GEOCHEMISTRY
Pathways
and Processes

STEVEN M. RICHARDSON
Iowa State University

HARRY Y. McSWEEN, JR.
University of Tennessee, Knoxville

Prentice Hall, Englewood Cliffs, New Jersey 07632

Library of Congress Cataloging-in-Publication Data

Richardson, Steven McAfee.
    Geochemistry : pathways and processes / Steven M. Richardson,
Harry Y. McSween, Jr.
      p. cm.
    Bibliography
    Includes index.
    ISBN 0-13-351073-5
    1. Geochemistry. I. McSween, Harry Y. II. Title
QE515.R53 1989   88-4567CIP
551.9—dc19

Editorial/production supervision: Karen Winget/Wordcrafters
Cover design: Diane Saxe
Manufacturing buyer: Paula Benevento

The program diskette that accompanies this text is available
in your instructor's copies or from Prentice-Hall College Software.

Printed in the United States of America
10 9 8 7 6 5 4 3 2 1

ISBN 0-13-351073-5

PRENTICE-HALL INTERNATIONAL (UK) LIMITED, *London*
PRENTICE-HALL OF AUSTRALIA PTY. LIMITED, *Sydney*
PRENTICE-HALL CANADA INC., *Toronto*
PRENTICE-HALL HISPANOAMERICANA, S.A., *Mexico*
PRENTICE-HALL OF INDIA PRIVATE LIMITED, *New Delhi*
PRENTICE-HALL OF JAPAN, INC., *Tokyo*
SIMON & SCHUSTER ASIA PTE. LTD., *Singapore*
EDITORA PRENTICE-HALL DO BRASIL, LTDA., *Rio de Janeiro*

GEOCHEMISTRY
Pathways
and Processes

For Cathy and Sue

Contents

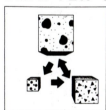

CHAPTER 2

A First Look at Thermodynamic Equilibrium 16

CHAPTER 3

How to Handle Solutions 46

CHAPTER 4

The Oceans and Atmosphere as a Geochemical System   82

CHAPTER 5

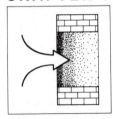

CHAPTER 6

CHAPTER 7

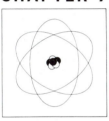

Using Stable Isotopes   **208**

CHAPTER 8

CHAPTER 9

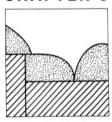

CHAPTER 10

Kinetics and Crystallization   **296**

CHAPTER 11

CHAPTER 12

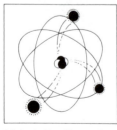

CHAPTER 13

Stretching Our Horizons: Cosmochemistry   **413**

APPENDIX A
Mathematical Methods   455

APPENDIX B
Finding and Evaluating Geochemical Data   465

Preface

The various operations of Nature, and the changes which take place in the several substances around us, are so much better understood by an attention to the laws of chemistry, that in every walk of life the chemist has a manifest advantage over his illiterate neighbor.—Samuel Parkes (1816), *The Chemical Catechism.*

This is a book about chemistry, written expressly for geologists. To be more specific, it is a text that we hope will be used to introduce advanced undergraduate and graduate students to the basic principles of geochemistry. With luck, it may inspire some to explore further—perhaps even to become geochemists themselves. Whether we achieve that goal or not, our major objective in writing this book is to help geology students gain that manifest advantage over their illiterate neighbors by showing that concepts from chemistry have a place in *all* corners of geology. The ideas in this book should be useful to all practicing geologists, from petrologists to paleontologists.

Students entering a new discipline should expect to spend a considerable amount of time assimilating its vocabulary and theoretical underpinnings. It isn't long, though, before most students begin to concede that theory is elegant, and start to ask, "What can I *use* it for?" Geologists, being practical people, reach this point perhaps earlier than most. In this book, therefore, we have tried not to just talk about geochemistry, but to show how its principles can be used to solve problems. In each chapter, we have integrated a number of Worked Problems, solved step-by-step to demonstrate the way ideas can be put to practical use. As often as possible, we have devised problems that are based on published research, so that the student can relate abstract principles to the real concerns of geologists. These are key parts of the book, and we encourage our readers to think about them. Working the problems at the end of each chapter will further reinforce what has been covered in the text.

This attention to problem-solving reflects our feeling that geochemistry, beyond the most rudimentary level, is quantitative. To answer real questions about the chemical behavior of the earth, a geologist must be prepared to perform calculations. We address this philosophy directly at the end of Chapter 1. It should be understood at the outset, however, that most of the questions at the level of this book can be answered by anyone with a standard undergraduate background in the sciences. We assume only that the student has had a year of college-level chemistry and a year of calculus. Methods for handling a few specialized concepts, such as partial derivatives, are outlined in an appendix. It is important for students to realize that most of the routine approaches to solving geochemical problems are already within their reach.

Beyond this gentle word of encouragement, we have taken two major steps to inspire students to be creative problem-solvers. First, we find that the ready availability of desktop computers has made it possible to involve students in solving problems which are highly illuminating, but would be far too tedious or time-consuming to do by hand. Because a high proportion of today's students has easy access to computers, we have made available a diskette of original programs. Far from being parlor tricks to amaze the neophytes, these programs should invite students to address challenging problems in geochemistry. They are introduced as supplements to problem-solving in the text, and are discussed collectively in an appendix. We hope that they can help students explore a wider range of topics than is usually possible in an introductory course, and that they prove useful in later research applications. For some, they may even serve as an inspiration to write programs of their own.

To Use The Disk

1. MAKE A BACKUP COPY FIRST! It is easiest to do this with the DOS copy utility. Read your system manual for details.
2. Insert the disk in your A: drive and type A: ⟨return⟩ to make this your default drive.
3. Type WELCOME ⟨return⟩.

General Advice

1. Some programs on the disk need to update data files. You have two options: (1) Remove the protection tab so your computer can write on the disk; (2) Copy data files onto a separate disk. We recommend option 2.
2. All BASIC programs on the disk call WELCOME.BAS as they run. If you see the message "File not found in 5", you have ignored instruction #2 above. If your BASIC interpreter is on a different disk drive, you must type its label when you run a BASIC program. For example, if your interpreter is on the C: drive and you want to run REGULAR.BAS, you should type C:BASICA REGULAR.
3. If you have problems or suggestions, write to S. M. Richardson, Geochemical Programs, Dept. of Earth Sciences, 253 Science I, Iowa State Univ., Ames, IA 50011.

Our second incentive for the student consists of highlighted boxes designed to present thought-provoking ideas that in many cases fall outside the standard narrative of introductory geochemistry. They offer a glimpse at topics that challenge the professional geochemist, faced with the world beyond textbook problems. We hope that these appeal particularly to students who want to become geochemists.

The contents of this book have been organized along lines that may seem unorthodox to some readers. As the title suggests, we have tried from the beginning to balance the traditional thermodynamic perspective with observations from a kinetic viewpoint, emphasizing both pathways and processes in geochemistry. Because a substantial amount of theory is involved, we decided to focus initially on processes in which temperature and pressure are nearly constant. Therefore, after an abbreviated introduction to the laws of thermodynamics and to fundamental equations for flow and diffusive transport, we spend the first half of the book investigating solution chemistry in natural waters, diagenesis, and chemical weathering. We then return for a second look at thermodynamics and kinetics as they apply to systems undergoing changes in temperature and pressure during magmatism and metamorphism, which command our attention for the second half of the book. Chapters 4, 11, and 13 serve to focus broad themes in the text by emphasizing, respectively, the ocean-atmosphere, the mantle-crust, and the solar system as geochemical systems. This loose division of geochemical topics has served us well in the classroom. Among other advantages, it spreads out the burden of learning basic theory and allows us to address the practical side of geochemistry before students' eyes begin to glaze over.

Experienced readers may recognize the invisible hands of J. B. Thompson, J. F. Hays, H. D. Holland, and J. A. Wood, each of whom shaped our early perceptions of geochemistry at Harvard. We owe them a great deal for sharing what they know and for giving us an appreciation for the breadth of geochemistry. A. C. Lasaga, J. F. Kasting, R. A. Berner, and many others have introduced us to unfamiliar corners of the field, in person or through their research, and have influenced the style of this text. Although each of these illustrious geochemists has contributed to our pleasure in the business, however, none of them is at all to blame for the way we have assembled ideas in writing this book. We have tried to present basic concepts in a way that makes sense to us and that we hope will inspire students.

In the preparation of this book, we have had many helpful suggestions from R. T. Dodd, L. S. Land, W. G. Ernst, R. Yuretich, M. D. Feigenson, C. K. Richardson, D. I. Siegel, L. K. Benninger, S. Willson, T. R. Nathan, T. C. Labotka, and X. Cao. Rusty Freeman shared his talents in the artwork at the beginning of each chapter. Our greatest thanks, though, go to Holly Hodder at Prentice-Hall. As editor and friend, she has directed traffic for us and offered repeated words of enthusiasm to shape all of these ideas into a book.

Steven M. Richardson
Harry Y. McSween, Jr.

GEOCHEMISTRY

Pathways
and Processes

Chapter 1

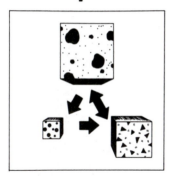

Introducing Concepts
in Geochemical Systems

OVERVIEW

In this introductory chapter our goal is to examine the range of problems that interest geochemists, and to compare two fundamental approaches to geochemistry. Thermodynamics and kinetics, as we attempt to illustrate, are complementary ways of viewing chemical changes that may take place in nature. By learning to use both approaches, you will come to appreciate the similarities among geochemical processes and be able to follow the many pathways of change. To develop some skills in using the tools of geochemistry, we also examine the limitations of thermodynamics and kinetics, and discuss practical considerations in problem solving.

WHAT IS GEOCHEMISTRY?

In studying the earth, geologists use a number of borrowed tools. Some of these come from physicists and mathematicians, others were developed by biologists, and still others by chemists. When we use the tools of chemistry, we are looking at the world as geochemists. From this broad perspective, the distinction between geochemistry and some other disciplines in geology, like metamorphic petrology or crystallography, is a fuzzy and artificial one.

A Historical Overview

The roots of geochemistry belong to both geology and chemistry. Many of the practical observations of Agricola (Georg Bauer), Nicolas Steno, and other Renaisssance geologists helped to expand the knowledge of the behavior of the elements and their occurrence in nature. These observations formed the basis from which both modern chemistry and geology grew in the late eighteenth century. The writings of Lavoisier and his contemporaries, during the age when modern chemistry began to take shape, were filled with conjectures about the oceans and atmosphere, soils, rocks and the processes that modify them. In the course of separating and characterizing the properties of the elements, Lavoisier, Davy, Dalton, and others also contributed to the growing debate among geologists about the composition of the earth. Similarly, De l' Isle and Hauy made the first modern observations of crystals, leading very quickly to advances in mineralogy and in structural chemistry.

The term *geochemistry* was apparently used for the first time by the Swiss chemist Schönbein in 1838. The emergence of geochemistry as a separate discipline, however, came with the establishment of major laboratories at the U.S. Geological Survey (U.S.G.S.) in 1884, at the Carnegie Institution of Washington in 1904, and in several European countries, notably in Norway and the U.S.S.R., between roughly 1910 and 1925. From these came some of the first systematic surveys of rock and mineral compositions, and some of the first experimental studies in which the thermodynamic conditions of mineral stability were investigated. F. W. Clarke's massive work, *The Data of Geochemistry*, published in 1908 and revised several times in the next 16 years, summarized analytical work performed at the U.S.G.S. and elsewhere, allowing geologists to estimate the average composition of the crust. The phase rule, which was first suggested through theoretical work by J. Willard Gibbs in the 1880s, was applied to field studies of metamorphic rocks by B. Roozeboom and P. Eskola, who thus established the chemical basis for the metamorphic facies concept. Simultaneously, A. L. Day and others at the Carnegie Institution of Washington Geophysical Laboratory began a program of experimentation focused on the processes which generate igneous rocks.

During the twentieth century, the course of geochemistry has been guided by several technological advances. The first of these was the discovery by F. Laue in 1912 that the internal arrangement of atoms in a crystalline substance can serve as a diffraction grating to scatter a beam of X-rays. Within a short time, W. L. Bragg used this technique to determine the structure of halite. In the 1920s, V. M. Goldschmidt and his associates at the University of Oslo determined the structures of a large number of common minerals and from these structures formulated principles regarding the distribution of elements in natural compounds. Their papers, published in the series *Geochemische Verteilungsgesetze der Elemente*, were perhaps the greatest contributions to geochemistry in their time.

During the 1930s and into the years surrounding World War II, new steel alloys were developed which permitted experimental petrologists to investigate processes at very high pressures for the first time. P. Bridgeman, N. L. Bowen, and an increasing army of colleagues built apparatus with which they could synthesize mineral assemblages at temperatures and pressures similar to those in the lower crust and upper mantle. With these technological changes, geochemists were finally able to investigate the chemistry of inaccessible portions of the earth.

The use of radioactive isotopes in geochronology began at the end of the nineteenth century, following the dramatic discoveries of E. Rutherford, H. Bequerel, and the Curies. The greatest influence of nuclear chemistry on geology, however, came with the development of modern, high-precision mass spectrometers in the late 1930s. Isotope chemistry gained rapid visibility through the careful mass spectroscopic measurements of A.O.C. Nier; between 1936 and 1939, he determined the isotopic compositions of 25 elements. Nier's creation in 1947 of a simple but highly precise mass spectrometer brought the technology within the budgetary reach of many geochemistry labs and opened a field that, even today, is still one of the most rapidly growing areas of geochemistry. Beginning with the work of H. Urey and his students in the late 1940s, stable isotopes have become standard tools of economic geology and environmental geochemistry. In recent years, precise measurements of radiogenic isotopes have also proved increasingly valuable in tracing chemical pathways in the earth.

The growth of geochemistry has been paralleled by the development of cosmochemistry, and many scientists have spent time in both disciplines. From the uniformitarian point of view, the study of terrestrial pathways and processes in geochemistry is merely an introduction to the broader search for principles that govern the general behavior of planetary materials. V. M. Goldschmidt, for example, refined many of his notions about the affinity of elements for metallic, sulfidic, or siliceous substances by studying meteorites. In more recent years, our understanding of crust-mantle differentiation in the early earth has been enhanced by studies of lunar samples.

This rapid historical sketch has unavoidably bypassed many significant events in the growth of geochemistry. It should be apparent from even this short treatment, however, that the boundary between geology and chemistry has been a porous one. For the generations of investigators who have worked at the boundary during the past two centuries, it has been an exciting and challenging time. This period has also been marked by recurring cycles of interest in data-gathering and the growth of basic theory, to which both chemists and geologists have contributed. Each advance in technology has invited geochemists to scrutinize the earth more closely, has expanded the range of information available to us, and has led to a new interpretation of fundamental processes.

Beginning Your Study of Geochemistry

In this book, we explore many ideas that geology shares with chemistry and, in doing so, offer a perspective on geology that may be new to you. You need not be a chemist to use the techniques of geochemistry, nor be a physicist to understand the constraints of seismic data, nor be a biologist to benefit from the wealth of paleontology. On the other hand, the language and methods of geochemistry are different from much that is familiar in standard geology. You will have to assimilate a new vocabulary and learn some fundamental concepts of chemistry to appreciate the role geochemistry plays in interpreting geological events and environments. As we use these new concepts to interpret environments from magma chambers to pelagic sediment columns, you will find that your understanding of other disciplines in geology is enriched.

As a student of geology, you have already been introduced to a number of unifying concepts. Plate tectonics, for example, provides a physical framework in which seemingly independent events such as island arc volcanism, earthquakes, and blueschist metamorphism can be interpreted as the products of large-scale movements of lithospheric plates. As geologists fine-tune the theory of plate tectonics, its potential for letting us in on broader secrets of how the earth works is even more tantalizing than its promise of showing us how the individual parts behave.

In the same way, as we try and answer the question "What is geochemistry?" in this book, we look for unifying concepts by exploring a wide variety of geologic environments and seeing how they can be interpreted with the tools of geochemistry. In Chapter 2, for example, we introduce the concept of chemical potential and demonstrate its role in characterizing directions of change in chemical systems. We then encounter chemical potential in various guises in later chapters as we examine topics in chemical weathering, diagenesis, stable isotope geothermometry, magma fractionation, and condensation of gases in the early solar system, to mention a few. In this way, you should come to recognize that chemical potential is a common strand in the fabric of geochemistry. We offer you the continuing challenge of recognizing other unifying concepts as we proceed through this book.

A glance at the Table of Contents shows you that geochemistry reaches into virtually all corners of geology. Because our interest is in identifying and learning to work with broadly applicable ideas in geochemistry, we do not concentrate exclusively on problems from one corner. Instead, we choose problems that are of interest to many kinds of geologists and help you find ways to approach and (we hope) solve them.

GEOCHEMICAL VARIABLES

Most geochemical processes can be described in rather broad, qualitative terms if we are presenting them for a lay audience or if we are trying to establish for each other a "big picture" of how things work. As a rule, however, qualitative discussions are only credible if they are backed up by more rigorous, quantitative lines of study. It is thus desirable to identify a set of properties (variables) sufficient to account for all relevant changes in whatever geochemical system we are considering.

We will begin by defining two types of variables, which we will refer to as *extensive* and *intensive*. The first of these is a measure of the size or *extent* of the system in which we are interested. For example, *mass* is an extensive variable because (all other properties being held constant) the mass of a system is a function of its size. If we choose to examine a block of galena that is 20 cm on a side, we expect to find that it has a different mass than a block that measures only 5 cm on a side. *Volume* is another familiar extensive property, clearly a measure of the size of a system. A characteristic of extensive variables that should be obvious from these examples is that extensive quantities are additive. That is, the mass of galena in the two blocks described above is equal to the sum of their individual masses. Volumes are also additive.

Intensive variables, in contrast, have values that are independent of the size of the system under consideration. Two of the most commonly described intensive properties are *temperature* and *pressure*. To show that these are independent of the size of a system, imagine that one block of galena is found to be uniformly at room temperature, roughly 25°C. Experience tells us that if we do nothing to the block except cut it in half, we will then have two blocks of galena, each of which is still at 25°C. Pressure is similarly independent of the dimensions of a system, provided that it too is uniform.

It will be desirable in many geochemical problems to cast results in terms of intensive rather than extensive properties of the system. By doing so, we free ourselves from caring about the size of a system and worry simply about its behavior. If we want to determine, for example, whether the mineral assemblage consisting of quartz and olivine is more or less stable than an assemblage of pyroxene, the answer will be most useful if it is cast in terms of energy *per unit of matter* or *per unit of volume*.

The conversion from extensive to intensive properties involves normalizing to some convenient measure of the size of the system. For example, we may choose to describe a block of galena by its mass (an extensive property) normalized by its volume (another extensive property). The resulting quantity, *density*, would be the same regardless of which block we choose to consider, and is therefore an intensive property of the system. We might choose to normalize instead by multiplying the mass of the block by 6.02×10^{23} (Avogadro's number: the number of formula units per mole of any substance, and in this case an appropriate scale factor) and then dividing by the number of PbS formula units in it. This exercise should yield the *molecular weight*, another intensive property. In general, the ratio of two extensive variables is an intensive variable, as is the derivative of an extensive variable with respect to either an extensive or an intensive variable.

GEOCHEMICAL SYSTEMS

Several times in the preceding discussion, we used the word *system*. By this, we mean the portion of the universe that is of interest for a particular problem, whether it is as small as an individual clay particle or as large as the protosolar nebula.

Because the results of geochemical studies will usually depend strongly on how a system is defined, it is always important to begin by defining the boundaries of the system. There are three generic types of systems, each defined by the condition of its boundaries. Systems are *isolated* if they cannot exchange either matter or energy with the universe beyond their limits. A perfectly insulated thermos bottle would be such a system. The simplifying assumption that systems are isolated is often made because it is easier to keep track of the condition of a system both physically and mathematically if we can draw a well-defined wall around it. In Chapter 2, we also find that the theoretical existence of isolated systems enables us to put limits on the behavior of fundamental thermodynamic functions. In the real universe, however, truly isolated systems cannot exist because there are no perfect insulators to prevent the exchange of energy across system boundaries.

In nature, systems can either be *closed* or *open*. A closed system is one that can exchange energy, but not matter, across its boundaries. An open system can exchange both energy and matter. In fact, this distinction is somewhat artificial, although it is a practical one in most cases. Provided that the permeability of beds confining an aquifer is acceptably low, for example, we may consider the aquifer to be a closed system for the purposes of a problem concerning relatively rapid chemical reactions within it. On the other hand, if we were considering a toxic waste disposal problem, the definition of "acceptably low" permeability would probably be different, and the same aquifer might be considered an open system. Often, then, the definition of a system will have to include a realistic assessment of the rates of transfer across its boundaries. Provided that the system is contained within limits of acceptably low transfer rates. we can usually justify treating it as either a closed or an isolated system. Time is an important part of this way of defining systems.

THERMODYNAMICS AND KINETICS

There are two different sets of conditions under which we might study a geochemical system. First, we might consider a system in a state of *equilibrium*, a consequence of which is that observable properties of the system undergo no change with time. (A more exact definition in terms of *thermodynamic* conditions will be investigated in Chapter 2.) In the case of a system at equilibrium, we will not care what its history has been and using the tools of thermodynamics would not permit us to ask questions about its evolutionary path anyway. However, with the appropriate equations, we will be able to estimate what the stable system will look like under any environmental conditions we choose. Because most geological systems are able to exchange energy with their surroundings, environmental parameters such as temperature and pressure are imposed on them from the outside. The methods of thermodynamics allow us to use these con-

straits, plus the bulk composition of the system, to predict which phases will be stable and in what relative amounts they will be present. In this way, the thermodynamic approach to a geochemical system can help us measure its stability and predict the direction in which it will change if its environmental parameters are changed.

The alternative to studying a system in terms of thermodynamic conditions would be to examine the *kinetics* of the system; that is, to study each of the pathways along which the system may evolve with time as it moves toward some state of thermodynamic equilibrium and to determine the rates of change of system properties along that pathway. Most geochemical systems can change between equilibrium states along a variety of different pathways, some of which are more efficient than others. In a kinetic study, the task is often to determine which of the competing pathways is dominant. In other words, while the focus of thermodynamics is upon the end states of a system, before and after a change has taken place, a kinetic treatment of geochemical systems sheds light on what happens in between. As geochemists, we have an interest in determing not only what *should* happen in real systems (the thermodynamic answer) but also *how* it is most likely to happen (the kinetic answer). By combining these two approaches, we can hope to interpret the large number of systems in transition that we encounter.

AN EXAMPLE: COMPARING THERMODYNAMIC AND KINETIC APPROACHES

Figure 1.1a is a schematic drawing of a hypothetical and somewhat frivolous system that illustrates these two approaches. It is a closed system consisting of five subsystems or *reservoirs* (indicated by boxes) that are connected by *pathways* (labeled by arrows). To make it interesting, we will also redraw this sytem in Figure 1.1b as a sketch of the real system from which the schematic diagram was abstracted, the floor plan of a small house. We can use this familiar setting to demonstrate many of the principles that apply to dynamic geochemical systems.

Assume that you, the owner of the house, have invited several of your friends to a party, and that they are all hungry. As we take our first look at the party, people are scattered unevenly in the living room, the dining room, the kitchen, the bedroom, and the bathroom. People are free to move from one room to another, provided that they use doorways and do not break holes through the walls, but there are some rules of population dynamics (which we will describe soon) which govern the rates at which they move.

From a thermodynamic point of view, the system we have described is initially in a state (i.e., hungry) which we would have a hard time describing mathematically, but which is clearly different from the final state that you, as host, would like to approach (i.e., well-fed). It is also clear that if you were to begin serving food to your friends, they would gradually move from the hungry state to the well-fed state. If people were as well-behaved as chemical systems, we ought to be able to use an approach analogous to thermodynamics to calculate the "energy" difference between the two states (Just *how* hungry are your guests?) and therefore predict the amount of food you will have to provide to let everyone go home feeling satisfied. The mathematics necessary to do this calculation might involve writing a differential equation for each room, expressing variations in some function W (an "inverse hunger" function) in terms of variations in the amounts of ham, cole slaw, and beer:

$$dW = (\partial W/\partial Ham)dHam + (\partial W/\partial Slaw)\, dSlaw + (\partial W/\partial Beer)\, dBeer.$$

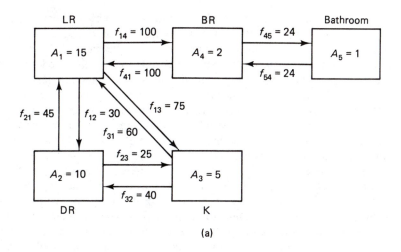

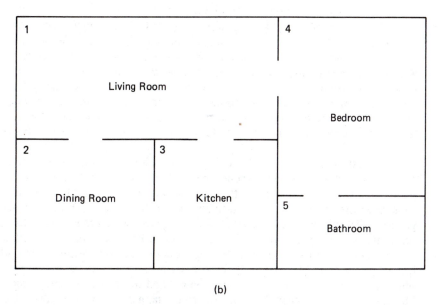

(b)

Figure 1.1 (a) Schematic diagram for a house with five rooms. The number of people in each room is given by A_i, and their rates of movement between rooms by f_{ij}, where i is a room being vacated and j is a destination. The units of f_{ij} are people hour^{-1}. (b) Floor plan for the house described schematically in (a).

This equation is in the form of a total differential, described in Appendix A. It will become familiar in later chapters.

Thermodynamics, then, allows us to recognize that the initial hungry state of your guests is less stable than the final well-fed state, and provides a measure of how much change to expect during the party. Even with an adequate description of a system, however, thermodynamics can give no guarantee that the real world will reach the equilibirium configuration we calculate on paper. Despite our best thermodynamic calculations, for example, the people at our party may remain indefinitely in a hungry state. (Maybe they don't like your cooking, or the large dog in the corner of your kitchen repels them.)

In the geologic realm, minerals or mineral assemblages may appear to be stable simply because we observe them over times that are short compared to the rates of reactions that alter them. Such materials are said to be *metastable*. Thus, granites and limestones are found at the earth's surface despite the fact that we know them to be less stable than clays and salt-bearing ground waters. The aragonitic tests that are typically excreted by modern shelled organisms are less stable than calcite, which thermodynamics tells us should be present, yet aragonite is only slowly converted to calcite. Basalts commonly contain glass, which strict thermodynamic calculations predict should have devitrified, instead of crystalline phases. Diamonds, stable only under the high-pressure conditions of the upper mantle, do not spontaneously restructure themselves to form graphite, even over very large spans of time. In each of these cases, the reaction path between the metastable and stable states involves intermediate steps that are difficult to perform and therefore take place very slowly or not at all. We will refer to these intermediate steps as *kinetic barriers* to reaction. It would be most accurate to say that, given an adequate description of the compositional and environmental constraints on a system, thermodynamics can tell us what that system *should* look like if it is given infinite time in which to overcome kinetic barriers.

We encounter many geologically important systems in transition between equilibrium states. Despite our reservations about systems with kinetic barriers, is it possible to apply thermodynamic reasoning to systems in transition? It is common to find that very large systems like the ocean are in a *state of dynamic balance* or *steady state*, in which environmental conditions are different from one part of the system to another, but the system may remain with no net change in its appearance for an indefinite period of time. Under these conditions, thermodynamics remains a useful approach, although it should be clear that no system-wide state of thermodynamic equilibirum can be described.

To return to the example in Figure 1.1, let us assume that randomly-moving guests will nibble some food whenever they find a serving table. Further, suppose that people in any particular room at the party always move around more rapidly within the room than they do from one room to another. If that is so, then the proportion of hungry to moderately well-fed people at different places *within* that room will be fairly uniform any time we peek into it. Different rooms, however, may have different proportions of hungry and well-fed people, depending on the size of the room and whether it has a serving table in it. Each of the rooms, in other words, is always in a local equilibrium state different from the other rooms, and the entire house is a mosaic of equilibirum states. If we knew how the "thermodynamics" of people works, then we could apply it within rooms to help describe how hungry the guests in them were at any time. Furthermore, despite the fact that the rooms are not in equilibrium with each other, consideration of the mosaic provides valuable information about the direction of overall change in the house.[1]

This is the basis of the concept of *local* or *mosaic equilibrium* that is often used in metamorphic petrology to discuss mineral assemblages that are well-defined on the scale of a thin section, but which may differ from one hand sample to another across an outcrop. Within small parts of the system, rates of reaction may often be rapid enough that equilibrium is maintained and thermodynamics can comment on the existence and

---

[1] Every example has its limitations. The statistics of small numbers, in this case, make it difficult to take this observation literally. The more people we invite to the party, however, the more uniform the population is in any given room.

relative abundance of minerals in which we are interested. Provided that we can agree on the definition of the mineral assemblages, and therefore the sizes of the regions of mosaic equilibrium, we can use thermodynamics to analyze them individually and to predict directions of change in the system as a whole.

The point here is that thermodynamics remains a useful part of our tool kit even if we examine a system that is far from equilibrium. We cannot use thermodynamics, however, to answer questions such as how rapidly the party will approach its well-fed state, or how people may be distributed throughout the house at any instant in time. In the world of geochemistry, we may be able to describe deviations from equilibrium by applying thermodynamics, but we will find ourselves incapable of predicting which of many pathways a system may follow toward equilibrium or how rapidly it gets there. Those are kinetic problems.

Suppose we address one of the kinetic problems for the party system: how are people distributed among the rooms as a function of time? To answer the question, let's choose one room at random and look at the operations that will cause its population to change with time. Briefly, we do this by taking an inventory of the rates of input to and output from the room, and comparing them. We can write this as a differential equation:

$$dA_i/dt = \text{total inputs} - \text{total outputs}$$

$$= \sum_j f_{ji} - \sum_j f_{ij}. \tag{1-1}$$

Here we are using the symbol A_i to indicate the number of people in the room we have chosen (room i, where i is a number from one to five). The quantities labeled f are *fluxes*; that is, numbers of people moving from one room to another per unit of time. The order of subscripts on f is such that the first one identifies *from* where people are moving, and the second tells us *to* where they are moving. This notation is used to label the pathways in Figure 1.1a. Room j in each case is some room other than i.

We will have to write an expression of this type for each room in the example (for dA_1/dt, dA_2/dt, and so forth). These *flux equations*, then, may be solved simultaneously for any time during the party, providing that we define the initial values for A_i and f_{ij}. This step is often intimidating, but only occasionally impossible.

For now, we will not try to solve the equations, but look at a few concepts we can derive from them. (See the accompanying box for a discussion of methods of solution, if you are interested.) First, we note in passing that kinetic models generally consider the distribution of mass (in this case, numbers of people) in a system rather than the distribution of energy. The driving force that causes the contents of a reservoir to change with time is a difference in energy levels (in this case, inverse hunger, the "thermodynamic" quantity that establishes the *direction* of change). The *rate* of change, however, depends on interactions between individual atoms, or molecules, or (in this example) people and serving tables, and will therefore depend on how many individuals there are. Consequently, the appropriate units in most problems in geochemical kinetics are units of concentration.

It is useful to reexamine the concept of a steady state for our system. It can now be seen to be a condition in which each of the flux equations is equal to zero. If there is such a state, the population in each room will not change with time. Notice that this does not imply that the *fluxes* are zero, but that the inputs and outputs from each room are balanced. That is, steady state may also be defined as a condition in which the time derivatives df_{ij}/dt are equal to zero.

Worked Problem 1-1

In Figure 1.1a, we have shown numerical values for each of the nonzero fluxes in the party example. How can we verify, using these constant values for f_{ij}, that the five flux equations for this system describe a system in steady state?

The flux equations are

$$dA_1/dt = f_{21} + f_{31} + f_{41} - f_{12} - f_{13} - f_{14}$$

$$= 45 + 60 + 100 - 30 - 75 - 100 = 0$$

$$dA_2/dt = f_{12} + f_{32} - f_{21} - f_{23}$$

$$= 30 + 40 - 45 - 25 = 0$$

$$dA_3/dt = f_{13} + f_{23} - f_{31} - f_{32}$$

$$= 75 + 25 - 60 - 40 = 0$$

$$dA_4/dt = f_{14} + f_{54} - f_{41} - f_{45}$$

$$= 100 + 24 - 100 - 24 = 0$$

$$dA_5/dt = f_{45} - f_{54}$$

$$= 24 - 24 = 0$$

Because each of the time derivatives is equal to zero, the reservoir contents, A_i, must be fixed at the values shown in Figure 1.1a. Therefore, despite the fact that people are free to move between rooms, the party is in a steady state.

Notice also, without proof for the moment, that there is no guarantee that all systems will have a steady state, particularly if they are open. A system may continually accumulate or lose mass through time. A dissolving crystal might be one such system. At one time, it was believed that the salinity of the oceans was a unidirectional function of time, and that the accumulation of salt in seawater could be used to infer the age of the oceans. In that view, the ocean was not at or approaching a steady state. It is also possible that a system may have more than one steady state. Systems which involve interacting fluxes for large numbers of species, or which have an oscillatory behavior (due, perhaps, to the passage of seasons) often have multiple steady states.

When we introduced the party example, we said that it would become necessary to define some laws of population dynamics. Those "laws" are a set of phenomenological statements that tell us how fluxes, f_i in the set of equations we are developing, change with time. In this problem, it is fair to assume that the rate at which people leave any given room to move to another is directly proportional to the number of people in the room. The more crowded the room becomes, the more likely people are to leave it. That is,

$$dA_i^{out}/dt = \sum_j k_{ij} A_i = A_i \sum_j k_{ij}, \tag{1-2}$$

in which the quantities k_{ij} are simple rate constants expressing migrations from room i to room j. The k_{ij} have units of inverse time. This "law" is an assumption of *first order* kinetics, which is commonly justifiable in geochemical problems as well. The rate of removal of magnesium from seawater at midocean ridges appears to be directly proportional to the Mg^{2+} concentration of seawater, for example. The rate of formation of halite evaporites is probably, in a similar way, a first-order function of the concentration of $NaCl$ in the oceans.

By introducing rate laws such as the first-order expression applied here, we have recognized that fluxes, in general, are not constant in systems away from their steady state. If we assume a first-order law, $f_{ij} = k_{ij}A_i$. In the example at hand, this means that we could induce changes in the rates of movement from one room to another either by varying the rate constants k_{ij} or by changing the population in one or more of the rooms. We might change the rate constants in any number of ways, such as putting out more food and drink in the dining room, putting on some dance music in the living room, or simply waiting for people to get full and mellow as the evening wears on. Any of these will violate the assumption of simple first-order kinetics and cause a shift in population with time. On the other hand, if more people were suddenly to arrive at the party, for the moment overcrowding the living room, we might expect the rate of movement of people into other rooms to increase for a while. This change in the system fluxes would be accomplished without changing any of the rate constants.

Generalizing from our example, we may say that the answer we obtain from a kinetic analysis of a geochemical system does not depend simply on how well we have identified mass reservoirs and instantaneous fluxes between them at the beginning and end of the problem. Kinetics is path-dependent, and the answer we get can be only as good as our knowledge of the chemical path between beginning and end. It is not appropriate to assume that all geologic processes follow first-order rate laws. Many do not. Other rate laws may be appropriate for systems that exhibit seasonal variability or in which the rate of removal of material from reservoir i is a nonlinear function of A_i. Complex processes of detachment at the surfaces of crystals, for example, may cause their rates of solubility to be very nonlinear functions of undersaturation. Determination of appropriate rate laws and constants is, in most cases, a difficult business at best. For this reason, you will find that many potentially interesting kinetic problems have been discussed in schematic terms by geochemists, but solved only in a qualitative or greatly simplified form.

HOW CAN SYSTEMS OF FLUX EQUATIONS BE SOLVED?

The flux equations introduced in this chapter describe a set of relationships among the reservoir contents, written in differential form as dA_i/dt. Our task in "solving" this set of equations is to extract from them a set of values of A_i that will tell us the reservoir contents at any time we choose. We are searching, in other words, for a set of equations to define $A_i(t)$.

Problems of this type can be solved by a variety of mathematical techniques. Perhaps the simplest, conceptually, is to begin by substituting initial values for A_i and f_{ij} into each of the flux equations to calculate instantaneous values of dA_i/dt for each of the reservoirs at time zero (t^0). If we assume that these time derivatives will be nearly constant for a short time interval, Δt, then we can calculate new reservoir contents at the end of the interval from

$$A_i(t^0 + \Delta t) = A_i(t^0) + (dA_i/dt)\Delta t.$$

New values of f_{ij}, therefore, can be calculated as well. These updated values of A_i and f_{ij}, in turn, serve as inputs to another interval of change. By keeping Δt relatively small, we can calculate successive steps in time, tracking linear changes in reservoir contents explicitly. Clearly, this is a job for a computer. It is also a potentially slow (and therefore expensive) process if we are forced to make a large number of very small steps.

Most modern computer installations support software for "implicitly" solving sets of differential equations. One such routine is the subroutine DIFSUB written by Gear (1971). It is impractical here to discuss the details of such programs, and it is not truly necessary either. Most users treat routines like DIFSUB successfully as "black boxes" into which they feed data and out of which answers are extruded. In broad terms, though, implicit methods predict the behavior of time derivatives on the basis of results in previous time steps and then decide how large or small to make Δt in later steps. Because Δt can, therefore, be made very large as a system of equations approaches steady state, computing time is reduced dramatically.

A third method uses methods of linear algebra and can be applied to systems of first-order flux equations to generate an analytical (that is, a closed form) solution. It has been described by Lasaga (1980, 1981) and is the basis for the program labeled CYCLE.EXE on the diskette available for this text. In this method, the first-order rate constants k_{ij} are reassembled in a matrix K defined by

$$K = \begin{cases} K_{ij} = k_{ji} & \text{for} \quad i \neq j \\ K_{ij} = -\sum_l k_{il} \ (i \neq l) & \text{for} \quad i = j. \end{cases}$$

The eigenvalues E_i and eigenvectors Ψ_i of this matrix are the basis of a solution for the set of flux equations, which takes the form

$$A_i(t) = \sum_j a_j \Psi_i \exp (E_j t),$$

in which the coefficients a_j contain information about the initial values of A_i. Intimidating as this method may appear if you are unfamiliar with matrix manipulation, it is surprisingly straightforward and has the advantage of producing a set of concise equations for $A_i(t)$. Using these equations, we can calculate the reservoir contents at any time of interest directly, without need for iterative computation. Relevant principles of matrix algebra and an introduction to eigenvalue problems are discussed in Appendix A, and documentation for CYCLE.EXE is in the file CYCLE.DOC on the program diskette available for this text.

NOTES ON PROBLEM-SOLVING

In this chapter, we have tried to illustrate the range of problems that interest geochemists and to compare in broad terms the thermodynamic and kinetic approaches to them. In the remaining chapters we look in detail at selected geochemical topics and develop specific techniques for analyzing them. As we progress, you should recall several of the points we have just made.

First, predictions made from thermodynamic or kinetic calculations cannot be any better than our description of the system allows them to be. In the frivolous example we have been using, we cannot expect thermodynamics to comment on events outside of the house, or on the state of hunger being felt by guests lurking in undescribed closets. We also cannot comment on the relative efficiency of kinetic pathways we have failed to include in the model, such as extra doors between rooms.

It may seem obvious that this limitation applies to real geochemical problems as well, but it is surprising how easy this is to overlook. If we are examining a system consisting of metamorphosed shales, for example, and forget to include in our description any chemical reactions that produce biotite, there is no way that we can expect thermodynamics to predict its existence. This omission has a serious effect. For a mineral like biotite, which may be a volumetrically significant phase in some metapelites, this error in description will render meaningless any thermodynamic analysis of the system as a whole. If we decide to omit a minor phase like apatite, on the other hand, the effect on thermodynamic calculations should be minimal. Thermodynamics will not be able to comment on its stability, but our overall description of the system should not be seriously in error.

Because geochemical reactions commonly proceed along several parallel pathways, a valid kinetic treatment depends critically on how well we have identified and characterized those pathways. Most systems of even moderate complexity, however, include pathways that are easy to underestimate or to overlook completely. In Chapter 4, for example, we consider the effect of hydrothermal circulation at midoceanic ridge crests on the mass balance of calcium and magnesium in the oceans. Until the mid-1970s, few data suggested that this was a significant pathway in the geochemical cycle for magnesium. Kinetic models, therefore, regularly overestimated the importance of other pathways, such as glauconite formation in marine sediments, in an attempt to solve the "magnesium problem."

It is clear, then, that the credibility of any statements we make regarding the appearance of phases, the stability of a system, or rates of change depends in large part on how carefully we have described the system *before* applying the tools of geochemistry. This is, perhaps, the single greatest source of difficulty you will experience in applying geochemical concepts to real-world problems.

Second, there are few concepts that are the exclusive property of igneous or sedimentary petrologists, of oceanographers or soil chemists, of economic geologists or cosmochemists. The breadth of geochemistry can be overwhelming. However, it can also be advantageous when we try to solve practical problems. A familiarity with the data and methodology used in one area of geochemistry will often allow formulation of alternative solutions to a problem in another area. Many of the fundamental quantities of thermodynamics and of kinetics should gradually become familiar as they appear in different contexts. Where possible, we try to point out areas of obvious overlap.

Finally, we encourage the development of the simplest geochemical models that will adequately answer the questions posed about the earth. All too often, problems appear impossible to solve because we have made them unnecessarily complex and begun treating them as abstract mathematical exercises rather than as investigations of the real earth. In most cases, a careful selection of variables and a set of physically justifiable simplifying assumptions can make a problem tractable without sacrificing much accuracy. What you already know as a geologist about the earth's properties and behavior should always serve as a guide in making such choices.

Having just said this, we should also emphasize that mathematics is an essential tool for the quantitative work of geochemistry. Now would be a good time to familiarize yourself with the contents of the appendixes at the opposite end of this book, before we begin our first excursion into thermodynamics. The first of these appendixes is an abbreviated overview of mathematical concepts, emphasizing methods that may not have been part of your first year of college-level calculus. Among these are principles of matrix algebra and partial differentiation, and an introduction to numerical methods for fitting functions to data and finding roots. Refer to Appendix A as often as necessary to make the mathematical portions of this book more palatable.

Some introductory texts include a short table of preferred values for data needed in problem-solving. We have decided not to do this. Instead, Appendix B is a reference list of standard sources for geochemical data with which you should become familiar. This is unquestionably less convenient than a compact data table would have been. We hope, though, that you will quickly become accustomed to finding and evaluating data for yourself, and that our list will be more useful in the long run than a table might have been.

Breaking only slightly with this precedent, Appendix C contains the values of fundamental constants and conversion factors. The literature of geochemistry has been slow to adopt SI units. Resistance to the use of pascals instead of bars or dm^3 in place of liters, for example, is very strong in some camps, and may well remain so. We have made no attempt to favor one system over another in this book. Our practical opinion is that because data are reported in several conventional forms, you should be conversant with them all and should know how to shift readily from one set of units to another. Thus, the need for conversion factors.

In the final appendix, we introduce you to the use of computer programs for addressing geochemical problems. In particular, we discuss the programs included on the diskette which is available with this textbook. This is a departure from standard geochemistry books, to make it possible for you to work on problems that have always been considered too time-consuming or too complex to present to beginning students. We hope that you find these programs a challenging supplement to the text, and that you regularly use them to solve problems at the ends of chapters.

SUGGESTED READINGS

These first references will illustrate the breadth of geochemistry. They also provide a wealth of data that may be most useful to professionals.

FAIRBRIDGE, R.W., ed. 1972. *The encyclopedia of geochemistry and environmental sciences.* New York: Van Nostrand Reinhold Co.. (A frequently referenced source book for geochemical data.)

GOLDSCHMIDT, V.M. 1954. *Geochemistry.* Oxford: Clarendon Press. (This book is now dated, but remains a classic in the field.)

MASON, B., and MOORE, C. 1982. *Principles of geochemistry.* New York: John Wiley & Sons. (Probably the most readable survey of the field of geochemistry now available.)

WEDEPOHL, K.H., ed. 1969–78. *Handbook of geochemistry.* 5 Vols. Berlin: Springer Verlag. (A five-volume reference set with ring bound pages that are designed for convenient updating. A definitive source book on the occurrence and natural chemistry of the elements.)

These two papers establish the theoretical basis for the matrix method of solving sets of first-order flux equations. See the box on page 11.

LASAGA, A.C. 1980. The kinetic treatment of geochemical cycles. *Geochim. Cosmochim. Acta* 44: 815–28. (A modern classic that describes the role of kinetics in global geochemistry.)

LASAGA, A.C. 1981. Dynamic treatment of geochemical cycles: Global kinetics. In *Kinetics of geochemical processes,* ed. A.C. Lasaga and R.J. Kirkpatrick, 69–110. Reviews in Mineralogy 8. Washington, D.C.: Mineral. Soc. of America. (An excellent teaching tool. This article provides a tutorial for the kinetics of global cycling.)

This reference discusses the details of the DIFSUB routine described in the box on page 11, as well as other initial value problems.

GEAR, C.W. 1971. *Numerical initial value problems in ordinary differential equations.* Englewood Cliffs, N.J.: Prentice-Hall. (A challenging book, even for professionals; a marvel of logic, but not for bedtime reading.)

PROBLEMS

1. You are already accustomed to describing geologic materials and environments in terms of measurable variables like temperature or mass. Make a list of 10 such variables and indicate which ones are intensive or extensive properties.

2. Consider the mass balance and the pathways for change in a residual soil profile.
   (a) Sketch a diagram similar to Figure 1.1a on which you indicate the way in which major species such as Fe, Ca, or organic matter may be redistributed with time.
   (b) Write a complete set of schematic flux equations to describe mass balance in your system, following the example of Worked Problem 1-1.

3. It has been said that diagenetic reactions generally occur in an open system, whereas metamorphic reactions more commonly take place in a closed system. Explain what this geochemical difference means and suggest a reason for it.

The following problems for math review test your skills with concepts we will be using from time to time:

4. Write the derivative with respect to t for each of the following expressions:
   (a) $\ln (t)^3$
   (b) $\exp (-2\,a/\alpha t)$
   (c) $\sin (\alpha t^2)$
   (d) $\log_{10}(\alpha/t)$
   (e) $[t \ln t + (1 - t) \ln (1 - t)]$
   (f) $\Sigma_i \exp (-E_i/Rt)$
   (g) $\int_0^{\alpha t} \sin (t - z)dz$

5. Write down the total differential, df, for each of the following expressions:
   (a) $f(x, y, z) = x^2 + y^2 + z^2$
   (b) $f(x, y, z) = \sin (xyz)$
   (c) $f(x, y, z) = x^3 y - xy^2$

6. Determine which of the following expressions are exact differentials:
   (a) $df = (y + z)dx + (z + y)dy + (x + y)dz$
   (b) $df = 2xy\,dx + 2yz^2 dy - (1 - x^2 - 2y^2 z)dz$
   (c) $df = z\,dx + x\,dy + y\,dz$
   (d) $df = (x^3 + 3x^2 y^2 z)dx + (y + 2x^3 yz)dy + (x^3 y^2)dz$

7. Consider the two differentials, $df = y\,dx - x\,dy$ and $df = y\,dx + x\,dy$. One of them is exact; the other is not. Evaluate the line integral of each over the following paths described by Cartesian coordinates (x,y):
   (a) A straight line from $(1,1)$ to $(2,2)$.
   (b) A straight line from $(1,1)$ to $(2,1)$, then another from $(2,1)$ to $(2,2)$.
   (c) A closed loop consisting of straight lines from $(1,1)$ to $(2,1)$ to $(2,2)$ and then to $(1,1)$.

8. Perform the following matrix operations:
   (a) $\begin{pmatrix} 1 & 2 & 3 \\ -1 & 2 & 0 \\ 3 & -2 & 1 \end{pmatrix}\begin{pmatrix} 3 \\ -1 \\ 2 \end{pmatrix} =$

   (b) $5\begin{pmatrix} -4 & 3 & 2 \\ 1 & 5 & 1 \\ -3 & 2 & 5 \end{pmatrix} + \begin{pmatrix} 1 & 2 & 6 \\ 0 & -3 & 2 \\ 5 & 10 & -1 \end{pmatrix} =$

   (c) $\begin{pmatrix} 2 & x & -x^2 \\ 1 & -2x & x^2 \\ 3 & 4x & -2x^2 \end{pmatrix}\begin{pmatrix} 1 \\ 2x \\ x^2 \end{pmatrix} =$

   (d) $\det \begin{pmatrix} 2 & -3 & 4 & 1 \\ -2 & 4 & -1 & 0 \\ 3 & 2 & 1 & -4 \\ 0 & 1 & 2 & 4 \end{pmatrix} =$

   (e) $\begin{pmatrix} 1 & 2 & 1 \\ 0 & 2 & -1 \\ 2 & 0 & 1 \end{pmatrix}^{-1} =$

9. Change units on the following quantities as directed:
   (a) Convert 1.5×10^3 bars to GPa.
   (b) Convert 3.5 g cm^{-3} to g bar cal^{-1}.
   (c) Convert 1.987 cal deg^{-1} to kJ deg^{-1}.
   (d) Convert 5.0×10^{-13} cm^2 sec^{-1} to m$^2(10^6$yr$)^{-1}$.

Chapter 2

A First Look at Thermodynamic Equilibrium

OVERVIEW

The foundations of thermodynamics are introduced in this chapter. A major goal is to establish what is meant by the concept of equilibrium, which was described informally in Chapter 1. The idea of equilibrium is an outgrowth of our understanding of three laws of nature that describe the relationship between heat and work and identify a sense of direction for change in natural systems. These are the heart of the subject of thermodynamics, which we will apply later to a variety of geochemical problems.

As we discuss the three laws of thermodynamics, it is necessary to examine more closely some familiar concepts like temperature, heat, and work. It is also necessary to introduce many new quantities like entropy, enthalpy, and chemical potential. The search for a definition of equilibrium also reveals a set of four fundamental equations which describe potential changes in the energy of a system in terms of temperature, pressure, volume, entropy, and composition.

TEMPERATURE AND EQUATIONS OF STATE

It is part of our everyday experience to arrange items on the basis of how hot they are. In a colloquial sense, our notion of temperature is associated with this ordered arrangement, so that we speak of items having higher or lower temperatures than other items, depending on how "hot" or "cool" they feel to our senses. In the more rigorous terms we must use as geochemists, however, this casual definition of temperature is inadequate for two reasons. First of all, a practical temperature scale should have some mathematical basis so that successive changes in temperature can be related in a rational way to other continuous changes in system properties. It is hard to use the concept of temperature in a predictive way if it relies on discrete, subjective observations. Second, and more important, the common perception of temperature just described is difficult to separate from the more elusive notion of "heat," the quantity that is transferred from one body to another to cause changes in temperature. It would be helpful to develop a definition of temperature that does not depend on our understanding of heat, but instead relies on thermodynamic properties that are easier to handle.

To do this imagine two closed systems, each of which is homogeneous; that is, each consists of a single, chemically uniform substance (to which we will refer from now on as a *phase*) that is not undergoing any chemical reactions. If this condition is met, the thermodynamic state of either system can be completely described by defining the values of any two of its intensive properties. These might be pressure and viscosity, for example, or acoustic velocity and molar volume. In order to study the concept of temperature, we will examine the pressure and the molar volume, using the symbols p and $\bar{v}$ in one system, and P and $\bar{V}$ in the other. (Notice that volume has now been made an intensive property as the result of dividing by the number of moles.)

If we place the two containers in contact, changes of pressure and molar volume will take place spontaneously in each system until a sufficiently long period of time has passed after which no further changes occur. At this point, the two systems are said to be in *thermal equilibrium*. Let's now measure their pressures and molar volumes and label them P_1, $\bar{V}_1$, and p_1, $\bar{v}_1$. There is no reason to expect that $P_1 = p_1$ or $\bar{V}_1 = \bar{v}_1$; in general the systems will not have identical properties.

If we separate the systems for a moment, we will find it possible to change one of them so that it is described by new values P_2 and $\bar{V}_2$ that still lead to thermal equilibrium with the system at p_1 and $\bar{v}_1$. There are, in fact, an infinite number of possible combinations of P and $\bar{V}$ that satisfy this condition. These combinations define a curve on a graph of P versus $\bar{V}$ (Figure 2.1a). Each combination of P and $\bar{V}$ describes a state

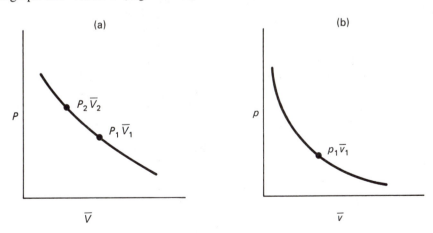

Figure 2.1 (a) All states of a system for which $f(P,\bar{V}) = T$ are said to be in thermal equilibrium. A line connecting all such states is called an *isothermal*. (b) If another system contains a set of states lying on the isothermal $f(p,\bar{v}) = t$, and if $t = T$, then the two systems are in thermal equilibrium.

Temperature and Equations of State

of the system in thermal equilibrium with p_1 and $\bar{v}_1$. All points on the curve are, therefore, also in equilibrium with each other.[1] It should be clear that we could just as well have varied pressure and molar volume in the *other* system and found an infinite number of values of p and $\bar{v}$ that lead to thermal equilibrium with the system at P_1, $\bar{V}_1$. These values define a curve on a graph of p versus $\bar{v}$, shown in Figure 2.1b. We refer to curves like these as *isothermals* and say that *states of a system have the same temperature if they lie on an isothermal, and systems have the same temperature if their states on their respective isothermals are in thermal equilibrium.*

In an algebraic, rather than a grapical way, what we have demonstrated is that the state in each of the two systems in our example can be described by a function

$$f(p, \bar{v}) = t \quad \text{or} \quad F(P, \bar{V}) = T \tag{2-1}$$

and that the systems are in thermal equilibrium if $t = T$. Functions like those in equation 2.1, which define the interrelationships among intensive properties of a system, are called *equations of state*.

Because many real systems behave in roughly similar ways, several standard forms of the equation of state have been developed. We find, for example, that many gases at low pressure can be described adequately by the ideal gas law:

$$P\bar{V} = RT, \tag{2-2}$$

in which R is the gas constant, the value of which is given in Appendix C. Other gases, particularly those containing many atoms per molecule or those at high pressure, are characterized more appropriately by expressions like the Van der Waals equation:

$$(P + a/\bar{V}^2)(\bar{V} - b) = RT, \tag{2-3}$$

in which a and b are empirical constants.[2] Thermodynamic states for liquids and solids can also be described approximately by equations in a number of standard forms. Despite the similarities that allow us to derive standard forms for the equation of state, however, real systems are rarely identical in detail, and we should not expect isothermals like those in Figure 2.1a and 2.1b to have the same shape.

Our notion of temperature, therefore, depends on the conventions we choose to follow in writing equations of state. A practical temperature scale can be devised simply by choosing a well-studied reference system (a *thermometer*), writing an arbitrary function that remains constant (that is, generates isothermals) for various states of the system in thermal equilibrium, and agreeing on a convenient way to number selected isothermals. In principle, this means that we are free to use any equation of state that does not lead to an ambiguous temperature scale. A gas that is adequately described by equation 2-2, for example, would be described equally well by

$$P\bar{V} = -cT \quad \text{or} \quad 1/(P\bar{V}) = cT \quad \text{or} \quad e^{P\bar{V}} = cT,$$

in which c is any constant, although each of these would lead to a highly unfamiliar temperature scale. Later in this chapter, we will find that there is a thermodynamic rationale for choosing an *absolute temperature scale* consistent with equation 2-2, rather than making one of these unconventional choices.

---

[1] This observation has been called the *zeroth law of thermodynamics: If A is in equilibrium with B, and B is in equilibrium with C, then A is also in equilibrium with C.*

[2] The empirical basis for the van der Waals equation of state is discussed in Chapter 3.

In general terms, *work* is performed whenever an object is moved by the application of a force. An infinitesimal amount of work, dw, is therefore described by writing

$$dw = F\,dx, \tag{2-4}$$

in which F is a generalized force and dx is an infinitesimal displacement. By convention, we define dw so that it is positive if work is performed by a system on its surroundings, and negative if the environment performs work on the system. We know from experience that an object may be influenced simultaneously by several forces, however, so it is more useful to write equation 2-4 as

$$dw = \sum_i F_i\,dx_i. \tag{2-5}$$

The forces may be hydrostatic pressure, P, as described in the earlier paragraphs of this chapter, or may be directed pressure, surface tension, or electrical or magnetic potential gradients. Since all forms of work are equivalent, the total work performed on or by a system can be calculated by including all possible force terms in the summation of equation 2-5. The thermodynamic relationships that build on equation 2-5 do not depend on the identity of the forces involved.

In practice, this broad understanding of work should be tempered by two considerations. First, work is performed in most geochemical systems when a volume change dV is generated by application of hydrostatic pressure:

$$dw = P\,dV. \tag{2-6}$$

This is the equation we will most often encounter, and geochemists commonly speak of work as if equation 2-6, rather than equation 2-5, were the fundamental definition of work. Usually, the errors introduced by this simplification are small. It is always wise, however, to examine each new problem to see whether work resulting from other forces is significant. Later chapters of this book will discuss some conditions in which this is necessary.

Second, keep in mind that equation 2-5 is a differential equation. The total amount of work performed in a process is the integral of that equation between the initial and final states of the system. Because forces in geologic environments rarely remain constant as a system evolves, the integral becomes extremely difficult to evaluate if we do not specify that the process is a slow one. The work performed by a gas expanding violently, as in a volcanic eruption, is hard to estimate because pressure is not the same at all places in the gas, and because the gas is not expanding in mechanical equilibrium with its surroundings.

THE FIRST LAW OF THERMODYNAMICS

During the 1840s, a series of fundamental experiments were performed in England by the chemist James Joule. In each of them, a volume of water was placed in an insulated container, and work was performed on it from the outside. Some of these experiments are illustrated in Figure 2.2. The paddle wheel, iron blocks, and other mechanical devices are considered to be parts of the insulated container. The temperature of the water was monitored during the experiments, and Joule reported the surprising result

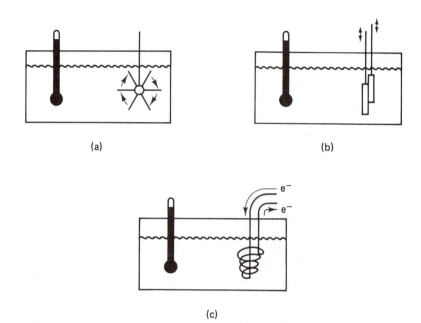

(a)

(b)

(c)

Figure 2.2 English chemist James Joule performed experiments to relate mechanical work to heat. (a) A paddlewheel is rotated and a measurable amount of work is performed on an enclosed water bath. (b) Two blocks of iron are rubbed together. (c) Electrical work is performed through an immersion heater.

that a specific amount of work performed on an insulated system, by any process, always results in the same change of temperature in the system.

What changed inside the container, and how do we explain Joule's results? In physics, equation 2-4 is often introduced in the context of a potential energy function. A block of lead may be lifted from the floor to a tabletop by applying an appropriate force to it. When this happens, we recognize not only that an amount of work has been expended on the block, but also that its potential energy has increased. Usually, problems of this type assume that no change takes place in the internal state of the block (that is, its temperature, pressure, and composition remain the same as the block is lifted), so the change in potential energy is just a measure of the change in the block's position.

By analogy, we may reason that when Joule performed work on the insulated container of water, some energy function associated with the water changed. In this case, however, it was not the position of the water that changed, but its internal state, as measured by the change in its temperature. The energy function used to describe this situation is called the *internal energy* (symbolized by E) of the system, and Joule's results can be expressed by writing

$$dE = -dw. \tag{2-7}$$

It should now be clear that there are two ways to change the internal energy of a system. First, as has just been demonstrated, the system and its surroundings can perform work on each other. Second, as we implied while considering the meaning of temperature, energy can be passed directly through system walls if those walls are noninsulating. If Joule's containers had been noninsulators, he could have produced temperature changes without performing any work, simply by lighting a fire under them. Energy transferred by this second mode is called *heat*.

In order to investigate these two modes of energy transfer, it is necessary to distinguish between two types of walls that can surround a closed system. What we have spoken of loosely as an "insulated" wall is more properly called an *adiabatic* wall. *Adiabatic processes* are ones which, like those in Joule's experiments, do not involve any transfer of heat between a system and its surroundings. A *nonadiabatic* wall, on the other hand, allows the passage of heat. Perfect adiabatic processes are seldom seen in the real world, but adiabatic behavior is often assumed as an approximation. A rapidly moving magma body, for example, may ascend almost adiabatically. It is possible to perform work on a system that is bounded by either type of wall.

Equation 2-7 is an extremely useful statement, referred to as the *First Law of Thermodynamics*. In words, it tells us that *the work done on a system by an adiabatic process is equal to the increase in its internal energy, a function of the state of the system*. We may conclude further that if a system is isolated, rather than simply closed behind an adiabatic wall, no work can be performed on it from the outside, and its internal energy must remain constant.

The total energy associated with a body, then, consists of its potential energy and its kinetic energy, plus its internal energy. In most of what follows in this book, our attention will be focused on internal energy and on other functions which relate to the internal state of a system. Most systems of geochemical interest stay in one place during a process, so it is reasonable to consider internal energy separately from any functions which relate only to the position or velocity of a body. Keep in mind, however, that this is strictly possible only if a system is isolated and that, as discussed in Chapter 1, it is vitally important to keep track of the extent and nature of system boundaries. Meteorite impact, lithospheric plate movement, and planetary core formation are good examples of geologic processes in which potential or kinetic energy are exchanged for internal energy. Depending on the nature of the question being asked, it may be necessary to consider more than thermodynamics in order to understand the energetics of these processes.

Equation 2-7 can be expanded to say that for any system, a change in internal energy is equal to the sum of the heat gained (dq) and the work performed on the system ($-dw$):

$$dE = dq - dw. \tag{2-8}$$

Notice that dE is a small addition to the amount of internal energy already in the system. Because this is so, the total change in internal energy during a geochemical process is equal to the sum of all increments dE:

$$\int_{E_{initial}}^{E_{final}} dE = E_{final} - E_{initial} = \Delta E. \tag{2-9}$$

The value of ΔE, in other words, does not depend on how the system evolves between its end states. In fact, if the system were to evolve along a path that eventually returns it to its initial state, there would be no net change whatever in its internal energy. This is an important property of functions of state.

In contrast, the values of dq and dw describe the amount of heat or work expended across system boundaries, rather than increments in the amount of heat or work "already in the system." When we talk about heat or work, in other words, the emphasis is on the process of energy transfer, not the state of the system. For this reason, the integral of either quantity depends on which path the system follows from its initial to its final state. Heat and work, therefore, are not functions of state.

The First Law of Thermodynamics

Worked Problem 2-1

Consider a system, illustrated in Figure 2.3, whose initial state is described by a pressure P_1 and a temperature T_1. This might be, for example, a portion of the atmosphere. Recall that the equation of state that we use to define isothermals for this system tells us the molar volume $\bar{v}_1$, under these conditions. For ease of calculation, assume that the equation of state for this system is $\bar{v} = RT/P$. Compare two ways in which the system might slowly evolve to a new state in which $P = P_2$ and $T = T_2$. In the first process, let pressure increase slowly from P_1 to P_2 while the temperature remains constant, then let temperature increase from T_1 to T_2 at constant pressure P_2. In the second process, let temperature increase first at constant pressure P_1 and then let pressure increase from P_1 to P_2. (If this were, in fact, an atmospheric problem, the isobaric segments of these two paths might correspond to rapid surface warming on a sunny day, and the isothermal segments might reflect the passage of a frontal system.) Is the amount of work done on these two paths the same?

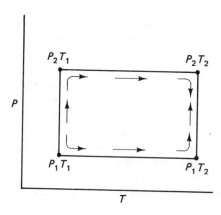

Figure 2.3 As a system's pressure and temperature are adjusted from $P_1 T_1$ to $P_2 T_2$, the work performed depends on the path the system follows.

If the only work performed on the system is due to pressure-volume changes, then the work along each of the paths is defined by the integral of equation 2-4. For path 1,

$$w_1 = \int_{P_1 T_1}^{P_2 T_1} P\, dV + \int_{P_2 T_1}^{P_2 T_2} P\, dV.$$

Along path 2,

$$W_2 = \int_{P_1 T_1}^{P_1 T_2} P\, dV + \int_{P_1 T_2}^{P_2 T_2} P\, dV.$$

In order to evaluate these four integrals, find an expression for $d\bar{V}$ by expressing the equation of state as a total differential of $\bar{V}(T,P)$, as follows:

$$d\bar{V} = (\partial\bar{V}/\partial T)_P dT + (\partial\bar{V}/\partial P)_T dP = (R/P)dT - (RT/P^2)dP,$$

and substitute the result into W_1 and W_2 above. The result, after integration, is that the work performed on path 1 is

$$W_1 = R[(T_2 - T_1) + T_1 \ln(P_1/P_2)],$$

but the work performed on path 2 is

$$W_2 = R[(T_2 - T_1) + T_2 \ln(P_1/P_2)],$$

clearly different. From the discussion of equations 2-8 and 2-9, it should follow that q_1 and q_2, the amounts of heat expended along these paths, must also be different.

Taken by itself, equation 2-8 tells us that heat and work are equivalent means for changing the internal energy of a system. In developing that expression, however, we have engaged in a little slight-of-hand that may have led you, quite incorrectly, to another conclusion as well. By stating that heat and work are equivalent modes of energy transfer, the First Law may have left you with the impression that heat and work can be freely exchanged for one another. This is not the case, as a few commonplace examples will show.

A glass of ice water left on the kitchen counter gradually gains heat from its surroundings, so that the water warms up and the room temperature drops ever so slightly. In the process, there has been an exchange of heat from the warm room to the relatively cool water. As Joule showed, the same transfer could have been accomplished by having the room perform work on the ice water. It is clearly impossible, however, for a glass of water at room temperature to cool spontaneously and begin to freeze, although we can certainly remove heat by transferring it first from the water to a refrigeration system. In a similar way, a lava flow gradually solidifies by transferring heat to the atmosphere, but this natural process cannot reverse itself either. It is impossible to melt rock by transferring heat to it directly from cold air.

If we were shown films of either of these events, we would have no trouble recognizing whether they were being run forward or backward. Equation 2-8, however, does not provide us with a means of making this decision theoretically. On the basis of experience with natural·processes, therefore, we are driven to formulate a *Second Law of Thermodynamics*. In its simplest form, first stated by Rudolf Clausius in the middle of the last century, the Second Law says that *heat cannot spontaneously pass from a cool body to a hotter one.* Another way of stating this, which may also be useful, is that *any natural process involving a transfer of energy is inefficient, with the result that a certain amount is irreversibly converted into heat that cannot be involved in further exchanges.* The Second Law, therefore, is a recognition that natural processes have a sense of direction.

Just as the internal energy function was introduced to present the First Law in a quantitative fashion, it is necessary to define a thermodynamic function that expresses the sense of direction embodied in the Second Law. This new function, known as *entropy* and given the symbol S, is defined in differential form by

$$dS = (dq/T)_{\text{rev}}, \qquad (2\text{-}10)$$

in which dq is an infinitesimal amount of heat gained by a body at temperature T in a reversible process. Because real processes are never perfectly reversible, we will find that entropy is, therefore, a measure of the degree to which a system has irretrievably lost heat and therefore some of its capacity to do work. Many people find entropy an elusive concept to master at first, so it is best to become familiar with the idea by discussing examples of its use and properties.

Figure 2.4 illustrates a potentially reversible path for a system enclosed in a nonadiabatic wall. Suppose that the system expands isothermally from A to B. In doing so, it performs an amount of work on its surroundings which can be calculated by the integral of PdV. Graphically, this integral is represented by the area AA'B'B. Because this is an isothermal process, the internal energy of the system must remain constant (recall the discussion at the beginning of this chapter). Therefore, the work performed by the system must be balanced by an equivalent amount of heat gained, according to equation 2-8. Suppose, now, that it were possible to compress the system isothermally, thus re-

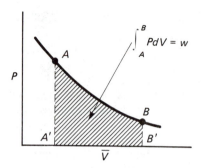

Figure 2.4 The work performed by isothermal compression from state A to state B is equal to the area AA'B'B. Because $dE = 0$, an equivalent negative amount of heat is gained by the system. If this change is reversed precisely, the net change in entropy, $\oint dS$, is zero.

versing along the path B → A without any interferences due to friction or other real-world forces. We would find that the work performed *on* the system (the heat *lost* by the system) would be numerically equal to the work on the forward path, but with a negative sign. That is, the net change in entropy around the closed path is

$$\oint dS = \int_A^B dq/T - \int_B^A dq/T = 0. \tag{2-11}$$

The entropy function is, therefore, a function of state for systems following reversible pathways, just like the internal energy function.

Next, consider an isolated system in which there are two bodies separated by a nonadiabatic wall (Figure 2.5). Let the two bodies initially be at different temperatures T_1 and T_2. If we allow them to approach thermal equilibrium, there must be a transfer of heat dq_2 from the body at temperature T_2. Because the total system is isolated, an equal amount of heat dq_1 must be gained by the body at temperature T_1. The net change in entropy for the system is

$$dS_{sys} = dS_1 + dS_2$$

or

$$dS_{sys} = dq_1((1/T_1) - (1/T_2)).$$

The Second Law tells us that this result is only valid if $T_2 > T_1$. Therefore, the entropy of an isolated system can only increase as it approaches internal equilibrium:

$$dS_{sys} > 0. \tag{2-12}$$

Finally, look at the closed system illustrated in Figure 2.6. It is similar to the previous system in all respects, except that is bounded by a nonadiabatic wall, so that the two bodies can exchange amounts of heat dq_1' and dq_2' with the world outside. As they approach thermal equilibrium, the heat exchanged by each is given by

$$(dq_1)_{total} = dq_1 + dq_1'$$

or

$$(dq_2)_{total} = dq_2 + dq_2'.$$

Notice that any heat exchanged *internally* must still show up either in one body or the other, so as before

$$dq_1 = -dq_2.$$

The net change in system entropy is, therefore,

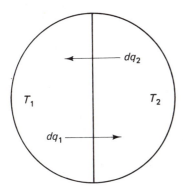

Figure 2.5 Two bodies in an isolated system are separated by a nonadiabatic wall and may, therefore, exchange quantities of heat dq_1 and dq_2 as they approach thermal equilibrium.

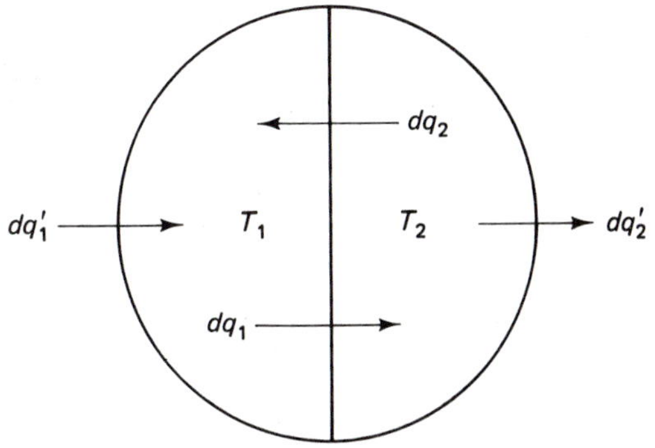

Figure 2.6 The system in Figure 2.5 is allowed to exchange heat with its surroundings. The net change in internal entropy depends on the values of d_1' and dq_2'.

$$dS_{sys} = (dS_1)_{total} + (dS_2)_{total}$$

$$= (dq_1'/T_1) + (dq_2'/T_2) + dq_1((1/T_1) - (1/T_2)).$$

It has already been shown that the Second Law requires the last term of this expression to be positive. Therefore,

$$dS_{sys} > (dq_1'/T_1) + (dq_2'/T_2). \tag{2-13}$$

Although the entropy change due to internal heat exchanges in a closed system will always be positive, notice that we have no way of predicting the direction in which the system will evolve unless we also have information about dq_1', dq_2', and the temperature in the world outside the closed system.

To appreciate this ambiguity, think back to the glass of ice water and compare the situation in which the glass is on the kitchen counter with the later situation in which it has been placed in the freezer. In both cases, internal heat exchange results in an increase in system entropy. When the glass is refrigerated, however, the terms on the right side of equation 2-13 become potentially large negative quantities, with the result that the change due to internal processes is overwhelmed and S_{sys} decreases. The Second Law, therefore, does not rule out the possibility that thermodynamic processes can be reversed. It does tell us, though, that they can be reversed only by transferring heat to some external system. Careful reexamination of this last example should convince you that a local decrease in entropy must be accompanied by an even larger increase in entropy in the world at large. Therefore, except in those cases where a system evolves gradually enough that we are justified in approximating its path as a reversible one, its change in entropy must be defined not by equation 2-10, but by

$$dS > (dq/T)_{irrev}. \tag{2-14}$$

ENTROPY AND DISORDER

Entropy is commonly described in elementary texts as a measure of the "disorder" in a system. This description is sometimes less than satisfying, because the mathematical definitions in equations 2-10 and 2-14 do not seem to address entropy in those terms.

To see why it is valid to think of entropy as an expression of disorder, consider this common problem.

Worked Problem 2-2

Two adjacent containers are separated by a removable wall. Into one of them, we introduce 1 mole of nitrogen gas. We fill the other with 1 mole of argon. If the wall is removed, we expect that the two gases will mix spontaneously rather than remaining in their separate ends of the system. It would require considerable work to separate the nitrogen and argon again. By randomizing the positions of gas molecules and contributing to a loss of order in the system, we have therefore caused an increase in entropy.

Suppose that the mixing takes place isothermally, and that the only work involved is mechanical. Assume also, for the sake of simplicity, that both argon and nitrogen are ideal gases. How much does entropy increase in the mixing process? First, write equation 2-8 as

$$dq = dE + PdV.$$

Because we agreed to carry out the experiment at constant temperature, dE must be equal to zero. Using the equation of state for an ideal gas, we can see that dq now becomes

$$dq = (nRT/V)\, dV,$$
$$dS = dq/T = (nR/V)\, dV.$$

(Notice that this problem is cast in terms of *volume*, rather than *molar volume*, for reasons that will be apparent soon.) We can think of the current problem as one in which each of the two gases has been allowed to expand from its initial volume (V_1 or V_2) into the larger volume $V_1 + V_2$. For each, then, the entropy change due to isothermal expansion is

$$\Delta S_i = \Delta n_i R \ln ((V_1 + V_2)/V_i),$$

and the combined entropy change is

$$\Delta S_1 + \Delta S_2 = n_1 R \ln ((V_1 + V_2))/V_1) + n_2 R \ln (V_1 + V_2))/V_2).$$

This can be simplified by defining the *mole fraction*, X_i, of either gas as $X_i = n_i/(n_1 + n_2) = V_i/V_1 + V_2)$. The combined entropy is then expressed by:

$$\Delta \bar{S} = -R(X_1 \ln X_1 + X_2 \ln X_2),$$

where the bar over $\Delta \bar{S}$ indicates that it has been normalized by $n_1 + n_2$ and is now a molar quantity. The entropy of mixing will always be a positive quantity, since X_1 and X_2 are always less than 1.0. The answer to our question, therefore, is that entropy increases by

$$2 \times \Delta \bar{S} = 2 \times -(1.987)(0.5 \ln 0.5 + 0.5 \ln 0.5)$$
$$= 2.76 \text{ cal deg}^{-1} \text{ or } 11.55 \text{ J deg}^{-1}$$

ENTROPY AND DISORDER: WORDS OF CAUTION

The description of entropy as a measure of disorder is probably the most often used and abused concept in thermodynamics. In many applications, the idea of "disorder" is coupled with the inference from the Second Law that entropy should increase through time. Like all brief descriptions of complex ideas, though, the statement that "entropy measures system disorder" must be accepted with some care. It is common practice among chemical physicists to estimate the entropy of a system by calculating its statistical degrees of freedom—computing, in effect, the number of different ways that atoms can be configured in the system, given its bulk conditions of temperature, pressure, and other intensive parameters. A less rigorous description of disorder, however, can lead to very misleading conclusions about the entropy of a system and thus the course of its evolution.

Creationists, for example, have grasped at the idea of entropy as disorder as a vindication of their belief that biological evolution is impossible. In their view, "higher" organisms are more ordered than more primitive organisms from which biologists presume that they evolved. Because this interpretation presents evolution as a historical progression from a less ordered to a more ordered state, it describes an apparent violation of the Second Law. There are so many flaws in this argument that it is difficult to know where to begin to discuss it, and we will mention only a few.

The first is a question of the meaning of "order" as applied to organisms or families of organisms. This is far from a semantic problem, since it is by no means clear that higher organisms are more ordered than primitive ones, from the point of view of thermodynamics. The biological concept of order, although potentially connected to thermodynamics in the realm of biochemistry, is based largely on functional complexity, not equations for energy utilization. The definition becomes even cloudier when applied to groups of organisms, rather than to individuals.

Furthermore, living organisms cannot be viewed as isolated systems, separated from their inorganic surroundings. As we have tried to emphasize, the definition of system boundaries is crucial if we are to interpret system changes by the Second Law. A refrigerator, for example, might be mistakenly seen to violate the Second Law if we failed to recognize that entropy in the kitchen around it increases even as the contents of the refrigerator become more ordered.

Finally, we note that there may be instances in which entropy and disorder in the macroscopic sense can be decoupled, even in a correctly defined isolated system. For example, if two liquids are mixed in an adiabatic container, we might expect that they will mix to a random state at equilibrium. We know, however, that other system properties may commonly preclude a random mixture. Oil and vinegar in a salad dressing unmix readily, as a manifestation of differences in their bonding properties. If the attractive force between similar ions or molecules is greater than between dissimilar ones, then complete random mixing would imply an increase, rather than a decrease, in both internal energy and entropy. A macroscopically ordered system, in other words, can easily be more stable than a disordered one without shaking our faith in the Second Law of Thermodynamics.

REPRISE: THE INTERNAL ENERGY FUNCTION MADE USEFUL

With the First and Second Laws in hand, we may now return to the subject of equilibrium, which was defined informally at the beginning of this chapter as a state in which no change is taking place. The First Law restates this condition as one in which dE is equal to zero. From the Second Law, we find that entropy is always maximized during the approach to equilibrium, and that dS becomes equal to zero once equilibrium is reached.

The results expressed in equations 2-10 and 2-14 can be used to recast equation 2-8 as

$$dE \leq TdS - dw. \qquad (2\text{-}15)$$

Under most geological conditions, mechanical (pressure-volume) work is the only significant contribution to dw, and it is reasonable to make a final substitution of equation 2-6 in equation 2-15:

$$dE \leq TdS - PdV. \qquad (2\text{-}16)$$

This is a practical form of equation 2-8, though only correct under the conditions just discussed. In most geologic environments, we will assume that change takes place very slowly, so that systems can be regarded as following nearly reversible paths. Generally, therefore, the inequality in equation 2-16 can be disregarded, even though it is strictly necessary.

Because much of the remaining discussion of thermodynamics in this book derives from equation 2-16, we should recognize three other relationships that follow directly from it and from the fact that the internal energy function is a function of state. First, equation 2-16 is a differential form of $E = E(S, V)$. Assuming that changes take place reversibly, it can be written as a total differential,

$$dE(S, V) = (\partial E/\partial S)_V \, dS + (\partial E/\partial V)_S \, dV. \qquad (2\text{-}17)$$

Comparison with equation 2-16 yields the following two statements:

$$T = (\partial E/\partial S)_V \quad \text{and} \quad P = -(\partial E/\partial V)_S. \qquad (2\text{-}18)$$

In other words, the familiar variables temperature and pressure can be seen as expressions of the manner in which the internal energy of a system responds to changes in entropy (under constant volume conditions) or volume (under adiabatic conditions). Second, because $E(S, V)$ is a function of state, equation 2-17 is a perfect differential (see Appendix A). That is,

$$(\partial^2 E/\partial S \, \partial V) = (\partial^2 E/\partial V \, \partial S)$$

or

$$(\partial T/\partial V)_S = -(\partial P/\partial S)_V. \qquad (2\text{-}19)$$

This last equation is known as one of the Maxwell relationships. These and other similar expressions to be developed shortly can be used to investigate the interactions of S, P, V, and T. Some of these will be discussed more fully in Chapter 8 when we look in more detail at the effects of changing temperature or pressure in the geologic environment.

AUXILIARY FUNCTIONS

Equation 2-16 is adequate for solving most thermodynamic problems. In many situations, however, it is possible to use equations of greater practical interest, which can be derived by imposing environmental constraints on the problem.

Enthalpy

The first of these equations can be derived by writing equation 2-16 as

$$dE + PdV \leq TdS.$$

If we restrict ourselves by looking only at processes that take place at constant pressure, this is the same as

$$d(E + PV) \leq TdS. \qquad (2\text{-}20)$$

The quantity $E + PV$ is a new function, called *enthalpy* and commonly given the symbol H. Its primary value is as a measure of heat exchanged under isobaric conditions, where $dH = dq$. Reactions which evolve heat, and therefore have a positive change in enthalpy, are *exothermic*. Those which result in a decrease in enthalpy are *endothermic*.

The left side of equation 2-20 can be expanded to reveal a differential form for dH:

$$dH = d(E + PV) = dE + PdV + VdP$$

or

$$dH \leq TdS + VdP. \tag{2-21}$$

Relationships similar to equation 2-18 can be written by inspection of the total differential $dH(S, P)$ and equation 2-21 under reversible conditions:-

$$T = (\partial H/\partial S)_P \quad \text{and} \quad V = (\partial H/\partial P)_S. \tag{2-22}$$

Enthalpy, like internal energy and entropy, can be shown to be a function of state, since it experiences no net change during a reversible cycle of reaction paths. This makes it possible, among other things, to determine the amount of heat that would be exchanged in geologically important reactions, even when those reactions may be too sluggish to study directly at the low temperatures where they take place in nature. The careful determination of such values is the business of *calorimetry*.

Because H is a function of state, it is also possible to extract one more Maxwell relation from its cross-partial derivatives, following the example of equation 2-19:

$$(\partial T/\partial P)_S = (\partial V/\partial S)_P. \tag{2-23}$$

The Helmholtz Function

By analogy with the way we introduced enthalpy, we can discover another useful function by writing equation 2-16 as

$$dE - TdS \leq -PdV.$$

Under isothermal conditions, this expression is equivalent to

$$d(E - TS) \leq -PdV. \tag{2-24}$$

The function $F = E - TS$ is best referred to as the *Helmholtz Function*, although you may see it referred to elsewhere as *Helmholtz Free Energy* or the *work function*.[3] Because it might be inadvertently mistaken for Gibbs Free Energy, which is ultimately a more useful function in geochemistry, it is best to avoid the first of these alternatives. The name *work function*, however, is informative. Inspection of equation 2-24 shows that the integral of dF at constant temperature is equal to the work performed on a system. This expression has its greatest application in mechanical engineering.

From the left side of equation 2-24, the differential form dF can be easily derived:

$$dF = d(E - TS) = dE - TdS - SdT$$

or

$$dF \leq -SdT - PdV. \tag{2-25}$$

---

[3] Unfortunately, a variety of symbols have been used for the internal energy, enthalpy, and Helmholtz functions, as well as the Gibbs Free Energy, which will be discussed shortly. This has led to some confusion in the literature. In the United States, F is often used to designate Gibbs Free Energy (for example, in thermodynamic publications of the Bureau of Standards). The International Union of Pure and Applied Chemistry (IAUPAC), however, has recommended that G be the standard symbol for Gibbs Free Energy. This usage, if not yet standard, is at least widespread. In the United States, those who use F for the Gibbs Free Energy generally use the symbol A for the Helmholtz function. To complete the confusion, internal energy, which we have identified with the symbol E, is frequently referred to as U, in order to avoid confusion with total (internal plus potential plus kinetic) energy. It is always prudent to be sure you know whose symbols you are using. N. Vanserg has written an excellent article on the subject, which is listed among the references at the end of this chapter.

The total differential $dF(T, V)$, when compared with equation 2-25, yields the useful expressions

$$S = -(\partial F/\partial T)_V \quad \text{and} \quad P = -(\partial F/\partial V)_T. \tag{2-26}$$

From the fact that F is a function of state and that $dF(T, V)$ is therefore a perfect differential, we also gain another Maxwell relation:

$$(\partial S/\partial V)_T = (\partial P/\partial T)_V. \tag{2-27}$$

Although the Helmholtz function is rarely used in geochemistry, this last relationship is quite widely used. When we return for a second look at thermodynamics in Chapter 8, equation 2-23 will be discussed as the basis of the *Clapeyron equation*, a means for describing pressure-temperature relationships in the geochemical world.

Gibbs Free Energy

Perhaps the most frequently used quantity in thermodynamics can be derived by writing equation 2-16 in the form

$$dE - TdS + PdV \leq 0.$$

Under conditions in which both temperature and pressure are held constant, this expression becomes

$$d(E - TS + PV) \leq 0. \tag{2-28}$$

In the now familiar fashion in which H and F have already been defined, we designate the quantity $E - TS + PV$ with the symbol G and call it the *Gibbs Free Energy* function (in colloquial usage, simply *free energy*), in honor of the nineteenth century theoretician Josiah Willard Gibbs.[4]

The differential on the left side of equation 2-28 can be written

$$dG = d(E - TS + PV)$$
$$= dE - TdS - SdT + PdV + VdP$$
$$= -SdT + VdP. \tag{2-29}$$

As with the previous fundamental equations, we can apply our knowledge of the total differential $dG(T, P)$ of this state function to find that

$$-S = (\partial G/\partial T)_P, \qquad V = (\partial G/\partial P)_T, \tag{2-30}$$

and

$$-(\partial S/\partial P)_T = (\partial V/\partial T)_P. \tag{2-31}$$

It can be seen from equation 2-29 that the Gibbs free energy function is the first of our fundamental equations written solely in terms of the differentials of intensive parameters. This, and the fact that both temperature and pressure are usually measured quite easily, contribute to the great practical utility of this function.

What, however, is "free" about the Gibbs Free Energy function? Consider the intermediate step in equation 2-29. At constant temperature and pressure, this reduces to

$$dG = dE - TdS + PdV.$$

---

[4] As a chemistry professor at Yale University, Gibbs wrote a classic series of papers in which virtually all of the fundamental equations of modern thermodynamics appeared for the first time. His papers provided a theoretical framework in which the experimental work of Thomson, Carnot, Clausius, Joule, and other earlier scientists could be related.

Substituting $dE = dq - dw$, we have

$$dG = dq - dw - TdS + PdV. \tag{2-32}$$

If the quantity of heat dq is transferred to the system isothermally and any changes are reversible, then $dq = TdS$ and $dw = dw_{rev}$, so equation 2-32 can be rewritten as

$$-dG = dw_{rev} - PdV. \tag{2-33a}$$

Therefore, the decrease in free energy of a system undergoing a reversible change at constant temperature and pressure is equal to the nonmechanical (i.e., non-pressure-volume) work that can be done by the system. If the condition of reversibility is relaxed, then

$$-dG > dw_{irrev} - PdV. \tag{2-33b}$$

In either case, the change in G during a process is a measure of the portion of the system's internal energy that is "free" to perform nonmechanical work.

The free energy function provides a valuable practical criterion for equilibrium. At constant temperature, the net change in free energy associated with a change from state 1 to state 2 of a system can be calculated from equation 2-28:

$$\int_1^2 dG = \int_1^2 d(E - TS + PV) = \int_1^2 d(H - TS),$$

or

$$\Delta G = \Delta H - T\Delta S, \tag{2-34}$$

where $\Delta G = G_2 - G_1$, $\Delta H = H_2 - H_1$, and $\Delta S = S_2 - S_1$.

Equations 2-33a and 2-33b make it clear that energy is available to produce a spontaneous change in a system as long as ΔG is negative. This can be accomplished, according to equation 2-34, under any circumstances in which $\Delta H - T\Delta S$ is negative. Most exothermic changes, therefore, are spontaneous. Endothermic processes (those in which ΔH is positive) can also be spontaneous, but only if they are associated with a large positive change in entropy.[5] A change will proceed as long as there can be a further decrease in free energy. Once free energy has been minimized ($dG = 0$), the system has attained equilibrium.

Worked Problem 2-3

Consider the sublimation of a solid at constant temperature in a closed container. The solid and vapor phases will be in equilibrium proportions when the system attains some minimal value of the Gibbs free energy, G_{min}. How is this value related to the enthalpy and entropy functions for the system?

As shown in Figure 2.7, the enthalpy of the system increases in a linear fashion as vaporization takes place. This can be rationalized by noting that heat must be added to the system to break molecular bonds in the solid. Therefore, if enthalpy were the only factor involved in the process the system would be most stable when it was 100% solid, since that is where H is minimized. Entropy, on the other hand, is maximized when the system is 100% vapor, because the degree of system randomness is greatest there. It can be shown from statistical arguments, however, that the increase in S follows a logarithmic function, rising most rapidly when the vapor fraction in the system is low (Denbigh, 1968, p. 485). Therefore, the difference, $G = H - TS$, is less than the free energy of the pure solid until

---

[5] This possibility was not appreciated at first by physical chemists. In fact, in 1879 the French thermodynamicist Berthelot used the term *affinity*, defined by $A = -\Delta H$, as a measure of the direction of spontaneous change. According to his reasoning, a spontaneous chemical reaction could only occur if $A > 0$. If an endothermic reaction turned out to be spontaneous, then some unobserved mechanical work must have been done on the system.

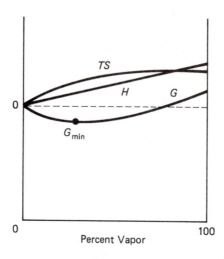

Figure 2.7 The quantity $G = H - TS$ varies with the amount of vapor as sublimation takes place in an enclosed container. The free energy when vapor and solid are in equilibrium is G_{min}. Energy units on the vertical axis are arbitrary. (Modified from Denbigh 1968.)

a substantial amount of vapor has been produced. Sublimation occurs spontaneously. At some intermediate vapor fraction, $H - TS$ reaches a minimum and solid and vapor are in equilibrium at G_{min}. If the vapor fraction is increased further, the difference $H - TS$ exceeds G_{min} again and condensation occurs spontaneously.

CLEANING UP THE ACT: CONVENTIONS FOR E, H, F, G, AND S

Except in the abstract sense that we have just used G in Worked Problem 2-3, there is no way we can talk about absolute amounts of energy in a system. To say that a beaker of reagents contains 100 kilojoules of internal energy or 55 kilocalories of free energy is meaningless. Instead, the fundamental quantities we have introduced so far are compared to their values in pure elements at the same temperature and at one atmosphere of pressure. It is customary, then, to speak of ΔH_f^o or ΔG_f^o, by which we mean the enthalpy or free energy of formation from the elements. The ΔH_f^o or ΔG_f^o for a system composed of a single element at any temperature is equal to zero. The same convention is observed for internal energy and the Helmholtz function as well, although they are less frequently encountered in geochemistry.

In the discussions so far, we may have given the impression that whatever thermodynamic data we might need to solve a particular problem will be readily available. In fact, the references in Appendix B do include data for a large number of geologic materials. We have, however, largely ignored the thorny problem of where the data come from and where we turn for data on materials that have not yet appeared in the tables. As an example of how thermodynamic data are acquired, we will now show how *solution calorimetry* is used to measure enthalpy of formation. In Chapter 9, we will examine how thermodynamic data can also be extracted from phase diagrams.

It is not necessary (or even possible) to make an absolute determination of enthalpy for individual phases. Instead, we measure the heat evolved or absorbed during a specific reaction between phases. Because most reactions of geologic interest are abysmally slow at low temperatures, this enthalpy change may be almost impossible to measure directly in some cases. However, the additive properties of this extensive variable permit us to employ some slight-of-hand: we can measure the heat lost or gained when the products and reactants are dissolved in some solvent at low temperature, and the heats of solution for the various products and reactants can be combined to obtain an equivalent value for the heat of reaction. Solvents suitable for silicate minerals include such corrosive fluids as hydrofluoric acid and molten lead borate.

Worked Problem 2-4:

Consider the following laboratory exercise. We wish to determine the molar enthalpy for the following reaction:

$$2MgO + SiO_2 \longleftrightarrow Mg_2SiO_4,$$
$$\text{periclase} \quad \text{quartz} \qquad \text{forsterite}$$

to which we will assign the value ΔH_1. This can be determined by measuring the heats of solution for periclase, quartz, and forsterite in HF at some modest temperature:

$$2MgO + SiO_2 + HF \longleftrightarrow (2MgO, SiO_2)_{solution},$$

which gives ΔH_2, and

$$Mg_2SiO_4 + HF \longleftrightarrow (2MgO, SiO_2)_{solution},$$

which gives ΔH_3. If the solutions are identical, we can see that the first of these reactions is mathematically equivalent to the second minus the third, so that

$$\Delta H_1 = \Delta H_2 - \Delta H_3.$$

In this way, calorimetry can be used to determine the enthalpy of formation for forsterite from its constituent oxides at low temperature.

One minor complication is that published tables of calorimetric data in many cases are given in terms of elements instead of oxides. If we wished to solve the problem above using such tabulated data, we could recognize that

$$2Mg + O_2 \longleftrightarrow 2MgO,$$

which gives ΔH_4, and

$$Si + O_2 \longleftrightarrow SiO_2,$$

which gives ΔH_5. Therefore, for the enthalpy ΔH_6 of the net reaction

$$2Mg + 2O_2 + Si \longleftrightarrow Mg_2SiO_4,$$

we obtain

$$\Delta H_6 = \Delta H_1 + \Delta H_4 + \Delta H_5.$$

The value ΔH_6 in this case is the enthalpy of formation for forsterite from the elements, ΔH_f.

Although we have now demonstrated that most thermodynamic functions are best defined as relative quantities like ΔH_f^o, entropy is a major exception. The convention observed most commonly is an outgrowth of the *Third Law of Thermodynamics*, stated here in the following paraphrase from Lewis and Randall (1961): *If the entropy of each element in a perfect crystalline state is defined as zero at the absolute zero of temperature, then every substance has a nonnegative entropy: at absolute zero, the entropy of* all *perfect crystalline substances becomes zero.* This statement, which has been tested in a large number of experiments, provides the rationale for choosing the absolute temperature scale as our standard for scientific use, as we discussed in the context of equation 2-2. For now, the significance of the Third Law is that it defines a state in which the *absolute* or *third law entropy* is zero. Because this state is the same for all materials, it makes sense to speak of S^o, *rather than* ΔS_f^o.

With this new perspective, return for a moment to equation 2-34. It should now be clear that the symbol Δ has a different meaning in this context. In fact, although it is rarely done, it would be less confusing to write equation 2-34 as

$$\Delta(\Delta G) = \Delta(\Delta H) - T\Delta S,$$

in which the deltas outside the parentheses and on S refer to a change in state (that is, a change in these values during some reaction), and the deltas inside the parentheses refer to values of G or H relative to some reference state. We will examine this concept more fully in later chapters.

Although we have not felt it necessary to prove, it should be apparent that each of the functions E, H, F, G, and S is an extensive property of a system. Therefore, the amount of energy or entropy in a system depends on the size of the system. In most cases, this is an unfortunate restriction, because we either don't know or don't care how large a natural system may be. For this reason, it is customary to normalize each of the functions by dividing them by the total number of moles of material in the system, thus making each of them an intensive property. To recognize molar quantities, look for a bar over the symbol, as in $\Delta \overline{G}^{\circ}_f$, the standard molar free energy of formation from the elements.

COMPOSITION AS A VARIABLE

Up to now, the functions we have considered all assume that a system is chemically homogenous. Most systems of geochemical interest, however, consist of more than one phase. The problem most commonly faced by geochemists is that the bulk composition of a system can be packaged in an extremely large number of ways, so that it is generally impossible to tell by inspection whether the collection or *assemblage* of phases actually found in a system is the most likely one. To answer questions dealing with the stability of multiphase systems, we need to write a separate set of equations for E, H, F, and G for each phase, and to apply criteria for solving them simultaneously. We will do this job in two steps.

Components

In order to describe the possible variations in the compositions and proportions of phases, it is necessary to define a set of *thermodynamic components* which satisfy the following rules:

1. The set of components must be sufficient to describe all of the compositional variations allowable in the system.
2. Each of the components must be *independently* variable in the system.

As long as these criteria are met, the specific set of components chosen to describe a system is arbitrary, although there often may be practical reasons for choosing one set rather than another.

These rules set stringent restrictions on the way components can be chosen, so it is a good idea to spend time examining them carefully. First, notice that the solid phases we encounter most often in geochemical situations have compositions that are either fixed or are variable only within bounds allowed by stoichiometry and crystal structure. This means that if we are asked to find components *for a single mineral*, they must be defined in such a way that they can be added to or subtracted from the mineral without destroying its identity. For example, consider the system consisting of isostructural carbonates with compositions between calcite ($CaCO_3$) and magnesite ($MgCO_3$).

It cannot be described by entities like Ca^{2+}, MgO, or CO_2, because none of them can be independently added to or subtracted from the carbonates without violating both their structure and their stoichiometry. The most obvious, though not unique, choice of components in this case would be $CaCO_3$ and $MgCO_3$.

When a system consists of more than one phase, it is common to find that some components selected for individual phases are redundant in the system as a whole. This occurs because it is possible to write stoichiometric relationships that express a component in one phase as some combination of those in other phases. For each stoichiometric equation, therefore, it is possible to remove one phase component from the list of system components, and thus to arrive at the independently variable set required by rule 2.

It is also possible, and often desirable, to choose system components that could not serve as components for any of the individual phases in isolation. Such a choice is compatible with the selection rules if the amount of the component in the system as a whole can be varied by changing proportions of individual phases. Petrologists usually refer to nonaluminous pyroxenes, for example, in terms of the components $MgSiO_3$, $FeSiO_3$, and $CaSiO_3$, even though $CaSiO_3$ cannot serve as a component for any pyroxene considered by itself.

Worked Problem 2-5

Olivine and orthopyroxene are both common minerals in basic igneous rocks. For the purposes of this problem, assume that olivine's composition varies only between Mg_2SiO_4(Fo) and Fe_2SiO_4(Fa), and that orthopyroxene is a solid solution between $MgSiO_3$ (En) and $FeSiO_3$ (Fs). What components might be used to describe each of them individually, and how might we select components for a rock containing both minerals?

The simplest components for the *minerals* are the end-member compositions themselves. The end-member compositions are completely independent of one another in each mineral and, as can be seen at a glance in Figure 2.8, any mineral composition in either solid solution can be formed by some linear combination of them. Therefore, compositions corresponding to Fo, Fa, En, and Fs satisfy our selection rules.

It is instructive to see that a choice like FeO or MgO would *not* be valid as a mineral component, because neither one could be added to or subtracted from olivine or orthopyroxene unless it were accompanied by a stoichiometric amount of SiO_2. Changing FeO or MgO alone would produce system compositions that lie off the solid solution lines in Figure 2-8.

If olivine and orthopyroxene are not isolated phases, but are constituents of a rock, different components are appropriate. The mineral components can still be used, but one of them is now redundant and can be eliminated by writing a stoichiometric relationship involving the other three. For example:

$$MgSiO_3 = 0.5Mg_2SiO_4 + FeSiO_3 - 0.5Fe_2SiO_4.$$

Notice that this is a mathematical relationship between abstract quantities, not a chemical reaction between phases. Because components are abstract constructions, we are free to use positive or negative amounts of them to arrive at the stoichiometry of any phase in the system. We have tried to emphasize this by using an "equals" sign rather than an arrow.

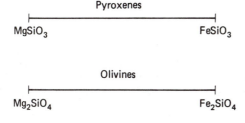

Pyroxenes

$MgSiO_3$ $FeSiO_3$

Olivines

Mg_2SiO_4 Fe_2SiO_4

Figure 2.8 Ferromagnesian olivine and pyroxene solid solutions can each be represented by two end-member components.

The solid triangle in Figure 2.9 illustrates the system as defined by the components on the right side.

As required by the negative amount of Fe_2SiO_4 in the equation above, magnesium-bearing orthopyroxenes have compositions outside the triangle defined by the system components. There is nothing wrong with this representation, but it is awkward for most petrologic applications. A more conventional selection of components is shown in Figure 2.10. The mineral compositions at the ends of the olivine solid solution are still retained as system components, but the third component, SiO_2, does not correspond to either mineral in the system.

It is very important to recognize that components are abstract means of characterizing a system. They do not need to correspond to substances that can be found in nature or manufactured in a laboratory. Orthopyroxenes, for example, are frequently described by the components $MgSiO_3$ and $FeSiO_3$, despite the fact that there is no natural phase with the composition $FeSiO_3$. Monatomic components like F, S, or O are also legal, even though fluorine, sulfur, and oxygen invariably occur as molecules containing two or more atoms. For some petrological applications, it makes sense to use components like $CaMg_{-1}$, which clearly do not exist as real substances. In fact, the carbonates discussed above can be characterized quite well by $MgCO_3$ and $CaMg_{-1}$, as can be seen from the stoichiometric relationship

$$MgCO_3 + CaMg_{-1} \longrightarrow CaCO_3.$$

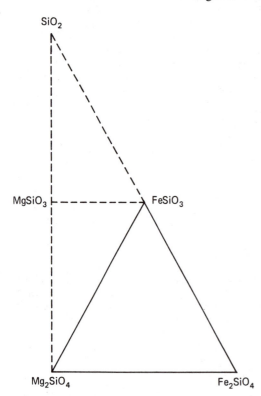

Figure 2.9 If we choose to describe a system containing olivines *and* pyroxenes, we need three components. Phase compositions lying outside the triangle of system components require negative amounts of one or more components.

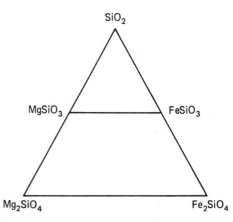

Figure 2.10 An alternative selection of components for the system in Figure 2.9. All olivine and pyroxene compositions now lie within the triangle.

Components of this type, known as *exchange operators*, have been used to great advantage in describing many metamorphic rocks (Thompson et al., 1981; Ferry, 1982).

Worked Problem 2-6

The ratio K/Na in coexisting pairs of alkali feldspars and alkali dioctahedral micas can be used to infer pressure and temperature conditions for rocks in which they were formed. What selection of components might be most helpful if we were interested in $K \longleftrightarrow Na$ exchange reactions between these two minerals?

The exchange operator KNa_{-1} is a good choice for a system component in this case. Feldspar compositions can be generated from

$$NaAlSi_3O_8 + x\, KNa_{-1} = K_xNa_{1-x}AlSi_3O_8$$

for any x between zero and 1. Similarly, mica compositions can be derived from

$$NaAl_3Si_3O_{10}(OH)_2 + y\, KNa_{-1} = K_yNa_{1-y}Al_3Si_3O_{10}(OH)_2.$$

If we also select the two sodium end-member compositions as components, all possible compositional variations in the system can be described. Figure 2.11a shows one way of illustrating this selection. A more subtle diagram, using the same set of components, is presented in Figure 2.11b.

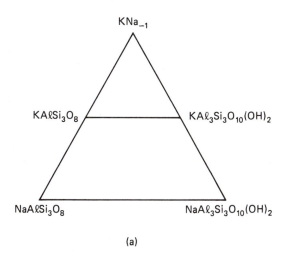

(a)

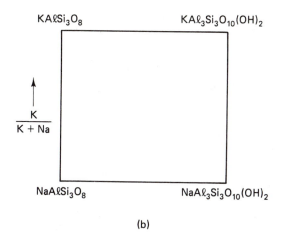

(b)

Figure 2.11 (a) Alkali micas and feldspars can be described by a component set that includes KNa_{-1}. (b) An orthogonal version of the diagram in 2.11a. Its vertical edges both point to KNa_{-1}.

Changes in E, H, F, and G because of Composition

Consider an open system containing only magnesian olivine, a pure phase. The internal energy of the system is equal to $\bar{E} = E_{Fo} n_{Fo}$, where $\bar{E}_{Fo}$ is the molar internal energy for pure forsterite and n_{Fo} is the number of moles of the forsterite component in the system. Suppose that it were possible to add a certain number of moles of the fayalite component, n_{Fa}, to the one-phase system without causing any increase in energy as a result of the mixing itself. Because E is an extensive property, the internal energy of the phase would then be equal to

$$E = \bar{E}_{Fo} n_{Fo} + \bar{E}_{Fa} n_{Fa}. \tag{2-35}$$

Further infinitestimal changes in the amounts of forsterite or fayalite in the system would result in a small change in E, namely

$$dE = \bar{E}_{Fo} \, dn_{Fo} + \bar{E}_{Fa} \, dn_{Fa}. \tag{2-36}$$

This differential equation can be written in the form of a total differential, from which it can be seen that

$$\bar{E}_{Fo} = (\partial E / \partial n_{Fo})n_{Fa} \qquad \text{and} \qquad \bar{E}_{Fa} = (\partial E / \partial n_{Fa})n_{Fo}. \tag{2-37}$$

This is an idealized process, of course, chosen to demonstrate that the internal energy of a phase can be changed by varying its composition, in addition to the ways we have already discussed. It is more realistic to recognize that the mixing process as described involves increases in both entropy and volume. Notice also that the total internal energy of the phase, E, is *not* a molar quantity, because it has not been divided by the total number of moles in the system. The relationships $dE_{olivine} = \bar{E}_{Fo} \, dn$ or $dE_{olivine} = \bar{E}_{Fa} \, dn$ can only be valid if the olivine is either pure forsterite or pure fayalite. In between, as we will see in later chapters, dE takes a nonlinear and generally complicated form. The quantities in equation 2-37 are more useful if written with these restrictions in mind:

$$\bar{E}_i = (\partial E / \partial n_i)_{S,V,nj \neq i}. \tag{2-38}$$

We have now identified a new thermodynamic quantity, the *partial molar internal energy*, which describes the way in which total internal energy for a phase responds to a change in the amount of component i in the phase, all other quantities being equal. It may be thought of, if you like, as a chemical "pressure" or a force for energy change in response to composition, in the same way that pressure is a force for energy change in response to volume. In order to emphasize the importance of this new function, it has been given the symbol μ_i and is called the *chemical potential* of component i in the phase.

The internal energy of a phase, then, is

$$E = E(S, V, n_1, n_2, \ldots, n_i),$$

and equation 2-16 can be rewritten to include the new-found chemical potential terms:

$$dE \leq TdS - PdV + \mu_1 dn_1 + \mu_2 dn_2 + \ldots + \mu_i dn_i$$

$$\leq TdS - PdV + \sum_i \mu_i dn_i \quad . \tag{2-39a}$$

In the same way, it can be shown that the auxiliary functions H, F, and G are also functions of composition:

$$dH \leq TdS + VdP + \mu_1 dn_1 + \mu_2 dn_2 + \ldots + \mu_i dn_i$$

$$\leq TdS + VdP + \sum_i \mu_i dn_i \quad , \tag{2-39b}$$

$$dF \leq -SdT - PdV + \mu_1 dn_1 + \mu_2 dn_2 + \ldots + \mu_i dn_i$$

$$\leq -SdT - PdV + \sum_i \mu_i dn_i \quad , \tag{2-39c}$$

$$dG \leq -SdT + VdP + \mu_1 dn_1 + \mu_2 dn_2 + \ldots + \mu_i dn_i$$

$$\leq -SdT + VdP + \sum_i \mu_i dn_i \tag{2-39d}$$

Chemical potential, therefore, can be defined in several equivalent ways:

$$\mu_i = (\partial E / \partial n_i)_{S,V,n_{j \neq i}} \tag{2-40a}$$

$$= (\partial H / \partial n_i)_{S,P,n_{j \neq i}} \tag{2-40b}$$

$$= (\partial F / \partial n_i)_{T,V,n_{j \neq i}} \tag{2-40c}$$

$$= (\partial G / \partial n_i)_{T,P,n_{j \neq i}} \tag{2-40d}$$

and can be correctly described as the *partial molar enthalpy*, the *partial molar Helmholtz function*, or the *partial molar free energy*, provided that the proper variables are held constant, as indicated in equations 2-40a–d.

CONDITIONS FOR HETEROGENEOUS EQUILIBRIUM

We are now within reach of a fundamental goal for this chapter. Having examined the various ways in which internal energy is affected by changes in temperature, pressure, or composition, we may now ask: "What conditions must be met if a system containing several phases is in internal equilibrium?" This is a circumstance usually referred to as *heterogeneous equilibrium*.[6]

A system consisting of several phases can be characterized by writing an equation in the form of equation 2-39a for each individual phase:

$$dE_\phi \leq T_\phi dS_\phi - P_\phi dV_\phi + \sum_i \mu_{i\phi} dn_{i\phi},$$

where the subscript ϕ is being used to identify properties of the phase of interest. This can also be written as

$$dS_\phi \geq (1/T_\phi)dE_\phi + (P_\phi/T_\phi)dV_\phi - \sum_i (\mu_{i\phi}/T_\phi)dn_{i\phi}. \tag{2-41}$$

We will now proceed to examine the full set of these equations to see what thermodynamic conditions must be satisfied if all phases in the system are in equilibrium. Be-

---

[6] This derivation of conditions for heterogeneous equilibrium will provide a good test of your understanding of principles introduced in this chapter. It is important, therefore, that you fight the temptation to skip to the conclusions in equations 2–43 and 2–45.

cause we are only interested in equilibrium rather than the approach to equilibrium, however, it is proper to ignore the inequality in equation 2-41 for each phase.

In order to avoid worrying about what may happen to the heterogeneous system as a result of outside influences, we will consider it to be isolated. From equation 2-12, we recall that the total entropy change in an isolated system is

$$0 \leq dS_{sys} = \sum_\phi dS_\phi \,,$$

so the equations 2-41 for each of the phases can be combined to yield a master entropy equation:

$$0 \leq \sum_\phi (1/T_\phi)dE_\phi + \sum_\phi (P_\phi/T_\phi)dV_\phi - \sum_\phi \sum_i (\mu_{i\phi}/T_\phi)dn_{i\phi} \,. \qquad (2\text{-}42)$$

The inequality that appears here is not the one we discarded from equation 2-41. *This* inequality arises from equation 2-12. It recognizes that reactions carried out irreversibly among the various phases will increase the total entropy by more than the amount inherited from the individual phases considered separately. Because we are still interested in equilibrium among phases, a state that we have learned can only exist if entropy is maximized, we can safely ignore this inequality as well.

It is easiest to examine the equilibrium condition among phases if we consider only the simple (though geochemically unlikely) situation in which system components correspond one-for-one with components of the individual phases. Because the system is isolated, we know that its extensive parameters must be fixed:

$$dE_{sys} = \sum_\phi dE_\phi = 0 \qquad (2\text{-}43a)$$

$$dV_{sys} = \sum_\phi dV_\phi = 0 \qquad (2\text{-}43b)$$

$$dn_{1,sys} = \sum_\phi dn_{1,\phi} = 0 \qquad (2\text{-}43c)$$

$$dn_{2,sys} = \sum_\phi dn_{2,\phi} = 0 \qquad (2\text{-}43d)$$

$$\vdots$$

$$dn_{i,sys} = \sum_\phi dn_{i,\phi} = 0 \,. \qquad (2\text{-}43i)$$

There is one $dn_{i,sys}$ equation for each component in the system. Despite these equations, though, there is no constraint that prevents the relative values of E, V, or the various ns from readjusting themselves, as long as their totals remain zero. On the other hand, *because* their totals are constrained, the various individual values of E_ϕ, V_ϕ, and the $n_{i,\phi}$ are not independent. Whatever leaves one phase in the system must show up in at least one of the others. Equation 2-42, therefore, contains some redundant terms among its summations.

Eliminate the extra terms in the following way. First, select any phase at random, which will be identified by the subscript x. Multiply equation 2-43a by $(1/T_x)$ and subtract the quantity $\sum_\phi (1/T_x)dE_\phi$, which is equal to zero, from the master entropy equation, 2-42. This procedure eliminates the dE_x terms from the first summation. The remaining dE terms are now independent of each other. Next, repeat the procedure by choosing a phase to be labeled y at random (it could even be the same phase as x, if we wish). This time, multiply equation 2-43b by (P_y/T_y) and subtract the quantity $\Sigma_\phi(P_y/T_y)dV_\phi$ from the master entropy equation. This removes the dV_y terms and leaves the rest in that summation independent. Finally, repeat the procedure as many times as necessary to remove one set of $\Sigma_\phi(\mu_{iz}/T_z)dn_{i\phi}$ terms from the double summation in equation 2-42 for each of the mass constraints in equations 2-43c–i.

This looks like a cumbersome method mathematically, but its purpose is simply to produce a streamlined version of equation 2-42, from which we have removed one term for each of the constraints in equations 2-43a–i and left only independently variable terms:

$$0 = \sum_\phi (1/T_\phi - 1/T_x)dE_\phi + \sum_\phi (P_\phi/T_\phi - P_y/T_y)dV_\phi$$

$$- \sum_\phi \sum_i (\mu_{i\phi}/T_\phi - \mu_{iz}/T_{iz})dn_{i\phi}. \qquad (2\text{-}44)$$

The right side of equation 2-44 is still equal to dS_{sys}, because we have done nothing more than repackaging equation 2-42 to get here. Remember that although equations 2-43a–i are each equal to zero, there is no reason why the internal energies, the volumes, and the number of moles of the various components in the system cannot vary independently. Therefore, the only way to guarantee that the summation terms in equation 2-44 also remain zero is to be sure that the terms in parentheses are all equal to zero. This leads finally to the set of equilibrium conditions among intensive parameters that we have been seeking:

$$T_x = T_A = T_B = \cdots = T \qquad (2\text{-}45a)$$

$$P_y = P_A = P_B = \cdots = P \qquad (2\text{-}45b)$$

$$\mu_{1z} = \mu_{1A} = \mu_{1B} = \cdots = \mu_1 \qquad (2\text{-}45c)$$

$$\mu_{2z} = \mu_{2A} = \mu_{2B} = \cdots = \mu_2 \qquad (2\text{-}45d)$$

$$\vdots \qquad \vdots \qquad \vdots \qquad \vdots \qquad \vdots$$

$$\mu_{iz} = \mu_{iA} = \mu_{iB} = \cdots = \mu_i \qquad (2\text{-}45i)$$

We already knew or suspected the first two of these conclusions: In order for all phases in a system to be in equilibrium, they must be at the same temperature (thermal equilibrium) and under the same pressure (mechanical equilibrium). The unfamiliar constraints are the ones relating to chemical potentials. Because these are so crucial in geochemistry, we will emphasize them here: *At equilibrium, the chemical potential of any component must be the same in all phases in a system.*

The results in equations 2-45a–i do not depend on the decision to consider only the case in which system components and phase components match one-for-one. For the more common situation, in which system components differ from phase components, there are stoichiometric equations relating them. These have the following form:

One unit of phase component $i = \sum_j m_{ji}$ units of system component j,

in which the coefficients m_{ji} tell us how many units of each system component are involved in making one unit of phase component i. This set of relationships can be solved for each of the system components:

$$n_{j,sys} = \sum_j n_{j\phi} = \sum_\phi \sum_i m_{ji} n_{i\phi}.$$

The chemical constraints represented by equations 2-43c-i are now interpreted as constraints on the system components j, and have become

$$dn_{1,sys} = \sum_\phi \sum_i m_{1i} dn_{i\phi} = 0$$

$$dn_{2,sys} = \sum_\phi \sum_i m_{2i} dn_{i\phi} = 0$$

$$\vdots$$

$$dn_{j,sys} = \sum_\phi \sum_i m_{ji} dn_{i\phi} = 0.$$

To get the repackaged entropy equation equivalent to equation 2-44, we follow the same procedure as before, except that the composition terms chosen for elimination are all ones in which a system component is always a phase component in some real or potential phase z. The final equation, then, is:

$$0 = \sum_\phi (1/T_\phi - 1/T_x) dE_\phi + \sum_\phi (P_\phi/T_\phi - P_y/T_y) dV_\phi$$

$$- \sum_\phi \sum_i (\mu_{i\phi}/T_\phi - \sum_j (\mu_{jz}/T_z) m_{ji}) dn_{i\phi}. \tag{2-46}$$

The equilibrium conditions in equations 2-45a–i follow directly from this equation in exactly the same way they did from equation 2-44.

To see whether the implications of this section are clear, study the following problem closely.

Worked Problem 2-7

An experimental igneous petrologist, working in his laboratory, has produced a run product which consists of quartz and a pyroxene intermediate in composition between $FeSiO_3$ and $MgSiO_3$. Assuming that the two minerals were formed in equilibrium, what conditions must have been satisfied?

The most obvious conditions are those of thermal and mechanical equilibrium, which we expressed in equations 2-45a and 2-45b:

$$T_{qz} = T_{cpx} = T$$

$$P_{qz} = P_{cpx} = P$$

The compositional conditions are less obvious, particulary if you are still uncertain about what components and chemical potential are. As presented in equations 2-40a–d, a chemical potential is a measure of the way in which the energy of a phase changes if we change the amount of a component in the phase. To find the compositional conditions, then, we must first choose a set of components for the system. Several selections are possible, some of which we considered in Worked Problem 2-5. This time, let's choose the end-member mineral compositions $FeSiO_3$, $MgSiO_3$, and SiO_2. From equations 2-45c–i, it follows that

$$\mu_{SiO_2,qz} = \mu_{SiO_2,cpx} = \mu_{SiO_2}$$

$$\mu_{FeSiO_3,qz} = \mu_{FeSiO_3,cpx} = \mu_{FeSiO_3}$$

$$\mu_{MgSiO_3,qz} = \mu_{MgSiO_3,cpx} = \mu_{MgSiO_3}.$$

Notice that the chemical potentials of $FeSiO_3$ and $MgSiO_3$ are defined in quartz, despite the fact that they are not components of the mineral quartz itself, and μ_{SiO_2} is defined in pyroxene, although SiO_2 is not a component of pyroxene! This may seem contradictory, but is quite normal. All three are *system* components, and the various chemical potentials express the relative sensitivity of each of the phases to changes in system composition. The statement about $FeSiO_3$, for example, should be read to mean "At constant temperature and pressure, the derivative of free energy with respect to the number of moles of $FeSiO_3$ is identical in pyroxene and in quartz." Therefore, if it were possible to add the same infinitesimal amount of $FeSiO_3$ to each, the free energies of the two phases would each change by the same amount:

$$dG = \mu_{FeSiO_3}dn_{FeSiO_3}.$$

THE GIBBS-DUHEM EQUATION

One final, very useful relationship can be derived from this discussion of chemical potentials. Consider equation 2-39a for a phase that is in equilibrium with other phases around it. It is possible to write an integrated form of equation 2-39a:

$$E = TS - PV + \sum_i \mu_i n_i.$$

Therefore, the differential energy change, dE, that takes place if the system is allowed to leave its equilibrium state by making small changes in any intensive or extensive properties, is

$$dE = TdS + SdT - PdV - VdP + \sum_i \mu_i dn_i + \sum_i n_i d\mu_i. \qquad (2\text{-}47)$$

In order to see how the intensive properties alone are interrelated, subtract equation 2-39a, which was written in terms of variations in extensive parameters alone, from equation 2-47 to get

$$0 = SdT - VdP + \sum_i n_i d\mu_i. \qquad (2\text{-}48)$$

This expression is known as the *Gibbs-Duhem* equation. It tells us that there can be no net gradients in intensive parameters at equilibrium. Of more specific interest in geochemical problems, at constant temperature and pressure there can be no net gradient in internal energy as a result of composition in a phase in internal equilibrium. This can be seen as another way of stating the results of equations 2-45c–i.

SUMMARY

What is equilibrium? Our discussion of the three laws of thermodynamics has led us to discover several characteristics that answer the question. First, systems in equilibrium must be at the same temperature; this is the condition of *thermal equilibrium*. Second, provided that work is exclusively defined as the integral of PdV, there can be no pressure gradient between systems in equilibrium; this is the condition of *mechanical equilibrium*. Finally, the chemical potential of each system component must be the same in all phases at equilibrium. This is the least obvious of the conditions we have discussed, and will be the focus of many discussions in chapters to come.

We have also repackaged the internal energy function in several ways, and have examined the role of entropy, which will always be maximized during the approach to equilibrium. In order to demonstrate that each of the equilibrium criteria is a direct consequence of the three laws, we have used the internal energy function, $E(S,V)$, as the basis of our discussion. As we progress, though, we will quickly abandon E in favor of the Gibbs Free Energy function. Any of the energy functions, however, can be used to clarify the conditions of equilibrium.

With the ideas we have introduced in this chapter, we are now ready to begin exploring geologic environments. This has been only a first look at thermodynamics, however. We will return to that topic many times.

SUGGESTED READINGS

There are many good introductory texts in thermodynamics, though most are written for chemists, rather than for geochemists. Below are listed some of the best—or at least most widely used—texts for geochemists.

ATKINS, P. W. 1984. *The second law.* New York: W. H. Freeman. (An extremely lucid discussion of entropy and its implications, presented with almost no mathematics.)

DENBIGH, K. 1968. *The principles of chemical equilibrium.* London: Cambridge Univ. Press. (Heavy going, but almost anything you want to know is in here somewhere.)

FRASER, D. G., ED. 1977. *Thermodynamics in geology.* Dordrecht, Holland: D. Reidel. (Topical chapters showing both fundamentals and applications of thermodynamics.)

GREENWOOD, H. J., ED. 1977. *Short course in application of thermodynamics to petrology and ore deposits.* Ottawa, Mineral. Assoc. of Canada. (A very readable book, with many sections of interest to economic geologists.)

KERN, R., and WEISBROD, A. 1967. *Thermodynamics for geologists.* San Francisco: Freeman, Cooper, & Co. (Perhaps the simplest treatment of thermodynamics written specifically for the geologist, although a bit dated by now.)

LEWIS, G. N., and RANDALL, M. 1961. *Thermodynamics.* Revised by K. S. Pitzer and L. Brewer. New York: McGraw-Hill. (A classic text still in widespread use.)

NORDSTROM, D. K., and MUNOZ, J. 1985. *Geochemical thermodynamics.* Menlo Park, Calif.: Benjamin Cummings. (An excellent, highly readable text for the serious student.)

SMITH, E. B. 1982. *Basic chemical thermodynamics.* Oxford: Oxford University Press. (A short, easy book with well-chosen problems, well-suited to the beginning student.)

TUNELL, G. 1985. *Condensed collection of thermodynamic formulas for one-component and binary systems of unit and variable mass.* Washington, D.C.: Carnegie Institute Pub. No. 408B. (Primarily for the professional, but a gold mine for any chemist who finds it necessary to derive arcane thermodynamic quantities from experimental data.)

The following articles provide an excellent though somewhat difficult discussion of compositional variables in geochemistry:

FERRY, J. M., ED. 1982. *Characterization of metamorphism through mineral equilibria.* Reviews in Mineralogy. Washington, D.C.: Mineral. Soc. of Amer. (An enlightening selection of articles designed for classroom use. Chapters 1–4 are particularly relevant to our first discussion of thermodynamics. Be prepared to stretch.)

THOMPSON, J. B.; LAIRD, J.; and THOMPSON, A. B. 1982. Reactions in amphibolite, greenschist, and blueschist. *J. Petrol.* 23: 1–27. (A good study for the motivated student. Reactions in amphibole-bearing assemblages are discussed in terms of exchange operators.)

The historical development of our notions of heat and temperature is discussed in chapter 4 of this thought-provoking book:

GOLDSTEIN, M., and GOLDSTEIN, I. F. 1978. *How we know: an exploration of the scientific process.* New York: Plenum.

This article does not deal with thermodynamics directly, but is a commentary on the use of symbols. Its author was better known by his real name, Hugh McKinstry, Professor of Geology at Harvard for many years.

VANSBERG, N. 1958. Mathmanship. *Amer. Scientist* 46: 94A–98A.

PROBLEMS

1. Refer to Figs. 2.9 and 2.10. Notice that by adding or subtracting SiO_2, we change the proportions of olivine and pyroxene in a rock, but not their compositions. Changing the amount of any other system component in Figure 2.9 or 2.10, however, alters both the compositions and the proportions of minerals. Can you find another set of components that includes one that, when varied, changes the compositions but not the proportions of olivine and pyroxene?

2. Show that the work done by 1 mole of gas obeying Van der Waal's equation (2-3) during isothermal expansion from $\bar{V}_1$ to $\bar{V}_2$ is equal to
$$w = RT \ln ((\bar{V}_2 - b)/(\bar{V}_1 - b)) + a((1/\bar{V}_2) - (1/\bar{V}_1)).$$

3. How much work can be obtained from the isothermal, reversible expansion of one mole of chlorine gas from 1 to 50 liters at 0°C, assuming ideal gas behavior (equation 2-2)? What if chlorine behaves as a Van der Waal's gas (equation 2-3)? (Use the result of problem 2. Insert the values $a = 6.493$ l^2 atm mol^{-2}, $b = 0.05622$ l mol^{-1}.)

4. What is the maximum work that can be obtained by expanding 10 grams of helium, an ideal gas, from 10 to 50 liters at 25°C? Express your answer in
   (a) calories,
   (b) Joules.

5. What conditions must be satisfied if hematite, magnetite, and pyrite are in equilibrium in an ore assemblage?

6. Construct a triangular diagram for the alkali feldspar-mica system similar to Figure 2.11a but in which you swap the positions of $NaAl_3 Si_3 O_{10}(OH)_2$ and $KAlSi_3 O_8$. What component must now take the place of KNa_{-1} at the top corner of the diagram? Where would this component have been plotted on Figure 2.11a? Where does KNa_{-1} plot on your diagram?

7. Using the Maxwell relations, verify that $-(\partial V/\partial T)_p/(\partial P/\partial T)_V = (\partial V/\partial P)_T$.

Chapter 3

How to Handle
Solutions

OVERVIEW

This is a chapter about solutions, important to us because most geological materials are composed of more than one substance. They are, in fact, mixtures at the submicroscopic scale among idealized end-member substances like albite, water, grossular, dolomite, or carbon dioxide, which lose their molecular identities in the mixture. In order to see how thermodynamics can be used to predict the equilibrium state in a system dominated by phases with variable compositions, we examine first the structure of solutions of solids, liquids, and gases, and discuss ways in which this architecture is reflected in mole fractions of end-member components.

In Chapter 2, we developed thermodynamic equations (2–43a–i and 2–45a–i) that describe the state of equilibrium for a heterogeneous system. Among these are constraints which imply that the compositions of phases in equilibrium are controlled by the system's overall drive toward a minimum in ΔG. Our appreciation for the structure of solutions allows us to look explicitly at the compositions of phases in equilibrium and to devise ways of using them to apply the thermodynamic principles in Chapter 2. Specifically, we begin to develop equations that relate mole fractions and other mixing parameters (which we introduce) to ΔG.

The results of this investigation can be applied to many problems in the realm of geochemistry. In this chapter, we begin applying them by studying aqueous fluids and the phenomenon of solubility. This study will not only present an opportunity to use the equations we develop, but also prepare the way for our later discussions of the oceans, diagenesis, and weathering reactions.

WHAT IS A SOLUTION?

Most phases of geochemical interest do not have fixed compositions. With the conspicuous exception of quartz, each of the major rock-forming minerals is a solid solution between two or more end-member molecules. The range of allowable compositions in any solution is dictated by rules of structural chemistry that consider both crystal architecture and the size and electronic configuration of ions. For many minerals, these rules permit liberal replacement of iron by magnesium or calcium, sodium by potassium, and silicon by aluminum. To the uninitiated eye, the degree of flexibility that we observe in mineral compositions gives them an appearance of randomness and reduces any hope that they might be used for thermodynamic modeling. Liquids and gases, of course, follow fewer structural guidelines and have even more diverse compositions. Can the principles of equilibrium thermodynamics we have just explored make sense in the more complicated world of these *real* phases? To answer, we first need to know more about solutions.

Crystalline Solid Solutions

Most geologically important materials are crystalline solids; that is, solids in which atoms are arranged in a periodic three-dimensional array that maintains its gross structural identity over large distances. A "large" distance is not easily defined, however. If we see a mineral grain in thin section that has continuous optical properties, it is clearly large enough. Even submicroscopic grains are judged to be crystalline if their structures are continuous over distances sufficient to yield an X-ray diffraction pattern. For most geochemical purposes, this distance is small enough, although modern electron imaging techniques allow us to see periodic structure even at the unit cell level.

The thermodynamic behavior of crystalline materials is determined almost entirely by their average properties over distances of thousands of unit cell repeats. There are circumstances in which this is not true, some of which we investigate in Chapter 10. Among them are extremely fine-grained crystalline aggregates, in which surface properties make a major contribution to free energy; highly stressed materials, in which dislocations or defects constitute a significant volume fraction of the material; and intimately intergrown materials, like those in Figure 3.1, in which two or more phases are structurally compatible and randomly intermixed. In this chapter, however, we will disregard these complications.

Regardless of the scale of structural continuity, the rules which dictate where atoms sit in a crystalline substance are quite flexible. Because oxygen and silicon are the most abundant elements in the crust, most rock-forming minerals consist of a framework of oxygen and silicon atoms. In the following brief discussion, therefore, we refer to silicate minerals, although it should be understood that the principles may be applied to any mineral.

The atoms in silicates are arranged in such a way that each silicon is surrounded by four oxygens to form a tetrahedral unit with a net electrical charge of -4. These units are connected to each other and to other atoms by bonds that are largely ionic in nature. The manner in which the units are linked together, and the identity of other atoms which occupy sites between them, determines the mineral's structure and establishes its bulk chemical identity.

Two major factors govern which atoms may occupy a given site: their ionic charges and their sizes. Charge is important because of the need to obtain electrical

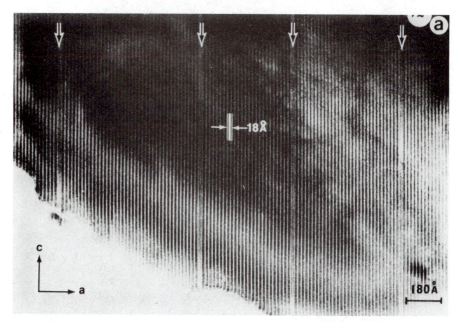

Figure 3.1 High-Resolution Transmission Electron Microscope (HRTEM) image of an intergrowth of ortho- and clinopyroxenes. This material is crystalline, but the identity of the structural unit changes after a random, rather small number of repeats. Thermodynamic characterization of such material is very difficult. (Courtesy of P. N. Buseck.)

neutrality over the structure. Size is important because it influences the degree of "overlap" between the orbitals of valence electrons of the ion and those which surround it. As is apparent from Figure 3.2, however, several ions available to a growing crystal will satisfy these constraints. In addition, the charge balance requirement can be satisfied by the formation of randomly distributed vacancies or by inserting ions into defects or into positions that are normally unoccupied. Therefore, local charge imbalances caused by placing an inappropriately charged ion on a site can be averaged out over the structure as a whole. As a result, minerals are most properly viewed as crystalline solutions in which many competing ions substitute freely for one another. As we will find in later sections of this chapter, both the bulk composition of a solution and the way in which it is mixed will influence its thermodynamic stability.

To illustrate, let us examine the clinopyroxene solid solution $CaMgSi_2O_6$ (diopside) $-$ $NaAlSi_2O_6$ (jadeite). As shown in Figure 3.3, these are C2/c pyroxenes, in which cations occupy three types of sites.[1] Silicon atoms are in SiO_4 tetrahedra, linked by corners to form long, single chains. Between these chains are two other types of cation sites, designated M1 and M2. Of these, M1 is surrounded by 6 oxygens and is nearly a regular octahedron. It is occupied by the relatively small cations Mg^{2+} and Al^{3+}. The M2 site is 8-coordinated, 20–25% larger, and irregular in shape. Ca^{2+} and Na^+, relatively large cations, prefer this site.

If we consider only the pure *end-member* pyroxenes, we have no trouble deciding how sites are occupied, because there is only one type of cation available for each site. Once we investigate intermediate compositions, however, two extreme possibilities, in-

---

[1] For a detailed but very readable description of pyroxene crystal chemistry, see Cameron and Papike (1980) or Cameron and Papike (1981).

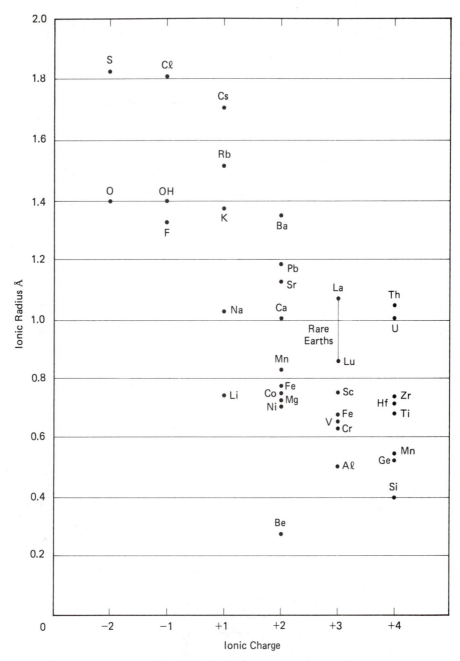

Figure 3.2 The relationship between ionic radius and charge for some common elements. In general, an ion has a larger radius if it has (a) a higher number of electron shells (that is, a higher principal quantum number), or (b) a larger number of electrons in valence orbitals. Compare this diagram with a periodic table to verify these qualitative rules. (After B. Mason and C. B. Moore, *Principles of geochemistry, 4th Edition*. Copyright 1982 by John Wiley and Sons. Reprinted with permission.)

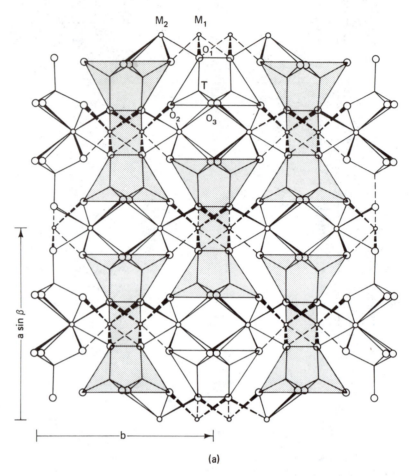

a sin β

b

(a)

Figure 3.3 The crystal structure of C2/c pyroxene, showing the geometry of M1 and M2 cation sites, each of which is surrounded by six oxygen atoms, and their relationship to tetrahedral (T) sites which are largely occupied by silicon atoms. These diagrams are drawn to emphasize the geometry of cation sites. Thus, atoms are only indicated schematically, or not at all. (a) Projection down the [001] crystallographic axis. (b) Projection onto the (100) crystallogrphic plane, shown on facing page. (After M. Cameron and J. J. Papike, Crystal chemistry of silicate pyroxenes. In *Pyroxenes,* ed. C. T. Prewitt, 5–92. Reviews in Mineralogy 7. Copyright 1980 by the Mineralogical Society of America. Reprinted with permission.)

dicated schematically in Figure 3.4, suggest themselves. The first, which we may call *molecular mixing,* assumes that charge balance is always maintained over very short distances. That is, calcium and magnesium ions substitute for sodium and aluminum ions as a coupled pair: each time we find a Ca^{2+} in an M2 site, there is a Mg^{2+} in an adjacent M1 site. The end-member pyroxenes mix as discrete molecules and complete short-range order results. In this case, the mole fraction of $CaMgSi_2O_6$ in the solid solution ($X_{CaMgSi_2O_6,cpx}$) is equal to the proportion of the diopside molecule present. Molecular mixing has been widely assumed by petrologists. The other possibility, known as *mixing on sites,* assumes that Ca^{2+} and Na^+ are randomly distributed on M2 sites, unaffected by whether adjacent M1 sites are filled with Mg^{2+} or Al^{3+} ions. This style of mixing may produce local charge imbalances, but the average stoichiometry is satisfied even in the absence of short-range order. Because each of the sites, according

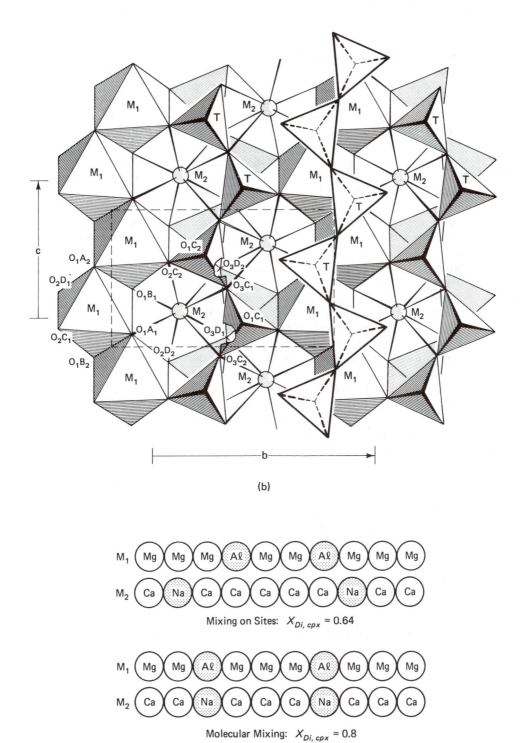

Figure 3.4 A schematic representation of a Di-Jd pyroxene in which M1 is 80% Mg, 20% Na, and M2 are 80% Ca, 20% Al. Using mixing on sites, $X_{Di,cpx} = 0.64$ and $X_{Jd,cpx} = 0.04$. Using a molecular mixing model, $X_{Dicpx} = 0.8$ and $X_{Jd,cpx} = 0.2$.

What Is a Solution? 51

to this model, behaves independently of the others, the mole fraction $X_{CaMgSi_2O_6,cpx}$ must now be calculated as the product of $X_{Ca,M2,cpx}$ and $X_{Mg,M1,cpx}$. This follows from a basic principle of statistics: If the probability of finding a Mg^{2+} ion in M1 is some value x and the probability of finding a Ca^{2+} in M2 is y, then the combined probability of finding both ions in the structure is xy. Thus, if 80% of the ions in the M1 site are Mg^{2+} and 80% of the M2 atoms are Ca^{2+}, $X_{CaMgSi_2O_6}$ is equal to 0.64. If the tetrahedral site were partially occupied by something other than silicon, we would have to multiply by $X_{Si,T,cpx}$ as well.

In a qualitative sense, we may infer that the way in which atoms are distributed on sites will affect the molar entropy of a phase, and therefore each of the energy functions E, H, F, and G. In theory, it should be possible to use an atom-by-atom approach to calculate the relative stabilities of clinopyroxene solutions based on these two extreme models. To do this would require calculating bond energies from electrostatic equations and detailed crystallographic data. Large-scale computer models exist for this purpose (for example, Busing, 1981), but are successful only with very simple structures. Cohen (1986), however, has evaluated a large body of calorimetric data and phase equilibrium studies for aluminous clinopyroxenes in light of these alternate mixing models and concludes that the macroscopic behavior of pyroxene solid solutions provides no observational justification for molecular mixing. His conclusions suggest that other common silicate solutions (feldspars, micas, amphiboles) behave similarly. We may assume, therefore, that short-range ordering is not significant unless crystallographic evidence suggests otherwise, and thus that mixing takes place on sites.

In this discussion of clinopyroxenes, we have assumed that each of the cations occupies only one type of site. In fact, this is not the case. Although Mg^{2+} prefers the smaller M1 site, it may also be present in M2. Fe^{2+} and Mn^{2+} may also be found in either M1 or M2. As a result, models of site occupancy rely strongly on observations made by spectroscopic and X-ray diffraction methods, and on our understanding of the rules of crystal chemistry that govern element partitioning. In practice, there is considerable uncertainty in this exercise, so most calculations can be performed only by making simplifying assumptions. Other possible sources of uncertainty include the stabilizing effect felt by certain transition metal cations as a result of site distortions (more about this in Chapter 11), and the degree of covalency of bonds in the structure (particularly in sulfide solid solutions). Because we generally have no way of testing each of the various mixing models directly, thermodynamic interpretation of crystalline solutions is an empirical process, based largely on the macroscopic behavior of materials.

Worked Problem 3-1

As an illustration of the way in which we can calculate mole fractions in a disordered mineral structure, consider the following analysis of a C2/c omphacite, which was first reported by Clark et al. (1969) and then included in the discussion by Cameron and Papike (1980). The abundances are recorded as numbers of cations per 6 oxygens in the clinopyroxene unit cell.[2]

Atom	Abundance
Si	1.995
Al	0.238
Fe^{3+}	0.123
Fe^{2+}	0.116
Mg	0.582
Ca	0.583
Na	0.325

[2] The analysis in Clark et al. (1969) also includes 0.002 Ti^{3+}, which we have omitted for simplicity. Including Ti^{3+} would cause an insignificant change in the contents of the M1 site.

To estimate the site occupancy, we first assume that all tetrahedral sites are filled with either Si or Al (the only two atoms in the analysis that regularly occupy tetrahedral sites). Because the ratio of tetrahedral cations to oxygens in a stoichiometric pyroxene should be 2/6, we see that 0.005 Al must be added to the 1.995 Si to fill the site. An alternative, which might also be justifiable, would be to assume that Si is the only tetrahedral cation, and simply normalize the Si in the analysis to 2.00.

Of the remaining cations, we may assume that Ca and Na (both of which are large) are restricted to the M2 site, while Al and Fe^{3+} (both relatively small) prefer the more compact M1 site. In addition, X-ray structure refinement by Clark et al. indicates that the ratio Fe/Mg in M1 is 0.40 and in M2 is 0.54. Using these pieces of information, we proceed as follows:

1. The amount of Fe^{2+} in M1 is some unknown quantity we may call x. The amount of Mg in M1, also unknown, we may call y. We know, however, that $(Fe^{3+} + x)/y = 0.4$.

2. The amount of Fe^{2+} in M2 is unknown, but must be equal to $0.116 - x$, just as the amount of Mg in M2 must be $0.582 - y$. We know, therefore, that $(0.116 - x)/(0.582 - y) = 0.54$.

3. If we solve the two equations in x and y simultaneously, we conclude that $x = 0.092$ and $y = 0.538$.

4. The total cation abundance in M1 (ΣM1) is, therefore,

$$\Sigma M1 = (Al) + (Fe^{3+}) + x + y$$
$$= 0.233 + 0.123 + 0.092 + 0.538 = 0.986.$$

The total cation abundance in M2 (ΣM2), similarly, is

$$\Sigma M2 = (Ca) + (Na) + (0.116 - x) + (0.582 - y)$$
$$= 0.583 + 0.325 + 0.024 + 0.044 = 0.976.$$

5. In order to calculate the mole fractions on each site, we normalize their respective contents by $1/\Sigma M1$ or $1/\Sigma M2$ to get

M1		M2	
Cation	$X_{,M1}$	Cation	$X_{,M2}$
Al	0.236	Ca	0.598
Fe^{3+}	0.125	Na	0.333
Fe^{2+}	0.093	Fe^{2+}	0.025
Mg	0.546	Mg	0.045

One way to model this pyroxene is to choose likely molecular end members such as jadeite ($NaAlSi_2O_6$), diopside ($CaMgSi_2O_6$), aegirine ($NaFe^{3+}Si_2O_6$), among others. We might then calculate mole fractions of each on the assumption that an M1 site occupied by Al is always adjacent to an M2 site containing Na, an M2 occupied by Na is either adjacent to an Al or an Fe^{3+} in M1, and so forth. We have argued that this is generally a poor way to proceed, however. Instead, if we wish to consider a reaction in which the $NaAlSi_2O_6$ component of this pyroxene is involved, we calculate its mole fraction as the cumulative probability of finding both Al and Na at random on their respective sites:

$$X_{Al,M1,cpx} X_{Na,M2,cpx} = (0.236)(0.333) = 0.079.$$

Likewise,

$$X_{Aeg,cpx} = X_{Fe^{3+},M1,cpx} X_{Na,M2,cpx} = (0.125)(0.333) = 0.042$$

and

$$X_{Di,cpx} = X_{Mg,M1,cpx} X_{Ca,M2,cpx} = (0.546)(0.598) = 0.327.$$

What Is a Solution?

Amorphous Solid Solutions

High-resolution electron imagery and the variation of physical properties with composition suggest that glasses and other amorphous geological materials have structures which differ from crystalline materials only in their degree of coherency. An amorphous solid may consist of a large number of very small-scale domains, each of which has an ordered structure but which, taken together, display no long-range structural continuity. Many of the concerns we had about mixing when discussing crystalline solids are also valid for amorphous materials. Their greatest importance for geochemists, however, is that they provide a conceptual bridge toward our understanding of liquids.

Melt Solutions

The only geologically significant melt systems that we understand in any detail are silicate magmas. Our knowledge of silicate melt structures has been acquired primarily since the late 1970s, with the help of improved X-ray diffraction and spectroscopic methods. It is now understood that silicon and oxygen atoms continue to associate in SiO_4 tetrahedral units beyond the melting point, and that the units tend to form polymers. Depending on the composition of the melt, these may be chain polymers similar to the ones found in pyroxenes, or may be branched structures of various sizes—the remnants (or precursors) of sheet and framework crystalline structures.

The degree of polymerization depends on temperature, but is most strongly influenced by the proportion of *network-formers* like silicon and aluminum and *network-modifiers* like Fe^{2+}, Mg^{2+}, Ca^{2+}, K^+, and Na^+. In silica-rich melts, the highly covalent nature of the Si-O bond tends to stabilize structures in which a large number of oxygen atoms are shared between SiO_4 units. As a result, these melts are highly viscous and have low electrical conductivities. The addition of even a small proportion of Na_2O or another network modifier, however, reduces viscosity drastically and increases melt conductivity. As in crystalline silicates, the bonds between oxygen and the larger cations are ionic. These break more easily than the Si-O bonds, thus reducing the melt's overall tendency to form large, tenacious polymeric structures. Water also acts as a network modifier in silica-rich melts, apparently by reacting with the bridging oxygens in a polymer. This reaction may be described generically by

$$H_2O + \ \overset{|}{\underset{|}{-Si}}-O-\overset{|}{\underset{|}{Si}}- \ \longleftrightarrow \ 2(-\overset{|}{\underset{|}{Si}}-OH)$$

Figure 3.5 illustrates the qualitative relationship between viscosity and composition to suggest the degree to which melt structure influences thermodynamic properties.

Worked Problem 3-2

Before much of the modern work by Raman spectroscopy and nuclear magnetic resonance was begun that examined the relationship between melt structure and viscosity, Bottinga and Weill (1972) developed a simple, empirical method for predicting melt viscosity. They noted that the natural logarithm of viscosity (η) may be successfully approximated by a function of composition that is linear within restricted intervals of SiO_2 content. Specifically, they showed that the empirical coefficients D_i in the equation

$$\ln \eta = \sum_i X_i D_i$$

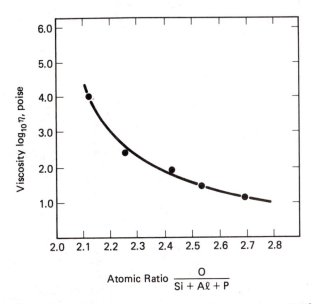

Figure 3.5 Relationship between melt viscosity and composition at 1150°C. The variable on the horizontal axis is the atomic ratio of oxygen atoms to network-formers (Si + Al + P). The dots represent compositions of natural magmas.

are constants at a given temperature and a given value of X_{SiO_2}. In the table below are values of D_i at 1400°C for melts with two different silica contents, calculated statistically by Bottinga and Weill from the measured viscosities of a large number of silicate melts. You will notice from the arithmetic signs on D_i that some melt components tend to increase viscosity (that is, they act as network-formers), while others tend to decrease it (by acting as network-modifiers). Notice also that some components, like FeO, behave differently in silica-rich melts than in silica-poor ones.

	$0.75 < X_{SiO} < 0.81$	$0.55 < X_{SiO_2} < 0.65$
Component	D_i	D_i
SiO_2	10.50	9.25
TiO_2	−3.61	−4.26
FeO	6.17	−6.64
MnO	−6.41	−5.41
MgO	−2.23	−4.27
CaO	−3.61	−5.54
$MgAl_2O_4$	−4.82	−1.71
$CaAl_2O_4$	−1.74	−0.22
$NaAlO_2$	15.1	7.48
$KAlO_2$	15.1	7.48

To illustrate this method, we will consider two different melt compositions, a granite and a diabase. For each, we have tabulated a bulk analysis in weight percent of oxides and a recalculated analysis in mole percent. In the recalculation, all iron has been converted to FeO, phosphorous has been added to SiO_2, water has been omitted (a serious omission, but at least we can compare anhydrous melt viscosities this way), and the total has been adjusted to 100 mole percent.

What Is a Solution?

Component	Granite Wt. %	Diabase Wt. %
SiO_2	70.18	50.48
TiO_2	0.39	1.45
Al_2O_3	14.47	15.34
Fe_2O_3	1.57	3.84
FeO	1.78	7.78
MnO	0.12	0.20
MgO	0.88	5.79
CaO	1.99	8.94
Na_2O	3.48	3.07
K_2O	4.11	0.97
H_2O	0.84	1.89
P_2O_5	0.19	0.25

Component	Mole %	Mole %
SiO_2	80.12	58.45
TiO_2	0.38	1.26
FeO	1.83	10.83
MnO	0.11	0.12
MgO	1.02	9.95
CaO	0.00	4.76
$MgAl_2O_4$	0.47	0.00
$CaAl_2O_4$	2.41	6.28
$NaAlO_2$	7.68	6.85
$KAlO_2$	5.97	1.43

For the granite,

$$\ln \eta = \sum_i X_i D_i = (.8013)(10.5) + (.0038)(-3.61) + (.0183)(6.17)$$
$$+ (.0011)(-6.41) + (.0102)(-2.23) + (.0047)(-4.82)$$
$$+ (.0241)(-1.74) + (.0768)(15.1) + (.0597)(15.1) = 10.48.$$
$$\eta = 3.56 \times 10^4 \text{ Poise} = 3.56 \times 10^3 \text{ kg m}^{-1} \text{ sec}^{-1}.$$

For the diabase,

$$\ln \eta = \sum_i X_i D_i = (.5845)(9.25) + (.0126)(-4.26) + (.1083)(-6.64)$$
$$+ (.0019)(-5.41) + (.0995)(-4.27) + (.0476)(-5.54)$$
$$+ (.0628)(-0.22) + (.0685)(7.48) + (.0143)(7.48) = 4.54.$$
$$\eta = 9.37 \times 10^1 \text{ Poise} = 9.37 \text{ kg m}^{-1} \text{ sec}^{-1}.$$

To an even greater extent than is true for crystalline solids, our limited understanding of melt structures makes it impossible to calculate thermodynamic quantities from models that consider the potential energy contributions of individual atoms. Enthalpies and free energies of formation for melts are determined on macroscopic systems, and therefore represent an average of the contributions from an extremely large number of local structures. Thermodynamic models that make simple assumptions about how mixing takes place between melts of different end-member compositions are surprisingly successful, but tend to be more descriptive than predictive. We will discuss these more fully in Chapter 8.

Electrolyte Solutions

An *electrolyte solution* is one in which the dissolved species (the *solute*) are present in the form of ions in a host fluid that consists primarily of molecular species (the *solvent*). In the geological world, these are usually aqueous fluids, although CO_2-rich solvents may be increasingly important with depth in the crust and upper mantle. Electrolyte solutions do not exhibit the large-scale periodic structures found in either melts or solids. The ions and solvent molecules, however, cannot generally be treated as a mechanical mixture of independent species. Electrostatic interactions among solute and solvent species place limits on the concentration of free solute ions in solution and can produce complexes which influence its thermodynamic behavior. These effects are most obvious as the concentration of dissolved species increases, and require a rather complicated mathematical treatment. Many of the later sections of this chapter are devoted to understanding the behavior of aqueous electrolyte solutions.

Gas Mixtures

Gases at low pressure mix nearly as independent molecules. In this way, they are perhaps the simplest of geologic materials and therefore the best to begin studying. Except for transient species in high-energy environments, most gas species are electrically neutral. That characteristic and the relatively large distances between molecules at low pressure allow us to treat atmospheric gases and most volcanic gases as mechanical mixtures. As such gas mixtures interact with other geological materials (say, in weathering reactions), the thermodynamic influence of each of the individual gas species is directly proportional to its abundance in the mixture. It is only at elevated pressures that gas molecules interfere with each other and begin to alter the thermodynamic behavior of the mixture.

SOLUTIONS THAT BEHAVE IDEALLY

For each type of solution we have considered in this brief discussion, there is some range of composition, pressure, and temperature over which the end-members mix as nearly independent species. For many, this range is distressingly narrow, but it provides us a simplified system from which we can extrapolate to examine the behavior of "real" solutions. When the end-member constituents of solutions act nearly as if they were independent, we refer to their behavior as *ideal*.

In this section and the following one on nonideal solutions, we will begin with a discussion of gases. This is done partly because gases are a ubiquitous part of the geological environment, but largely because their behavior is the easiest to model.

We have already seen (equation 2-29) that the Gibbs Free Energy function for a one-component system can be written as $G(T, P)$. In differential form:

$$dG = -SdT + VdP. \tag{3-1}$$

Furthermore, dG can be recognized as $dG = n\, d\overline{G} = n\, d\mu$. Suppose we are interested in the difference in chemical potential between some reference state for the system (for which we will continue to use the superscript ^{0}) and another state in which temperature or pressure is different. To extract this information, we must integrate equation 3-1:

$$n \int_{\overline{G}^0}^{\overline{G}} d\overline{G} = n \int_{\mu^0}^{\mu} d\mu = - \int_{T^0}^{T} S \, dT + \int_{P^0}^{P} V \, dP. \qquad (3\text{-}2)$$

The integration, unfortunately, is harder than it looks. It is permissible to move the variable n outside of the integrals for $\overline{G}$ and μ because the amount of material in a system is not a function of its molar free energy. We could not do the same with S or V, however, because they are, respectively, functions of T and P. In order to make the problem solvable, we need to supply $S(T)$ and $V(P)$. The first of these is difficult enough to define that we will avoid it for now by declaring that we will only be interested in isothermal processes. The integral of $S \, dT$, therefore, is equal to zero. Later, in Chapter 8, we will consider temperature variations.

The pressure-volume integral is more tractable. The function $V(P)$ is obtained from an equation of state, the most familiar of which is the ideal gas equation: $V = nRT/P$. Equation 3-2 for an isothermal ideal gas, therefore, is reduced to

$$n \int_{\mu^0}^{\mu} d\mu = nRT \int_{P^0}^{P} (dP/P), \qquad (3\text{-}3)$$

from which we see that

$$\mu - \mu^0 = RT \ln P/P^0. \qquad (3\text{-}4)$$

It is standard practice to choose the reference pressure as unity, so that equation 3-4 becomes

$$\mu = \mu^0 + RT \ln P. \qquad (3\text{-}5)$$

(Notice that because P in equation 3-5 really stands for P/P^0, it is a dimensionless number.) We have now derived an expression for the chemical potential of a single component in a one-component ideal gas at a specified temperature and pressure.

Next, let us generalize the problem by considering a phase that is a multicomponent mixture of ideal gases. There must be a stoichiometric equation that describes all of the potential interactions between species ($A_1, A_2, A_3, \ldots, A_i$) in the gas:

$$\nu_{j+1} A_{j+1} + \cdots + \nu_{i-1} A_{i-1} + \nu_i A_i = \nu_1 A_1 + \nu_2 A_2 + \cdots + \nu_j A_j.$$

The quantities ν_1 through ν_i are stoichiometric coefficients, arranged here so that 1 through j refer to product species and $j + 1$ through i refer to reactants. (If the gas were a two-component system of carbon and oxygen species, for example, and if we were to ignore all possible species except for CO_2, CO, and O_2, then the stoichiometric equation would be

$$CO + 1/2 \, O_2 = CO_2.$$

Counting the product species first, CO_2, CO, and O_2 are A_1, A_2, and A_3, respectively, and their stoichiometric coefficients are $\nu_1 = \nu_2 = 1$ and $\nu_3 = 0.5$.) The expanded form of equation 3-1 that describes this multicomponent phase is

$$dG = -SdT + VdP + \sum_{k=1}^{j} \mu_k \, dn_k - \sum_{k=j+1}^{i} \mu_k \, dn_k, \qquad (3\text{-}6)$$

in which the chemical potentials are properly defined as partial molar free energies (Equation 2-40d):

$$\mu_i = (\partial G/\partial n_i)_{T,P,n_{j \neq i}} = \overline{G}_i.$$

The summation terms in equation 3-6 do not contradict equation 2-39d, where positive and negative compositional effects on free energy were combined implicitly in a single term. Here, we separate them explicitly merely to emphasize that quantities associated with product species (1 through j) are meant to be added to the free energy of the system; those associated with reactant species ($j + 1$ through i) are subtracted.

It may bother you that we began this discussion by using the subscript i to identify real chemical species and have now reverted to the convention of using i to identify components. This is allowable because we recognize that the practice of using i in equation 3-6 to count species implies that there are a number of stoichiometric equations that relate species compositions to system components. Therefore, it is fair to include μ_{CO}, μ_{O_2}, and μ_{CO_2} in equation 3-6 for the two-component gas CO-O_2 because the Gibbs-Duhem equation (2-48) assures us that at equilibrium $\mu_{CO_2} = \mu_{CO} + 0.5\mu_{O_2}$. The individual quantities dn_i in equation 3-6, in other words, are not independent, but are related by

$$
\begin{aligned}
(dn_1/\nu_1) = (dn_2/\nu_2) = \cdots &= (dn_j/\nu_j) \\
&= -(dn_{j+1}/\nu_{j+1}) = \cdots = -(dn_{i-1}/\nu_{i-1}) \\
&= -(dn_i/\nu_i) = d\zeta.
\end{aligned}
\tag{3-7}
$$

The new variable, $d\zeta$, is a convenient measure of the extent of reaction, upon which we will not elaborate further. The important point here is that changes in the amount of each species in the system are proportional by a common factor to their stoichiometric amounts in the reaction.

With this insight, we can rewrite equation 3-6 at constant T and P as

$$
\begin{aligned}
dG &= (\nu_1\mu_1 + \nu_2\mu_2 + \cdots + \nu_j\mu_j - \nu_{j+1}\mu_{j+1} - \cdots - \nu_{i-1}\mu_{i-1} - \nu_i\mu_i)\, d\zeta \\
&= \left(\sum_{k=1}^{j} \nu_k\mu_k - \sum_{k=j+1}^{i} \nu_k\mu_k \right) d\zeta \\
&= \Delta G_r\, d\zeta.
\end{aligned}
\tag{3-8}
$$

The quantity ΔG_r is known as the *free energy of reaction* or the *chemical reaction potential*. At equilibrium, it has the value zero. If $\Delta G_r < 0$, then the reaction proceeds to favor the product species; if $\Delta G_r > 0$, then the reaction is reversed to favor reactant species. You may prefer to view this as an extension of the Gibbs-Duhem equation or, perhaps, as a quantitative explanation of LeChatlier's Principle, which states that an excess of species involved on one side of a reaction causes the reaction to proceed toward the opposite side.

Because this is still a discussion of *ideal* gas mixtures, the equation of state for the gas phase remains $V = nRT/P$, but the total amount of material in the system, n, is now equal to the sum of the amounts of species, n_i. The pressure exerted by any individual species in the mixture is, therefore, a *partial pressure,* defined by

$$
P_i = n_i RT/V = Pn_i/n = PX_i.
\tag{3-9}
$$

We can now write an expression like equation 3-3 for each gas species, and follow a derivation identical to the one for a one-component gas to conclude that

$$
\mu_i = \mu_i^0 + RT \ln P_i.
\tag{3-10}
$$

For the mixture as a whole, then, equations 3-8 and 3-10 tell us that the free energy of reaction can be divided into two parts, one of which refers to the free energy in some standard state (often one in which solutions with variable compositions consist of a limiting end-member) and the other of which tells us how free energy changes as a result of compositional deviations from the standard state. The free energy of reaction is written as

$$\Delta G_r = \sum_i \nu_i \mu_i$$

$$= \sum_i \nu_i \mu_i^0 + RT \sum_i \ln P_i$$

$$= \Delta G^0 + RT \sum_i \ln P_i^{\nu_i}$$

$$= \Delta G^0 + RT \ln \frac{P_1^{\nu_1} P_2^{\nu_2} \cdots P_j^{\nu_j}}{P_{j+1}^{\nu_{j+1}} \cdots P_{i-1}^{\nu_{i-1}} P_i^{\nu_i}}. \tag{3-11}$$

If the gas phase is in internal (homogeneous) equilibrium, then $\Delta G_r = 0$ and

$$\frac{-\Delta G^0}{RT} = \ln \frac{P_1^{\nu_1} P_2^{\nu_2} \cdots P_j^{\nu_j}}{P_{j+1}^{\nu_{j+1}} \cdots P_{i-1}^{\nu_{i-1}} P_i^{\nu_i}}$$

or

$$\exp\left(-\Delta G^0/RT\right) = \frac{P_1^{\nu_1} P_2^{\nu_2} \cdots P_j^{\nu_j}}{P_{j+1}^{\nu_{j+1}} \cdots P_{i-1}^{\nu_{i-1}} P_i^{\nu_i}} = K_{eq}. \tag{3-12}$$

The new term, K_{eq}, is called the *equilibrium constant,* and equation 3-12 is the first answer to our question, "What controls the compositions of phases at equilibrium?"

Worked Problem 3-3

A gas at equilibrium at 327°C contains the species CO, O_2, and CO_2. If the total gas pressure is 1 bar and P_{O_2} is 0.8 bar, what are the partial pressures of carbon monoxide and carbon dioxide, assuming that each behaves as an ideal gas?

The first step is to write a balanced chemical reaction among the species:

$$CO + 1/2\, O_2 \longleftrightarrow CO_2.$$

Equation 3-12, therefore, should be written

$$K_{eq} = \exp\left(-\Delta G^0/RT\right) = \frac{P_{CO_2}}{P_{CO} P_{O_2}^{0.5}}.$$

The value of ΔG^0 can be calculated from the standard free energies of formation of each of the reactant and product species from the elements at 327°C (600 K):

$$\Delta G^0 = \Delta G_{f,CO_2} - \Delta G_{f,CO} - 0.5(\Delta G_{f,O_2})$$
$$= -94.445 - (-39.308) - 0.5(0.0) = -55.137 \text{ kcal mol}^{-1}.$$

The value of K_{eq}, therefore, is

$$K_{eq} = \exp\left(\frac{55.137 \text{ kcal mol}^{-1}}{(0.001987 \text{ kcal deg}^{-1} \text{ mol}^{-1})(600 \text{ deg})}\right) = 1.22 \times 10^{20}.$$

We are given that $P_{CO_2} + P_{CO} + P_{O_2} = 1$ bar. From this and the expression for K_{eq}, we calculate that

$$1.22 \times 10^{20}\, P_{CO} P_{O_2}^{0.5} + P_{CO} + P_{O_2} = 1 \text{ bar}$$

so

$$P_{CO} = 1.833 \times 10^{-21} \text{ bar}$$

and

$$P_{CO_2} \simeq 0.2 \text{ bar.}$$

With only small modifications, this derivation can be used to produce a similar expression that can be used for ideal liquid or solid solutions. If we stipulate that pressure, as well as temperature, is constant for these solutions, then

$$dP_i = d(PX_i) = P \, dX_i,$$

and equation 3-3 for any individual species i becomes

$$d\mu_i = (RT/P_i) \, dP_i = (RT/PX_i)P \, dx_i = (RT/X_i) \, dX_i.$$

The equation analogous to 3-10 is, therefore,

$$\mu_i = \mu_i^* + RT \ln X_i, \tag{3-13}$$

and the free energy of reaction is

$$\Delta G_r = \Delta G^* + RT \ln \frac{X_1^{\nu_1} X_2^{\nu_2} \cdots X_j^{\nu_j}}{X_{j+1}^{\nu_{j+1}} \cdots X_{i-1}^{\nu_{i-1}} X_i^{\nu_i}}. \tag{3-14}$$

The superscript * on both μ^* and ΔG^* is a reminder that each is a function of temperature and pressure, unlike μ^0 and ΔG^0, which depend only on temperature. In Chapter 8, when we discuss the effects of pressure on mineral equilibria, we will see that it is necessary to add an extra term to the free energy equations for solid and liquid phases to allow for compressibility. For gases, that adjustment is made in the $RT \ln K_{eq}$ term.

Because equations 3-11 and 3-14 are similar in form, they can be combined to formulate a general equation that relates the compositions of several ideal solutions or gas mixtures in a geochemical system to their free energies, as follows:

$$\Delta G_r = (\Delta G^0 + \Delta G^*) + RT \ln \frac{X_1^{\nu_1} X_2^{\nu_2} \cdots P_{j-1}^{\nu_{j-1}} P_j^{\nu_j}}{X_{j+1}^{\nu_{j+1}} \cdots P_{i-1}^{\nu_{i-1}} P_i^{\nu_i}}. \tag{3-15}$$

What we are doing here is similar to the step we took in Chapter 2 when we discussed conditions for homogeneous equilibrium and then developed equation 2-46 to describe heterogeneous systems. As we pointed out in the context of equation 3-7, however, the existence of stoichiometric relationships among species (indicated here by the coefficients ν_i) reduces any apparent excess in phase components. It makes no difference, then, whether equations 3-11 and 3-14 are written for a homogeneous system or a heterogeneous one, and we are free to combine them in equation 3-15. Symbols become a bit clumsy at this point, however. Each of the free energies, mole fractions, and partial pressures should be understood to carry a subscript ($\emptyset$) to identify the phase to which they refer. Thus, the mole fraction of $CaAl_2Si_2O_8$ in plagioclase should be shown as $X_{CaAl_2Si_2O_8, PL}$. Except where they are necessary to avoid ambiguity, however, we will follow standard practice and omit phase subscripts.

Worked Problem 3-4

During greenschist facies metamorphism (~400°C), rocks containing crysotile, calcite, and quartz commonly react to form tremolite, releasing both CO_2 and water. Assume for the moment that calcite is pure $CaCO_3$, that the other two solids are pure magnesian end-members of crystalline solutions, and that water and CO_2 mix as an ideal gas phase. This

Solutions that Behave Ideally

reaction will be strongly affected by total pressure, and we have not yet discussed adjustments for pressure. Suppose, however, that you tried to simulate this reaction by heating crysotile, calcite, and quartz to 400°C in an open crucible. What would ΔG_r be under these circumstances?

The balanced reaction for this example is

$$5Mg_3Si_2O_5(OH)_4 + 6CaCO_3 + 14SiO_2 \longleftrightarrow$$
$$3Ca_2Mg_5Si_8O_{22}(OH)_2 + 6CO_2 + 7H_2O.$$

The data relevant to this problem, available in Helgeson, 1969, are

	Tremolite	CO_2	H_2O	Crysotile	Calcite	Quartz
$\Delta \bar{H}^0$ (cal mol^{-1})	-2894140	-87451	-54610	-1012150	-279267	-212480
$\bar{S}^0$ (cal deg^{-1} mol^{-1})	258.8	65.91	51.97	118.55	41.65	20.87

The partial pressure of CO_2 in the atmosphere is $\sim 3.2 \times 10^{-4}$ atm, and the partial pressure of H_2O, although variable, is roughly 1×10^{-2} atm. If the crucible is open to the atmosphere and mixing is rapid, then these partial pressures should remain roughly constant.

The free energy of reaction can be calculated from

$$\Delta G_r = \Delta \bar{H}^0 - T\Delta \bar{S}^0 + RT \ln K_{eq},$$

in which

$$K_{eq} = \frac{X^3_{Ca_2Mg_5Si_8O_{22}(OH)_2, trem} P^6_{CO_2} P^7_{H_2O}}{X^5_{Mg_3Si_2O_5(OH)_4, cry} X^6_{CaCO_3, cc} X^{14}_{SiO_2, qz}}.$$

From the data above, we calculate

$$\Delta \bar{H}^0 = 3(-2894140) + 6(-87451) + 7(-54610) - 5(-1012150) -$$
$$6(-279267) - 14(-212480) = 121676 \text{ cal mol}^{-1},$$

and

$$\Delta \bar{S}^0 = 3(258.8) + 6(65.91) + 7(51.97) - 5(118.55) - 6(41.65) - 14(20.87)$$
$$= 399.32 \text{ cal deg}^{-1} \text{ mol}^{-1}.$$

Because we assumed that the solid phases have end-member compositions, each of the mole fractions is equal to 1. To reach the final answer, then, we calculate

$$\Delta G_r = 121256 - 673(399.32) + 1.987(673)(6 \ln (3.2 \times 10^{-4}) + 7 \ln (1 \times 10^{-2}))$$
$$= -255161 \text{ cal mol}^{-1}.$$

The reaction, therefore, favors the products under the conditions specified.

Suppose that we performed the same experiment, but used an impure calcite with a composition $(Ca_{0.8}Mn_{0.2}CO_3)$. How would you expect the free energy of reaction to change, assuming that Ca-Mn-calcite behaves as an ideal solid solution?

The expressions for ΔG_r and K_{eq} remain the same, as do the values for ΔH^0 and ΔS^0. The mole fraction of $CaCO_3$ in calcite is no longer 1.0, however, so the value of $(-6 \ln X_{CaCO_3})$ is no longer zero. We now find that

$$\Delta G_r = 121256 - 673(399.32) + 1.987(673)(6 \ln (3.2 \times 10^{-4}) +$$
$$7 \ln (1 \times 10^{-2}) - 6 \ln (0.8)) = -253371 \text{ cal mol}^{-1}.$$

The reaction still favors the products, but ΔG_r has increased by 1790 cal mol^{-1}.

In many cases, it is appropriate to assume that real gases follow the ideal gas equation of state. More generally, however, interactions between molecules make it necessary to modify the ideal gas law by adding "correction" terms to adjust for nonideality. Before we worry about the precise nature of these adjustments, let us illustrate how nonideality complicates matters. We shall do this by writing the corrected equation of state as a generic power function

$$PV = RT + BP + CP^2 + DP^3 + EP^4 + \cdots, \tag{3-16}$$

in which the coefficients B, C, D, and so forth are empirical functions of T alone. This expression can now be inserted in equation 3-2. Assuming isothermal conditions, the integrated form analogous to equation 3-5 is

$$\mu = \mu^0 + RT \ln P + B(P - 1) + C(P^2 - 1)/2 + \cdots. \tag{3-17}$$

This is a clumsy equation to handle, particularly if we are headed for a generalized form of equation 3-15, and it is only valid for the empirical equation of state we chose. The way we avoid this problem is to repackage all the terms containing P and introduce a new variable, f, so that

$$\mu = \mu^0 + RT \ln (f/f^0). \tag{3-18}$$

This new quantity is known as *fugacity,* and it will serve in each of the equations as a "corrected" pressure; that is, as a pressure that has been adjusted for the effects of nonideality. Notice that equation 3-18 is identical in form to equation 3-4. The reference fugacity, f^0, used for comparison is usually[3] taken to be the fugacity of the pure substance at 1 atmosphere total pressure and at a specified temperature. For a pure (one-component) gas, $f^0 = P^0 = 1$ atm. The ratio f/f^0 is given the special name *activity,* for which we use the symbol a. The activity of a gas is, therefore, a dimensionless value, numerically equal to its fugacity.

Changing variables certainly simplifies equation 3-17, but we have not improved our understanding of nonideal gases unless we can evaluate fugacity. To see how this is done, differentiate equation 3-18 with respect to P at constant temperature:

$$(\partial\mu/\partial P)_T = RT (\partial \ln a/\partial P)_T.$$

(Remember that μ^0 is a function of T only, so that $(\partial\mu^0/\partial P)_T$ is equal to zero.) This can be recast another way by recalling that $(\partial\mu/\partial P)_T = (\partial\overline{G}/\partial P)_T = \overline{V}$:

$$RT \, d \ln a = \overline{V} \, dP.$$

Subtract the quantity $RT \, d \ln P$ from both sides to get

$$RT \, d \ln (a/P) = \overline{V} \, dP - RT \, d \ln P$$
$$= (\overline{V} - (RT/P)) \, dP,$$

and then integrate the result between the limits zero and P. This procedure yields the expression

---

[3] A variety of different standard states are possible. For any given problem, the one we choose is partly a matter of convenience and partly determined by convention. In Chapter 8, we will introduce and discuss a number of alternate standard states.

$$\ln a = \ln P + \int_o^P \left(\frac{\bar{V}}{RT} - \frac{1}{P} \right) dP, \qquad (3\text{-}19)$$

which is just what we were looking for. The adjustment for nonideality is expressed as an integral that can be evaluated by replacing $\bar{V}$ with an appropriate equation of state, $\bar{V}(P)$. Notice that if $\bar{V}(P) = RT/P$ (the ideal gas equation), then the integral has the value zero at any pressure. The quantity

$$\gamma = \exp \left[\int_o^P \left(\frac{\bar{V}}{RT} - \frac{1}{P} \right) dP \right] \qquad (3\text{-}20)$$

defines the *activity coefficient*, γ, which is equal to the ratio a/P.

ACTIVITY COEFFICIENTS AND EQUATIONS OF STATE

One good way to illustrate the nonideal behavior of real gases with increasing pressure is to plot values of pressure against the experimentally determined quantity $P\bar{V}/RT$, which serves as a measure of compressibility. For an ideal gas, $P\bar{V}/RT$ will, of course, be equal to 1.0 at all pressures; values below 1.0 indicate that a gas is more compressible than an ideal gas; values above 1.0 indicate that it is less compressible. In Figure 3.6, we have shown the behavior of molecular hydrogen and oxygen on this type of plot.

At low pressures, most gases occupy less volume than we would expect from the ideal gas equation of state, suggesting that attractive forces between gas molecules reduce the effective mean distance between them. In 1879, the Dutch physicist van der Waals recognized this phenomenon and adjusted the equation of state by adding a term $a/\bar{V}^2$ to the observed pressure, where a is an empirically determined constant for the gas. This modification correctly describes the deviation shown in Figure 3.6 at very low pressures, although research has shown that a actually is not a constant but depends on both pressure and temperature.

This correction for intermolecular attractions, however, predicts that real gases will become even more compressible as pressure is increased. This is clearly not the case. Consequently, the full form of van der Waals' equation includes a second empirical adjustment on the molar volume:

$$(P + (a/\bar{V}^2))(\bar{V} - b) = RT.$$

As pressure is increased, an increasingly significant volume in the gas is occupied by the

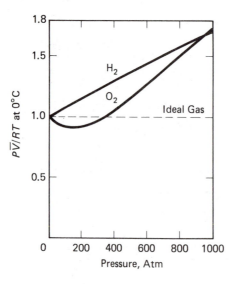

Figure 3.6 Total pressure vs. $P\bar{V}/RT$ for H_2 and O_2. A gas for which $P\bar{V}/RT < 1$ is more compressible than an ideal gas; if $P\bar{V}/RT > 1$, the gas is less compressible. Deviations from ideality at low pressure are described by the $a/\bar{V}^2$ attractive term in the van der Waals equation. At higher pressures, the positive deviations are described by b, the excluded volume.

molecules themselves. The factor b, therefore, can be thought of as a measure of the *excluded volume*, which already contains molecules. The remaining volume, in which molecules are free to move, is much less than the total volume, and pressure is higher than it would be for an ideal gas. The full equation, therefore, reflects a balance between attractive and repulsive intermolecular forces that affect the molar volume of a real gas.

In order to calculate an activity coefficient using the van der Waals equation of state, we need to rearrange it in terms of V:

$$\bar{V}^3 - \bar{V}^2(b + RT/P) + \bar{V}(a/P) - ab/P = 0.$$

This cubic equation can be shown to have three real roots at temperatures and pressures below the critical point,[4] of which the largest root is the true molar volume for the gas. It is this solution that should replace $\bar{V}$ in equation 3-20. In practice, equation 3-20 is solved numerically by computing molar volumes iteratively over the range of pressures from zero to P as part of the algorithm which solves the integral. This is a reasonably straightforward and reliable procedure at moderate temperatures and pressures.

At more elevated pressures where many igneous or metamorphic reactions may involve gas phases, the van der Waals equation of state is less successful. Redlich and Kwong (1949) found that they could only predict the nonideal behavior of gases at very high pressures by modifying the attractive force term. Their equation,

$$P = RT/(\bar{V} - b) - a/(\sqrt{T}\, \bar{V}(\bar{V} + b)),$$

has been widely used in petrology, and has been modified further in many studies. Good reviews of these modifications, and of the use of Redlich-Kwong type equations of state in calculating activity coefficients, can be found in Holloway (1977) and in Kerrick and Jacobs (1981).

Activities of species in nonideal gas mixtures are defined in a fashion similar to the one we followed in deriving equations 3-10 and 3-11, with the exception that all partial pressures P_i must now be multiplied by appropriate activity coefficients γ_i. The resulting equilibrium constant is given by

$$
\begin{aligned}
K_{eq} = \exp\left(-\Delta G^\circ/RT\right) &= \frac{a_1^{\nu_1} a_2^{\nu_2} \cdots a_j^{\nu_j}}{a_{j+1}^{\nu_{j+1}} \cdots a_{i-1}^{\nu_{i-1}} a_i^{\nu_i}} \\
&= \frac{P_1^{\nu_1} P_2^{\nu_2} \cdots P_j^{\nu_j}}{P_{j+1}^{\nu_{j+1}} \cdots P_{i-1}^{\nu_{i-1}} P_i^{\nu_i}} \frac{\gamma_1^{\nu_1} \gamma_2^{\nu_2} \cdots \gamma_j^{\nu_j}}{\gamma_{j+1}^{\nu_{j+1}} \cdots \gamma_{i-1}^{\nu_{i-1}} \gamma_i^{\nu_i}}.
\end{aligned}
\tag{3-21}
$$

Figure 3.7 summarizes graphically the various contributions to ΔG_r.

In a similar way, nonideal liquid or crystalline solutions can be treated by modifying equation 3-14. As with gases, activity is defined as the ratio f/f°. Most species in liquid or solid solution, however, have negligible vapor pressures at 1 atmosphere total pressure, so f° is rarely equal to 1. Furthermore, because fugacities in solution are so small, it is impractical to use the fugacity ratio to measure quantities like the activity of Fe_2SiO_4 in a magma. Instead, it is common to write

$$
\begin{aligned}
K_{eq} = \exp\left(-\Delta G*/RT\right) &= \frac{a_1^{\nu_1} a_2^{\nu_2} \cdots a_j^{\nu_j}}{a_{j+1}^{\nu_{j+1}} \cdots a_{i-1}^{\nu_{i-1}} a_i^{\nu_i}} \\
&= \frac{X_1^{\nu_1} X_2^{\nu_2} \cdots X_j^{\nu_j}}{X_{j+1}^{\nu_{j+1}} \cdots X_{i-1}^{\nu_{i-1}} X_i^{\nu_i}}
\end{aligned}
\tag{3-22a}
$$

[4] We will not define the critical point until Chapter 9. For now, think of it as a condition in temperature and pressure beyond which there is no physical distinction between the liquid and vapor states. Virtually all near-surface environments are below the critical point.

Solutions that Behave Nonideally

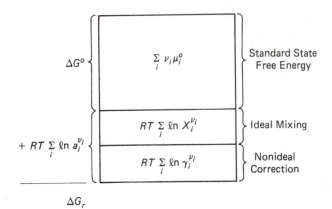

Figure 3.7 Schematic representation of the various contributions to the free energy of a nonideal solution.

in which we declare that $a_i = \gamma_i X_i$. In practice, it is common to specify, further, that a_i is equal to 1 for a pure substance at any chosen temperature. More specifically,

$$a_i \longrightarrow X_i \qquad \text{as} \qquad X_i \longrightarrow 1. \tag{3-22b}$$

Following this convention and applying equation 3-22, the activity of MgSiO$_3$, for example, would be equal to 1 in pure enstatite ($X_{\mathrm{MgSiO_3}} = 1$). Other conventions are also common, however. In very dilute aqueous solutions, we usually define the activity of solute species by the relation $a_i = \gamma_i m_i$, in which m_i is the concentration of species i in molal units. This leads to the expression

$$K_{\mathrm{eq}} = \exp\left(-\Delta G^*/RT\right) = \frac{a_1^{\nu_1} a_2^{\nu_2} \cdots a_j^{\nu_j}}{a_{j+1}^{\nu_{j+1}} \cdots a_{i-1}^{\nu_{i-1}} a_i^{\nu_i}}$$

$$= \frac{m_1^{\nu_1} m_2^{\nu_2} \cdots m_j^{\nu_j}}{m_{j+1}^{\nu_{j+1}} \cdots m_{i-1}^{\nu_{i-1}} m_i^{\nu_i}} \frac{\gamma_1^{\nu_1} \gamma_2^{\nu_2} \cdots \gamma_j^{\nu_j}}{\gamma_{j+1}^{\nu_{j+1}} \cdots \gamma_{i-1}^{\nu_{i-1}} \gamma_i^{\nu_i}}. \tag{3-23a}$$

We generally apply this expression by specifying that

$$a_i \longrightarrow m_i \qquad \text{as} \qquad m_i \longrightarrow 0. \tag{3-23b}$$

According to either of these conventions, as well as others which we have not mentioned, solutions approach ideal behavior ($\gamma_i \to 1$) as their compositions approach a single-component end member. The greatest difficulty arises when we must consider nonideal solutions that are far from a pure end-member composition. We will consider that problem for non-electrolyte solutions and will examine standard states for activity more fully in Chapter 8.

ACTIVITY IN ELECTROLYTE SOLUTIONS

Species in aqueous solutions, such as seawater or hydrothermal fluids, present an unusual challenge. Although natural fluids commonly contain some associated (that is, uncharged) species, it is much more common for them to contain free ions. In a typical problem, for example, we may be asked to evaluate the solubility of sodium sulfate in seawater. Ignoring the fact that crystalline sodium sulfate usually takes the form of a hydrate, Na$_2$SO$_4\cdot n$H$_2$O, the relevant chemical reaction is

$$Na_2SO_4 \longleftrightarrow 2Na^+ + SO_4^{2-},$$

which we always write with the solid phase as a reactant and fully dissociated ions as products. For this type of reaction, we refer to the equilibrium constant as an equilibrium *solubility product constant, K_{sp}*:

$$K_{sp} = (a_{Na^+})^2 a_{SO_4^{2-}} / a_{Na_2SO_4}.$$

In general, ionic species in solution do not behave ideally unless they are very dilute, because ions tend to interact electrostatically. They also generally associate with water molecules to produce "hydration spheres" in which the chemical potential of H_2O is different from that in pure water. The result is that the free energies of both the solvent (H_2O) and the solute (dissolved ions) differ from their standard states, and their activities differ from their concentrations.

Unfortunately, there is no way to measure the activity of a dissolved ion independently. Because charge balance must always be maintained, we cannot vary the concentration of a cation like Na^+ without also adjusting the anions in solution. It is impossible, for example, to determine how much of the potential free-energy change during evaporation of seawater is due to increasing a_{Na^+} and how much is the result of parallel increases in a_{Cl^-}, $a_{SO_4^{2-}}$, and other anion activities.

The Mean Salt Method

Several approaches have been taken to solve this problem. One is to define a *mean ionic activity*, given by

$$a_\pm = (a_+^{\nu_+} a_-^{\nu_-})^{1/\nu}, \tag{3-24}$$

in which a_+ and a_- are the individual activities of cations and anions, respectively, and the stoichiometric coefficients

$$\nu = \nu_+ + \nu_-$$

count the number of each formed per dissociated molecule of solute. The mean ionic activity of $MgCl_2$ in aqueous solution, calculated from equation 3-24, would be equal to

$$a_\pm(MgCl_2) = (a_{Mg^{2+}}(a_{Cl^-})^2)^{1/3}.$$

Mean activities, unlike the activities of charged ions, are readily measurable. From these, it is possible to calculate mean ion activity coefficients by using the relationship

$$\gamma_\pm = a_\pm / m_\pm. \tag{3-25}$$

The quantity $m_\pm$ is the *mean ionic molality,* which is related to the molal concentration of total solute m by

$$m_\pm = m(\nu_+^{\nu_+} \nu_-^{\nu_-})^{1/\nu}. \tag{3-26}$$

Again, for $MgCl_2$ in aqueous solution,

$$m_\pm(MgCl_2) = (m_{MgCl})_{total}((2)^2(1)^1)^{1/3}.$$

To calculate *individual ion activity coefficients,* γ_+ and γ_-, it is customary to choose a standard univalent electrolyte in a solution of the same effective concentration, and to assume that $\gamma_+ = \gamma_-$. Several studies have suggested that KCl is well-behaved and, thus, that

$$\gamma_\pm(KCl) = (\gamma_{K^+}\,\gamma_{Cl^-})^{1/2} = \gamma_{K^+} = \gamma_{Cl^-}.$$

If we wanted to determine an individual ion activity coefficient for Mg^{2+}, then, we could calculate that

$$\gamma_\pm(MgCl_2) = (\gamma_{Mg^{2+}}(\gamma_{Cl^-})^2)^{1/3} = (\gamma_{Mg^{2+}}(\gamma_\pm KCl)^2)^{1/3}$$

and, therefore, that

$$\gamma_{Mg^{2+}} = \gamma_\pm(MgCl_2)^3/\gamma_\pm(KCl)^2.$$

This method, known as the *mean salt method*, can be used to estimate activity coefficients for anions as well as cations.

The Debye-Hückel Method

An alternative means of calculating ion activity coefficients in dilute solutions is provided by Debye-Hückel theory, which attempts to calculate the effect of electrostatic interactions among ions on their free energies of formation. To evaluate the cumulative effect of attractive and repulsive forces, we define first a charge-weighted function of species concentration known as *ionic strength* (I):

$$I = \tfrac{1}{2} \sum_i m_i z_i^2, \tag{3-27}$$

in which z_i is the charge on an ion of species i. In this way, we recognize that ions in solution are influenced by the total electrostatic field around them, and that polyvalent ions ($|z| > 1$) exert more electrostatic force on their neighbors than monovalent ions. A 1 m solution of $MgSO_4$, for example, has an ionic strength of 4, while a 1 m solution of NaCl has an ionic strength of only 1.

In its most commonly used form, the Debye-Hückel equation takes into account not only the ionic strength of the electrolyte solution, but also the effective size of the hydrated ion of interest, thus estimating the number of neighboring ions with which it can come into contact. Values of the size parameter, $\mathring{a}$, for representative ionic species, are given in Table 3-1, along with values of empirical parameters A and B, which are functions only of pressure and temperature. We calculate γ_i, the activity coefficient for ion i, from

$$\log_{10} \gamma_i = \frac{-Az_i^2\sqrt{I}}{1 + B\mathring{a}\sqrt{I}}. \tag{3-28}$$

Pytkowicz (1983) has discussed the theoretical justification for this equation and a critique of its use in aqueous geochemistry.

Worked Problem 3-5

What is the activity coefficient for Ca^{2+} in a 0.05 m aqueous solution of $CaCl_2$ at 25°C? How closely do answers obtained by the mean salt method and the Debye-Hückel equation agree?

TABLE 3-1 VALUES OF CONSTANTS FOR USE IN THE DEBYE-HÜCKEL EQUATION[*]

Temperature, °C	A	$B (\times 10^{-8})$	$\mathring{a} \times 10^8$	Ion
0	0.4883	0.3241	2.5	Rb^+, Cs^+, NH_4^+, Tl^+, Ag^+
5	0.4921	0.3249	3.0	K^+, Cl^-, Br^-, I^-, NO_3^-
10	0.4960	0.3258	3.5	OH^-, F^-, HS^-, BrO_3^-, IO_4^-, MnO_4^-
15	0.5000	0.3262	4.0–4.5	Na^+, HCO_3^-, $H_2PO_4^-$, HSO_3^-, Hg_2^{2+}, SO_4^{2-}, $SeO_4^{2-} \cdot CrO_4^{2-}$, HPO_4^{2-}, PO_4^{3-}
20	0.5042	0.3273	4.5	Pb^{2+}, CO_3^{2-}, SO_3^{2-}, MoO_4^{2-}
25	0.5085	0.3281	5.0	Sr^{2+}, Ba^{2+}, Ra^{2+}, Cd^{2+}, Hg^{2+}, $S^{2-} \cdot WO_4^{2-}$
30	0.5130	0.3290	6	Li^+, Ca^{2+}, Cu^{2+}, Zn^{2+}, Sn^{2+}, Mn^{2+}, Fe^{2+}, Ni^{2+}, Co^{2+}
35	0.5175	0.3297	8	Mg^{2+}, Be^{2+}
40	0.5221	0.3305	9	H^+, Al^{3+}, Cr^{3+}, trivalent rare earths
45	0.5271	0.3314	11	Th^{4+}, Zr^{4+}, Ce^{4+}, Sn^{4+}
50	0.5319	0.3321		
55	0.5371	0.3329		
60	0.5425	0.3338		

[*]Data from Garrels and Christ (1982).

To calculate $\gamma_{Ca^{2+}}$ by the Debye-Hückel method, we first determine the ionic strength of the solution:

$$I = 0.5 \sum_i m_i z_i^2 = 0.5((0.05)(2)^2 + (0.1)(-1)^2) = 0.15.$$

Then, using the data in Table 3-1, we find that

$$\log_{10} \gamma_{Ca^{2+}} = \frac{-A\, z_i^2 \sqrt{I}}{1 + B \mathring{a} \sqrt{I}}$$

$$= \frac{-(0.5085)(2)^2(0.15)^{1/2}}{1 + (0.3281 \times 10^8)(6 \times 10^{-8})(0.15)^{1/2}} = -0.447.$$

So,

$$\gamma_{Ca^{2+}} = 0.357.$$

The mean ionic activity coefficient for $CaCl_2$ in 0.05 m aqueous solution has been reported by Goldberg and Nuttall (1978) to be 0.5773. To use the mean salt method, we need to know the mean ion activity coefficient for KCl under similar conditions. A KCl solution of ionic strength 0.15, however, is 0.15 molal. Robinson and Stokes (1949) report a value of 0.744 in 0.15 m aqueous KCl. From these, we calculate that

$$\gamma_\pm(CaCl_2) = (\gamma_{Ca^{2+}}(\gamma_{Cl^-})^2)^{1/3} = (\gamma_{Ca^{2+}}(\gamma_\pm KCl)^2)^{1/3}.$$

Therefore,

$$\gamma_{Ca^{2+}} = \gamma_\pm(CaCl_2)^3/\gamma_\pm(KCl)^2 = (0.577)^3/(0.744)^2 = 0.347.$$

This is very close to the value calculated by Debye-Hückel theory.

In Figure 3.8, we have repeated this pair of calculations for a range of ionic strengths from 0.05 to 3.0. It is apparent that the methods agree at ionic strengths less than about 0.3, but diverge strongly in more concentrated solutions as Debye-Hückel theory fails to model changes in the structure of the solute adequately. Other calculation schemes have been used to model highly concentrated solutions (see, for example, Harvie et al., 1984), but most researchers continue to use values obtained by some variant of the mean salt method.

Activity in Electrolyte Solutions

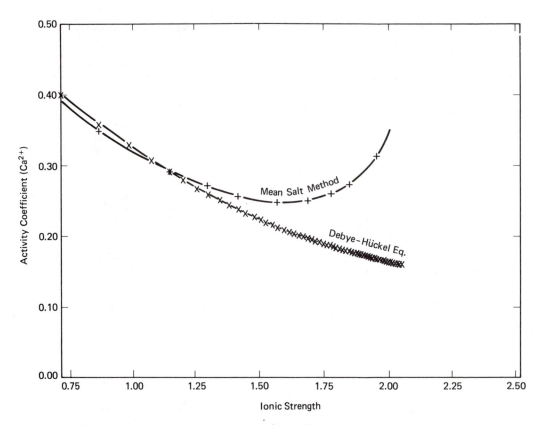

Figure 3.8 Ion activity coefficient for Ca^{2+} as a function of ionic strength, calculated from the Debye-Hückel equation and by the mean salt method. Notice the deviations at higher ionic strengths.

SOLUBILITY

Many of the interesting questions facing aqueous geochemists have to do with mineral solubility. How much sulfate is in groundwater that is in equilibrium with gypsum beds? How much fluoride can be carried by an ore-forming hydrothermal fluid? What sequence of minerals should be expected to form during evaporation of seawater? What processes can cause deposition of ore minerals from migrating fluids? How are the compositions of residual (zonal) soils affected by the chemistry of soil water? How might a chemical leakage from a waste disposal site affect the stability of surrounding rocks and soils? In order to address these questions and similar ones that appear in coming chapters, we must look first at broad conditions that affect solubility.

In the following series of worked problems, we will make successive refinements in our approach to a typical problem, illustrating by stages the major thermodynamic factors influencing solubility.

Worked Problem 3-6

What is the solubility of barite in water at 25°C? The simplest answer can be calculated from data for the free energies of formation of species in the reaction:

$$BaSO_4 \longleftrightarrow Ba^{2+} + SO_4^{2-}.$$

We find the following values in Parker et al. (1971):

	BaSO$_4$	Ba^{2+}	SO$_4^{2-}$
$\Delta\overline{G}_f$ (kcal mol^{-1})	-325.6	-134.02	-177.97

From these, the standard free energy of reaction is

$$\Delta\overline{G}_r^\circ = -134.02 + (-177.97) - (-325.6) = 13.61 \text{ kcal mol}^{-1}.$$

If the fluid is saturated with respect to BaSO$_4$, then the reaction above is perfectly balanced and $\Delta\overline{G}_r = 0$. Therefore,

$$\log_{10} K_{sp} = -\Delta\overline{G}_r^\circ/2.303RT$$
$$= -13.61/((298)(2.303)(0.001987)) = -9.98.$$
$$K_{sp} = 1.04 \times 10^{-10}.$$

If we assume that the solution behaves ideally, then the solubility of barite can be calculated by recognizing that the charge balance in the fluid requires that $m_{Ba^{2+}} = m_{SO_4^{2-}}$, and that

$$m_{Ba^{2+}} m_{SO_4^{2-}} = K_{sp}.$$

From this, we get

$$m_{Ba^{2+}} = m_{SO_4^{2-}} = 1.02 \times 10^{-5} \text{ mol kg}^{-1}.$$

Other common ways of presenting this result would be

$$m_{Ba^{2+}} = 1.02 \times 10^{-5} \text{ mol kg}^{-1} \times 137.34 \text{ g Ba mol}^{-1}$$
$$= 1.40 \times 10^{-3} \text{ g Ba kg}^{-1}$$
$$= 1.40 \text{ ppm Ba}.$$

or

$$= 1.02 \times 10^{-5} \text{ mol kg}^{-1} \times 233.40 \text{ g BaSO}_4 \text{ mol}^{-1}$$
$$= 2.39 \times 10^{-3} \text{ g BaSO}_4 \text{ kg}^{-1}$$
$$= 2.39 \text{ ppm BaSO}_4.$$

The result in Worked Problem 3-6 is remarkably close to the value of 1.06×10^{-5} mol kg^{-1} measured in experiments by Blount (1977), despite the fact that we assumed the fluid was ideal. This agreement may be fortuitous, however, because we should anticipate some nonideal behavior. Let's do the problem again.

Worked Problem 3-7

What are the activity coefficients for Ba^{2+} and SO$_4^{2-}$ in a solution saturated with respect to barite, and how much does the calculated solubility change if we allow for nonideal behavior?

The easiest way to solve this problem is to design a numerical algorithm and iterate toward a solution by computer. In outline form, the procedure looks like this:[5]

1. Using the ideal solubility as a first guess, calculate the ionic strength of a saturated solution.

---

[5] The program SOLUBLE, on the diskette available for this text, follows this algorithm.

2. Calculate activity coefficients for Ba^{2+} and SO_4^{2-} from the Debye-Hückel equation (3-28), using constants in Table 3-1.

3. Using the activity coefficients and the value of K derived from $\Delta \overline{G}°$, calculate improved values of $m_{Ba^{2+}}$ and $m_{SO_4^{2-}}$.

4. Repeat steps 1 through 3 until $m_{Ba^{2+}}$ and $m_{SO_4^{2-}}$ do not change in successive iterations by more than some acceptably small amount.

When we did this calculation, it took four cycles to satisfy the condition $|m_{Ba}(\text{old}) - m_{Ba}(\text{new})| \leq 10^{-9}$, which is a calculation error of less than one part in 10^4 (much better than is justified by the uncertainties in $\Delta \overline{G}°$). Notice that charge balance constraints require that $m_{Ba^{2+}} = m_{SO_4^{2-}}$.

Iteration	I.S.	$\gamma_{Ba^{2+}}$	$\gamma_{SO_4^{2-}}$	$\gamma \pm$	$m_{Ba^{2+}}$
1	2.41×10^{-5}	0.9775	0.9774	0.9774	1.0443×10^{-5}
2	4.18×10^{-5}	0.9705	0.9704	0.9705	1.0519×10^{-5}
3	4.21×10^{-5}	0.9704	0.9703	0.9704	1.0509×10^{-5}
4	4.20×10^{-5}	0.9704	0.9703	0.9704	1.0509×10^{-5}

As we suspected in Worked Problem 3-6, a solution saturated with $BaSO_4$ is nearly ideal. By taking the slight nonideality into account, however, the computed solubility increases to within 1% error of the value measured by Blount (1977).

The Ionic Strength Effect

The reason that there is progressive improvement with each iteration in Worked Problem 3-7 is that each cycle calculates the ionic strength with an improved set of activity coefficients. Suppose, however, that there were other ions in solution as well as Ba^{2+} and SO_4^{2-}. Natural fluids, in fact, rarely contain only a single electrolyte. Most hydrothermal fluids or formation waters are dominated by NaCl. To calculate the ionic strength correctly, we must not only consider the amount of Ba^{2+} and SO_4^{2-}, but also include the concentrations of all other ions in solution. This, in turn, will lead to an adjustment in the solubility of $BaSO_4$.

Worked Problem 3-8

What effect do other solutes have on the solubility of barite? In particular, what would be the solubility of barite in a solution at 25°C that is already 0.2m in NaCl?

Assuming that there is little tendency for Ba^{2+} to associate with Cl^- (Sucha et al., 1975) or for Na^+ to associate with SO_4^{2-} (Elgquist and Wedborg, 1978), we should expect that barite solubility will change only as the result of increased ionic strength. To verify this, we repeat the calculation of Worked Problem 3-7, this time calculating ionic strength in each iteration by $I = 0.5(m_{Ba^{2+}} \times 4 + m_{SO_4^{2-}} \times 4 + m_{Na^+} + m_{Cl^-})$.

I.S.	$\gamma_{Ba^{2+}}$	$\gamma_{SO_4^{2-}}$	$\gamma_\pm$	$m_{Ba^{2+}}$
0.2001	0.2987	0.2671	0.2824	3.611×10^{-5}

These values, once again, compare favorably with Blount (1977), who measured barite solubility in 0.2m NaCl − H_2O at 25°C and found $m_{Ba^{2+}} = 3.7 \times 10^{-5}$ mol kg^{-1} and $\gamma_\pm = 0.2754$.

The Common Ion Effect

In many natural fluids, the solubility of minerals like barite increases with ionic strength, as we saw in the previous problem. This will not be the case, however, if ions produced by dissociation of the solute are also present from some other source in solution. For example, if barite is in equilibrium with a solution that contains both NaCl and Na_2SO_4, the charge balance constraint on the fluid becomes

$$2m_{Ba^{2+}} + m_{Na^+} = 2m_{SO_4^{2-}} + m_{Cl^-}.$$

Sulfate is a *common ion* in $BaSO_4$ and Na_2SO_4. We should expect that when each salt releases SO_4^{2-} to solution, it will affect the solubility of the other. To demonstrate this common ion effect, we offer the following problem.

Worked Problem 3-9

How does the addition of successive amounts of Na_2SO_4 to an aqueous fluid affect the solubility of barite?

The molal concentration of Ba^{2+} can be calculated from charge balance and the equilibrium constant, K:

$$K = m_{Ba^{2+}} m_{SO_4^{2-}} \gamma_{Ba^{2+}} \gamma_{SO_4^{2-}}$$

$$= \frac{m_{Ba^{2+}} \gamma_{Ba^{2+}} \gamma_{SO_4^{2-}} (2m_{Ba^{2+}} + m_{Na^+} - m_{Cl^-})}{2}$$

$$m_{Ba^{2+}} = \frac{2K}{\gamma_{Ba^{2+}} \gamma_{SO_4^{2-}} (2m_{Ba^{2+}} + m_{Na^+} - m_{Cl^-})}.$$

This equation can be solved by the same numerical procedure we applied in Worked Problems 3-7 and 3-8. Because the solubility of barite has been shown to be quite low, we can make the simplifying assumption that all SO_4^{2-} in solution is due to dissociation of Na_2SO_4, as long as we examine only solutions of moderately high $m_{SO_4^{2-}}$. The charge balance constraint, in other words, is approximately

$$m_{Na^+} = 2m_{SO_4^{2-}} + m_{Cl^-}.$$

The ionic strength calculated with this simplification will differ only very slightly from the "true" ionic strength.

In order to illustrate the effect of a *common ion* (SO_4^{2-} in this case), we have calculated $m_{Ba^{2+}}$ in a variety of $NaCl-Na_2SO_4-H_2O$ solutions with ionic strength equal to 0.2.

$m_{Ba^{2+}}$	$m_{SO_4^{2-}}$	m_{Na^+}	m_{Cl^-}
3.51×10^{-6}	3.71×10^{-4}	2.00×10^{-1}	1.99×10^{-1}
9.38×10^{-6}	1.38×10^{-4}	1.99×10^{-1}	1.97×10^{-1}
6.39×10^{-7}	2.04×10^{-3}	1.98×10^{-1}	1.94×10^{-1}
1.94×10^{-7}	6.71×10^{-3}	1.93×10^{-1}	1.80×10^{-1}
9.76×10^{-8}	1.33×10^{-2}	1.86×10^{-1}	1.60×10^{-1}
6.50×10^{-8}	2.00×10^{-2}	1.80×10^{-1}	1.40×10^{-1}

The addition of sulfate, even in very small amounts, dramatically lowers the solubility. Barite will precipitate if the product, $m_{Ba^{2+}} m_{SO_4^{2-}}$, equals or exceeds K_{eq} (1.31×10^{-9} at ionic strength 0.2). With increasing amounts of sulfate in solution, this condition is met with vanishingly small amounts of dissolved Ba^{2+}. The same effect would be observed if we were to provide barium ions from some source other than barite (for example, from barium chloride).

Complex Species

In stream waters (ionic strength <0.01 on average) and in most surficial environments on the continents, dissolved species are highly dissociated. As ionic strength increases, however, electrical interactions between ions also increase. The result is that many ions in concentrated solutions exist in groups or clusters known as *complexes*. For ease of discussion, these may be divided into three broad classes: *ion pairs, coordination complexes,* and *chelates.* By convention (not universally observed, unfortunately), the stability of any complex is expressed by a *stability constant*, K_{stab}, which is written for the reaction forming a complex species from simpler dissolved species. Thus, for example, Mg^{2+} and SO_4^{2-} ions can associate to form the neutral ion pair $MgSO_4^0$ by the reaction

$$Mg^{2+} + SO_4^{2-} \longleftrightarrow MgSO_4^0,$$

for which $K_{stab} = a_{MgSO_4}/(a_{Mg^{2+}} a_{SO_4^{2-}}) = 5.62 \times 10^{-3}$ at 25°C. The larger a stability constant, the more stable the complex it describes.

The distinctions among different types of complexes are not easy to define and, for thermodynamic purposes, not very relevant. Ion pairs are characterized by weak bonding; hence, small stability constants. The dominant complex species in seawater and most brines fall into this category. In Chapter 4, we will examine the effect of ion pair formation on the chemistry of the oceans.

Coordination complexes involve a more rigid structure of *ligands* (usually anions or neutral species) around a central atom. This well-defined structure contributes to greater stability. Transition metals like copper, vanadium, and uranium have been shown to exist in aqueous solution primarily as coordination complexes like $Cu(H_2O)_6^{2+}$, in which the negatively charged ends of several polar water molecules are attracted to the metal cation to form a *hydration sphere*. Other polar molecules commonly form coordination complexes with metal cations. One that may be familiar to you from analytical chemistry is ammonia, which combines readily with Cu^{2+} in solution to form the bright blue complex $Cu(NH_3)_4^{2+}$. The stability constant for this species is 2.13×10^{14}.

Chelates differ from coordination complexes only in their degree of structure. Molecules like water, ammonia, Cl^-, and OH^-, which form simple coordination complexes because they can only attach to one metal cation at a time, are called *unidentate* ligands. Other species, usually organic molecules or ions, are called *polydentate* ligands. Chelates, which form around these, have larger, more complicated structures that include several metal cations. Some of these, like the anion of ethylenediaminetetraacetic acid (mercifully known as EDTA), are widely used by analytical chemists to scavenge metals from very dilute aqueous solutions. This is a hexadentate ligand; that is, it can complex six cations at sites indicated by the arrows:

In nature, the vast number of humic and fulvic acids and other dissolved organic species also serve as chelating agents. In general, the larger and more complicated these are, the more stable the complexes they form. Because they have very high stability constants and can coordinate several cations at once, chelating agents can contribute significantly to the chemistry of transition metals and of cations like Al^{3+} in natural waters, even if both they and the metal cations are in very low concentration.

In normal operation, nuclear reactors produce a variety of radioactive waste products that are potentially dangerous. Because they commonly remain hazardous for very long periods of time, storage facilities for them must be designed which rely not only on the integrity of manufactured containers, but also on the behavior of the surrounding geologic environment. Waste containment, therefore, has become a cooperative effort for chemical engineers and geochemists.

Radioactive waste products are classified according to their potential for health and environmental hazard. Most of these produce only relatively low levels of radiation, and many decay to negligible concentrations in a few decades. For many years it was standard practice to dispose of these by leaching them with appropriate solvents and then burying the liquids in shallow landfills where they were stabilized by reactions with geologic materials. These disposal sites, of course, leak very slowly, but only at rates that are slow compared to the half-lives of many isotopes. As they are gradually released into the surroundings, therefore, their concentrations are low enough that they present no measurable health hazard.

Oak Ridge National Laboratory, a nuclear research facility, used a storage area of this type from 1951 to 1966 to contain over one million curies of waste. Since then, additional wastes have been stored only in deep well sites, but the old disposal sites have been continuously monitored to verify their integrity. Observations reported by Olsen et al. (1986) give some insight into the solution behavior of many dissolved radionuclides at the ORNL site. Of particular interest is their observation that ⁶⁰Co, a fission product with a half-life of 5.3 years, is much more mobile than had been predicted when the area was designed.

Cobalt can exist in aqueous solution as either a divalent or a trivalent cation. As with other transition metals, Co^{2+} is the most stable form under the conditions of acidity and oxidation state of most near-surface waters. At pH 9 (the approximate value measured in ground water around the disposal site), however, the concentration of either should be extremely low. Cobalt should prefer to precipitate as a carbonate or to be adsorbed on clay and oxide surfaces in the soil. The relatively high solubility at ORNL is made possible by the formation of organic chelates.

Measurements at the ORNL site by Olsen et al. (1986) and at a similar site in Ontario by Killey et al. (1984) suggest that up to 80% of the ⁶⁰Co in solution is present in weakly anionic organic complexes of low molecular weight and size. The dominant chelating agent at ORNL is probably EDTA, which was used in the original decontamination process and is present as a contaminant at the site. The stability constants for Co^{3+}-EDTA and Co^{2+}-EDTA are $10^{40.6}$ and $10^{16.3}$, large enough that a significant fraction of dissolved cobalt prefers to associate with the organic anion rather than to precipitate. Observations by Olsen et al. and by Killey et al. indicate that chelated ⁶⁰Co is also not taken up by plants as readily as ⁶⁰Co^{2+}. Other contaminant and natural organic species apparently contribute to the solubility as well as EDTA, although these are less well characterized. Polydentate ligands, therefore, play an important role in this geochemical environment and must be considered in the design of low-level waste disposal sites.

The formation of complex species increases the solubility of electrolytes by reducing the effective concentration of free ions in solution. In a sense, we can think of ions bound up in complexes as having become electrostatically "shielded" from oppositely charged ions, and therefore unable to combine with them. Because the activity product of free ions is lowered, equilibria that involve those ions favor the further dissolution of solid material. Thus, although the activity product of free ions at saturation will still be equal to K_{sp}, the total concentration of species in solution will be much higher than in a solution without complexing.

Worked Problem 3-10

Calcium and sulfate ions can associate in aqueous solution to form a neutral ion pair, $CaSO_4^0$. This is one of several ion pairs that affect the concentration of species in seawater, as we shall see in Chapter 4. How does it affect the solubility of gypsum?

The solubility product constant, K_{sp}, for gypsum is calculated from the free energy of the reaction

$$CaSO_4 \cdot 2H_2O \longleftrightarrow Ca^{2+} + SO_4^{2-} + 2H_2O.$$

In reasonably dilute solutions, the activity of water can be assumed to be unity. Gypsum, a pure solid, is in its standard state and also has unit activity. K_{sp} then is equal to $a_{Ca^{2+}} a_{SO_4^{2-}}$. At 25°C, it has the value 2.45×10^{-5}.

The stability constant, K_{stab}, for $CaSO_4^\circ$ is derived by experiment for the reaction

$$Ca^{2+} + SO_4^{2-} \longleftrightarrow CaSO_4^\circ.$$

$$K_{stab} = a_{CaSO_4^\circ}/(a_{Ca^{2+}} a_{SO_4^{2-}}) = 2.09 \times 10^2$$

at 25°C. By combining the two expressions for K_{sp} and K_{stab} and rearranging terms, we can calculate

$$a_{CaSO_4^\circ} = K_{stab} K_{sp} = 5.09 \times 10^{-3}.$$

Because $CaSO_4^\circ$ is a neutral species, its concentration in solution will be nearly the same as its activity. The concentration of the ion pair, therefore, is about 5 mmol kg^{-1}. It can be shown, using the iterative procedure in Worked Problem 3-7, that the concentration of Ca^{2+} in the absence of complexing is about 10 mmol kg^{-1}. Thus, the total calcium concentration in solution is approximately 15 mmol kg^{-1}, or about 50% higher than the concentration of Ca^{2+} alone.

A KINETIC APPROACH TO SOLUTION EQUILIBRIA

An experimental petrologist, following standard procedure, would verify the proportions and compositions of phases at equilibrium by allowing the reaction describing the system to proceed first in a "forward" direction until no further change was observed, and then repeating the experiment in a "backward" direction. It is an unfortunate fact of life, however, that many of the systems we wish to study never attain equilibrium during the time we have available to observe them. This is particularly true at low temperatures, where reaction rates are sluggish, but may also be the case at elevated temperatures if reactions involve the slow formation of intermediate compounds. Where they are available, kinetic data can be useful in describing such systems. What follows here is a very brief discussion of rate theory for chemical reactions, to illustrate an alternative approach to the thermodynamic equations developed in this chapter.[6]

For simplicity, we will consider only those processes which are said to follow a *first order* kinetic law like the one we discussed for the party problem in Chapter 1. In qualitative terms, such a relationship tells us that the probability that a reaction will occur between molecular or ionic species in a given time interval is directly proportional to their abundance. The rate of precipitation of $BaSO_4$ from solution, for instance, is governed by the probability that a Ba^{2+} ion and a SO_4^{2-} ion will be close enough to combine. The more Ba^{2+} and SO_4^{2-} the solution contains, the higher that probability is. For the general reaction

$$\nu_{j+1} A_{j+1} + \cdots + \nu_{i-1} A_{i-1} + \nu_i A_i \longleftrightarrow \nu_1 A_1 + \nu_2 A_2 + \cdots + \nu_j A_j,$$

the rate of forward reaction is given by

$$v_f = k_f a_{j+1}^{\nu_{j+1}} a_{j+2}^{\nu_{j+2}} \cdots a_i^{\nu_i} \qquad (3\text{-}29)$$

---

[6] This discussion and Worked Problem 3-9 are modified from Stumm and Morgan (1981).

and the rate of the reverse reaction by

$$v_r = k_r a_1^{\nu_1} a_2^{\nu_2} \cdots a_j^{\nu_j}. \tag{3-30}$$

The quantities labeled ν are stoichiometric coefficients and k_f and k_r are first-order rate constants.

At equilibrium, the forward and reverse reaction rates must be equal if there is to be no net change in the system. It should be clear, therefore, that there is a direct relationship between the rate constants and the equilibrium constant for the reaction

$$K_{eq} = k_f/k_r. \tag{3-31}$$

Furthermore, the degree of disequilibrium can be estimated from the ratio v_f/v_r and the activities of species in the system, because

$$\frac{k_f}{k_r} = \frac{v_f}{v_r} \frac{a_1^{\nu_1} a_2^{\nu_2} \cdots a_j^{\nu_j}}{a_{j+1}^{\nu_{j+1}} a_{j+2}^{\nu_{j+2}} \cdots a_i^{\nu_i}}. \tag{3-32}$$

This expression can be written, for greater clarity, in the form

$$\frac{k_f}{k_r} = K_{eq} = \frac{v_f}{v_r} Q. \tag{3-33}$$

This degenerates to equation 3-21 for the equilibrium case, where $v_f/v_r = 1$, but it tells us more than that. If $K_{eq}/Q > 1$, then the rate of the forward reaction exceeds the rate of the reverse reaction. If $K_{eq}/Q < 1$, the opposite is true. Furthermore, it can be seen that a reaction proceeds most rapidly (in either a forward or a backward direction) when it is farthest from equilibrium; that is, when K_{eq}/Q is very much greater or very much less than 1. For this reason, the direction and rate of reaction are often good indicators of how far a system may be from equilibrium. In the same sense that an experimental petrologist can perform experiments to "bracket" an equilibrium phase assemblage, we can use equation 3-33 to estimate the approach to equilibrium.

Worked Problem 3-11

Reactions involving carbon dioxide and water are extremely important in the oceans and in surficial environments, as we will find in later chapters. One important reaction in this system involves the equilibrium between dissolved CO_2 gas and the bicarbonate ion:

$$CO_2(aq) + H_2O \longleftrightarrow HCO_3^- + H^+. \tag{3-34}$$

This is, in fact, a two-stage reaction which proceeds first by the slow step of hydration

$$CO_2(aq) + H_2O \longleftrightarrow H_2CO_3$$

and then much more quickly by dissociation

$$H_2CO_3 \longleftrightarrow HCO_3^- + H^+.$$

Hydration is such a slow step, comparatively, that $CO_2(aq)/H_2CO_3$ is about 600 at 25°C. The combined reaction, therefore, follows approximately a first-order rate law. Can we use its kinetic behavior to describe the approach to equilibrium and to estimate the value of K_{eq}?

The rate constants for equation 3-34 (Kern, 1960) are $k_f = 3 \times 10^{-2}$ sec^{-1} and $k_r = 7.0 \times 10^4$ sec^{-1}. Figure 3.9a shows the progress of reaction in a system presumed to consist initially of 1×10^{-5} mol liter^{-1} $CO_2(aq)$ at $pH = 7(a_{H^+} = 1 \times 10^{-7})$. The concentrations of $CO_2(aq)$ and HCO_3^- could, hypothetically, be measured as functions of time, but this is a nearly impossible task in the CO_2–H_2O system. In constructing Figure 3.9a, therefore, the concentrations were calculated algebraically from the experimentally determined rate constants. For details of the method, see Stumm and Morgan (1981).

A Kinetic Approach to Solution Equilibria

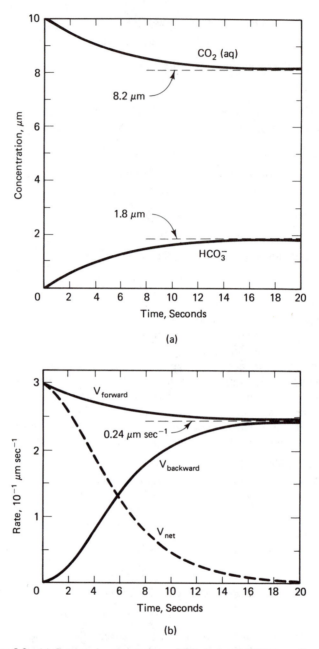

Figure 3.9 (a) Computed concentrations of CO_2 (aq) and HCO_3^- as a function of time for the reaction CO_2 (aq) $+ H_2O \longleftrightarrow HCO_3^- + H^+$ at 25°C in a closed aqueous system. The total concentration of carbon species remains constant at 1×10^{-5} m. (b) Computed reaction rates for the forward and reverse directions of the approximate first-order reaction illustrated in Figure 3.9 as functions of time. At equilibrium, both reaction rates approach 2.4×10^{-4} mole sec^{-1}. (After W. Stumm and J. J. Morgan, *Aquatic Chemistry*. Copyright 1981 by John Wiley and Sons. Reprinted with permission.)

Here we are interested in the rates of reaction as equation 3-34 approaches equilibrium. These can be calculated at any given time from

$$v_f = k_f a_{CO_2}(aq)$$

and

$$v_r = k_r a_{HCO_3^-} a_{H^+}.$$

The results are shown in Figure 3.9b. Notice that the rate profiles for the forward and reverse reactions are not mirror images of each other. During the approach to equilibrium, $v_f/v_r > 1$, so $K_{eq}/Q > 1$. Both ratios decrease with time, however, and within roughly 20 seconds v_f and v_r have approached the same value (2.4×10^{-4} mole sec^{-1}) to less than 1% error. The dashed line in Figure 3.9b indicates the progress of the net reaction rate, $v_f + v_r$, which approaches zero. The value of K_{eq}, calculated from the rate constants by equation 3-31, is 4.3×10^{-7}. This is virtually the same as the value (4.4×10^{-7}) calculated from the free energies of formation of dissolved species.

SUMMARY

We see now that the structure and stoichiometry of solutions play a key role in determining how they act in chemical reactions. The free energy of any reaction can be expressed in terms of a standard-state free energy and a contribution brought about by the mixing of end-member species in solutions. Relatively few solutions of geochemical interest are ideal; in most, interactions among dissolved species or among atoms on different stuctural sites produce significant departures from ideality. In many cases, however, we gain valuable first impressions of a system by assuming ideal mixing and then adding successive corrections to describe its actual behavior.

We will discuss solution models more fully in Chapters 8 and 9 where, in particular, we will address the problem of calculating activity coefficients for nonideal systems of petrologic interest. In the next few chapters, however, we will make use of the Debye-Hückel equation and the mean salt method, which yield the activity coefficients we need to explore dilute aqueous solutions. Solubility equilibria will dominate our discussion of the oceans, of chemical weathering, and of diagenesis.

SUGGESTED READINGS

Several excellent textbooks are available for the student interested in the thermodynamic treatment of solutions. The ones listed here are a selection of some which the authors have found particularly useful.

DREVER, J. I. 1982. *The geochemistry of natural waters.* Englewood Cliffs, N.J.: Prentice-Hall. (This is an excellent book for the student who is new to aqueous geochemistry. There are quite a few carefully written case studies to show the use of geochemistry in the real world.)

GARRELS, R. M., and CHRIST, C. L. 1982. *Solutions, minerals, and equilibria.* San Francisco: Freeman Cooper. (Perhaps the most often-used text in its field. Any geochemist who works with solutions should own a copy of this book.)

POWELL, R. 1978. *Equilibrium thermodynamics in petrology.* London: Harper and Row. (A good text for classroom use. Chapters 4 and 5 consider the calculation of mole fractions in crystalline and liquid solutions.)

ROBINSON, R. A., and STOKES, R. H. 1968. *Electrolyte solutions.* London: Butterworths. (A reference for chemists rather than geologists, but widely quoted by aqueous geochemists. Highly theoretical, but not difficult to follow.)

PYTKOWICZ, R. M. 1983. *Equilibria, nonequilibria, and natural waters, Vol. I.* New York: J. Wiley and Sons. (An exhaustive treatment of the theoretical basis for the thermodynamics of aqueous solutions. Chapter 5 has over 100 pages devoted to the determination of activity coefficients.)

STUMM, W., and MORGAN, J. J. 1981. *Aquatic chemistry.* New York: John Wiley and Sons. (A detailed, yet highly readable text with an emphasis on the chemical behavior of natural waters. Chapter 6, dealing with complexes, is particularly well-written.)

The following articles were referenced in this chapter. The interested student may wish to explore them more thoroughly.

BLOUNT, C. W. 1977. Barite solubilities and thermodynamic quantities up to 300°C and 1400 bars. *Amer. Mineral.* 62: 942–57.

BOTTINGA, Y., and WEILL, D. F. 1972. The viscosity of magmatic silicate liquids: A model for calculation. *Amer. J. Sci.* 272: 438–75.

BUSING, W. R. 1981. WMIN, A computer program to model molecules and crystals in terms of potential energy functions. Oak Ridge National Laboratory, NTIS Document ORNL-5747.

CAMERON, M. and PAPIKE, J. J. 1980. Crystal chemistry of silicate pyroxenes. In *Pyroxenes,* ed. C. T. Prewitt, 5–92. Reviews in Mineralogy 7. Washington, D.C.: Mineral. Soc. of Amer.

CAMERON, M., and PAPIKE, J. J. 1981. Structural and chemical variations in pyroxenes. *Amer. Mineral.* 66: 1–50.

CLARK, J.; APPLEMAN, D. E.; and PAPIKE, J. J. 1969. Crystal-chemical characterization of clinopyroxenes based on eight new structure refinements. In *Special Paper* 2. 31–50. Washington. D.C.: Mineral. Soc. of Amer.

COHEN, R. E. 1986. Thermodynamic solution properties of aluminous clinopyroxenes: Non-linear least-squares refinements. *Geochim. Cosmochim. Acta* 50: 563–76.

ELGQUIST, B., and WEDBORG, M. 1978. Stability constants of $NaSO_4^-$, $MgSO_4$, MgF^+, $MgCl^+$ ion pairs at the ionic strength of seawater by potentiometry. *Marine Chemistry* 6: 243–52.

GOLDBERG, R. N., and NUTTALL, R. L. 1978. Evaluated activity and osmotic coefficients for aqueous solutions: The alkaline earth halides. *J. Phys. Chem. Ref. Data* 7: 263.

HARVIE, C. E.; MOLLER, N.; and WEARE, J. H. 1984. The prediction of mineral solubilities in natural waters: The Na-K-Mg-Ca-H-Cl-SO₄-OH-HCO₃-CO₃-CO₂-H₂O system to high ionic strengths at 25°C. *Geochim. Cosmochim. Acta* 48: 723–52.

HELGESON, H. C. 1969. Thermodynamics of hydrothermal systems at elevated temperatures and pressures. *Amer. J. Sci.* 267: 729–804.

HOLLOWAY, J. R. 1977. Fugacity and activity of molecular species in supercritical fluids. In *Thermodynamics in geology,* ed. D. G. Fraser, 161–80. Boston: D. Reidell.

KERRICK, D. M., and JACOBS, G. K. 1981. A modified Redlich-Kwong equation for H_2O, CO_2, and H_2O-CO_2 mixtures at elevated pressures and temperatures. *Amer. J. Sci.* 281: 735–67.

KERN, D. B. 1960. The hydration of carbon dioxide. *J. Chem. Ed.* 37: 14–23.

KILLEY, R. W. D.; McHUGH, J. O.; CHAMP, D. R.; COOPER, E. L.; and YOUNG, J. L. 1984. Subsurface Cobalt-60 migration from a low-level waste disposal site. *Environ. Sci. Technol.* 18: 148–57.

MASON, B., and MOORE, C. B. 1982. *Principles of geochemistry.* New York: John Wiley and Sons.

OLSEN, C. R.; LOWRY, P. D.; LEE, S. Y.; LARSEN, I. L.; and CUTSHALL, N. H. 1986. Geochemical and environmental precesses affecting radionuclide migration from a formerly used seepage trench. *Geochim. Cosmochim. Acta* 50: 593–608.

REDLICH, O., and KWONG, J. N. S. 1949. An equation of state: Fugacities of gaseous solutions. *Chem. Rev.* 44: 233–44.

ROBINSON, R. A., and STOKES, R. H. 1949. Tables of osmotic and activity coefficients of electrolytes in aqueous solutions at 25°C. *Trans. Faraday Soc.* 45: 612.

SUCHA, L.; CADEK, J.; HRABEK, K.; and VESELY, J. 1975. The stability of the chloro complexes of magnesium and of the alkaline earth metals at elevated temperatures. *Coll. Czechoslov. Chem. Commun.* 40: 2020–24.

PROBLEMS

1. Use the Debye-Hückel equation to calculate values for the ion activity coefficients for Cl^- and Al^{3+} at 25°C in aqueous solutions with ionic strengths ranging from 0.01 to 0.2. Plot these on a graph of γ vs log I. What physical differences between these two ions account for the differences in their activity coefficients?

2. Using the procedure introduced in Worked Problem 3-6, calculate the solubility of gypsum in water at 25°C.

3. To produce a solution that is saturated with respect to fluorite at 25°C, you need to dissolve 6.8×10^{-3} g of CaF_2 per 0.25 liter of pure water. Assuming that CaF_2 is 100% dissociated, and that the solution behaves ideally, calculate the K_{sp} for fluorite. What is the free energy for the dissociation reaction?

4. The K_{sp} for celestite is 7.6×10^{-7} at 25°C. How many grams of $SrSO_4$ can you dissolve in a liter of 0.1 m $Sr(NO_3)_2$? Assume that the activity coefficients for all dissolved species are unity, and that no complex species are formed in solution.

5. Repeat the calculation in problem 4, again assuming that complex species are absent, but

correcting for the ionic strength effect by calculating activity coefficients with the Debye-Hückel equation.

6. The following analysis of water from Lake Nipissing in Ontario was reported by Livingstone (1963):

HCO_3^-	26.2 ppm	SO_4^{2-}	8.5 ppm
Cl^-	1.0 ppm	NO_3^-	1.33 ppm
Ca^{2+}	9.0 ppm	Mg^{2+}	3.6 ppm
Na^+	3.8 ppm		

What is the ionic strength of this water? (Recall that concentrations in ppm are equal to concentrations in mmol kg^{-1} multiplied by formula weight.)

7. Calculate the mole fractions of forsterite and fayalite in an olivine with the composition $Mg_{0.7}Fe_{1.3}SiO_4$.

8. A garnet has been analyzed by electron microprobe and found to have the following composition:

SiO_2	39.08 wt%
Al_2O_3	2.66 wt%
FeO	29.98 wt%
MnO	1.61 wt%
MgO	6.67 wt%

Calculate the mole fractions of almandine, pyrope, and spessartine in this garnet.

9. [7]Calculate the mole fractions of enstatite, ferrosilite, diopside, $CaCrAlSiO_6$ (Cr-CATS), and $Mg_{1.5}AlSi_{1.5}O_6$ (pyrope) in the orthopyroxene analysis below. Assume complete mixing on sites in such a way that all tetrahedral sites are filled by either Si or Al; M1 sites by Al, Cr, Fe^{3+}, Fe^{2+}, or Mg; and M2 sites by Ca, Fe^{2+}, or Mg. Assume also that the ratio of X_{Fe} to X_{Mg} is the same in M1 and M2 sites. (This problem will require a fair amount of effort.)

SiO_2	57.73 wt%	Cr_2O_3	0.46 wt%
Al_2O_3	0.95 wt%	FeO	3.87 wt%
Fe_2O_3	0.42 wt%	MgO	36.73 wt%
		CaO	0.23 wt%

[7] Modified from an example by Powell (1978, Chapter 4).

Chapter 4

The Oceans and Atmosphere as a Geochemical System

OVERVIEW

In this chapter we introduce three broad problems that have occupied the attention of geochemists who deal with the ocean-atmosphere system. The first of these concerns the composition of the oceans and the development of what has been called a "model" for seawater. This is a comprehensive description of dissolved species, constrained by mass and charge balance and the principles of electrolyte solution behavior we have just discussed in Chapter 3.

Closely linked to this problem are questions about the dynamics of chemical cycling between the ocean-atmosphere and the solid earth. Aside from the geologically rapid processes that control the electrolyte species in seawater, there are large-scale processes that maintain its bulk chemistry by balancing inputs from weathering sources against a variety of sinks. Although many of these inputs and outputs will be discussed in Chapters 5 and 6, in this chapter we consider the global dynamics of the sodium, magnesium, and carbon cycles as examples of this class of problem.

Finally, to assure you that oceanic geochemistry has room for visionaries, we discuss several ideas relating to the history of seawater and air. This third area of study draws on our understanding of electrolyte chemistry and our understanding of cycling processes to help interpret ancient marine environments and events, and to gain some perspective on the evolution of life.

COMPOSITION OF THE OCEANS

A Classification of Dissolved Constituents

The oceans provide geochemists with a vast laboratory in electrolyte solution chemistry. In one sense, this laboratory is boringly uniform: its composition is dominated by six major elements (Na, Mg, Ca, K, Cl, and S) whose abundances, relative to one another, are very nearly constant despite the fact that the total dissolved salt content (the *salinity*) of the open ocean varies from 33 ‰ to 38 ‰[1]. Of these elements, only calcium has been shown to vary in its ratios to the other five by a small but measurable amount. Constituents whose relative concentrations remain constant in seawater are identified as *conservative*. Variations in their absolute abundances can be attributed solely to the addition or subtraction of pure water to the oceans.

Attention is focused most commonly on the minor constituents of seawater. Some of them (like boron, lithium, antimony, uranium, and bromine) are also conservative, but many vary dramatically, relative to local salinity. These dissolved species are, therefore, *nonconservative*. Most of these variations are associated with the oceans' major role as a home for living organisms. Plants live only in near-surface waters, where they can receive enough sunlight to permit photosynthesis. They and animals that feed on them deplete those waters in nutrient species such as phosphate, nitrate, dissolved silica, and dissolved inorganic carbon ($\Sigma \ CO_2 = CO_2 + HCO_3^- + CO_3^{2-}$), which are converted to organic and skeletal matter. As these materials settle toward the ocean floor in dead organisms or in fecal matter, the hard body parts redissolve and tissues are rapidly consumed by bacteria, causing a relative enrichment of nutrients in bottom water. Of the elements associated with ocean life, the only significant exception to this pattern is oxygen, which is released by photosynthesis at the surface and consumed by respiration in deep water. Since the mid-1970s, it has gradually become apparent that many trace elements, such as strontium, cadmium, barium, and chromium, also follow abundance patterns that are nutrient-correlated, although it is not clear in many cases how these elements are involved in the oceans' biochemical mechanisms. From a geochemical viewpoint, these measurable gradients are essential to our ability to trace the pathways and processes of biological, sedimentary, and circulatory behavior in the oceans.

In Table 4-1, we have indicated the abundance and behavior of selected elements in the ocean, dividing them into conservative and nonconservative groups. Dissolved gases, except for those (like oxygen, CO_2, H_2S, and to some degree N_2 and N_2O) that are involved in metabolic processes, have abundance patterns that are controlled largely by their solubility. To a good first approximation, concentration in surface waters is fixed by the temperature of the water and the atmospheric partial pressure of the gas. Because atmospheric gases are also trapped in bubbles during stirring by waves, however, their measured abundances may be expected to deviate from this solubility-controlled concentration by an unpredictable (but usually small) amount. We also have indicated that human activity contributes in a major way to the abundance or distribution of a few elements. Lead, injected into the ocean-atmosphere system by automobile emissions, is a good example.

The central role of organisms in determining the distribution of elements in the oceans, aside from the six major conservative elements we have already mentioned,

---

[1] It is common practice to report salinity measurements in parts per thousand by weight, or *per mil* (‰). Thus, seawater that has a total dissolved salt content of 3.45 weight percent has a salinity of 34.5‰.

TABLE 4-1 ABUNDANCE AND BEHAVIOR OF SELECTED ELEMENTS IN SEAWATER[*]

Element	Surface Conc.	Conc. at Depth	Behavior
Li	178 μg/kg		Conservative
B	4.4 mg/kg		Conservative
C	2039 μmol/kg	2264 μmol/kg (714m)	Nutrient
N (N_2)		560 μmol/kg (995m)	Non-nutrient gas
(NO_3^-)	$< 3 \times 10^{-7}$ mol/kg	41 μmol/kg (891m)	Nutrient
O (O_2)	205 μmol/kg	47 μmol/kg (891m)	Nutrient-related
F	1.3 mg/kg		Conservative
Na	10.781 g/kg		Conservative
Mg	1.284 g/kg		Conservative
Al	1.0 μg/kg	0.6 μg/kg (1000m)	?
Si	2.0 μmol/kg	100.8 μmol/kg (891m)	Nutrient
P	0.08 μmol/kg	2.84 μmol/kg (891m)	Nutrient
S	2.712 g/kg		Conservative
Cl	19.353 g/kg		Conservative
K	399 mg/kg		Conservative
Ca	417.6 mg/kg	413.6 mg/kg (1000m)	Conservative[***]
Cr	268 ng/kg	296 ng/kg (1000m)	Nutrient-related
Mn	34 ng/kg	38 ng/kg (985m)	Nutrient-related
Fe	8 ng/kg	45 ng/kg	Nutrient-related
Ni	146 ng/kg	566 ng/kg (985m)	Nutrient-related
Cu	34 ng/kg	130 ng/kg (985m)	Nutrient-related
Zn	6–7 ng/kg	438 ng/kg (985m)	Nutrient-related
As	1.1 μg/kg	1.8 μg/kg (1000m)	Nutrient-related
Br	67 mg/kg		Conservative
Rb	124 μg/kg		Conservative
Sr	7.404 mg/kg	7.720 mg/kg (700m)	Nutrient-related
Cd	0.3 ng/kg	117 ng/kg (985m)	Nutrient-related
Sn	1.0 ng/kg		Anthropogenic
I	48 μg/kg	60 μg/kg (1036m)	Nutrient-related
Cs	0.3 μg/kg		Conservative
Ba	4.8 μg/kg	13.3 μg/kg (997m)	Nutrient-related
Hg	3 ng/kg	4 ng/kg (1000m)	Nutrient-related
Pb	13.6 ng/kg	4.5 ng/kg (1000m)	Anthropogenic
U	3.2 μg/kg		Conservative

[*]Modified from Quimby-Hunt and Turekian (1983).

[**]These values are for illustration only. They are not intended to be taken as average values, which are in most cases poorly known.

[***]Although these data indicate that typical surface and deep waters have similar calcium contents, calcium is commonly removed from surface water by biological precipitation of carbonates. The net loss of Ca, however, is a small fraction of the total amount present and is not easily discerned from global values. Therefore, Ca is nearly conservative.

should be apparent from the last column of Table 4-1. Of the elements listed as nutrients or nutrient-related, five are particularly noteworthy. Phosphate, nitrate, silica, zinc, and cadmium are nearly absent in surface waters, and are often referred to as *biolimiting* constituents of seawater because they are scavenged so efficiently by microorganisms. When depleted, these elements place a limit on the biomass of surface waters. Only one of these five is universally biolimiting, however. The supply of silica limits only the population of organisms (diatoms and radiolaria) that form hard body parts from SiO_2, but not the vast population of those that build skeletons with calcium carbonate. Nitrate is a limiting nutrient for a great many species, but not for the large family of blue-green algae, which can fix nitrogen directly from dissolved N_2. Phosphorus (as phosphate) alone is required in the metabolic cycles of all organisms, and therefore is the ultimate biolimiting element. Cadmium and zinc are apparently depleted drastically in surface waters only because plants mistakenly incorporate them as phosphorus.

One element that is conspicuously absent from the list of biolimiting constituents is carbon, about which we will have much more to say as this chapter unfolds. Notice that although surface waters have a lower concentration of total dissolved carbon species than deep waters, the depletion is on the order of 10%—not startlingly large. Surface-dwelling organisms, therefore, leave roughly 90% of the dissolved carbon around them untouched.

One way to appreciate more fully the nonlimiting nature of carbon is to compare its abundance with elements that *are* biolimiting. As organic tissue is formed in surface waters, its atomic ratio of phosphorus to nitrogen to carbon is very nearly constant at 1 : 15 : 105, a proportion called the *Redfield ratio,* for Alfred C. Redfield, the oceanic geochemist who first determined it. As this organic matter settles into the deep oceans, it is almost entirely consumed and returned to solution as dissolved species. It is not surprising, then, that the atomic ratio P : N in deep waters is 1 : 15. The deep ratio of P : C, however, is roughly 1 : 1000. Approximately 90% of the dissolved carbon in deep water, therefore, cannot have been carried down in organic matter. Some was carried down as particulate carbonate, but this is a rather small fraction. For every four atoms of carbon fixed in organic matter, approximately one is combined with a calcium ion and precipitated biochemically as carbonate. (This process accounts, qualitatively, for the slightly nonconservative behavior of calcium, evident in Table 4-1. We will elaborate on this relationship in a later section.)

We must conclude finally that most of the dissolved carbon in the deep ocean is not there as a direct result of the downward rain of dead organisms. For each phosphorus atom in the ocean, there are roughly 870 carbon atoms in excess of the number necessary to satisfy the growth demands of the biomass. If we were skeptical before, it should now be clear that carbon is by no means biolimiting, despite its central role in the architecture of living matter.

Chemical Variations with Depth

From our discussions so far, it should be clear that the major compositional gradients in the oceans are vertical and are driven by the biological cycling process we have just outlined. As a way of building a greater appreciation of these gradients, we have presented in Figure 4.1 several sets of graphical data from the Pacific Ocean gathered by the Geochemical Ocean Sections program (GEOSECS) during the 1970s.

The first two of these, showing vertical gradients in temperature and salinity, are important for our understanding of oceanic environments, but are controlled by physical rather than geochemical processes. The thermal gradient shown in Figure 4.1a is the result of solar heating at the surface. Water is sufficiently opaque to visible and infrared radiation that sunlight penetrates only a few tens of meters before it is absorbed almost completely. This, and thermal exchange with the atmosphere, result in significant warming of the top of the ocean. Mixing by winds and currents causes limited downward heat transport, but the water temperature throughout most of the deep ocean is nearly constant at 2°C.[2] It is convenient, therefore, to describe the ocean in

---

[2] Seawater experiences a minor temperature increase with depth (~0.1°C/km) due to compression, but this is an adiabatic effect. No net internal energy change would result, therefore, if the water were slowly brought to the surface again. For that reason, water temperatures at depth are generally reported as *potential temperatures* from which the adiabatic heating has been subtracted.

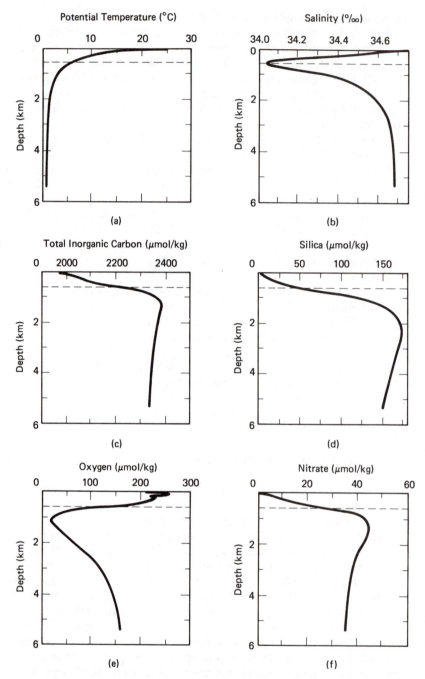

Figure 4.1 Variations in chemical and physical properties of the ocean with depth at GEOSECS station 214 in the North Pacific Ocean. (a) Potential temperature, (b) Salinity, (c) Total dissolved carbon, (d) Silica, (e) Oxygen, (f) Nitrate. (After W. S. Broecker and T-H. Peng, *Tracers in the sea*. Copyright 1982 by Wallace S. Broecker and Tsung-Hung Peng. Reprinted with permission.)

thermal terms as if it consisted of two zones: a warm surface zone, mixed by winds to an average depth of 70 meters, and a deep-water zone of constant cold temperatures that begins roughly 1 km below the surface. The transitional region between them is called the *thermocline*. Depending on distance from the equator and on the time of year, the thermocline may be very thin and close to the surface or (as in Figure 4.1a) much thicker and deeper. Because warm water is, in general, less dense than cold water, the shape of the temperature profile suggests that the oceans should be stable against thermal convection.

Salinity profiles are more complex than temperature profiles. At most latitudes, solar heating causes enough evaporation from the surface to increase the salinity of waters above the thermocline. Despite their higher salt content, these waters are warm and thus less dense than the less saline waters below. Near the poles, however, evaporation is minimized and the production of sea ice tends to leave cold, dense saline water, which descends to the ocean floor. Global circulation draws large amounts of this cold bottom water (largely from the North Atlantic Ocean and from the Weddell Sea off Antarctica) into more temperate latitudes, where it gradually displaces local bottom water and causes a gentle upward circulation. In large parts of the world, this pattern is complicated further by the lateral transport of cold, low-salinity melt waters (as in the North Pacific, for which the profile in Figure 4.1b was constructed) or warm, high-salinity waters from shallow seas (as in the Mid-Atlantic, where Mediterranean waters enter through the Straits of Gibraltar). As a result, salinity profiles may commonly have maxima and minima where waters of varying density are injected into the water column by global circulation.

Although Figures 4.1a and 4.1b are important to our appreciation of the physical behavior of the oceans, the remaining profiles (Figures 4.1c through 4.1f) are of greater geochemical interest because they help to illustrate the effects of the biological "pumping" mechanism that carries nutrients between surface and deep waters. Total inorganic carbon and silica show depth profiles that are roughly mirror images of the thermal profile. Together, these delineate the zone of photosynthesis and surficial biologic activity. The only difference between them is that silica, being biolimiting, begins at nearly zero concentration at the surface, while carbon begins at about 90% of its value in bottom water, as we discussed earlier. Many other elements, like barium, follow profiles that resemble the profile for silica qualitatively, but are not biolimiting. Like carbon, therefore, they do not reach zero concentration in surface waters. It is not yet clear how these bio-related elements are incorporated into organic matter or skeletons, but their association with constituents like silica (in hard body parts) and carbon (largely in tissues) is quite apparent from their similar depth profiles.

Oxygen and nitrate deserve particular attention, because their profiles demonstrate the influence of both biological mechanisms and global circulation. Near the surface, the oxygen content of seawater is constrained by its solubility in thermodynamic equilibrium with the atmosphere. In fact, surface waters generally exceed this level as the result of local photosynthesis. Just below the thermocline, however, the oxygen level drops precipitously as O_2 is used to oxidize organic matter raining down from above. Nitrate is released during this process, so its profile is almost a mirror image of that for oxygen. Dissolved oxygen content has not been found to reach zero at any place in the modern open oceans, but there are inland seas like the Black Sea which are anoxic below the thermocline, and the sedimentological record indicates that anoxia was more common in pre-Cenozoic oceans.

Composition of the Oceans

In polar waters, the story is somewhat different. Because of the low temperatures, organisms do not consume all of the nutrient nitrate and phosphate at the surface. The rain of organic matter toward the bottom is, consequently, less intense. As a result, the oxygen minimum below the thermocline is less pronounced and the nutrient content of polar bottom water is slightly higher than is observed at the base of the thermocline in warmer latitudes. When this water is transported by the deep global currents we discussed earlier, the result is that average bottom water is significantly more oxygen-rich and somewhat more nutrient-rich than average seawater. This is illustrated in Figures 4.1e and 4.1f. A profile of phosphate would be qualitatively similar to that for nitrate.

COMPOSITION OF THE ATMOSPHERE

Before we proceed with a more quantitative look at geochemical processes in the oceans, let us consider briefly the composition of the atmosphere. The three major reactive components of the atmosphere—oxygen, water, and carbon dioxide—are readily exchanged between the air and the oceans. Equilibrium or kinetic factors that influence their distribution, therefore, have an important effect on marine chemistry. Further down the chain of geochemical events (discussed later in this chapter and in Chapter 6), we will investigate how the atmosphere engenders weathering on land and thus provides further controls on the bulk composition of streams and the ocean.

Although some photochemical reactions that take place in the upper atmosphere may affect geochemical processes on the earth, we can largely disregard the composition of air above about 12 km. The lower portion of the atmosphere, known as the *troposphere,* contains the bulk of its mass, and is vigorously stirred by convection because of solar heating at the planetary surface. As a result, its composition, presented in Table 4-2, is nearly uniform. Water vapor, of course, is a major exception. Because atmospheric P_{H_2O} is, on average, very close to the saturation vapor pressure, relatively small local changes in temperature can make the difference between evaporation and precipitation. The conversion of liquid water to vapor and back again is a major means for energy transport at the earth's surface.

In addition to the species listed in Table 4-2, there are many gases with concentrations less than 1 ppm. Variations in some of these, like ozone (O_3), can have a

TABLE 4-2 THE COMPOSITION OF THE ATMOSPHERE AT GROUND LEVEL*

Constituent	Concentration (ppm by volume)
Nitrogen (N_2)	$78.084 \, (\pm 0.004) \times 10^4$
Oxygen (O_2)	$20.946 \, (\pm 0.002) \times 10^4$
Argon	$9340 \, (\pm 10)$
Water	40 to 4×10^4
Carbon Dioxide	320
Neon	$18.18 \, (\pm 0.04)$
Helium	$5.24 \, (\pm 0.04)$
Methane (CH_4)	1.4
Krypton	$1.14 \, (\pm 0.01)$

*Modified from Holland (1978).

significant effect on organisms, but they rarely have a measurable role in controlling geochemical environments.

The rare gases are all chemically inert, so they have no tendency to form compounds during chemical weathering or rock-forming processes. To a first approximation, then, they can be seen as gradually accumulating in the atmosphere with time. The picture is much more complicated than this, however. Helium, for example, is light enough to escape from the top of the atmosphere in a geologically short time, argon is continually produced in the crust by radioactive decay, and all of the rare gases have been shown to bond weakly to clay surfaces and undergo limited reburial during sedimentation. With allowances for these complexities, rare gas abundances have long been used to help interpret the history of the atmosphere and of the earth itself, precisely because they are nonreactive. This will be a topic of discussion near the end of this chapter.

Nitrogen is also nearly inert, although a limited amount of N_2 is fixed directly by microorganisms and incorporated by growing plants. Estimates in Holland (1978) suggest that of the 4700×10^{12} g of nitrogen used in marine photosynthesis each year, only 85×10^{12} g—less than 2%—is converted from atmospheric N_2. Therefore, most nitrogen in the nutrient cycles of the ocean or continents is recycled within the biosphere. The rates of nitrate burial in sediments and nitrogen release by weathering and volcanism are very small and nearly balanced, although estimates are difficult to verify. Additional values from Holland indicate that less than one part of the atmospheric mass of N_2 in 2×10^8 is lost to sedimentation each year. As a result, it is a very good approximation to call the atmospheric mass of nitrogen inert.

The two remaining constitutents of interest in this quick survey are oxygen and carbon dioxide, both of which owe their abundance in the atmosphere to processes of the biosphere. Each year, roughly 83×10^{15} g of carbon (12% of the atmospheric reservoir) is used in photosynthesis. The process liberates 2.2×10^{17} g of O_2. The total mass of atmospheric O_2 is estimated to be 1.2×10^{21} g. By dividing the mass liberated annually by the total mass, it is easy to calculate that the biosphere could regenerate the entire mass of the atmospheric O_2 once every 5500 years. The shorter time in which all CO_2 could theoretically be consumed by photosynthesis, approximately 9 years, is largely a function of the relative amounts of both gases in the atmosphere. In fact, the biogeochemical cycles for O_2 and CO_2 are very nearly balanced. The mass of carbon fixed in organic matter by photosynthesis each year is almost quantitatively oxidized by the yearly mass of "new" O_2 and returned to the atmosphere. Neither gas, in any case, can be regarded nonreactive, in the style of nitrogen and the noble gases.

Despite the dominance of biologic processes in controlling the fates of oxygen and CO_2, both gases also play a major role in the inorganic cycle of weathering and sedimentation. In a year, about 3.6×10^{14} g of carbon (as CO_2) are consumed in weathering continental rocks. This amounts to less than 0.5% of the mass of carbon fixed annually in organic matter. The O_2 consumed in all weathering reactions is about 4×10^{14} g yr^{-1}—again, a very small mass compared to the annual amount linked to photosynthesis. These small amounts, however, dominate the weathering process.

It is almost true that the biological cycle for these two gases is separate from the weathering-sedimentation cycle. Less than 0.1% of the carbon fixed yearly in organic matter is trapped in sediments rather than immediately reoxidized, and an equivalent amount is weathered out of continental rocks. The constituents of both cycles meet in the oceans, however, and are a major source of its chemical stability. In the following section, we will concentrate on the chemistry of marine carbonate species.

CARBONATE AND THE GREAT MARINE BALANCING ACT

Some First Principles

A good way to begin is by considering a simplified "ocean" that consists only of pure water in equilibrium with carbon dioxide. We can then write four major reactions relating species in the system to one another:

$$CO_2 \text{ (gas)} \longleftrightarrow CO_2 \text{ (aq)} \qquad (4\text{-}1)$$

$$CO_2 \text{ (aq)} + H_2O \longleftrightarrow H_2CO_3 \qquad (4\text{-}2)$$

$$H_2CO_3 \longleftrightarrow HCO_3^- + H^+ \qquad (4\text{-}3)$$

$$HCO_3^- \longleftrightarrow CO_3^{2-} + H^+ \qquad (4\text{-}4)$$

As we demonstrated at the end of Chapter 3, the first two of these reactions can easily be combined into one. Because the rate of reaction 4-1 is more than two orders of magnitude faster than reaction 4-2, the combined reaction follows a kinetic expression that is nearly first order in P_{CO_2}. It is customary, therefore, to use the expression

$$CO_2 \text{ (gas)} + H_2O \longleftrightarrow H_2CO_3 \qquad (4\text{-}5)$$

to describe the hydration of CO_2 and to speak of all CO_2 (aq) as H_2CO_3. No thermodynamic validity is lost by this simplification, because the equilibrium constant for reaction 4-5 is simply the product of the constants for reactions 4-1 and 4-2. The set of relevant relationships, therefore, consists of a solubility expression for CO_2 and two dissociation reactions. At 25°C, the equilibrium constants for these reactions have the values

$$K_1 \text{ (for reaction 4-3)} = a_{HCO_3^-} a_{H^+} / a_{H_2CO_3} = 10^{-6.35} \qquad (4\text{-}3a)$$

$$K_2 \text{ (for reaction 4-4)} = a_{CO_3^{2-}} a_{H^+} / a_{HCO_3^-} = 10^{-10.33} \qquad (4\text{-}4a)$$

and

$$K_{CO_2} \text{ (for reaction 4-5)} = a_{H_2CO_3} / P_{CO_2} = 10^{-1.47}. \qquad (4\text{-}5a)$$

We have now written three equations to relate five unknowns (P_{CO_2} and the activities of H_2CO_3, HCO_3^-, CO_3^{2-}, and H^+). If we are given any two other pieces of nonredundant information about the system, therefore, it should be possible to determine the activities of all five species. The other two pieces of information will vary from one problem to another, because they are conditions of the particular environment we wish to study.

Worked Problem 4-1

During this century, atmospheric P_{CO_2} has changed significantly as a result of fossil fuel combustion, so any published value is likely to be out of date. For dramatic verification of this statement, see Figure 4.2. If we wanted to know the current atmospheric P_{CO_2}, then, how could we measure it?

One way is to collect a beaker of pure rainwater[3] and measure its *pH*, on the reasonable assumption that rainwater is in equilibrium with atmospheric CO_2. The activity of H^+, therefore, is one piece of information we did not have before. It can be expressed in the form of a very simple equation; for example, $a_{H^+} = 10^{-5.65}$. Because it is necessary to

---

[3] We ignore the existence of other atmospheric gases that, when dissolved in rainwater, will also affect its *pH*. These constituents of "acid rain" can have serious environmental consequences, but are beyond the scope of this simple problem.

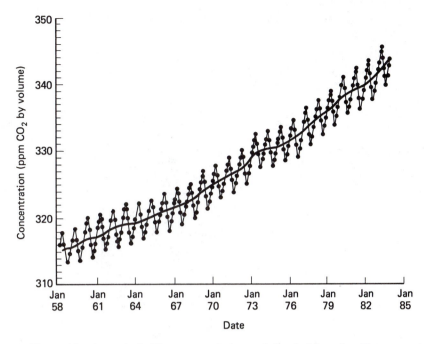

Figure 4.2 Atmospheric CO_2 concentrations recorded at the Mauna Loa Observatory, Hawaii. Annual oscillations follow the growing cycle in the Northern Hemisphere. (After E. T. Sundquist, Geological perspectives on carbon dioxide and the carbon cycle. In *The carbon cycle and atmospheric CO_2: Natural variations Archean to present,* ed. E. T. Sundquist and W. S. Broecker, 5–60. Monograph 32. Copyright 1985 by the American Geophysical Union. Reprinted with permission.)

maintain electrical neutrality, we can often write a charge balance equation that provides another constraint. In this case, that equation is

$$m_{H^+} = m_{HCO_3^-} + 2m_{CO_3^{2-}}.$$

With these two new expressions, plus equations 4-3a through 4-5a, we now have five equations in five unknowns.

To calculate P_{CO_2}, we solve the five equations simultaneously. Rainwater is a very dilute solution, so we will simplify the calculation by assuming that the activities of all aqueous species are equal to their concentrations and that the activity of water is equal to 1:

1. From reaction 4-4, we know that $m_{CO_3^{2-}} = K_2 m_{HCO_3^-}/m_{H^+}$. By combining it with the charge balance expression and performing a little algebra, we find that

$$(m_{H^+})^2 = m_{HCO_3^-}(m_{H^+} + 2K_2).$$

2. Reaction 4-3 gives us a way to express $m_{HCO_3^-}$ as a function of $m_{H_2CO_3}$ and m_{H^+}. If we combine it with the expression for K_{CO_2}, the result is

$$m_{HCO_3^-} = P_{CO_2} K_{CO_2} K_1/m_{H^+}.$$

3. We can finally substitute this expression into the result of step one. A little more algebraic manipulation yields the answer

$$P_{CO_2} = (m_{H^+})^3/(K_{CO_2} K_1(m_{H^+} + 2K_2)).$$

To calculate the atmospheric partial pressure of CO_2, then, we insert the measured

$pH(= -\log_{10} a_{H^+})$. If we had measured a pH of 5.65, for example, we would conclude that $P_{CO_2} = 331$ ppm.

It is instructive, while we are here, to consider the inverse of this problem. Suppose that we had measured atmospheric P_{CO_2} and wanted to know what rainwater pH to expect. The final equation we produced above cannot be rewritten to get a_{H^+} by itself on one side, so we are forced to extract pH by solving a cubic equation. Given a little time or a computer, it is not difficult to solve the problem by sequential approximations.[4]

A much more common method, however, involves using our knowledge of the H_2O–CO_2 system to simplify the problem. When carbon dioxide is dissolved in water at room temperature, the resulting solution always has a pH less than 7. As we will show shortly, this implies that $a_{CO_3^{2-}}$ is negligibly small. (In fact, it is roughly 2×10^{-5} times $a_{HCO_3^-}$ under the conditions we are using.) If this is the case, the charge balance equation is very nearly

$$a_{H^+} = a_{HCO_3^-}.$$

The expression for K_1, therefore, may be written as

$$(a_{H^+})^2 = K_1 a_{H_2CO_3}.$$

By inserting the expression for K_{CO_2}, we get

$$(a_{H^+})^2 = K_1 K_{CO_2} P_{CO_2},$$

or

$$pH = -0.5 \log_{10} (K_1 K_{CO_2} P_{CO_2}).$$

Using a little chemical sophistication, therefore, we have derived an approximate analytical solution that is easy to use with paper, pencil, and a pocket calculator.

How do we know when it is safe to make simplifying approximations like the one we just used in Worked Problem 4-1? Figure 4.3a is a plot illustrating the relative importance of H_2CO_3, HCO_3^-, and CO_3^{2-} in solutions of varying pH but a fixed total activity of dissolved carbon species. This sort of diagram was introduced in 1914 by the Swedish ocean chemist N. Bjerrum. From it, we see that there are two pH values at which the activities of a pair of carbon species are equal. We can derive these easily. Equation 4-3a can be reorganized to give a ratio of $a_{H_2CO_3}/a_{HCO_3^-}$:

$$a_{H_2CO_3}/a_{HCO_3^-} = a_{H^+}/K_1.$$

This ratio will be equal to 1.0 if $a_{H^+} = K_1$. Similarly, equation 4-4a can be written in the form

$$a_{HCO_3^-}/a_{CO_3^{2-}} = a_{H^+}/K_2,$$

from which we see that the activities of bicarbonate and carbonate will be equal if $a_{H^+} = K_2$. At 25°C, the magic pH values are 6.35 and 10.33. Because these crossover points on the Bjerrum plot are several pH units apart, we are justified in neglecting $a_{CO_3^{2-}}$ at pH values much below 8 or $a_{H_2CO_3}$ at pH values much above 9.

Because the activities of two dissolved carbon species are roughly the same in the vicinity of the crossover points, a CO_2-H_2O solution is said to have a high *buffering capacity* in these regions. By this, we mean that the solution pH is extremely resistant to change when we add an acid or a base to it. We can explain this feature of solution chemistry qualitatively by recognizing that hydrogen ions added around pH 6.35 or pH 10.33 are likely to be involved in reaction 4-3, rather than adding to the free hydrogen

---

[4] The program RAINWATR on the diskette available for this book solves for pH numerically, using a Newton-Raphson approach described in Appendix A.

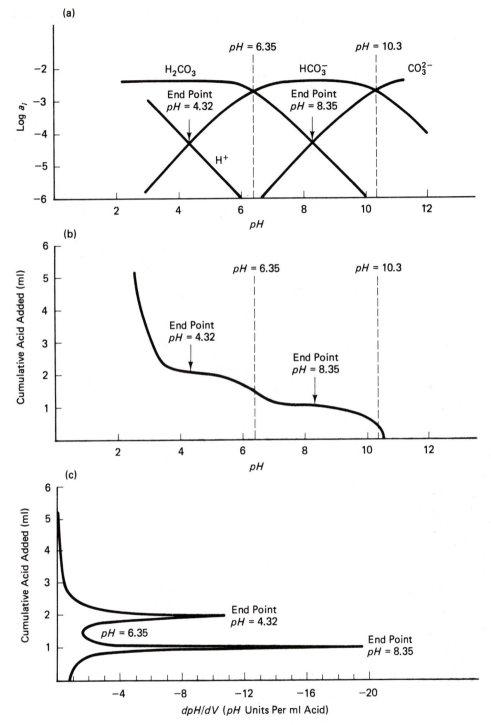

Figure 4.3 (a) Bjerrum plot for a solution with $CO_2 = 1 \times 10^{-3}$ m; (b) Titration curve showing the effect on pH of adding measured volumes of 1.0m HCl to the solution in (a); (c) $d(pH)/dV$ vs. volume of added acid for the titration in (b). Notice the changes in each curve at the titration end points.

Carbonate and the Great Marine Balancing Act

ions in solution. Charge balance is maintained in the solution, even as hydrogen ions (or OH^-) are added, simply by adjusting the ratio of H_2CO_3 to HCO_3^-. In a similar way, solutions with *pH* below 10 involve HCO_3^- and CO_3^{2-} in roughly the same concentration, and solutions below *pH* 3 have H_2CO_3 and H^+ in nearly the same concentration also. This qualitative explanation is incomplete, however, since the various carbonate reactions must consume or release hydrogen ions in *any pH* region. To see why buffering is strongest around *pH* 6.35 and 10.33, we will first consider a rather detailed problem. If you have the program diskette, you may wish to examine the BASIC source code for the program called TITRATE, which can be used to perform the following calculations.

Worked Problem 4-2

Imagine a closed container, filled with one liter of a 10^{-3} m Na_2CO_3 solution. How will the *pH* in the container change as we add successive small amounts of 1.0 m HCl? To make the calculation easier, assume that there is never any gas phase in the container, and pretend that the volume of solution always remains 1.0 l. Furthermore, let's avoid reality and assume that all species in this system behave ideally. If we are ever concerned about this last approximation, we can use the procedure of Worked Problem 3-7 to calculate the necessary activity coefficients.

First, what is the *pH* before we add any acid? There are four unknown quantities in this phase of the problem: $m_{CO_3^{2-}}$, $m_{HCO_3^-}$, $m_{H_2CO_3}$, and *pH*. We therefore need to find four equations to define the system. Two are the defining equations (4-3a and 4-4a) for K_1 and K_2 (we cannot use K_{CO_2} because no gas phase is present).

Charge balance must be maintained, so we can write

$$m_{Na^+} + m_{H^+} = m_{HCO_3^-} + 2m_{CO_3^{2-}} + m_{OH^-}.$$

This equation can be simplified somewhat if we realize that the initial solution will be quite alkaline. Before we begin adding acid, therefore, m_{H^+} is negligibly small and we can write as our third equation

$$m_{Na^+} = m_{HCO_3^-} + 2m_{CO_3^{2-}} + m_{OH^-}. \tag{4-6}$$

(The appearance of OH^- should not be disturbing. We have, of course, added another unknown species, but we have also implicitly gained another equation, describing the dissociation of water:

$$K_w = a_{H^+}a_{OH^-}/a_{H_2O}$$

[or, in this problem, $K_w = m_{H^+}m_{OH^-} = 10^{-14}$]. We therefore include the activity of OH^- and K_w explicitly as a fifth unknown and a fifth constraint, without making the problem any more difficult.)

The final equation could be, as in Worked Problem 4-1, a statement fixing the value of $a_{H_2CO_3}$, $a_{HCO_3^-}$, or $a_{CO_3^{2-}}$ at some measured value. In this case, however, we would find it difficult to measure any of these three activities independently. Instead, we write a statement of mass balance. Because the system is closed, the only possible source of any dissolved carbon species is the Na_2CO_3. We know, therefore, that

$$\Sigma\, CO_2 = m_{H_2CO_3} + m_{HCO_3^-} + m_{CO_3^{2-}}. \tag{4-7}$$

This, too, can be simplified because we know that $m_{H_2CO_3}$ must be negligible in alkaline solutions. Therefore, the fourth equation is

$$\Sigma\, CO_2 = m_{HCO_3^-} + m_{CO_3^{2-}}. \tag{4-8}$$

We know both $\Sigma\, CO_2$ and m_{Na^+} from the stoichiometry of Na_2CO_3 and the initial condition of the problem we are solving: $\Sigma\, CO_2 = \frac{1}{2}m_{Na^+} = 10^{-3}$ m.

These four equations (4-3a, 4-4a, 4-6, and 4-8), solved simultaneously, yield a quadratic expression in m_{H^+}:

$$m_{Na^+}(m_{H^+})^2 - 2K_w(m_{H^+} + K_2) = 0.$$

By inserting numerical values for m_{Na^+}, K_2, and K_w, we can solve for m_{H^+}. The initial pH turns out to be 10.56. In retrospect, this is clearly high enough to justify the simplifications we made in writing equations 4-7 and 4-8.

Now, back to the original question: *what happens when we start to add 1.0m HCl?* To begin, the charge balance equation must be rewritten to accommodate the addition of Cl^-. Also, we should no longer ignore m_{H^+}, which will become increasingly significant as more acid is added. Therefore,

$$m_{Na^+} + m_{H^+} = m_{HCO_3^-} + 2m_{CO_3^{2-}} + m_{OH^-} + m_{Cl^-}. \tag{4-9}$$

If we use the symbol V to represent the volume of acid added by the end of any stage in the experiment, then m_{Cl^-} is given by

$$m_{Cl^-} = V \text{ (ml)} \times 1.0 \text{ (mol } 1^{-1})/10^3 \text{ (ml)} = 10^{-3}V \text{ (mol } 1^{-1}).$$

Equation 4-9, therefore, can be recast in terms of V instead of m_{Cl^-}. This is a convenient substitution, because the volume of added acid is the control variable we want to have left when we reach the end of this calculation. Equation 4-9, therefore, has become

$$10^{-3}V = m_{Na^+} + m_{H^+} - m_{HCO_3^-} - 2m_{CO_3^{2-}} - m_{OH^-}. \tag{4-10}$$

By substituting the expressions for K_1, K_2, and K_w into 4-10 to reduce the number of compositional variables, we achieve the intermediate result

$$10^{-3}V = m_{Na^+} + m_{H^+} - (K_1 m_{H_2CO_3}/m_{H^+}) - (2K_2 m_{HCO_3^-}/m_{H^+}) - (K_w/m_{H^+}). \tag{4-11}$$

Next, the mass balance among dissolved carbon species must be written to include $m_{H_2CO_3}$, because it will be increasingly important as we add acid to the system. Therefore, we manipulate the equations for K_1 and K_2 once more to derive expressions for $m_{HCO_3^-}$ and $m_{CO_3^{2-}}$ in terms of $m_{H_2CO_3}$, m_{H^+}, and the equilibrium constants. We substitute the results in equation 4-6 to get a new mass balance equation:

$$\sum CO_2 = m_{H_2CO_3} + (K_1 m_{H_2CO_3}/m_{H^+}) + (K_1 K_2 m_{H_2CO_3}/(m_{H^+})^2). \tag{4-12}$$

The last step is to eliminate H_2CO_3 by combining equations 4-11 and 4-12 to get

$$10^{-3}V = m_{Na^+} - \frac{\sum CO_2 K_1(m_{H^+} + 2K_2)}{(m_{H^+})^2 + K_1 m_{H^+} + K_1 K_2} - K_w/m_{H^+} + m_{H^+}. \tag{4-13}$$

This equation, describing a *titration curve* for the Na_2CO_3 solution, is the answer we have been looking for. We can examine it graphically by inserting the values for m_{Na^+} and $\sum CO_2$ and solving for V at a number of selected values of pH. The result, calculated by the program TITRATE, is shown in Figure 4.3b. Below it, in Figure 4.3c, is a plot of $d(pH)/dV$.

While the experience of having derived and plotted equation 4-13 is still fresh, we can ask again how pH buffering in the carbonate system works. We see now that in the immediate vicinity of pH 6.35, m_{H^+} is very nearly the same as K_1. Equation 4-13, therefore, can be greatly simplified:

$$10^{-3}V \simeq m_{Na^+} - \frac{\sum CO_2(m_{H^+} + 2K_2)}{(2m_{H^+} + K_2)} - \frac{K_w}{m_{H^+}} + m_{H^+}$$

or, since $K_2 \ll m_{H^+}$ and $K_w \ll m_{H^+}$ under these circumstances,

$$10^{-3}V \simeq m_{Na^+} - \frac{\sum CO_2}{2} + m_{H^+}.$$

Carbonate and the Great Marine Balancing Act

Around *pH* 6.35, therefore, m_{H^+} is almost directly proportional to V. As we move away from *pH* 6.35 in either direction, however, this approximation breaks down, and the quadratic behavior of m_{H^+} in the denominator of the second term gradually speeds up the way in which *pH* changes with V. From a mathematical perspective, then, the buffering region around *pH* 6.35 arises because equation 4-13 becomes steeper and more nearly linear when $m_{H_2CO_3}/m_{HCO_3^-} = m_{H^+}/K_1 = 1$. A similar situation arises when *pH* is around 10.3, where the solution is buffered by nearly equal amounts of carbonate and bicarbonate, so that $m_{H^+} \simeq K_2$. Beyond *pH* 11, the largest changes in equation 4-13 are due to the K_w/m_{H^+} term; below about *pH* 3, the last term is most significant. In both regions, *pH* is again relatively less affected by the second term and, therefore, changes slowly with the addition of acid.

Notice, on the other hand, that there are two regions in which the pH of the solution changes very rapidly with only a small addition of acid. These regions are centered on the *end points* of the titration at *pH* 8.35 (where HCO_3^- is the dominant species) and 4.32 (where H_2CO_3 is dominant). At these points, the inverse quadratic behavior of the second term in equation 4-13 has a significant effect on the shape of the titration curve. It is common to say that in these regions all of the dissolved carbon species have been titrated to form either H_2CO_3 or HCO_3^-. In fact, that is not true, because we can always find all three species at any *pH*. It is true, however, that one dissolved carbon species is significantly more abundant than the other two.

Earlier, we used the terms *conservative* and *nonconservative* to classify elements in seawater on the basis of whether their relative abundances remain constant from place to place in the ocean. Now we see that the same concept can be applied to ions as well, and that it has a special significance when we consider the ocean's response to processes that generate or consume hydrogen ions. Most ions in seawater, such as Na^+, Ca^{2+}, K^+, Mg^{2+}, SO_4^{2-}, NO_3^-, and Cl^-, are unaffected by changes in acidity, because they are not associated with species (like bicarbonate) which contain one or more hydrogen atoms. These are conservative ions. A handful of others, including not only HCO_3^- and CO_3^{2-} but also $B(OH)_4^-$, $H_3SiO_4^-$, and a variety of organic ions, are nonconservative. Their relative abundances, therefore, can vary (sometimes considerably) from one part of the ocean to another.

To understand further how the carbonate system helps maintain charge balance in the oceans, let us sum the charge equivalents in average seawater due to conservative cations and compare them with the summed charge equivalents due to conservative anions:

Ion	Conc. (mol/kg)	Charge (eq/kg)
Na^+	0.470	0.470
Mg^{2+}	0.053	0.106
K^+	0.010	0.010
Ca^{2+}	0.010	0.020
Sum of cation charges =		0.606
Cl^-	0.547	0.547
SO_4^{2-}	0.028	0.056
Br^-	0.001	0.001
Sum of anion charges =		0.604

We commit only a small error by ignoring the remaining conservative ions, which are much less abundant. The difference between our totals, therefore, is 0.002 eq kg^{-1}. This charge deficit, referred to as the *alkalinity* of the seawater, must be balanced by some combination of nonconservative anions. To a good first approximation, this means that

$$\text{alkalinity} = m_{\text{HCO}_3^-} + 2m_{\text{CO}_3^{2-}}, \tag{4-14}$$

since these two ions make up the bulk of the remaining dissolved species in seawater. More correctly, this quantity is known as the *carbonate alkalinity*. Unless we are making very careful measurements, the only difference between it and the true alkalinity is the omission of $m_{\text{B(OH)}_4^-}$, which is a small but not negligible quantity in the oceans.

With this perspective, it should now be apparent why the titration we described at length in Worked Problem 4-2 is often called an *alkalinity titration*. The amount of acid necessary to bring a solution's *pH* to its first end point (8.35) is a measure of $m_{\text{CO}_3^{2-}}$, because that is the *pH* at which "all" CO_3^{2-} has been converted to bicarbonate. Titration to the second end point (4.32) gives us an indirect measure of $\Sigma\,\text{CO}_2$, because at this point "all" of the original CO_3^{2-} and HCO_3^- have been converted to H_2CO_3. Because $\Sigma\,\text{CO}_2 = m_{\text{CO}_3^{2-}} + m_{\text{HCO}_3^-}$, we can now use equation 4-14 to calculate the alkalinity of the solution. Full details of this procedure are discussed by Edmond (1970).

Why should we expect $m_{\text{HCO}_3^-}$ and $m_{\text{CO}_3^{2-}}$ in seawater to be different from one part of the ocean to another? Early in this chapter, we discussed how carbon is carried from surface to deep waters in the organic and inorganic remains of planktonic life. We now see that two sorts of changes take place in this process. First, photosynthesis allows plants to consume carbon from seawater and fix it in organic matter. As a result, $\Sigma\,\text{CO}_2$ is lower in the surface ocean than it is in the deep ocean. A very slight alkalinity increase results from the uptake of dissolved NO_3^- by tissues. Except for that minor contribution, however, the alkalinity of surface water is unaffected by the formation of organic matter, because the balance of other conservative ions is left intact. On the other hand, as carbonate shells form, both $\Sigma\,\text{CO}_2$ and the alkalinity of surface seawater change. Each mole of carbon that leaves the surface ocean as a constituent of calcite takes a mole of calcium ions with it, carrying two moles of positive charge. Formation of calcite, therefore, causes the alkalinity (in milliequivalents per liter) to change by exactly twice the amount that $\Sigma\,\text{CO}_2$ changes (in mmoles per liter).

To appreciate the combined effects of tissue and shell formation, recall that four out of every five carbon atoms transferred from surface to deep ocean waters is carried in organic matter. Therefore, if we momentarily ignore the alkalinity change due to nitrate, we find that every five millimoles of carbon leaving the surface ocean are accompanied by one mole of Ca^{2+} ions, thus reducing the alkalinity by two milliequivalents. Returning to take into account the slight charge increase due to nitrate reduction, we conclude that alkalinity decreases only 28% as fast as $\Sigma\,\text{CO}_2$ during growth of marine organisms. In deep ocean waters, the reverse takes place. As respiration, fermentation, and dissolution return the constituents of dead organisms to seawater, alkalinity increases at 28% of the rate of total carbon increase.

These relative changes can also be seen as changes in the ratio of carbonate to bicarbonate with depth. Equations 4-8 and 4-14 can be rearranged to let us express $m_{\text{CO}_3^{2-}}$ and $m_{\text{HCO}_3^-}$ in terms of $\Sigma\,\text{CO}_2$ and alkalinity:

$$m_{\text{CO}_3^{2-}} = \text{Alkalinity} - \sum \text{CO}_2$$

and

$$m_{\text{HCO}_3^-} = 2 \sum \text{CO}_2 - \text{Alkalinity}.$$

A little algebraic shuffling shows us, therefore, that the ratio of $m_{\text{CO}_3^{2-}}$ to $m_{\text{HCO}_3^-}$ increases as surface organisms consume carbon, nitrogen, and calcium ions. In deep waters, $m_{\text{CO}_3^{2-}}/m_{\text{HCO}_3^-}$ decreases by the same factor. Relative variations in dissolved carbon species with depth, then, are a direct result of the life-death cycle in the oceans. It should be apparent that these variations will occur laterally as well, as a result of differences in biologic activity from place to place on the globe, as illustrated in Figure 4.4.

Carbonate and the Great Marine Balancing Act

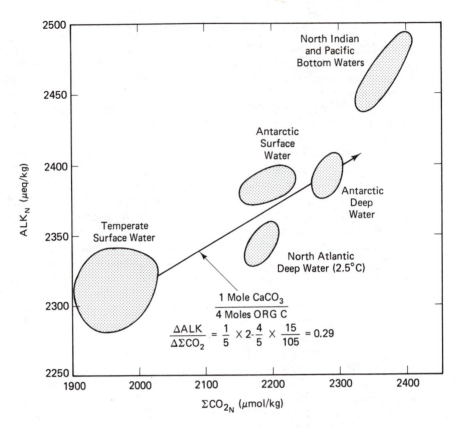

Figure 4.4 Plot of alkalinity vs. $\Sigma\ CO_2$ for major surface and deep waters in the world oceans. The slope of the solid line confirms our deduction that particulate carbon is transferred from the surface ocean to the bottom primarily in the form of organic debris. The ratio of $CaCO_3$ to organic carbon transported vertically is ¼. (Modified from Broecker and Peng 1982.)

Calcium Carbonate Solubility

By introducing carbon species from a source other than atmospheric CO_2, Worked Problem 4-2 brought us closer to understanding carbonate equilibria in the oceans, where the abundance of dissolved carbonate species depends not only on equilibrium with the atmosphere, but also on reactions involving continental silicates and marine carbonate sediments. Chemical weathering provides a continuous source of dissolved ions, including bicarbonate, to river water by processes we will discuss in greater detail in Chapter 6. These accumulate in seawater until they reach steady-state limits imposed by solubility or adsorption equilibria. In addition to reactions 4-3, 4-4, and 4-5, then, two solubility equilibria are important. These are

$$CaCO_3 \text{ (calcite)} \quad \longleftrightarrow \quad Ca^{2+} + CO_3^{2-} \qquad (4\text{-}15)$$

and

$$CaCO_3 \text{ (aragonite)} \quad \longleftrightarrow \quad Ca^{2+} + CO_3^{2-}. \qquad (4\text{-}16)$$

Carbonates of sodium, magnesium, and potassium, the other major cations in seawater, are significantly more soluble than either calcite or aragonite. For kinetic reasons, dolomite also does not precipitate in the open ocean. Reactions 4-15 and 4-16, therefore, dominate carbonate equilibria in most ocean chemistry problems. Because most $CaCO_3$ is precipitated from surface waters by plankton, the relative abundances of calcite and aragonite follow variations in the population of microbiota. Reactions 4-15 and 4-16, then, are each important in different marine settings. At 25°C, the solubility product constants for these reactions are:

$$K_{cal} = 10^{-8.34} \tag{4-15a}$$

and

$$K_{arag} = 10^{-8.16}. \tag{4-16a}$$

With this addition, problems in oceanic carbonate chemistry become only slightly more complicated. There are now six potentially variable quantities: P_{CO_2} and the activities of H_2CO_3, HCO_3^-, CO_3^{2-}, H^+, and Ca^{2+}. Their values can be determined, therefore, by the simultaneous solution of any six independent equations that relate them. Four of these are equations 4-3a, 4-4a, 4-5a, and either 4-15a or 4-16a (depending on which $CaCO_3$ polymorph is present). As in previous examples, the remaining two may include explicit values for one or two of the activities, or for some combination of two or more activities (like $\Sigma\, CO_2$, for example), or a charge balance expression.

Worked Problem 4-3

A sample of limestone consisting of pure calcite is placed in a beaker of water and allowed to reach equilibrium with the water and a CO_2 atmosphere above it. Suppose that we can control the partial pressure of CO_2 experimentally, and that $m_{HCO_3^-}$ can be measured. How does the solubility of limestone vary as a function of $m_{HCO_3^-}$ and P_{CO_2}?

We know that

$$K_1 = a_{H^+} a_{HCO_3^-} / a_{H_2CO_3}$$

and

$$K_2 = a_{H^+} a_{CO_3^{2-}} / a_{HCO_3^-}.$$

If we divide one by the other to eliminate a_{H^+}, and then rearrange the result to get an expression for the activity of CO_3^{2-}, we find that

$$a_{CO_3^{2-}} = \frac{K_2 a_{HCO_3^-}^2}{K_1 a_{H_2CO_3}}.$$

Because we also know that

$$K_{cal} = a_{Ca^{2+}} a_{CO_3^{2-}},$$

if the solid $CaCO_3$ is pure calcite, we can once again combine expressions to eliminate $a_{CO_3^{2-}}$:

$$a_{Ca^{2+}} = \frac{K_{cal} K_1 a_{H_2CO_3}}{K_2 a_{HCO_3^-}^2}.$$

Also, P_{CO_2} is related to $a_{H_2CO_3}$ by

$$K_{CO_2} = a_{H_2CO_3} / P_{CO_2},$$

so we can write

$$a_{Ca^{2+}} = \frac{K_{cal} K_1 K_{CO_2} P_{CO_2}}{K_2 a_{HCO_3^-}^2}.$$

Finally, we want an answer written in terms of the measurable quantities, $m_{Ca^{2+}}$ and $m_{HCO_3^-}$, so we must include the appropriate activity coefficients:

$$m_{Ca^{2+}} = \frac{K_{cal} K_1 K_{CO_2} P_{CO_2}}{K_2 m_{HCO_3^-}^2} \frac{1}{\gamma_{Ca^{2+}} \gamma_{HCO_3^-}^2}.$$

To obtain the most reliable answer, it is best to perform this calculation iteratively, as in Worked Problem 3-7, in order to refine $\gamma_{Ca^{2+}}$ and $\gamma_{HCO_3^-}$.

Worked Problem 4-4

As a first approximation, the *pH* in surface ocean waters is fixed by equilibrium between calcite, seawater, and atmospheric CO_2. If we assume that P_{CO_2} is 340 ppm (3.4×10^{-4} atm), what should be the equilibrium value of *pH*?

By combining the expressions

$$K_1 = a_{H^+} a_{HCO_3^-} / a_{H_2CO_3}$$

and

$$K_{CO_2} = a_{H_2CO_3} / P_{CO_2},$$

we find that

$$a_{HCO_3^-} = K_{CO_2} P_{CO_2} K_1 / a_{H^+}. \tag{4-17}$$

Because we also know that

$$K_2 = a_{H^+} a_{CO_3^{2-}} / a_{HCO_3^-},$$

we can eliminate $a_{HCO_3^-}$ by substitution to get

$$a_{CO_3^{2-}} = K_1 K_2 K_{CO_2} P_{CO_2} / a_{H^+}^2.$$

We can now eliminate $a_{CO_3^{2-}}$ by combining this expression with the solubility product constant for calcite:

$$a_{H^+}^2 = a_{Ca^{2+}} K_1 K_2 K_{CO_2} P_{CO_2} / K_{cal}. \tag{4-18}$$

The remaining unknown quantity is $a_{Ca^{2+}}$. The best method for isolating it involves the charge balance equation

$$2m_{Ca^{2+}} + m_{H^+} = 2m_{CO_3^{2-}} + m_{HCO_3^-} + m_{OH^-}.$$

Recall that we have defined a simplified "seawater" for the purposes of this problem. As you are aware from earlier problems, the bulk composition of natural seawater is dominated by many species that have not been included in this charge balance expression. We will address this shortcoming in a moment. It is intriguing, however, to observe that the *pH* of natural seawater can be estimated quite reliably by ignoring species other than the ones in our abbreviated expression.

We anticipate that the *pH* of ocean water will be somewhere between 7 and 9, so it is fair to simplify this expression by disregarding m_{H^+}, m_{OH^-}, and $m_{CO_3^{2-}}$, each of which should be much smaller than either $m_{Ca^{2+}}$ or $m_{HCO_3^-}$. Therefore,

$$2m_{Ca^{2+}} \simeq m_{HCO_3^-},$$

and

$$a_{Ca^{2+}} \simeq \frac{a_{HCO_3^-} \gamma_{Ca^{2+}}}{2\gamma_{HCO_3^-}}.$$

This, combined with equations 4-17 and 4-18, finally yields

$$a_{H^+}^3 = \frac{P_{CO_2}^2 K_1^2 K_{CO_2}^2 K_2 \gamma_{Ca^{2+}}}{2K_{cal} \gamma_{HCO_3^-}}, \tag{4-19}$$

from which we easily extract *pH*.

It is tempting to suggest that we iterate toward "best" values for $\gamma_{Ca^{2+}}$ and $\gamma_{HCO_3^-}$, as we have done before. In this case, however, the exercise should not be taken lightly. Each cycle of iteration requires us to calculate improved values not only for $m_{Ca^{2+}}$ and the concentrations of carbonate species, but also those of the other major ions and complexes in seawater, plus each of their activity coefficients. The product of this exercise, when it is applied to natural seawater rather than the simplified composition we have chosen, has been called a *chemical model for seawater*. Several have been produced and refined by geochemists, beginning with Garrels and Thompson (1962). Rather than expend the effort to construct and justify such a model, we can complete this Worked Problem satisfactorily by using activity coefficients derived from experiments with synthetic seawater (for example, those in Pytkowicz, 1983, Table 5.89), and inserting the results in equation 4-19. We calculate that if $\gamma_{Ca^{2+}} = 0.261$ and $\gamma_{HCO_3^-} = 0.532$, surface ocean water has a $pH = 8.39$. The presence of calcite in the ocean, therefore, maintains a pH very near to the HCO_3^-/CO_3^{2-} titration end-point we discussed in Worked Problem 4-2.

The solubility of calcite varies as a function of depth in the oceans, in ways that are directly related to the steady downward rain of dead organisms. To see how this works, consider our earlier discussion of alkalinity and $\Sigma\,CO_2$ variations with depth, in which we concluded that the ratio $m_{CO_3^{2-}}/m_{HCO_3^-}$ decreases from the surface downward. In fact, although the total concentration of dissolved carbon is 10% higher in bottom waters, $m_{CO_3^{2-}}$ is lower by almost two-thirds. Figure 4.5 illustrates a vertical profile of

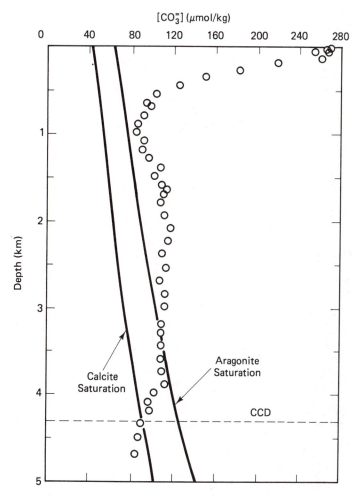

Figure 4.5 Carbonate ion variations with depth at a station in the western South Atlantic Ocean. The depth at which measured concentrations are the same as the theoretical saturation limit for calcite is the carbonate compensation depth (CCD). (Modified from Broecker and Peng 1982.)

Carbonate and the Great Marine Balancing Act

$m_{CO_3^{2-}}$ at a station in the South Atlantic Ocean, in which this decrease is apparent. Notice that the most dramatic change occurs within the wave-mixed surface zone, and that rather gradual changes continue below about 500 meters.

The two solid lines on Figure 4.5 indicate the theoretical values of $m_{CO_3^{2-}}$ in equilibrium with calcite and aragonite. The slight increase in each with depth is due primarily to pressure, but is also a function of decreasing temperature. To the right of either line, seawater is supersaturated with respect to a carbonate mineral; to the left, it is undersaturated. Notice that the measured profile, in this example, lies above both solid lines at depths less than about 3300 m. Therefore, a carbonate particle falling through the upper portion of the ocean should act as a nucleus for further growth. Below 3300 m, though, aragonite should dissolve in seawater. Below about 4300 m, calcite also is unstable. The lower of these two saturation levels has been called the *carbonate compensation depth* (CCD), representing the level below which no carbonate mineral is stable.

In practice, the CCD is generally defined in kinetic terms as the level at which the rates of carbonate growth and dissolution are balanced. Rates of reaction at 2°C are slow enough, in fact, that some carbonate persists in pelagic sediments even below the CCD. A great many studies, therefore, refer instead to the *lysocline,* the depth at which dissolution features are first observed in carbonate sediments. The position of the lysocline varies from place to place in the oceans, depending on the rate of carbonate supply and burial, its composition, and the local degree of undersaturation.

SEARCHING FOR THE CCD

At first glance, it would seem that the easiest way to locate the saturation depth in the oceans would be to look for dissolution features in pelagic sediment. The abyssal plains, however, have remarkably low relief. Except on the flanks of seamounts, much of the ocean floor is well below the CCD or the lysocline, and is therefore virtually carbonate-free.

Several people have devised clever ways to measure the CCD experimentally. The simplest of these was introduced by M. Peterson of Scripps Oceanographic Institute (1966), who hung preweighed spheres of calcite at various depths in the Pacific Ocean, suspending them on cables from a current-monitoring buoy. After 250 days, he retrieved the spheres and reweighed them to assess their net gain or loss.

Following Peterson's example, several scientific teams repeated and improved upon the original experiment. One of the most elegant was performed by Honjo and Erez (1978), who reasoned that there could be a significant uncertainty in Peterson's measurements if the carbonate samples had seen different current flow rates. Their solution was to use a small pump at each sample depth to pull water at a constant rate through a container with a preweighed mass of calcite or aragonite.

Direct measurements have also been made by what has been termed a *saturometer* (Weyl, 1961). In this apparatus, seawater is pumped at a high flow rate through a tube filled with calcite crystals and then stopped abruptly. The *pH* of water trapped in the tube is monitored by an embedded glass electrode. If the water is initially undersaturated, then the *pH* rises as its acidity is titrated by dissolving calcite. If it is initially supersaturated, then calcite precipitation causes a drop in *pH*. With time, the *pH* value in the tube approaches a saturation value asymptotically, and the initial CO_3^{2-} concentration can be calculated from

$$m_{CO_3^{2-}} \text{ (initial)} = \frac{m_{CO_3^{2-}} \text{ (final)} \times m_{H^+} \text{ (final)}}{m_{H^+} \text{ (initial)}}.$$

This apparatus has been modified since the mid-1970s, so that it is now possible to make *in situ* measurements by lowering a saturometer from a ship into the deep waters (see, for example, Ben-Yaakov, 1974). It is from measurements of this type that Figure 4.5 was constructed.

Chemical Modelling of Seawater: A Summary

Seawater is the most widely available natural electrolyte solution, and serves both to control and to record the effects of many global geochemical processes. As we have seen, biological processes cause significant variations in seawater chemistry from place to place, particularly in vertical profiles. It is not surprising, then, that geochemists have invested a considerable amount of work in understanding how ionic species are distributed in the oceans.

To this point in the chapter, we have emphasized the role of carbonate equilibria in controlling the chemistry of many natural waters, including seawater, through the mechanism of charge balance. It should be clear, however, that the other major ions in seawater (Mg^{2+}, Na^+, K^+, Cl^-, and SO_4^{2-}) may modify the simple equilibria we have described. In addition, the high ionic strength of the oceans (~ 0.7) favors the formation of ion pairs, which may influence solution chemistry significantly.

Garrels and Thompson (1962) recognized ten complexes whose association constants are large enough that they contribute significantly to the ionic strength of seawater: $MgHCO_3^+$, $MgCO_3^0$, $MgSO_4^0$, $CaHCO_3^+$, $CaCO_3^0$, $CaSO_4^0$, $NaHCO_3^0$, $NaCO_3^-$, $NaSO_4^-$, and KSO_4^-. The activity coefficients for these and for the six major ions were estimated by the mean salt method or, in the case of uncharged species, were assumed to be equal to unity. Then, by extension of the methods we have just discussed, they wrote a series of mass and charge balance equations. These, together with appropriate expressions for dissociation and association constants, were solved simultaneously to calculate the equilibrium abundances of all major dissolved species.

Many geochemists have applied Garrels and Thompson's method to other brine systems, or have developed more sophisticated means of estimating activity coefficients in seawater. Pytkowicz and co-workers have made some of the most impressive advances in the laboratory and in theoretical modelling of activity-concentration relationships. It is now apparent that other complex species (particularly those involving chloride) are also present in significant concentrations. Instead of the mean salt method, investigators now generally employ a variety of comprehensive theoretical equations (in the spirit of the Debye-Hückel equation, although different in form) to calculate activity coefficients. A good review of the current state of seawater models can be found in Chapter 5 of Pytkowicz (1983).

Despite the sophistication of these studies, a variety of problems are not easily interpreted by any thermodynamic model of seawater. The relative magnitudes of K_{cal} and K_{arag}, for example, suggest that aragonite should always be metastable relative to calcite. We know, however, that both carbonates are common in marine sediments and that aragonite is particularly abundant in tropical waters. To a certain degree, this observation can be explained by recognizing that Mg^{2+} enters most readily into the structure of calcite, where it increases its solubility slightly. Aragonite accepts magnesium ions only sparingly, and is therefore only marginally less stable than "impure" calcite.

That explanation is only part of an answer, however. Virtually all carbonate precipitated from the sea is nucleated biochemically, and aragonite depends on the peculiarities of ocean life for its tenuous stability. Some organisms, particularly among the scleractinian corals, pelecypods, and gastropods, build shells and skeletons exclusively with aragonite. Others precipitate only calcite, and some build with both polymorphs interchangeably. Once formed, aragonite persists metastably until it is gradually replaced in ancient carbonate rocks by calcite. This, then, is a kinetic problem outside the scope of thermodynamic models of seawater.

It is also observed that the concentration of CO_3^{2-} in the surface ocean is roughly five times greater than should be expected for seawater in equilibrium with calcite (see Figure 4.5). Carbonate precipitation by organisms, as we found earlier in this chapter, is limited by the availability of phosphate. Why, then, doesn't excess carbonate precipitate inorganically? Once again, the apparent answer lies with kinetics and biology, rather than with thermodynamics. Plankton are such voracious and nondiscriminating eaters that they ingest not only nutrients, but also a large fraction of the suspended submicroscopic calcite particles that might otherwise act as seeds for inorganic crystal growth. When these are later excreted by the plankton, they carry a thin coating of organic matter which is sufficient to retard further growth. Carbonate ions, therefore, accumulate in surface seawater well past the concentration at which K_{cal} is exceeded.

In summary, the distribution of major chemical species in the ocean can be understood, with an impressive degree of confidence, by recognizing the central role of carbonate equilibria and manipulating the proper thermodynamic equations. Few other geochemical systems have been as well described. As the two brief examples above show, however, there are still problems in seawater chemistry that defy solution by standard thermodynamic modelling.

GLOBAL MASS BALANCE AND STEADY STATE IN THE OCEANS

Examining the Steady State

Until now, this chapter has emphasized the thermodynamic behavior of nonconservative species in the oceans. Another class of geochemical problems is equally challenging, however. Put simply, these problems compel us to ask what kinetic factors control the abundance of conservative or nonconservative species. Because dissolved matter is constantly being supplied in river water, the influx of each conservative species must be exactly balanced by removal processes. Without such a balance, a species abundance cannot be in steady state, and it must either accumulate or vanish with the passage of time. In a similar way, nonconservative species like HCO_3^- and CO_3^{2-} may vary in concentration from one place to another in the ocean, but there are kinetic factors that operate to keep the HCO_3^-/CO_3^{2-} ratio nearly constant at each place. A steady state for nonconservative species is therefore maintained among various subreservoirs within the ocean. It is a matter of great importance, in other words, to identify each of the sources and sinks for dissolved material in seawater, and to estimate their magnitudes so that we can evaluate the chemical pathways for change in the oceans.

Worked Problem 4-5

What does the oceanic budget for sodium look like? At first glance, this appears to be an easy problem. Sodium is the most abundant cation in seawater (see Table 4-1). Its concentration, on average, is 10.78 g/kg. The total mass of sodium in the oceans, then, is

$$A_{Na^+} = 10.78 \text{ g/kg} \times 1.4 \times 10^{21} \text{ kg} = 15.1 \times 10^{21} \text{ g}.$$

The average concentration of Na^+ in river water is 6.9 mg/kg. Rivers supply approximately 4.6×10^{16} kg of water to the oceans each year, so the rate of sodium supply is

$$dA_{Na^+}/dt = 6.9 \text{ mg/kg} \times 4.6 \times 10^{16} \text{ kg/yr} \times 0.001 \text{ g/mg}$$
$$= 3.2 \times 10^{14} \text{ g/yr}.$$

At that rate, it would take the world's rivers only 4.7×10^7 years to carry the ocean's entire present mass of sodium. Using earlier and somewhat less accurate numbers, the nineteenth-century Irish geologist James Joly calculated a value of 9.0×10^7 years, which he then concluded was the age of the earth. This was a clever calculation, and the results were remarkably good for the period. His interpretation was flawed, however, because he failed to recognize the existence of sinks as well as sources for sodium.

We now know the age of the earth to be much greater than 47 or even 90 million years. Furthermore, the compositions of formation brines and the stratigraphic record of evaporite formation suggest that the sodium content of the oceans probably has not varied by more than 30% during the Phanerozoic (Holland, 1984). This implies that the river input of sodium is very nearly balanced by processes that remove sodium from the oceans; that is, the oceanic concentration of Na^+ has been at or very close to a steady state value during the past 600 million years. Holland has presented further observations which suggest that the concentration of NaCl in seawater has varied by less than a factor of ± 2 in the past 3.5×10^9 years. Instead of yielding the age of the oceans, as Joly supposed, the balanced source and loss rates give us a measure of the *mean residence time* for sodium in the ocean. The average sodium ion, in other words, spends 4.7×10^7 years in the ocean before it is removed by some loss process.

The best available data (Holland, 1978) suggest that a little more than 40% of the sodium in river water is derived from the solution of halite in evaporites, that roughly 35% is derived from weathering of all other rocks, and that the remaining 20 to 25% comes from sea spray carried inland by winds. The major removal processes from seawater include precipitation in evaporites, burial in pore waters in marine sediments, reaction with silicates in submarine geothermal systems, and atmospheric transport of sea spray. Holland (1978) estimates on the basis of experimental results reported from several labs that the oceans annually lose between 1.0 and 1.4×10^{14} g Na^+ as seawater circulates through midocean ridge systems. This corresponds roughly to 30% or 40% of the annual river flux. Burial of sodium in interstitial waters in the sediment column may account for another 15% of the annual river flux. Estimates of sea salt transport by winds are highly variable, but it is fairly clear that sodium derived from sea spray is carried onshore at roughly the same annual rate that rivers return it to the oceans. We reported this value above to be 20–25% of the river flux of sodium per year. The remainder, about 40% of the annual input of sodium, must be lost by formation of evaporites.

We present these estimates here and in Figure 4.6 as an abbreviated example of the sort of mass balance exercise that must be performed if we are to account for the prove-

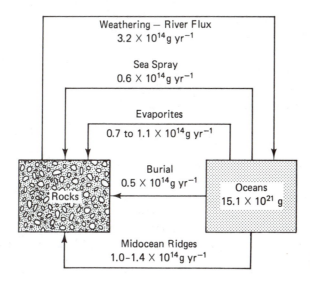

Weathering — River Flux
3.2×10^{14} g yr^{-1}

Sea Spray
0.6×10^{14} g yr^{-1}

Evaporites
0.7 to 1.1×10^{14} g yr^{-1}

Burial
0.5×10^{14} g yr^{-1}

Rocks

Oceans
15.1×10^{21} g

Midocean Ridges
1.0–1.4×10^{14} g yr^{-1}

Figure 4.6 A simple box model illustrating major reservoirs and pathways for sodium.

nance of each of the dissolved constituents in seawater and evaluate the geochemical pathways they follow. You should not be misled, however, by the apparent confidence with which we have reported magnitudes of the loss processes. The rates of these processes for most dissolved species are poorly known. By some estimates, for example, formation of evaporites may account for as much as 60% of the annual river flux of sodium. Consequently, this implies that the importance of other loss processes has been exaggerated. Even the oceanic budget for sodium, then, is not as simple as it looks.

On occasion, our ignorance about some fundamental pathways has led geochemists to overestimate greatly the importance of others. The mass balance cycle for magnesium is a good example. Rivers annually carry 1.4×10^{14} g of Mg^{2+} per year to the oceans (Meybeck, 1979), the result of weathering both silicates and dolomitic carbonates on the continents. If the magnesium budget of the oceans is in a steady state, then some combination of processes must remove an equal amount of magnesium each year. Until the early 1970s, geochemists believed that the most important loss process was the formation of authigenic clays (glauconite) in marine sediments. The only other known loss process was dolomitization, widely recognized as negligible in today's oceans.

As more analyses of marine clays became available, however, it quickly became apparent that authigenic clay formation is insufficient to balance the river flux. The "magnesium problem" became most acute with Drever's (1974) failure to identify any clearly authigenic magnesian marine clays. Laboratory experiments and the observations gathered during the Deep Sea Drilling Program have now shown that reactions between seawater and hot basalt, previously thought to be inconsequential, remove virtually all of the magnesium from seawater as it is cycled through midocean ridges. Estimates of the cycling rate itself depend on the inferred distribution of heat and its rate of loss around ridges. Because these involve significant uncertainties, calculations of the total amount of magnesium removed from the oceans by this process per year are still imprecise. It is now clear, however, that a major portion of the annual river flux may be balanced this way. Some observations suggest that a substantial amount of magnesium is also lost during alteration of basalts at bottom water temperatures (close to 0°C) distant from ridge systems. Minor quantities are also deposited in magnesian calcite or are involved in exchange reactions with marine clays, and some, in fact, may be lost during authigenic mineralization. Although uncertainties persist, in any event, the "magnesium problem" seems to have vanished.

Each of these two examples—the sodium and magnesium cycles—illustrates the difficulty of describing a steady state in a system of global scale. In most cases, we must accept estimates of reservoir contents and interreservoir fluxes that have large uncertainties, and we must be prepared to revise them frequently.

How Does the Steady State Evolve?

There is a level of interest in kinetic modelling that extends beyond understanding the steady state. Clearly, geological conditions change through time. Paleogeographic studies indicate that the rates of seafloor spreading and volcanism have not remained constant, and that climatic fluctuations have induced changes in rates of weathering, erosion, and evaporation. If the ocean has managed to remain nearly in steady state through all of these fluctuations, it is logical to ask what keeps the system under control. What sorts of feedback mechanisms allow seawater to recover from external perturbations, and how rapidly do they operate?

As an example of a familiar elemental cycle whose evolving steady state has been debated quite intensely, let us consider the fate of carbon. In a sense, this is an arbitrary choice, because our purpose here is to demonstrate how global-scale cycles stay in balance, not necessarily to study carbon. On the other hand, as we have already seen in earlier parts of this chapter, the carbon cycle is of pivotal importance to the chemistry of the oceans and the ocean biosphere. By looking closely at the carbon cycle, we accomplish two goals at once.

Since the late 1960s, ocean chemists have had a continuing interest in models which predict how the global carbon balance is likely to adjust to the increase in atmospheric P_{CO_2}, illustrated in Figure 4.2. These calculations often also attempt to estimate the climatic changes that may occur as a larger fraction of infrared radiation is absorbed by CO_2. Such predictions have great value for agricultural, economic, and political planners. A large number of these models have been produced, each differing from the others in either the number of reservoirs and pathways it considers or the numerical approach it follows in evaluating them (for a summary, see Sundquist, 1985). We will briefly consider three of these to demonstrate that no model is "best," but that each is appropriate for addressing a different problem.

A Short-Term Model

Figure 4.7 illustrates a model to evaluate the effect of twentieth century anthropogenic CO_2, one of several models constructed by Sundquist (1985). In all, 18 reservoirs for carbon are considered. Twelve of these represent chemically distinct vertical zones in four parts of the global ocean, characterized by differences in intensity and style of biological activity. Five are reservoirs on the continents, again characterized by differences in the type of dominant biological community and its metabolic role in the carbon cycle. (The terms *woody* and *nonwoody* refer to the branches and leaves of trees.) The remaining reservoir, the atmosphere, communicates between land and ocean reservoirs. Various pathways connect these reservoirs, as indicated by the arrows. The amounts of carbon in each reservoir (in gigatons) and the fluxes (in gigatons per year) are subject to some large uncertainties, but represent the best estimates available in the literature.

Two features of this model should be noted. Both are consequences of the time scale which Sundquist considers. First, we observe that changes in seawater chemistry that take place in less than a few thousand years (the time scale over which global stirring occurs) will not affect the entire ocean at once. In fact, the shorter our attention span is, the more the ocean behaves as a series of semiindependent reservoirs, and the the greater the number of reservoirs that must be considered in our model. The same also is true on the continents. The intimidating complexity of Figure 4.7, then, is a natural consequence of the type of problem Sundquist is trying to address with this model.

Second, over periods of a few years to a few hundred years, the contents of geochemical reservoirs are most sensitive to changes in biological processes rather than geological ones. Populations of living organisms respond rapidly to the environmental pressures of food supply and climate. In the twentieth century, for example, we have seen rapid environmental changes caused by human activity. By comparison, most geologic processes, particularly those which cause slow changes in the positions of conti-

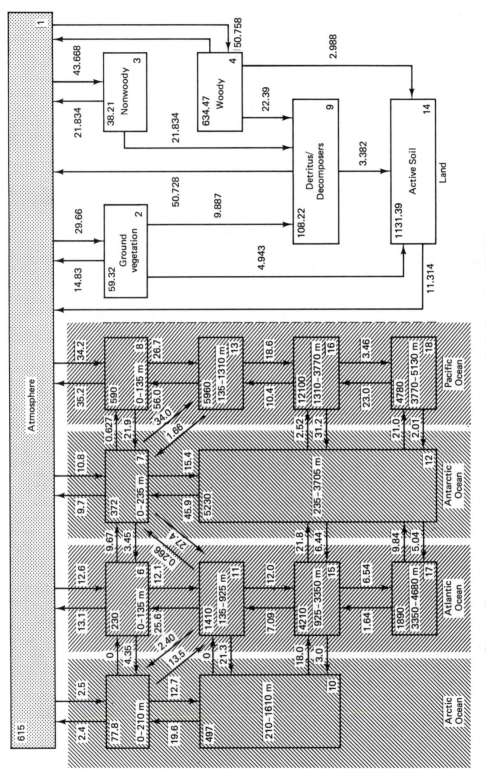

Figure 4.7 A box model for the short-term carbon cycle described by Sundquist (1985). In each box, the key number in the lower right corner identifies the reservoir, and the number in the upper left corner indicates the mass of carbon (in 10^{12} tons) in it. The range of water depths defining each of the oceanic boxes is shown in meters. Numbers along paths indicated by arrows show the flux of carbon (in 10^{12} tons) transferred per year. (After E. T. Sundquist, Geological perspectives on carbon dioxide and the carbon cycle. In *The carbon cycle and atmospheric CO_2: Natural variations Archean to present*, ed. E. T. Sundquist and W. S. Broecker, 5–60. Monograph 32. Copyright 1985 by the American Geophysical Union. Reprinted with permission.)

nents or the average height of mountains, have operated nearly at a constant rate during our lifetimes. Although rates of tectonic processes change, they do it so slowly that biological communities do not notice[5]. Therefore, only the concentrations of nutrient-related species vary significantly in the short term, and little geological insight is needed to interpret them.

The other side of this second observation is that the natural feedback mechanisms that tend to keep biological fluctuations from getting out of hand are based on processes that change only with geologic slowness. A planktonic community that overeats, for example, finds that it grows only until it reaches the limits of its food supply. The river supply of phosphate or nitrate is a function of the global weathering rate, which remains nearly constant over periods of hundreds or thousands of years. Biological systems may change rapidly, therefore, but their effects on the ocean/atmosphere system are severely limited by the slower kinetics of geologic processes.[6]

Worked Problem 4-6

How rapidly does Sundquist's model system adjust to change? To answer, we need to estimate the mean residence time and response time of each of the reservoirs in the system, as well as for the system as a whole. These are concepts we first encountered in Chapter 1.

In order to examine his model's sensitivity, Sundquist assumes that material is transferred out of any reservoir (i) according to a first-order kinetic law. That is, the rate of decrease in carbon content (A_i) in the reservoir is directly proportional to A_i itself. If the flux of carbon into the reservoir (F_{input}) is constant, then:

$$dA_i/dt = F_{input} - F_{output} = F_{input} - kA_i. \tag{4-20}$$

(Compare this with Equation 1-2.) For a single reservoir where A_i initially has a value A_i^0, the carbon content at any given time can therefore be calculated from the solution to Equation 4-20:

$$A_i(t) = \frac{F_{input}}{k} - \left(\frac{F_{input}}{k} - A_i^0\right) \exp{(-kt)}. \tag{4-21}$$

To get a feel for what this result means, consider first what it would look like if the reservoir were in steady state. F_{input}/k would then be equal to A_i^0, and the entire second term in Equation 4-21 would be equal to zero. If we were to dump an unexpected amount of carbon into the reservoir, its input and output rates would have to readjust until they were once again equal in a new steady state. The rate at which this readjustment occurs is given by the exponential factor ($-kt$). It is common practice to declare that a new steady state is reached when further changes in the second term of Equation 4-21 are less than $1/e$ (about 0.36) times $((F_{input}/k) - A_i^0)$. The time at which this condition is met is called the *response time* or, sometimes, the *e-folding time* for the reservoir. Numerically, it is equal to $1/k$. For a single reservoir, this is the same as the mean residence time for carbon, which we discussed earlier. As shown below, these values range from 274 years for moderately deep waters in the Pacific Ocean to less than a year for the nonwoody parts of trees (i.e., the leaves).

---

[5] In fact, some studies during the 1980s (for example, Brewer et al., 1983) have indicated that some measurable climatic changes may take place in association with shifts in deep ocean circulation over a span of a few decades. Similarly, volcanic eruptions like those at Mt. St. Helens during the early 1980s can produce short-term climatic changes. The effect of these processes on global geochemical cycles like the one we are discussing here, however, is trivial.

[6] Kasting et al. (1986), in fact, have demonstrated that the oceans recover from even a nonbiological short-term perturbation like the impact of an asteroid within a few tens of years because long-term geological conditions limit the possible range of values for alkalinity and total dissolved carbon. Dramatic departures from "normal" or steady state conditions, therefore, are precluded, and rapid kinetics within the biosphere easily reestablish ocean chemistry to its preimpact state.

MEAN RESIDENCE TIMES FOR CARBON IN THE SYSTEM
ILLUSTRATED IN FIGURE 4-7.

Reservoir #	t_{res} (yr)	Reservoir #	t_{res} (yr)
1	3.3	10	13.4
2	2.0	11	23.5
3	0.86	12	65.2
4	12.5	13	83.8
5	2.7	14	100.0
6	5.5	15	110.5
7	4.1	16	274.2
8	6.5	17	163.2
9	2.0	18	195.4

A measure of the response time for the entire system, as opposed to the isolated response time for individual reservoirs, can be obtained by eigensystem analysis of the matrix of rate constants (k), repackaged as described in Chapter 1. We can think of the eigenvalues as a set of "harmonic constants" for the system, taking into account not only the dynamics of individual reservoirs, but also the ways in which they affect each other. The basis for this method is discussed in Lasaga (1980, 1981).[7] Using it, we find that the system will reach a new steady state 234 years after it is perturbed. This system response time is less than the mean residence time for reservoir 16 (the slowest individual reservoir) because eigensystem analysis implicitly takes into account the interdependence of fluxes.

Whether we examine mean residence times or system response time, we conclude from this model that the chemistry of seawater adjusts to change on the order of 10^2 times less rapidly than the biosphere and atmosphere. The effects of twentieth century fossil-fuel burning, a biological perturbation to the system, will propagate through the world's oceans for at least the next 200 years.

As the time scale of interest becomes longer, biological factors have less effect than geological factors on system kinetics. Sundquist has shown, for example, that the model we have just described can be greatly simplified by lumping the atmosphere together with all surface ocean reservoirs, land plants, and detritus (boxes 1–9 in Figure 4.7) to make a super-reservoir. The mean residence times of these reservoirs are all so short that, from the perspective of several hundred years or more, they all appear to be in equilibrium. With increasing time scale, however, Sundquist finds that it is necessary to add reservoirs for marine and terrestrial sediments, which interact with the ocean/atmosphere system along weathering and diagenetic pathways. Response times for the modified geochemical cycles are appropriately longer as a result. The following model illustrates these points.

A Midrange Model

In 1982, Broeker suggested that variations in the extent of the late Pleistocene ice sheets may have been moderated by changes in the distribution of carbon and nutrients in the oceans. He reasoned that because organic carbon burial takes place almost exclusively on continental shelves, it should be least effective during a glacial epoch, when sea level is lowest. During deglaciation, however, sea level rises, the proportion of shelf area increases, and more organic carbon should be buried. As submarine shelf area fluctuated, therefore, the balance of carbon species in shallow and deep waters should also have shifted, as should the amount of CO_2 in the atmosphere. Because

---

[7] Lasaga's method is the basis for the program CYCLE, which is on the diskette with this text.

global temperatures follow P_{CO_2}, this constitutes a climatic feedback system. Carbon contents of pelagic and shelf sediments and of atmospheric gases trapped in ancient polar ice follow qualitative trends consistent with this scheme. We should wonder, though, whether these predictions are compatible with our understanding of the kinetic behavior of the ocean/atmosphere system. Can we estimate quantitatively how sensitive the oceans and the atmosphere are to a geological perturbation like glaciation?

Keir and Berger (1983, 1985) produced models of the global carbon cycle that examine the effects of glaciation during the past 120,000 years. Their model, shown in Figure 4.8, consists of five reservoirs. Three of these are peripheral to the ocean/atmosphere system, and are large enough that changes in their carbon contents over 10^5 years are almost certainly negligible: (a) continental rocks provide a contnuous flux (H_{riv}) of dissolved HCO_3^- to the oceans, (b) the shelves act as a sink for $CaCO_3$ (E_{carb}) and organic matter (E_{org}), and (c) pelagic sediments exchange both organically and inorganically derived carbon with seawater in the deep oceans. These last interactions in-

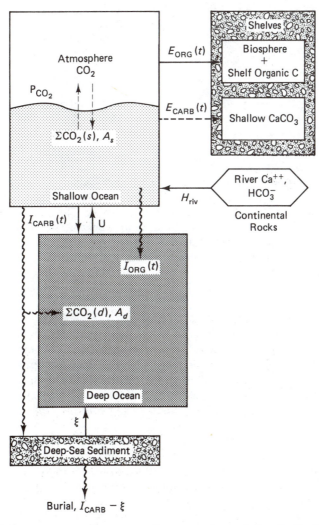

Figure 4.8 A box model for the intermediate-term carbon cycle described by Keir and Berger (1983).

Global Mass Balance and Steady State in the Oceans

volve inorganic (I_{carb}) and organic (I_{org}) carbon deposition in the ocean floor, and carbonate dissolution (ζ) from pelagic sediments.

In order to test Broecker's hypothesis, Keir and Berger set up four differential equations which examine the evolution of total dissolved carbon (Σ CO$_2$) and alkalinity (A) in surface and deep waters. The atmosphere and surface waters follow the relations:

$$\frac{d(V_s \Sigma CO_2(tot))}{dt} = u(\Sigma CO_2(d) - \Sigma CO_2(s)) - I_{org} - I_{carb} - E_{org} - E_{carb} + H_{riv}$$

$$(4\text{-}22)$$

and

$$\frac{d(V_s A_s)}{dt} = u(A_d - A_s) - 2I_{carb} - 2E_{carb} + \nu_o I_{org} + 2H_{riv}. \qquad (4\text{-}23)$$

In the deep ocean,

$$\frac{d(V_d \Sigma CO_2(d))}{dt} = -u(\Sigma CO_2(d) - \Sigma CO_2(s)) + I_{org} + \zeta(\Sigma CO_2(d), A_d, I_{carb})$$

$$(4\text{-}24)$$

and

$$\frac{d(V_d A_d)}{dt} = -u(A_d - A_s) - \nu_o I_{org} + 2\zeta. \qquad (4\text{-}25)$$

By inspection, you can see that each equation consists of source terms (with positive arithmetic signs) and loss terms (with negative signs), which may combine to produce a net increase or a decrease during any time interval of interest. These four equations, therefore, are no more difficult than Sundquist's equations (4-20). The total surface CO$_2$ (Σ CO$_2$ (tot)) is partitioned into an atmospheric fraction and a dissolved fraction (Σ CO$_2$ (s)) on the basis of solubility equilibria, and the rate of dissolution (ζ) is controlled by the solubility of calcite in deep waters and the rate of pelagic sedimentation. Other parameters appearing in equations 4-22 through 4-25 are the volumes of water in surface and deep ocean (V_s and V_d), the rate of physical overturning of deep and surface waters (u), and the ratio of alkalinity loss to CO$_2$ production due to oxidation of organic matter (ν_o). This last factor takes into account any changes in alkalinity due to conversions between NO$_3^-$ and organic N.

The glacial perturbations to equations 4-21 to 4-25 are introduced in E_{org} and I_{carb}, both of which are described as functions of sea level, as inferred from the geologic record. Given these functions and reasonable initial choices for each of the model parameters, Keir and Berger then solve the four differential equations repeatedly at closely spaced time steps to watch the way in which atmospheric P_{CO_2} and the preservation rate for pelagic carbonate evolve. Their numerical results are consistent with the carbonate record in deep sea drill cores and with the CO$_2$ trapped in air bubbles in polar ice caps.

We have presented Keir and Berger's model without any detailed discussion of its numerical approach, which is beyond the immediate scope of this chapter. The points to observe in their model and its results, however, are these:

First, although the alkalinity and carbon contents of surface and deep waters differ because of the biological pathways that dominate them, the physical rate of overturn in the oceans (the parameter u in equations 4-22 through 4-25) places a practical limit on the rate at which perturbations at one level can be communicated to the other. Short-term biological perturbations within surface and deep reservoirs, such as those addressed by Sundquist, constitute a background "noise" that is insignificant over the time scale that Keir and Berger consider. Their model, therefore, is largely geological.

Second, although the interactions among reservoirs are more pronounced and the equations describing each must therefore contain more terms, the system itself is physically simpler. Because the differences among Sundquist's 12 oceanic reservoirs are insignificant at this time scale, Keir and Berger can compress them into two.

A Long-Term Model

If we consider even longer time scales, structure in the ocean-atmosphere system can be ignored altogether. For example, Berner et al. (1983, 1985) have shown that variations in the sea floor spreading rate can be related to long-term fluctuations in atmospheric P_{CO_2} during the past 100 million years. Their model, in its simplest form, includes an undifferentiated ocean reservoir with no hint of biological control. Mass transfer between the ocean and various solid-earth reservoirs takes place only by the slow geologic processes of weathering, metamorphism, volcanism, and seawater cycling at ridge crests. The response of seawater to a perturbation, in this formulation, is ultimately governed by the mean residence time for bicarbonate (10^5 yr), calcium (10^7 yr), and magnesium (10^7 yr) in the oceans. Even the interglacial fluctuations considered by Keir and Berger would generate only negligible "noise" at this time scale, and they are therefore ignored.

As in Keir and Berger's model, the atmosphere follows changes in ocean chemistry very closely. Berner and associates, however, have incorporated an empirical equation in their model that describes the rate of chemical weathering as a nonlinear function of P_{CO_2}. As the amount of carbon dioxide in the atmosphere increases, both temperature and rainfall increase as well, and the rates of chemical weathering for continental rocks rise rapidly. This natural mechanism provides a negative feedback that prevents P_{CO_2} from varying by more than a factor of two or three. Calculated values of P_{CO_2} since the late Mesozoic era, from Berner et al. (1983), are shown in Figure 4.9. Kasting and Richardson (1986) have used this model's prediction of a rise in P_{CO_2} around 45 million years ago to account for the 2°–3°C global warming that is generally inferred to have taken place during the Eocene. Considered over the appropriate time scale, therefore, the kinetic behavior of the oceans can be exploited to show a causal link between plate tectonics and paleoclimate.

A Brief Summary

There are two insights to be drawn from the past few pages of discussion. One has to do with the chemistry of the ocean-atmosphere, and of carbon in particular: the chemistry of the oceans is resilient. If we examine the dynamic redistribution of an element like carbon, we find that its abundance in seawater is rarely at a true steady state value, but that it is also rarely far from one. Short-term perturbations have a limited effect on the ocean-atmosphere system because the response times of reservoirs dominated by biology are very short, and because long-term geological processes place stringent limits on the growth of biological communities. The geological processes that tend to disturb the oceans' chemistry are themselves gently cyclical and have periods that are long compared to the mean residence times for many dissolved species in the oceans. Therefore, their abundances and rates of movement into and out of neighboring reservoirs

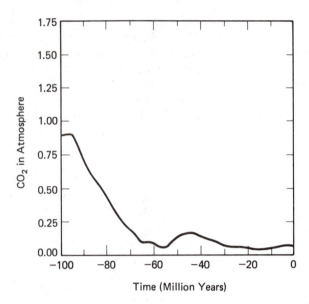

Figure 4.9 Variations in P_{CO_2} during the Cenozoic, as calculated by the model of Berner et al. (1983, 1985). (After R. A. Berner, A. C. Lasaga, and R. M. Garrels, An improved geochemical model of atmospheric CO_2 fluctuations over the past 100 years. In *The carbon cycle and atmospheric CO$_2$: Natural variations Archean to present,* ed. E. T. Sundquist and W. S. Broecker, 397–411. Monograph 32. Copyright 1985 by the American Geophysical Union. Reprinted with permission.)

follow gradual changes in global tectonics from one geologic period to another without deviating markedly from the evolving steady state.

The second insight concerns the models we use to describe the ocean-atmosphere system: no single model is appropriate for studying the full range of kinetic behavior in the system. Depending on the processes that interest us, time scales can range from less than a year to many tens of millions, and reservoirs can be grouped and regrouped in many ways. The ultimate tests of success are internal consistency and the ability to elucidate some facet of the system's behavior in a way that is compatible with the geologic record.

GRADUAL CHANGE: THE HISTORY OF SEAWATER AND AIR

Beyond the time scales we have just considered lie broader geochemical questions about the evolution of our planet. What did the earliest ocean and atmosphere look like? What pathways did they follow in arriving at their present compositions? How has evolution of the ocean-atmosphere system been linked with evolution of the crust and with the history of life on the earth? These are highly speculative realms, but by careful interpretation of the rock record and the application of thermodynamic constraints, geochemists have been able to peer through the fog and glimpse the outlines of the early state of the ocean-atmosphere.

Early Outgassing and the Primitive Atmosphere

Among the most useful bits of information we have to work with is the earth's *excess volatile inventory.* This tabulation, first suggested by Rubey (1951), is an estimate of the total amount of carbon, nitrogen, sulfur, and water that have been released from the solid earth since its time of accretion. A certain fraction of this inventory is now in the ocean-atmosphere system, but much of it has been recombined with crustal rocks during chemical weathering and now resides primarily in the sedimentary column as a constituent of carbonates, clay minerals, and fossil organic matter. In particular, roughly a third of the outgassed water and nitrogen now reside in the crust, along with almost all of the carbon and sulfur. These are regularly recycled through the surface reservoirs of the earth as a result of volcanism, metamorphism, and erosion.

Our understanding of the history of volatiles on the earth has been dominated by two schools of thought. According to one, the volatile inventory has accumulated gradually through geologic time as the result of volcanic emanations. The other holds that degassing of the planet was an early, catastrophic event associated with accretion and differentiation. As is often the case when scientific processes are interpreted from divergent points of view, the truth lies somewhere between them.

Clearly, some outgassing has continued throughout the earth's history. The best evidence we have comes not from the major volatiles listed in Table 4-3, but from the rare gases that are presumed to have followed similar outgassing histories. The advantage of working with noble gases is that they are chemically nonreactive and therefore accumulate in the atmosphere through time rather than being partially recycled into the crust. Abundance ratios of their stable and radiogenic isotopes can, in many cases, yield valuable clues to the history of outgassing. Atmospheric argon, for example, consists largely (99.6%) of the isotope ^{40}Ar, which is a natural decay product of ^{40}K. The long (1.25×10^9 yr) half-life for radioactive potassium assures us that ^{40}Ar production has continued in the solid earth since its formation. Let us suppose that today's atmospheric argon had been generated entirely during an early devolatilization event. To produce such a large amount of ^{40}Ar 4.5×10^9 years ago would have required a very high initial abundance of ^{40}K in the earth and should have led, even today, to an average crustal heat flow many times greater than we observe. Also ^{40}Ar should have continued to accumulate since the early Archean, with the result that there should now be far more argon in the solid earth than in the atmosphere. This possibility seems remote, considering the vigor with which the process of plate tectonics has recycled the upper portions of the planet. We conclude, therefore, that some outgassing has continued throughout earth's history. The exact amount, unfortunately, cannot be extracted directly from the ^{40}Ar data because too little is known about the concentration and distribution of ^{40}K in the crust and mantle.

Further convincing evidence has come from the revelation that ^{3}He is usually abundant in deep ocean waters, in sea floor basalts, and in the fluids released at hydrothermal vents on ridge crests (Clarke et al., 1969; Kurtz et al., 1982). Helium is light enough that its abundance in the atmosphere is controlled by a dynamic balance between outgassing and escape to space, the mean residence time of an atmospheric he-

TABLE 4-3 DISTRIBUTION OF CARBON, NITROGEN, SULFUR, AND WATER IN NEAR-SURFACE GEOCHEMICAL RESERVOIRS[*]

	Carbon ($\times 10^{18}$ g)	Nitrogen ($\times 10^{18}$ g)	Sulfur ($\times 10^{18}$ g)	Water ($\times 10^{21}$ g)
Atmosphere	0.69	3950	<0.001	≤0.2
Biosphere	1.1	0.02	<0.1	1.6
Hydrosphere	40	22	1300	1400
Crust	90,000	2000	15,000	600
Inventory	90,041.79	5972.02	16,300.1	2001.8

[*]After Holland (1984), Table 3.5.

Gradual Change: The History of Seawater and Air

lium atom being roughly 2×10^6 years. The primary source for helium is the radioactive decay of uranium and thorium, which yields the isotope ^4He, accounting for 99.99986% of the released helium. The remainder is ^3He, long believed to arise from the decay of tritium (^3H) in the solid earth and by interactions between cosmic rays and atmospheric gases. Repeated calculations, however, have shown that these mechanisms can only account for 10% or less of the ^3He lost annually from the upper atmosphere. It has now become apparent from studies of submarine volcanism that much of the outgassed ^3He is primordial, released from mantle sources at the rate of $3.0(\pm 1.0) \times 10^{19}$ atoms sec^{-1} (1.6×10^3 mole yr^{-1}) (Jenkins, 1980).

The release rates of these and other noble gas isotopes, therefore, have been used to estimate the outgassing history of the earth. Uncertainties in the data are frequently large, of course, and individual observations are subject to multiple interpretations. Most seriously, model calculations of outgassing history rely heavily on our understanding of the volatile content of the mantle, which is itself a subject of much debate. Still, many good attempts have been made. Representative of these is Figure 4.10, reproduced from Holland (1984), which illustrates outgassing curves that result from several different assumptions regarding the present volatile content and homogeneity of the mantle. From these, we may conclude that unless the mantle is currently more than 70% free of primordial gases, two-thirds or more of the planet's volatile inventory was released catastrophically during the earliest Archean era. This result is compatible with our modern conception of planetary formation, which will be discussed in Chapter 13. We should keep in mind, though, that it depends on many barely concealed uncertainties and may change dramatically as we continue to gather data.

In addition to the composition of the earth's excess volatile inventory and the rate at which it was released, a significant effort has been made to deduce the species abundances in the ocean-atmosphere through geologic time. Although some attempts have been made, geochemical cycling models such as those for the carbon cycle that we discussed earlier have not proven useful in pre-Phanerozoic studies. Our knowledge of initial reservoir contents and rate constants is too scanty. Also, we understand so little about indirect controls, like the configuration of continents and the average depth of seas, that it is very difficult to construct a reliable kinetic model. Instead, the chemical

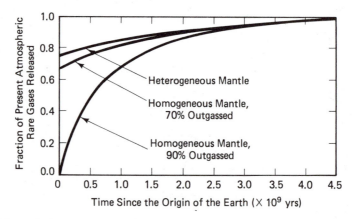

Figure 4.10 Accumulation of the noble gases in the atmosphere. The upper curve was calculated by assuming a heterogeneous mantle. Others were calculated by assuming that the mantle is homogeneous. Notice that the two models of mantle structure have indistinguishable effects unless the rare gas inventory in a homogeneous mantle is at least 70% outgassed. (Based on calculations by Holland 1984.)

history of the Precambrian oceans and atmosphere has been reconstructed by studying a succession of presumed equilibrium states through which they passed.

During planetary accretion, it seems likely that primordial gases were released from the solid earth in equilibrium with silicate magmas. The species abundances in the gas mixture, therefore, can be estimated theoretically by assuming that their bulk composition was that of the present excess volatile inventory and then using the tools of thermodynamics to find the mixture whose free energy, in association with the magma, is lowest. The major controlling parameter is oxygen fugacity, which is determined in most magmatic systems by the oxidation state of iron.

No samples of volcanic rock from the first half billion years have survived, so we have no direct information about their composition. It is almost certainly true, however, that the oxidation state of volcanic gases during the early Archean was no higher than it is today. Several studies (e.g., Gerlach and Nordlie, 1975 a, b, c) have shown that modern volcanic gases behave as if their oxidation state is controlled by the reaction

$$3SiO_2 + 2Fe_3O_4 \longleftrightarrow 3Fe_2SiO_4 + O_2,$$
$$\text{quartz} \quad \text{magnetite} \qquad \text{fayalite}$$

commonly referred to as the QFM oxygen fugacity buffer. This places an upper limit on the oxidation state of the primordial atmosphere. The lower limit is probably best defined by the reaction

$$SiO_2 + 2Fe + O_2 \longleftrightarrow Fe_2SiO_4,$$
$$\text{quartz} \quad \text{iron} \qquad\qquad \text{fayalite}$$

which may have been dominant before metallic iron was segregated to form the earth's core. The oxygen fugacity controlled by this assemblage, known as QFI, is about four orders of magnitude lower than QFM at any temperature of interest in a magmatic system (see Figure 4.11). As we will see in Chapter 13, prevailing ideas about the differentiation of the primitive earth favor conditions near this lower limit.

Using this information and the tabulated ΔG_f^0 values for a large number of possible gas species, we can compute equilibrium mole fractions for the system. We can also compute equilibrium fugacity ratios by manipulating the calculated mole fractions or simply by evaluating specific gas phase reactions.

Worked Problem 4-7

Consider a closed system containing a gas phase and the solid QFM oxygen buffer at 1100°C. The total pressure on the gas is 1.0 atm., and it consists of a mixture of C-H-O species. What are the partial pressures of each of those species if the system is at equilibrium and the bulk atomic ratio C : H is 0.05?

As phrased, this is a very difficult problem to do on paper, because it involves a great number of potentially stable species. To make it easier, let us assume for purposes of illustration that there are only four gases in the mixture, other than O_2: H_2, H_2O, CO, and CO_2. The partial pressures of these five, then, constitute the unknown values we must calculate.

To solve for the five unknowns, we must write five equations. Two of these follow directly from the bulk chemistry of the system. Because total pressure is 1.0 atm., we know that

$$pH_2 + pH_2O + pCO + pCO_2 + pO_2 = 1.0. \qquad (4\text{-}26)$$

Because the atomic ratio C : H is 0.05, we also know that

$$(pCO + pCO_2)/(pH_2 + pH_2O) = 0.1. \qquad (4\text{-}27)$$

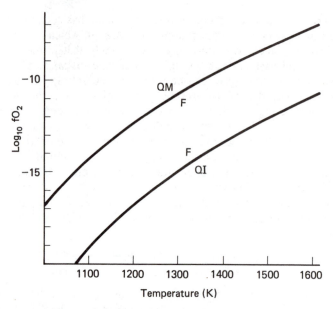

Figure 4.11 Calculated curves indicating the oxidation state of magmatic systems buffered by Quartz + Fayalite + Magnetite (QFM) or Quartz + Fayalite + Iron (QFI) as a function of temperature. The two curves are approximately four orders of magnitude apart over the temperature range for most volcanic events.

The partial pressure of O_2 is defined by the QFM buffer at the temperature specified. An empirical equation commonly used for this purpose (Eugster and Wones, 1962) is

$$\log f_{O_2} = 9.00 - (25738/T),$$

where T is the temperature in Kelvins. In this example, then,

$$f_{O_2} = 1.8 \times 10^{-10}. \tag{4-28}$$

Assuming ideality, this is also the partial pressure of O_2, in atmospheres.

The final two equations are based on the stoichiometry of the gas species. From the potential reaction

$$2CO_2 \longleftrightarrow 2CO + O_2,$$

we calculate that

$$f_{CO_2}/f_{CO} = (f_{O_2}/K_c)^{1/2}, \tag{4-29}$$

in which K_c is the equilibrium constant for this reaction between carbon species. It has a value, under these conditions, of 3.79×10^{-13}. The only other independent chemical reaction that can be written is

$$2H_2O \longleftrightarrow 2H_2 + O_2,$$

from which

$$f_{H_2O}/f_{H_2} = (f_{O_2}/K_h)^{1/2}. \tag{4-30}$$

The equilibrium constant, K_h, has the value of 8.82×10^{-14} at 1373 K.

Equations 4-26 through 4-30 can be solved simultaneously to yield

$$P_{H_2} = 1.97 \times 10^{-2} \text{ atm}$$
$$P_{H_2O} = 0.89 \text{ atm}$$
$$P_{CO} = 4.00 \times 10^{-3} \text{ atm}$$
$$P_{CO_2} = 8.70 \times 10^{-2} \text{ atm}$$

The method used in Worked Problem 4-7 becomes cumbersome if the potential number of gas species is large, so a computer is a more appropriate tool than pencil and paper. In order to investigate model atmospheres for the early Archean, we have calculated the mole fractions of 84 gas species in the system C-O-H-S, with its bulk composition derived from the volatile inventory in Table 4-3. The mixture, therefore, has atomic ratios $C/H = 0.032$, $C/N = 16.5$, and $C/S = 42.0$. We have assumed a total atmospheric pressure of 1.0 bars and a magma temperature of 1200°C, but have found that the results of the model calculation are very insensitive to changes in either parameter. The results are applicable, therefore, throughout the probable range of outgassing pressures and temperatures. Values were computed at both the QFM and QFI oxidation state limits. Here are the concentrations of the major species and some important abundance ratios:

	QFM	QFI
H_2O	0.91	0.30
H_2	0.02	0.64
CO_2	0.057	0.0097
CO	0.0031	0.051
SO_2	0.0013	2.0×10^{-7}
H_2S	1.3×10^{-5}	0.0014
N_2	0.0018	0.0018
O_2	3.4×10^{-9}	4.0×10^{-13}
CH_4	4.7×10^{-14}	2.7×10^{-7}
CO_2/CO	18.4	0.19
SO_2/H_2S	100.0	1.4×10^{-4}
H_2O/H_2	45.5	0.46

Today's volcanic gases are similar to the calculated mixture in the QFM column. They are dominated by H_2O. Among the carbon and sulfur species, CO_2 and SO_2 are most abundant, and N_2 is the major nitrogen species. If metallic iron were present in Archean magmas, however, we would expect to find H_2, a major reduced gas species, roughly twice as abundant as water, carbon monoxide five times as abundant as CO_2, and sulfur species heavily dominated by H_2S. In either case, partial pressures of atmospheric gas species should differ from those in volcanic gases because of reactions between gases during cooling or as the result of various loss processes. To the extent that thermodynamic rather than kinetic factors determined the state of the primitive atmosphere, therefore, we can estimate the relative abundances of gas species during the early Archean by allowing the mixture in the QFI column to reequilibrate at low temperature.

Carbon monoxide, for example, is not stable at low temperatures in a hydrogen-rich atmosphere, and should react to form CH_4 plus water or, as P_{H_2} drops below 10^{-4} atm, to form graphite plus water. Therefore, the dominant carbon species in the earliest Archean atmosphere should have been methane. Similarly, as long as P_{H_2} exceeded roughly 10^{-3} atm, the dominant nitrogen species at 25°C should have been NH_3 rather than N_2. Figure 4.12 illustrates the way in which each of these equilibria depends on the partial pressure of H_2. The actual equilibrium partial pressures of CH_4 and NH_3, of course, would also have depended on the amount of crustal outgassing, as suggested by the pair of curves in Figure 4.12b.

Hydrogen is lost from the upper atmosphere to space at a rate that is roughly proportional to its partial pressure and is rapid enough that the mean residence time for atmospheric H_2 today is four to seven years. In order to maintain the high initial partial

Gradual Change: The History of Seawater and Air   **119**

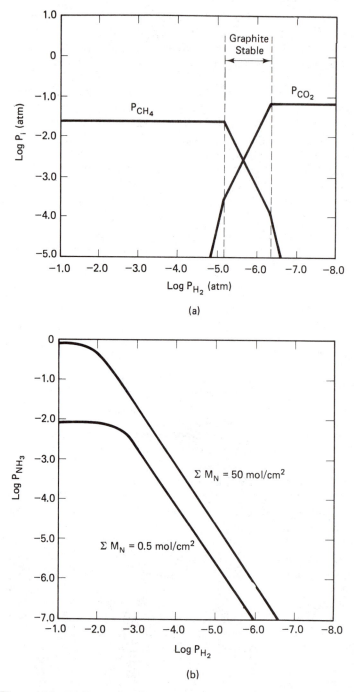

Figure 4.12 Equilibrium partial pressures of major carbon and nitrogen gases as functions of P_{H_2} at 25°C. (a) P_{CH_4} and P_{CO_2} calculated for a total atmospheric carbon concentration of 1.5 moles cm^{-2}; (b) P_{NH_3} calculated for two values of total nitrogen abundance. (After H. D. Holland, *The chemical evolution of the atmosphere and oceans*. Copyright 1984 by Princeton University Press. Reprinted with permission.)

pressure suggested by the results above, the rate of volcanic supply of H_2 would have had to meet or exceed its rate of loss by chemical reactions and escape to space. Following the early period of catastrophic outgassing, the rate of volcanic activity probably dropped sharply, and segregation of metallic iron into the core favored the exhalation of more oxidized volcanic gases. As a result, P_{H_2} decreased. This, in turn, would have led to progressively lower abundances of CH_4 and NH_3 with time. Hydrogen sulfide, easily destroyed by solar UV radiation or consumed in redox reactions with the crust, should also have become less abundant during this period. Also, as P_{H_2} continued to decrease, P_{CO_2} should have increased as a result of the reactions

$$CH_4 + 2H_2O \longleftrightarrow CO_2 + 4H_2$$

and then

$$C \text{ (graphite)} + 2H_2O \longleftrightarrow CO_2 + 2H_2,$$

as shown in Figure 4.12a. As the Archean period progressed, then, the oxidation state of the atmosphere should have gradually favored a mixture of CO_2 and N_2. Water vapor, of course, should have condensed and led to the early growth of the oceans.

It is important to reemphasize that this interpretation is based on thermodynamic arguments and may not bear any resemblance to the actual history of Archean atmosphere-ocean chemistry. Today's atmosphere, in fact, does not contain the mixture of NO_x gases that would be predicted by equilibrium thermodynamics. It is likely that the early atmosphere's composition, also, was controlled by reaction kinetics rather than thermodynamics. The true course of events may never be known. Regardless of the pathways, however, it is fairly certain that the atmosphere had become predominantly a CO_2-N_2 mixture by the time our stratigraphic record began 3.7×10^9 years ago. The earliest surviving sedimentary rocks already contain mature chemical weathering products, which are formed most readily by the interaction of CO_2-charged rainwater with crustal rocks.

The *greenhouse effect,* which arises because certain atmospheric gases are transparent to incoming shortwave solar radiation but are relatively opaque to radiation at the longer wavelengths generated by the earth's surface, may have been important in stabilizing surface temperatures. Of the likely atmospheric gases in the C-O-H-S system, only water and CO_2 are effective in this role. If either is present, outgoing radiation is partially trapped, and the surface temperature of the planet is raised. The degree of warming is a nonlinear function of the abundance of greenhouse gases and of the amount of incoming solar radiation. Using the estimated Pleistocene global surface temperatures as a guide, Kasting (1986) has calculated that a partial pressure of CO_2 near 0.05 bars, or roughly 100 times the present atmospheric level, would have been necessary to generate the same surface temperature by the greenhouse effect during the Huronian glacial event. By the Late Proterozoic, Kasting estimates that P_{CO_2} had fallen off to less than 3×10^{-3} bars, a decrease of roughly 10 times (see Figure 4.13). Like the equilibrium arguments we presented above, then, these results suggest that carbon in the atmosphere was predominantly in the form of CO_2 rather than CH_4, CO, or other reduced gases during most of the Precambrian.

As we will discuss in greater detail in Chapter 6, H_2CO_3 produced in CO_2-charged rainwater is one of the most common participants in reactions like

$$2KAlSi_3O_8 + 2H_2CO_3 + 9H_2O \longleftrightarrow$$
K-feldspar
$$2K^+ + 2HCO_3^- + Al_2Si_2O_5(OH)_4 + 4H_4SiO_4.$$
kaolinite

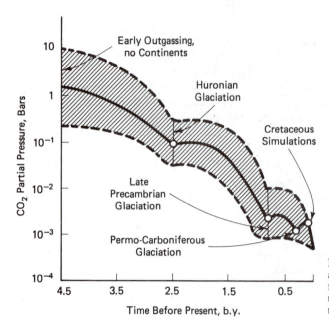

Figure 4.13 Limits on the evolution of atmospheric CO_2, inferred primarily from the greenhouse-supported temperatures necessary to avoid runaway glaciation. (Modified from Kasting 1986.)

ARCHEAN ATMOSPHERIC GASES AND THE ORIGIN OF LIFE

The composition of the early atmosphere is of interest not only to geochemists but also to biologists, who would like to know what environmental conditions were most likely to have prevailed when the first biological molecules were formed on the earth. The Russian biochemist A. I. Oparin opened discussion on the topic in the 1920s by theorizing that lightning or intense solar ultraviolet radiation could have produced amino acids and other organic building blocks from an early atmosphere rich in CH_4, H_2, and NH_3. These then accumulated in the oceans, forming a primordial "soup" in which more complex hydrocarbons such as proteins and nucleic acids finally developed. This notion was further reinforced as it became apparent that the atmospheres of Jupiter and the other outer planets were mixtures of these same reduced C-H-N gases, plus helium. Earth's present atmosphere was assumed by many scientists of the period to have evolved from a gravitationally accreted primordial one like Jupiter's.

The first modern experiments to shed light on this problem were performed by Stanley Miller (1953), who was then a graduate student in chemistry. Miller built a closed apparatus, illustrated in Figure 4.14, into which he introduced various mixtures of CH_4, H_2, and NH_3. Water was boiled in the small flask to add water vapor to the gas mixture and to encourage circulation through the larger flask, where the gases were subjected to a spark discharge. The resulting mixture was then condensed in a cold trap and fluxed back through the smaller flask. After a week of operation, during which nonvolatile products gradually accumulated in the small flask, samples were withdrawn and analyzed by gas chromatography.

Miller's experiments produced a number of fairly simple organic compounds, many of which are found in living organisms. In all, 15% of the carbon in the system had reacted to form amino acids, aldehydes, or polymeric hydrocarbons. This was a dramatic and somewhat surprising confirmation of Oparin's hypothesis, which seemed to confirm the importance of an early, highly reduced atmosphere.

Experiments like Miller's, but using a wide range of starting gas compositions and a number of different energy sources, have shown that prebiotic molecules can be formed from almost any nonoxygenated atmosphere. In Miller's highly reduced gas mixture, for example, glycine is probably produced by the reaction

$$NH_3 + 2CH_4 + 2H_2O \longleftrightarrow C_2H_5O_2N + 5H_2.$$
$$\text{glycine}$$

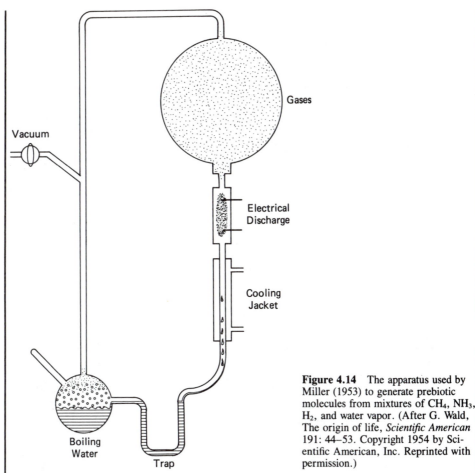

Vacuum

Gases

Electrical
Discharge

Cooling
Jacket

Boiling
Water

Trap

Figure 4.14 The apparatus used by Miller (1953) to generate prebiotic molecules from mixtures of CH_4, NH_3, H_2, and water vapor. (After G. Wald, The origin of life, *Scientific American* 191: 44–53. Copyright 1954 by Scientific American, Inc. Reprinted with permission.)

Goldsmith and Owen (1979) have also produced glycine from a less reduced mixture of N_2, H_2, H_2O, and CO by the reaction

$$3HCN + 2H_2O \longleftrightarrow C_2H_5O_2N + CN_2H_2.$$

glycine   cyanamide

The proportion and absolute abundance of amino acids decreases with an increasing oxidation state of the atmosphere, to the extent that none have been found in experiments on CO_2, N_2, H_2O mixtures. Even in these, however, some biologically significant aldehydes and simple molecules like HCN have been observed. Life may have evolved in a methane-ammonia atmosphere, as Oparin suggested, but that no longer seems to be a necessary condition. Geochemistry places more stringent constraints on models of the early atmosphere than biochemistry does.

Clay-forming reactions, therefore, would have served as a sink for CO_2 and water from the beginning, and would have increased in efficiency as P_{CO_2} increased. In the same way, weathering would have removed sulfur gases from the primitive "acid rain." Therefore, as chemical weathering progressed, a sizable portion of the excess volatile inventory was transferred to residual minerals and buried in sedimentary rocks. The mean residence times for most dissolved species in the oceans are short enough that these weathering reactions would probably also have led to seawater of nearly modern composition within a few hundred million years. As we remarked earlier, the earliest sedimentary record contains carbonates, evaporites, and shales that appear to have been deposited from an ocean similar to today's (Holland, 1984).

The Rise of Oxygen

The other major story we have to tell concerns the evolution of oxygen, which has been parallel to the evolution of life. Prebiotic O_2 levels were very low, partly because there were no volumetrically significant oxygen sources prior to photosynthesis and partly because crustal and atmospheric sinks were plentiful. The dissociation of H_2O by solar UV radiation, followed by the escape of hydrogen to space, would have provided a small but continuous supply of O_2 in the upper atmosphere. Molecular oxygen is easily consumed by back reaction with H_2, however, so much of the oxygen would have been consumed before it reached the earth's surface. Even as volcanic sources for hydrogen waned, other H_2 sources persisted in the Archean. The dominant ones were the photo-chemical reaction producing hydrogen peroxide from water vapor,

$$2H_2O \longleftrightarrow H_2O_2 + H_2,$$

and the photo-oxidation of ferrous iron in seawater,

$$Fe^{2+} + H^+ \longleftrightarrow Fe^{3+} + \tfrac{1}{2}H_2.$$

Kasting (1986) has shown that these reactions would have guaranteed a P_{H_2} high enough to preclude an O_2 abundance greater than about one part in 10^{-12} at the earth's surface during the prebiotic Archean.

The fossil evidence that would tell us when green-plant photosynthesis began is very equivocal. Stromatolites resembling modern algal mats, the product of blue-green algae, have been identified in rocks almost 3×10^9 years old. The earliest of these, however, may have been formed by photosynthetic bacteria, which use reduced compounds like H_2S and thiosulfate rather than H_2O as electron donors. In all likelihood, the transition to green-plant photosynthesis was very gradual, as early organisms developed the enzymatic mechanisms necessary to protect themselves from free oxygen.

Two sorts of observations tell us that this transition was well under way by about 2.5×10^9 years ago. On land, the oxidation state of various weathering products has been used to infer a gradual increase in P_{O_2} between 3.0 and 2.0×10^9 years ago. In marine environments, the appearance of thick deposits of banded iron formation during the same period has been used as evidence. We will examine each of these briefly.

Several rock units in Precambrian terrain, most notably the quartz-pebble conglomerates in the Witwatersrand Basin in South Africa and at Blind River and Elliot Lake in Canada, contain grains of uraninite and pyrite that are almost certainly detrital (see Figure 4.15). Over most of the natural range of pH and oxidation conditions today, pyrite is readily oxidized to form insoluble iron hydroxides. Uraninite, at pH values less than about 5, combines with O_2 and bicarbonate to form the soluble uranyl carbonate complex, $UO_2(CO_3)_2^{-2}$, and is thus easily destroyed. Between pH 4 and 6, reactions with O_2 and H^+ also yield $UO_2(OH)^+$. As a result, except in very arid regions or under conditions of unusually rapid erosion and burial, neither mineral commonly accumulates in modern sediments. The presence of both in 2.7 to 2.3×10^9 year-old conglomerates, but not in younger units, strongly suggests that this was the end of the "O_2-free" era. Holland (1984), using experimentally derived data on the dissolution rate of UO_2 (Grandstaff, 1980), has determined that the product, $P_{O_2}P_{CO_2}$, had a value of $10^{-4.9 \pm 0.8}$ during this period. Today it is close to 600.

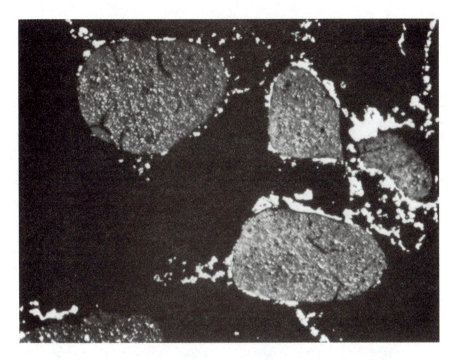

Figure 4.15 Uraninite grains in 2.3 b.y. old quartz-pebble conglomerate from the Witwatersrand Basin, South Africa. The degree of rounding suggests that these are detrital grains, deposited in a surficial environment with considerably less free O_2 than today. (After M. Schidlowski, Beiträge zur Kenntnis der radioactiven Bestandteile der Witwatersrand-Konglomerate: I. Uranpecherz in den Konglomeraten des Oranje-Freistaat-Goldfeldes, *Neues Jahrb. Mineral. Abh.* 105: 183–202. Copyright 1966 by E. Schweizerbart'sche Verlagsbuchhandlung, Stuttgart. Reprinted with permission.)

A considerable amount of attention has also been devoted to the few Late Archean and Early Proterozoic paleosols that have survived in the rock record and can be characterized despite the overprint of diagenetic and metamorphic events. The abundance of oxidized iron in a soil profile can be shown to depend on the relative availability of O_2 and acid weathering agents, primarily H_2CO_3. As the parent rock is decomposed, acids attacking iron-bearing silicates release Fe^{2+}, which may then be precipitated from soil water as Fe^{3+} by reaction with O_2. In most modern soil environments, sufficient oxygen is present to fix all of the mobilized iron as an oxide or hydroxide. If O_2 were exhausted from soil and ground water, however, Fe^{2+} would remain in solution and be washed out of the soil. Acids, of course, can be titrated by minerals other than iron silicates, and oxygen can be removed by reaction with sulfides and organic carbon as well as Fe^{2+}. Holland (1984) has argued, therefore, that the amount of iron retained in a soil depends not only on the atmospheric ratio P_{O_2}/P_{CO_2}, but also on the composition of the parent rock, as expressed by

$$R = \frac{\text{total oxygen demand}}{\text{total acid demand}}.$$

Figure 4.16, based on Holland's study of paleosols from 12 Precambrian localities, suggests the gradual increase in atmospheric oxidation state during the Proterozoic.

For the other part of our brief survey, we turn to the oceans. Because the atmosphere and the surface waters of the ocean are in contact, we may assume that a gradual

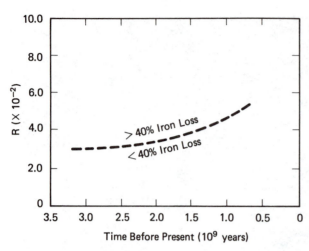

Figure 4.16 The function R, used in this figure, is a measure of the degree to which oxidizable components ("FeO," FeS_2, reduced carbon, and so forth) in a rock exceed those which are titrated by H_2CO_3 during weathering. Rocks with high R values should generally exhaust soil O_2 before soil CO_2, and should therefore lose a large fraction of their iron in solution as Fe^{2+}. Rocks with low R require less O_2 to convert all iron to Fe^{3+}, which is retained in the paleosol. This diagram suggests that the extent of iron-leaching decreased gradually through the late Precambrian era and thus that atmospheric P_{O_2}/P_{CO_2} gradually increased. (Modified from Holland 1984.)

increase in P_{O_2} during the Late Precambrian would have oxygenated surface waters first. The relatively slow rate at which deep ocean waters are overturned would itself have kept bottom waters anoxic for a long time. More importantly, oxygen carried downward in circulating surface waters should have been consumed rapidly by reduced species, largely Fe^{2+}.

There has been a long-standing debate concerning the source of iron in the Precambrian oceans and its role in regulating atmospheric oxygen. According to one view (for example, Holland, 1984), marine Fe^{2+} was derived from the global river flux, fed ultimately by the continental weathering reactions that led to iron depletion in paleosols. As green-plant photosynthesis gained dominance, the river supply decreased because iron was retained in the soils, and Fe^{2+} remaining in the deep ocean was gradually oxidized. The other possibility, articulated by Veizer (1983), is that iron was introduced into the oceans by seafloor volcanism, presumed to have been more vigorous during the first half of earth history than it is today. Its effectiveness as a sink for dissolved O_2, in this view, decreased as the earth's plate tectonic engine gradually slowed down.

Whether ferrous iron in the oceans was a cause or a side effect of low atmospheric P_{O_2}, it is clear that O_2 was a minor species in the atmosphere as long as the deep oceans were anoxic. Only after the supply of oxygen to the seafloor was able to exhaust the Fe^{2+} sink did it become possible for P_{O_2} to rise toward modern levels. The depositional history of banded iron formation is, therefore, a record of the transition to an oxygen-rich atmosphere.

SUMMARY

The field of ocean-atmosphere geochemistry is extensive, and we cannot claim to have given it more than a quick glimpse in this chapter. By example, however, we have shown three of the major directions geochemists have followed and have suggested others.

The composition of seawater, and its variations with depth and global location, are of great interest as indicators of environmental control. In particular, chemical species like HCO_3^- and CO_3^{2-} play a central role in the ocean's *pH* and alkalinity regulation and thus affect not only geologic but also biologic processes. The pathways along which change occurs, and the network of kinetic factors that define steady state condions, are our gateway to understanding the recent history of the ocean-atmosphere. Geochemical cycle models like the three we considered here can also help to elucidate the connections between geochemistry, biochemistry, and geophysics that force long-term change in the system. Finally, by examining the chemistry of seawater and air from their earliest days, we gain a perspective on earth history that bears on the evolution of life.

In presenting these three topics, we have also incidentally introduced subjects that will occupy the next two chapters: diagenesis and weathering. Each represents a boundary between the ocean-atmosphere system and the solid earth.

SUGGESTED READINGS

A great many books have been written in the field of ocean-atmosphere chemistry. The four we have selected here have been written by some of the unquestioned leaders of research in the area, and are impressive for their clarity of presentation.

BROECKER, W. S., and PENG, T-H. 1982. *Tracers in the sea.* New York: Eldigio Press, Columbia University. (Perhaps the most comprehensive book on chemistry of the oceans, written as a textbook, with many worked examples.)

HOLLAND, H. D.1978. *The chemistry of the atmosphere and oceans.* New York: Wiley Interscience. An exceptionally readable discussion of processes which relate the compositions of the ocean and atmosphere.)

HOLLAND, H. D. 1984. *The chemical evolution of the atmosphere and oceans.* Princeton N.J.: Princeton Univ. Press. (This is the best single volume of data and ideas regarding the history of the ocean-atmosphere system.)

WALKER, J. C. G. 1977. *Evolution of the atmosphere.* New York: Macmillan. (A particularly valuable reference for those interested in atmospheric chemistry.)

The following articles were referenced in this chapter. An interested student may wish to explore them further.

BEN-YAAKOV, S.; RUTH, E.; and I. R. KAPLAN 1974. Carbonate compensation depth: Relation to carbonate solubility in ocean waters. *Science* 184: 982–84.

BERNER, R. A.; LASAGA, A. C.; and GARRELS, R. M. 1983. The carbonate-silicate geochemical cycle and its effect on carbon dioxide over the past 100 million years. *Amer. J. Sci.* 283: 641–83.

BERNER, R. A.; LASAGA, A. C.; and GARRELS, R. M. 1985. An improved geochemical model of atmospheric CO_2 fluctuations over the past 100 years. In *The carbon cycle and atmospheric CO_2: Natural variations archean to present,* ed., E. T. Sundquist and W. S. Broecker, 397–41. Monograph 32. Washington, D.C.: Amer. Geophys. Union.

BREWER, P. G.; BROECKER, W. S.; JENKINS, W. J.; RHINES, P. B.; ROOTH, C. G.; SWIFT, J. H.; TAKAHASHI, T.; and WILLIAMS, R. T. 1983. A climatic freshening of the deep North Atlantic (north of 50°N) over the past 20 years. *Science* 222: 1237–39.

BROECKER, W. S. 1982. Glacial to interglacial changes in ocean chemistry. *Progr. Oceanogr.* 11: 151–97.

CLARKE, W. B.; BEG, M. A.; and CRAIG, H. 1969. Excess ^{3}He in the sea: Evidence for terrestrial primordial helium. *Earth Planet. Sci. Lett.* 6: 213–20.

DREVER, J. I. 1974. The magnesium problem. In *The sea,* Vol. 5, ed. E. D. Goldberg, 337–57. New York: Wiley Interscience.

EDMOND, J. M. 1970. High precision determination of titration alkalinity and total carbon dioxide content of sea water. *Deep Sea Res.* 17: 737–50.

EUGSTER, H. P., and WONES, D. R. 1962. Stability relations of the ferruginous biotite, annite. *J. Petrol.* 3: 82–125.

GARRELS, R. M., and THOMPSON, M. E.1962. A chemical model for sea water at 25° C and one atmosphere total pressure. *Amer. J. Sci.* 260: 57–66.

GERLACH, T. M., and NORDLIE, B. E. 1975a. The C-O-S-H gaseous system, Part I: Composition limits and trends in basaltic gases. *Amer. J. Sci.* 275: 353–76.

———, 1975b. The C-O-S-H gaseous system, Part II: Temperature, atomic composition, and molecular equilibria in volcanic gases. *Amer. J. Sci.* 275: 377–94.

———. 1975c. The C-O-S-H gaseous system, Part III: Magmatic gases compatible with oxides and sulfides in basaltic magmas. *Amer. J. Sci.* 275: 395–410.

GRANDSTAFF, D. E. 1980. Origin of uraniferous conglomerates at Elliot Lake, Canada, and Witwatersrand, South Africa: Implications for oxygen in the Precambrian atmosphere. *Precambrian Research* 13, 1–26.

HONJO, S., and EREZ, J. 1978. Dissolution rates of calcium carbonate in the deep ocean: An *in situ* experiment in the North Atlantic. *Earth Planet. Sci. Lett.* 40: 226–34.

KASTING, J. F. 1986. Theoretical constraints on oxygen and carbon dioxide concentrations in the Precambrian atmosphere. *Precambrian Research,* 34, 205–29.

KASTING, J. F.; POLLACK, J. B.; and CRISP, D. 1984. Effects of high CO_2 levels on surface temperature and atmospheric oxidation state on the early earth. *J. Atmos. Chem.* 1: 403–28.

KASTING, J. F., and RICHARDSON, S. M. 1986. Seafloor hydrothermal activity and spreading rates: The Eocene carbon dioxide greenhouse revisited. *Geochim. Cosmochim. Acta* 49: 2541–44.

KASTING, J. F.; RICHARDSON, S. M.; POLLACK, J. B.; and TOON, O. B.1986. A hybrid model of the CO_2 geochemical cycle and its application to large impact events. *Amer. J. Sci.* 286: 361–89.

KEIR, R. S., and BERGER, W. H. 1983. Atmospheric CO_2 content in the last 120,000 years: The phosphate extraction model. *J. Geophys. Res.* 88: 6027–38.

———. 1985. Late Holocene carbonate growth in the equatorial Pacific: Reef growth or neoglaciation? In *The carbon cycle and atmospheric CO_2: Natural variations Archean to present,* ed. E. T. Sundquist and W. S. Broecker, 208–20. Monoghraph 32, Washington, D.C.: Amer. Geophys. Union.

KURZ, M. D.; JENKINS, W. J.; SCHILLING, J. G.; and HART, S. R. 1982. Helium isotopic variations in the mantle beneath the central North Atlantic Ocean. *Earth Planet, Sci. Lett.* 58: 1–14.

LASAGA, A. C. 1980. The kinetic treatment of geochemical cycles. *Geochim. Cosmochim. Acta* 44: 815–28.

LASAGA, A. C. 1981. Dynamic treatment of geochemical cycles: Global kinetics. In *Kinetics of geochemical processes,* ed. A. C. Lasaga and R. J. Kirkpatrick, 69–110. Reviews in Mineralogy 8. Washington, D.C.: Mineral. Soc. of Amer.

MEYBECK, M. 1979. Concentration des eaux fluviales en éléments majeurs et apports en solution aux oceans. *Revue de Geol. Dynam. et de Geogr. Phys.* 21: 215–46.

MILLER, S. L. 1953. A production of amino acids under possible primitive Earth conditions. *Science* 117: 528–29.

MILLER, S. L. and ORGEL, L. E. 1974. *The origins of life on the earth.* Englewood Cliffs, N.J.: Prentice-Hall.

PETERSON, M. N. A. 1966. Calcite: Rates of dissolution in a vertical profile in the Central Pacific. *Science* 154: 1542–44.

PYTKOWICZ, R. M. 1983. *Equilibria, nonequilibria, & natural waters. Vol. 1.* New York: Wiley Interscience.

QUIMBY-HUNT, M. S. and TUREKIAN, K. K. 1983. Distribution of elements in sea water. *EOS, Trans. Amer. Geophys. Union* 64: 130–31.

RUBEY, W. W. 1951. Geologic history of seawater, an attempt to state the problem. *Bull. Geol. Soc. Amer.* 62: 1111–47.

SAGAN, C., and MULLEN, G. 1972. Earth and Mars: Evolution of atmospheres and surface temperatures. *Science* 177: 52–56.

SCHIDLOWSKI, M. 1966. Beiträge zur Kenntnis der radioactiven Bestandteile der Witwatersrand-Konglomerate: I. Uranpecherz in den Konglomeraten des Oranje-Freistaat-Goldfeldes. *Neues Jahrb. Mineral. Abh.* 105: 183–202.

SUNDQUIST, E. T. 1985. Geological perspectives on carbon dioxide and the carbon cycle. In *The carbon cycle and atmospheric CO_2: Natural variations Archean to present,* ed. E. T. Sundquist and W. S. Broecker, 5–60. Monograph 32, Washington, D.C.: Amer. Geophys. Union.

VEIZER, J. 1983. Geologic evolution of the Archean-early Proterozoic Earth. In *The earth's earliest biosphere: Its origin and evolution,* ed. J. W. Schopf, 240–259. Princeton, N.J.: Princeton Univ. Press.

WEYL, P. K. 1961. The carbonate saturometer. *J. Geology* 69: 32–43.

PROBLEMS

1. A naive geochemist might expect the nitrate concentration in surface waters of the ocean to be maintained at an equilibrium concentration for the reaction

$$N_2 + \tfrac{5}{2}O_2 + H_2O \longleftrightarrow 2H^+ + 2NO_3^-.$$

Use tabulated thermodynamic data and your knowledge of P_{O_2}, P_{N_2}, and seawater pH to estimate the expected oceanic NO_3^- concentration. How does this value compare with the concentration recorded in Table 4-1? What factors might account for this difference?

2. What is the pH of a 3×10^{-3} m Na_2CO_3 solution?

3. Using the procedure described in Worked Problem 4-2, construct titration curves to show the change in solution pH as a 3×10^{-3} m Na_2CO_3 solution is titrated by (a) 0.01m HCl, (b) 1.0m HCl, and (c) 10.0m HCl. Do the end points depend on the molality of the HCl solution? (This problem is most easily done with the program TITRATE.)

4. Suppose you were to add Fe^{2+} to the Na_2CO_3 solution in problem 2. How high could you raise the Fe^{2+} concentration before $FeCO_3$ began to precipitate? Assume that the K_{sp} for $FeCO_3$ is 2.0×10^{-11}, and that there are no kinetic barriers to precipitation.

5. How should the alkalinity of a beaker of seawater be affected by the addition of small amounts of the following substances? (a) $NaHCO_3$; (b) NaCl; (c) HCl; (d) $MgCO_3$?

6. In Worked Problem 4-1, we made the simplifying assumption that $a_{CO_3^{2-}}$ is negligible in rain water, and derived an expression to calculate pH, given P_{CO_2}. Use that expression to calculate rainwater pH if $P_{CO_2} = 380$ ppm. Repeat the calculation using the program RAINWATR on the diskette. Assuming that the numerical answer is more accurate, what is the percentage error in your approximate solution?

7. The program CYCLE.EXE on the program diskette uses Lasaga's (1980) eigenvalue method to study perturbations in a multireservoir geochemical system. Study the documentation file CYCLE.DOC by using TYPE or PRINT commands to display it on your monitor or printer. Then use CYCLE.EXE with the data file CARBON.DAT to determine how the global carbon cycle would be affected by (a) rapid injection of 1.1×10^{17}g of CO_2 into the atmosphere by fossil fuel burning; (b) instantaneous combustion of 2.0×10^{17}g of carbon in the terrestrial biosphere due to flash fires in a nuclear war. (Assume total conversion of the biospheric carbon to CO_2.)

8. In Worked Problem 4-7, we calculated the composition of a gas mixture under the moderately oxidizing conditions of the QFM buffer. Redo this calculation, using instead the oxygen fugacity defined by QFI. A useful empirical equation for this buffer is

$$\log f_{O_2} = 7.51 - 29382/T.$$

Chapter 5

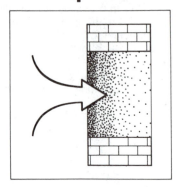

Diagenesis: A Study in Kinetics

OVERVIEW

Diagenesis embraces all of the changes that may take place in a sediment following deposition, except for those due to metamorphism or to weathering at the earth's surface. Intellectual battles have been waged over these two environmental limits to diagenesis. Rather than join these battles here, we focus on some of the geochemical processes that affect sediments after burial and consider some of the pathways along which they may change. This is a field of great interest in geology, especially among those who are interested in petroleum exploration.

Diagenetic changes take place slowly at low temperature. The assemblages we see in sediment, therefore, usually represent some transition between stable states. In previous chapters, we have begun to study thermodynamic principles that allow us to interpret those stable states. By studying the kinetics of transport and chemical reaction, it is possible to use those assemblages in transition to determine which of many chemical pathways through diagenesis a particular sediment may have followed.

Sediments at low temperature commonly contain organic compounds that can be useful in tracing the course of diagenesis. Of course, these are also interesting in their own right, as such compounds are the source of fossil fuels. We briefly survey the range of organic reactions that characterize the transition from living organic matter to kerogen, pointing out trends that can be used to indicate the degree of diagenesis a sample has attained.

WHAT IS DIAGENESIS?

Many diagenetic processes are associated with lithification. Among these are compaction, cementation, recrystallization, authigenic mineralization, and growth of concretions or nodules. It is through these mechanisms that unconsolidated sediment experiences a general reduction in porosity and develops a secondary framework that converts it to solid rock. Other changes also occur, often independent of the progress of lithification. Chief among these are the many processes that modify buried organic matter.

Each of these processes takes place, commonly, within a temperature range from about 20°C to 300°C, and close enough to the surface that total pressure is less than about 1 kbar. Almost invariably, diagenetic processes also involve the participation of fluids in the interstitial spaces between sediment particles. In freshly buried sediments, this fluid has the same bulk composition as the waters from which the sediments were deposited. As time proceeds, these *formation waters* accumulate and transmit the products of reactions within the sediment column. Many of the thermodynamic manipulations of earlier chapters are useful in examining these reactions. In most cases, however, sediments and fluids undergoing diagenesis are highly nonuniform; that is, their compositions vary from place to place within a distance of a few centimeters to a few meters. Because this is so, it is possible to extract kinetic information by studying concentration-distance profiles in a diagenetic environment.

KINETIC FACTORS IN DIAGENESIS

Transport of material can take place in either of two ways: by *diffusion* or by *advection*. When we speak of diffusive transport, we are considering the dispersion of ions or molecules through a medium that is not itself moving. The path followed by any single particle is random, but a net movement of material may result if there is a gradient in some intensive property such as temperature or chemical potential across the system.

Advective transport, on the other hand, takes place when the ions or molecules are carried in a medium, like an intergranular fluid, which *is* moving. The driving force in this case is simply a gradient in hydrostatic pressure.

In this section, we will develop a general expression to illustrate how the concentration of any mobile species in a diagenetic environment may change as the result of diffusive or advective transport, or as the result of chemical reactions between the fluid and solid phases in a sedimentary column.

Diffusion

In Chapter 2, we introduced thermodynamic functions which describe the behavior of systems in bulk, rather than developing a statistical method for describing systems at the atomic level. Such an approach is *empirical* rather than *atomistic*. In the same way, nineteenth-century scientists like Joseph Fourier and Georg Ohm described the transport of heat and electrical charge by assuming that they travel through a continuous medium rather than one consisting of discrete atoms. As with the perspective we have adopted for thermodynamics, this assumption leads to a description of transport properties by differential equations. Diffusive transport was first described in this way in the 1850s by Adolf Fick, a German chemist.

According to Fick's first law, the flux (J) of material along a composition gradient at any specific time (t) is directly proportional to the magnitude of the gradient. That is, if the ions or molecules in which we are interested move along a direction x parallel to a gradient in concentration, c,

$$J = -D(\partial c/\partial x)_t. \tag{5-1}$$

The quantity D is called the *diffusion coefficient*, and is commonly tabulated in units of $cm^2 sec^{-1}$. Equation 5-1 satisfies the empirical observation that flux drops to zero in the absence of a concentration gradient. It is applicable in any system experiencing diffusive transport, but is most appropriate for use when the system of interest is in a steady state (that is, when $(\partial c/\partial t)_x$ is a constant). It is similar to Ohm's law or to fundamental heat flow equations in that D can be shown experimentally to be independent of the magnitude of $(\partial c/\partial x)_t$. Because all of the quantities in equation 5-1 are positive numbers, we must place a negative sign on the right hand side to indicate that material diffuses in the direction of *decreasing* concentration.

If the concentration at a given point in the system is changing with time, equation 5-1 still holds, but it is not the most convenient way to describe diffusive transport. To derive a more appropriate differential equation, consider the schematic sediment column in Figure 5.1, in which we have measured the flux of some species at two depths x_1 and x_2, separated by a distance Δx, and have found J_1 to be different from J_2. So long as Δx is small, we can justify describing the relationship between J_1 and J_2 by

$$J_1 = J_2 - \Delta x(\partial J/\partial x).$$

Because fluxes at the two depths differ, the concentration of the species of interest must change in the sediment volume between them during any time interval we choose. If we assume a unit cross-sectional area for the column, then the affected volume is $1 \times \Delta x$ and the change can be written as

$$(J_1 - J_2) = \Delta x(\partial c/\partial t) = -\Delta x(\partial J/\partial x).$$

We can now substitute this result into equation 5-1 to obtain

$$\frac{\partial c}{\partial t} = \frac{\partial}{\partial x}\left(D\frac{\partial c}{\partial x}\right). \tag{5-2}$$

This equation is Fick's second law, sometimes referred to as a *continuity equation* because it arises directly from the need to conserve matter during transport.

As they are presented here, Fick's laws describe diffusion in a straight line and assume that the medium for diffusion is isotropic. In most diagenetic environments, where diffusion takes place in pore waters, this formulation is adequate. Lateral variations in the composition of interstitial waters in a sediment, in most cases, are much less pronounced than changes with depth, so the driving force for diffusion is primarily

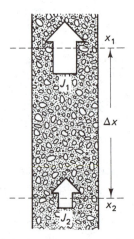

Figure 5.1 Sediment column with unit cross-sectional area. The flux (J_1) at depth x_1 is greater than the flux (J_2) at depth x_2.

one-dimensional. In general, however, diffusion takes place in three dimensions, (x, y, z), and equation 5-2 is more properly written as

$$\frac{\partial c}{\partial t} = \frac{\partial}{\partial x}\left(D\frac{\partial c}{\partial x}\right) + \frac{\partial}{\partial y}\left(D\frac{\partial c}{\partial y}\right) + \frac{\partial}{\partial z}\left(D\frac{\partial c}{\partial z}\right)$$

or, if you are familiar with vector notation, as

$$\frac{\partial c}{\partial t} = -\text{grad } J = \text{div}(D \text{ grad } c).$$

In this chapter, we will deal exclusively with the one-dimensional form. We will also generally assume that the diffusion coefficient, D, does not vary from place to place. These simplifications make mathematical modelling much easier, and in most cases we lose very little accuracy by adopting them.

If D is not a function of distance, then Fick's second law can be written more simply as

$$\frac{\partial c}{\partial t} = D\left(\frac{\partial^2 c}{\partial x^2}\right). \tag{5-3}$$

Solutions for this equation in a nonsteady state system (the most general situation) involve determining concentration as a function of distance and time, $c(x, t)$, for a given set of initial and boundary conditions. Even in the simplified form of equation 5-3, this task is often quite difficult. Many of the most useful solutions are described in Crank (1975). The two following problems illustrate common situations of interest in diagenetic environments.

Worked Problem 5-1

Suppose all of a potentially mobile species is initially concentrated in a very narrow layer. How can we describe its subsequent redistribution in surrounding sediments?

Imagine a column of lacustrine sediment in which the only inhomogeneity is an infinitesimally thin marker horizon at some distance x_0 below the sediment-water interface. At the start of the problem, this marker horizon contains a quantity, α, of some potentially mobile species which is not found elsewhere in the column. Such a situation is hard to imagine in a natural setting, because some diffusion into underlying sediments would have begun even before later sediments had buried the layer. A somewhat similar problem might occur, however, if waste from a chemical spill had spread evenly over a large part of the lake bottom and been buried rapidly by later sediments. This example is most valuable as an introduction to more complex ones.

The solution to equation 5-3 under the conditions we have described is

$$c(x, t) = \frac{\alpha}{2\sqrt{\pi Dt}} \exp\left(-\frac{x^2}{4Dt}\right), \tag{5-4}$$

where x is a distance either above or below the reference horizon x_c. This is a correct solution, because it can be differentiated to yield equation 5-3 again, and because it satisfies the initial conditions we specified:

$$\text{for } |x| > 0, \qquad c(x, t) \longrightarrow 0 \text{ as } t \longrightarrow 0$$

and

$$\text{for } x = 0, \qquad c(x_o, t) \longrightarrow \infty \text{ as } t \longrightarrow 0,$$

in such a way that the total amount of the mobile species is fixed:

$$\int_{-\infty}^{\infty} c(x, t)\, dx = \alpha.$$

Kinetic Factors in Diagenesis

Suppose, then, that the marker horizon contains 50 mg of Cu^{2+} per cm^2, and assume that the diffusion coefficient for Cu^{2+} in water is 5×10^{-6} cm^2 sec^{-1}. If diffusion proceeds outward from the marker horizon for 2 months, what will be the concentration of copper in pore waters 1 cm away? Applying equation 5-4, we find that

$$c(1 \text{ cm}, 5.1 \times 10^6 \text{ sec}) = 2.74 \text{ mg cm}^{-3}$$
$$\approx 2.74 \times 10^3 \text{ mg kg}^{-1}$$
$$\approx 2740 \text{ ppm}.$$

This particular example illustrates what is known as the *thin-film solution* to Fick's second law. The total mass of diffusing material is very small and is concentrated initially in a surface layer with "zero" thickness. In a real setting, of course, the source bed has a finite thickness. For that case, however, we would need to apply a slightly different solution to equation 5-4.

We can examine the general consequences of thin-film geometry by studying Figure 5.2, in which we have plotted $c(x, t)$ against x at successive times after diffusion outward from the marker horizon has begun. The area under the $c(x, t)$ curve remains constant, but the diffusing species spreads out symmetrically above and below the source layer as time advances. Referring to equation 5-4, we observe that the concentration at $x = 0$, $c(x_0, t)$, decreases at a rate proportional to $1/\sqrt{t}$. We also see that the curve has inflection points (where $\partial^2 c/\partial x^2 = 0$) above and below x_0, and that these separate a zone of solute depletion near the source bed from zones of solute enrichment at a distance. One of the other questions we might ask, then, is "How far will material have diffused by some time t?" This is a difficult question to answer, because $c(x, t)$ never reaches zero at *any* distance once diffusion has begun. As we saw in Chapter 4, though, one practical measure of the limit to a negative exponential function like equa-

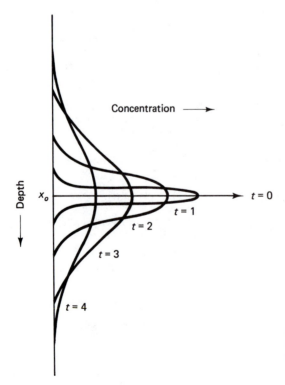

Figure 5.2 Diffusion away from a thin-film source creates a symmetrical profile that flattens and stretches out with time. The areas under the curves are constant.

Diagenesis: A Study in Kinetics   Chap. 5

tion 5-4 is the point at which it decays to $1/e$ ($\sim$0.36) times its value at x_0. In this example, that distance is equal to $x_p = 2\sqrt{Dt}$. For the numerical example above, $x_p = 10$ cm. This result is often generalized as a rule of thumb, called the Bose-Einstein relation, describing the *penetration distance* over which a diffusing species has traveled:

$$x_p = k\sqrt{Dt}. \tag{5-5}$$

The size of the proportionality constant, k, may vary from one problem to another, depending on the geometry of the system and how strictly we want to define the practical limits of the $c(x, t)$ curve.

Worked Problem 5-2

Suppose that the diffusing species is initially concentrated in one rather thick layer or a body of water, and then begins to travel across a boundary into a second layer. How does the concentration of diffusing ions vary in the second layer as a function of position and time?

Consider the situation illustrated in Figure 5.3a. Here, lake water containing a potentially mobile species at a concentration $c = c'$ overlies a sediment column which initially contains none. This is the configuration we might expect if, for example, acid mine drainage were suddenly diverted into a previously clean lake. The initial conditions are, therefore:

$$\text{for } x > 0 \text{ at } t = 0, c = 0$$
$$\text{for } x < 0 \text{ at } t = 0, c = c'.$$

In order to find a solution to equation 5-3 for this case, we recognize that the source for diffusing solute is no longer concentrated in a narrow plane, but is distributed throughout the volume of water above the sediment interface. We imagine, then, a column of lake water with unit cross-sectional area, divided into a set of n horizontal slices, each of thickness Δa. Each slice acts as a thin-film source for solute molecules that eventually reach a depth x_p in the sediment. The total concentration at x_p after a time t has passed can, therefore, be calculated by taking the sum of all the contributing thin-film solutions. From the previous Worked Problem, we know these to have the form

$$c(x, t) = \frac{c'\Delta a}{2\sqrt{\pi Dt}} \exp\left(-\frac{a_i^2}{4 Dt}\right),$$

where a_i is the distance from depth x to the center of the ith slice. Therefore, if each slice is infinitesimally thin and we treat the infinite sum of all slices as an integral,

$$c(x, t) = \frac{c'}{2\sqrt{\pi Dt}} \int_x^\infty \exp\left(-\frac{z^2}{4 Dt}\right) dz.$$

It is customary to rewrite this result in terms of the variable $\eta = z/(2\sqrt{Dt})$:

$$\begin{aligned}
c(x, t) &= \frac{c'}{\sqrt{\pi}} \int_{x/(2\sqrt{Dt})}^\infty \exp\left(-\eta^2\right) d\eta \\
&= \frac{c'}{\sqrt{\pi}}\left[1 - \int_0^{x/(2\sqrt{Dt})} \exp\left(-\eta^2\right) d\eta\right] \\
&= \frac{c'}{2}\left[1 - \operatorname{erf}\left(\frac{x}{2\sqrt{Dt}}\right)\right].
\end{aligned} \tag{5-6}[1]$$

Figure 5.3b illustrates the concentration profile generated by equation 5-6 for several times after the onset of diffusion. Notice that the profile is symmetrical around the sediment-water interface, where concentration at all times is equal to $c'/2$. Unless the water column is unusually stagnant, as in some swamps, we would not expect to see a mea-

---

[1] The error function, erf (x), is a standard mathematical function discussed briefly in Appendix A.

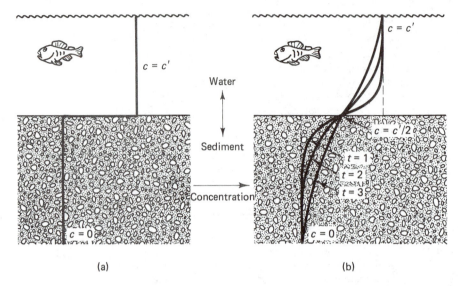

Figure 5.3 (a) At the beginning of Worked Problem 5-2, the concentration of the mobile species is equal to zero at all depths in the sediment, and is equal to c' throughout the overlying water. (b) Diffusion across the water sediment interface produces profiles described by equation 5-6. Notice that the concentration at the interface remains equal to $c'/2$ at all times after $t = 0$.

surable gradient like this above the interface. Instead, the lake water should be uniformly stirred by winds or thermal convection, with the result that the concentration for all values of $x \leq 0$ is c' at all times. Although we will not verify it mathematically, it can be shown that diffusion in the sediment column with this additional boundary condition follows the form of equation 5-6 very closely:

$$c(x, t) = c' \left[1 - \text{erf} \left(\frac{x}{2\sqrt{Dt}} \right) \right]. \tag{5-7}$$

Suppose, then, that we add enough Cu^{2+} to lake water to raise its concentration to 500 ppm. If the diffusion coefficient for Cu^{2+} in water is taken to be 5×10^{-6} cm^2 sec^{-1}, what should be the copper concentration in pore waters 10 cm below the water-sediment interface after a year? By applying equation 5-7, we get

$$c(10 \text{ cm}, 3.4 \times 10^7 \text{ sec}) = 500 \text{ ppm} \left[1 - \text{erf} \frac{10 \text{ cm}}{2\sqrt{(5 \times 10^{-6} \text{ cm}^2 \text{ sec}^{-1})(3.1 \times 10^7 \text{ sec})}} \right]$$

$$= 294 \text{ ppm}.$$

In the discussion so far, we have considered only the diffusion of dissolved material through water. This is appropriate, because diffusion through any solid medium is so much slower that it is negligible. Within the sediment column, however, this observation implies that the bulk diffusive flux depends on how much volume is *not* occupied by solid particles (that is, the porosity, ϕ). Diffusion may also be slowed significantly because of the geometrical arrangement of particles, since the effective path length for diffusion must increase as dissolved material detours around sediment particles. This second factor is known as the *tortuosity*, θ, defined by

$$\theta = dl/dx, \tag{5-8}$$

where dl is the actual path length a diffusing species must traverse in a given distance dx. Tortuosity cannot be measured directly, but is usually determined by comparing values of D determined in open water with those measured in pore water (Li and Gregory, 1974), or by comparing the electrical conductivities of sediments and pore waters (Manheim and Waterman, 1974). Taking both porosity and tortuosity into account, the diffusion coefficient in the sediment column, D_s, is related to the value of D in pure water by

$$D_s = \phi D/\theta^2. \tag{5-9}$$

Worked Problem 5-3

Measurements of Mg^{2+} concentration are made in the pore waters of a marine red clay, and it is found that Mg^{2+} decreases linearly from 1300 ppm at the sea floor to 1057 ppm at a depth of 1 m. Assuming that the profile is entirely due to diffusion, how rapidly is Mg^{2+} diffusing into the sediment?

The diffusion coefficient for Mg^{2+} in pure water close to 0°C is 3.56×10^{-6} cm^2 sec^{-1}, and the average porosity and tortuosity in marine red clay are 0.8 and 1.34, respectively (Li and Gregory, 1974). The effective diffusion coefficient, D_s, is therefore calculated from equation 5-9 to be

$$D_s = (0.8)(3.56 \times 10^{-6} \text{ cm}^2 \text{ sec}^{-1})/(1.34)^2$$
$$= 1.6 \times 10^{-6} \text{ cm}^2 \text{ sec}^{-1}.$$

From the fact that $\partial c/\partial x$ does not vary with depth, we conclude that the profile represents a steady-state flux of Mg^{2+} into the sediment. J at the interface is therefore easily calculated from Fick's first law. The concentration gradient is 243 ppm per meter, or 1.0×10^{-2} mol l^{-1} m^{-1}, or 0.1 μmol cm^{-4}. Therefore,

$$J = -D_s \frac{\partial c}{\partial x}$$
$$= -(1.6 \times 10^{-6} \text{ cm}^2 \text{ sec}^{-1})(0.1 \ \mu\text{mol cm}^{-4})$$
$$= -1.6 \times 10^{-7} \ \mu\text{mol cm}^{-2} \text{ sec}^{-1}$$
$$= -4.96 \ \mu\text{mol cm}^{-2} \text{ yr}^{-1}$$
$$= -0.12 \text{ mg cm}^{-2} \text{ yr}^{-1}.$$

In addition to porosity and tortuosity, other chemical factors will affect diffusion in a diagenetic environment. These arise because most material transported in pore fluids is in the form of charged particles. A flux of cations into a sediment column must be accompanied by an equal flux of anions, or must be balanced by a flux of cations of some other species in the opposite direction. Furthermore, we may expect that electrostatic attractions or repulsions between ions will affect their mobilities. The greater the charge on an ion, the more likely it is that it will be slowed by interaction with other ions. Figure 5.4 illustrates the magnitude of this effect for most simple aqueous ions by plotting the value of D against the charge/size ratio ($|z|/r$) for a variety of common ionic species. Lasaga (1979) has shown that ionic *cross-coupling* can change the value of D for some species by as much as a factor of 10.

Charged species not only interact with each other, but also interact with the sediment. Clay particles, in particular, develop a negatively charged surface when placed in water, as interlayer cations are lost into solution. Several important properties of clays arise from this behavior. Some cations from solution are adsorbed onto the clay surface to form a *fixed layer*, which may be held purely by electrostatic forces or may form complexes. Others form a *diffuse layer* at a greater distance from the clay-water interface, in which ions are only weakly bonded. This *double layer* of positively-charged ions repels similar structures around adjacent clay particles. In open water, these electrostatic effects support small particles in stable suspensions known as *colloids*, and prevent them from settling to the bottom. In the sediment column, they promote a very

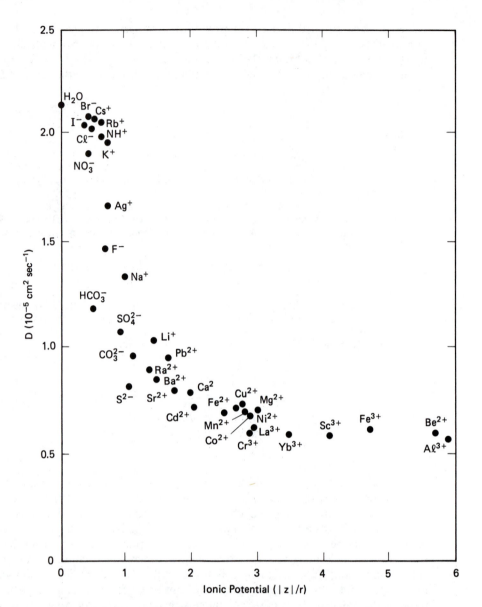

Figure 5.4 Tracer diffusion coefficients (D) for common ionic species as a function of ionic potential (Modified from Li and Gregory 1974, by K. Aplin, printed with permission.)

open structure, like a house of cards, and maintain a high porosity. In general, this tendency decreases with increasing ionic strength because the diffuse layer is overwhelmed by the abundance of free ions in solution. As a result, colloids carried stably in river water are commonly flocculated as they enter the ocean.

The way in which surface properties of clays affect ionic diffusion can be shown by considering the adsorption of a mobile ion A onto a clay particle as a chemical reaction:

$$A_{\text{free}} \longleftrightarrow A_{\text{bound}}.$$

Ignoring any effects of nonideality, the equilibrium constant for this reaction is given by

$$K = C/c,$$

where c is the concentration of A in solution and C is its concentration in bound form, in units of moles per liter of interstitial solution. Because the total amount of A is constant, the rate of change in c must be equal and opposite to the rate of change in C. For a rapid exchange between solution and solids, then, we can write

$$K(\partial c/\partial t) = -(\partial C/\partial t).$$

Equation 5-3, with this modification, becomes

$$\frac{\partial c}{\partial t} = D\frac{\partial^2 c}{\partial x^2} + \frac{\partial C}{\partial t}$$

$$= D\frac{\partial^2 c}{\partial x^2} - K\frac{\partial c}{\partial t},$$

or

$$\frac{\partial c}{\partial t} = \frac{D}{1 + K}\frac{\partial^2 c}{\partial x^2}. \tag{5-10}$$

The rate of diffusion, therefore, becomes much slower as the role of ion exchange or adsorption by clay surfaces increases. If we take chemical interactions with solids into account, the effective diffusion coefficient defined in equation 5-9 now becomes

$$D_s = \frac{\phi D}{\theta^2(1 + K)}. \tag{5-11}$$

Advection

Water can flow through pores in a sediment as the result of compaction or as the result of a regional hydraulic gradient. The first of these is most likely to dominate in a marine or lacustrine environment, and is therefore most commonly associated with early diagenesis. Regional flow, however, can be important where dissolved matter is transported and interacts with sediments in an aquifer, or where tectonic overpressure induces fluid flow in deep sediments.

As sediment accumulates, we can examine the progress of diagenesis in a given bed in either of two reference frames. In the simplest, we locate the bed by referring to its depth below the sediment-water interface. With continued deposition, the depth to the bed increases. Viewed in this reference frame, therefore, both the sediment and any interstitial fluid seem to move with time. This movement is only a *pseudo-advection*, however, because it results in no movement of sediment particles relative to each other. In the absence of compaction, there is also no movement of fluid relative to the bed. In a reference frame centered on the bed itself, this pseudo-advection disappears.

True fluid advection in marine or lacustrine sediments is most commonly driven by the gradual reduction in porosity that occurs as sediments are compacted with depth. If we choose a bed-centered reference frame, we may anticipate a gentle upward flow as water is squeezed out of sediments in deeper layers. This effect should be most prominent in clay-rich sediments, which initially have high porosities because of electrostatic repulsion between particles. Within a few tens of meters of the surface, these weak repulsive forces are overwhelmed by the weight of accumulating sediment, and porosity drops from 70 or 80% to less than 40%, gradually approaching an asymptotic limit within a few tens of meters, as is shown in the schematic profile in Figure 5.5.[2] Flow rates, however, are too low to measure, because clays typically accumulate and compact very slowly. In sandy sediments, particularly those with poor sorting, porosity is initially low and does not change much with depth. The angularity of most clastic particles makes reorientation difficult. Therefore, despite the fact that coarse sediments accumulate more rapidly than clays, fluid advection due to compaction in sands during marine or lacustrine diagenesis is even slower.

Because fluid flow is generally too slow to measure, we can only estimate its rate by constructing theoretical models of the compaction process. The simplest model we can devise (and the only one we will consider) assumes that the sedimentation rate is constant and that $(\partial \phi / \partial t) = 0$ at any depth in the sediment column. This is very nearly true within a few meters of the sediment-water interface.

Let us adopt the symbols ω and u to indicate the rates at which sediment and pore waters are buried, respectively. If there were no compaction, ω would be equal to u, and both would simply be equal to the rate of sediment accumulation. Under steady-state compaction, however, we find that as porosity decreases toward its asymptotic limit, ϕ_d, ω, and u each decrease as well, gradually approaching a limit where $\omega_d = u_d$. As long as the change in porosity with depth is fairly small (as it is below about ten meters in the sediment), we may conclude that

$$u\phi = \phi_d u_d = \phi_d \omega_d. \tag{5-12}$$

Assuming steady-state compaction, therefore, it is possible to calculate the rate of fluid burial at any depth if we have a porosity-depth profile and can estimate ω.

Clearly, however, at any depth less than the one where ϕ, u, and ω all approach their asymptotic limits, fluids are buried less rapidly than sediment. Expelled waters, therefore, flow upward at a rate (U) equal to

$$U = u - \omega.$$

If we multiply both sides of this expression by ϕ and combine it with equation 5-12, we see finally that it is possible to calculate the rate of upward flow from

$$U = \frac{\phi_d}{\phi} \omega_d - \omega. \tag{5-13}$$

Worked Problem 5-4

A team of researchers has measured porosity as a function of depth in a sediment column at the bottom of mythical Puzzle Pond. The data are plotted and it is found that they can be fit adequately by the mathematical function

$$\phi(x) = 0.5 \exp(-0.05x) + 0.2,$$

where x is depth in meters. How can we use this information to estimate the rate of compaction-driven fluid flow in the sediments? Assuming steady-state compaction, it can be shown (Berner, 1980) that

$$\frac{\partial[(1 - \phi)\omega]}{\partial x} = 0.$$

It follows, therefore, that

[2] This is an approximation. Porosity decreases very rapidly within a few tens of meters of the sediment surface, and then only very slowly for the next few thousand, as shown in Figure 5.5. For our purposes, it is reasonable to treat the porosity below a depth of about 30 to 50 meters as a constant.

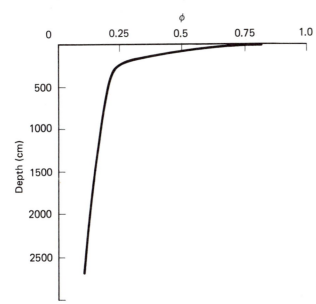

Figure 5.5 Porosity (ϕ) in clastic sediments decreases to a nearly asymptotic limit within 30 to 70 meters of the surface. This approximation makes it possible to estimate advective flow rates caused by compaction.

$$[1 - \phi(x)]\omega(x) = \text{constant}.$$

The measured value of ω at the sediment interface ($x = 0$) is equal to the sediment deposition rate. If we let $\omega(0)$ be designated as ω_0, then we can evaluate the constant by calculating $\phi(0)$ and inserting both numbers in this last expression to get

$$[1 - \phi(x)]\omega(x) = 0.3\omega_0.$$

The burial rate of the sediment in this column, therefore, is given by

$$\omega = \frac{3\omega_0}{[8 - 5\exp(-0.05x)]}.$$

In Puzzle Pond, the sedimentation rate is 0.2 cm yr^{-1}. Using these expressions, we verify that ϕ and ω approach nearly constant values ($\phi_d = 0.22$ and $\omega_d = 0.077$ cm yr^{-1}) about 60 meters down. These, then, can be inserted in equation 5-13 to yield

$$U = \frac{0.22}{\phi}(0.077 \text{ cm yr}^{-1}) - \omega.$$

The fluid advection rate at selected depths can be calculated by substituting appropriate values for ϕ and ω:

x (meters)	ϕ	ω	U (cm yr^{-1})
5	0.59	0.15	−0.117
10	0.50	0.12	−0.087
15	0.44	0.11	−0.068
20	0.38	0.097	−0.053
30	0.31	0.087	−0.033
40	0.27	0.082	−0.019
50	0.24	0.079	−0.009
60	0.22	0.077	−0.002

Notice that the advection rates are all negative, indicating upward flow.

Kinetic Factors in Diagenesis

Because these rates of fluid advection depend so strongly on grain size, sorting, and sedimentation rate, it is hard to devise reliable models that are more detailed than we have described above. Given a semi-empirical model like this, however, it is easy to compare rates of diffusion and advection in a sediment, and see which is the major mechanism for transport during diagenesis.

Worked Problem 5-5

How efficient are diffusion and advection at transporting Mg^{2+} interstitially in a typical marine mud? We have already seen (Worked Problem 5-3) that the effective diffusion coefficient (D_s) for Mg^{2+} under these circumstances is about 1.6×10^{-6} cm^2 sec^{-1}. As we found in Worked Problem 5-4, the fluid advection rate will vary with depth. Let's assume the same parameter values we used in that problem, so that we can apply those calculated flow rates here. To express D_s and U in compatible units, we will divide each calculated value of U by the conversion factor, 3.1×10^7 sec yr^{-1}. Because D_s has been adjusted for porosity, it will also be necessary to multiply each value of U by ϕ.

Diffusion and advection will be equally effective over a critical distance L such that

$$D_s/LU_s \simeq 1,$$

where $U_s = U\phi$ in cm sec^{-1}. The quantity on the left-hand side of this expression is known as the *Peclet number*. If its value is much less than 1, then advection dominates; if $D_s/LU_s \gg 1$, then diffusion is the more effective transport process. L, therefore, defines the boundary between diffusion-dominated and flow-dominated regimes. We can calculate the value of L for which the Peclet number is unity at any sediment depth. In this problem, we find:

x (meters)	L (meters)
5	7.2
10	11.3
15	16.8
20	24.3
30	48.6
40	99.4
50	234
60	1058

From these results, we conclude that because the actual distance over which Mg^{2+} is to be transported is always significantly less than the critical value of L, diffusion is the major transport process during diagenesis of marine muds. This result would not be significantly different if we had chosen to follow a different ion (see Figure 5.4).

To illustrate how much this inference depends on the value of U, consider the transport of Mg^{2+} in a sandy aquifer, where the rate of flow is imposed by some regional pressure gradient. To calculate such a rate, we might use a form of Darcy's law,

$$U_s = k(l/h),$$

where k is a measure of permeability and (l/h) is the local hydraulic gradient. It would not be unusual to find flow rates in excess of 100 cm yr^{-1}. Substituting in the expression above, we find that the Peclet number is equal to 1 for a distance of 0.5 cm. Not surprisingly, then, we verify that flow, not diffusion, is the major transport mechanism in an aquifer.

Kinetics of Chemical Reactions

In our discussion so far, we have considered changes in the sediment column from a mechanical point of view, in which diffusing or advecting ions are carried as inert particles. Variations in fluid composition during diagenesis, however, are not due to diffusion or advection alone. In addition, dissolved material can be added to solution (or lost from it) in a variety of chemical reactions. Reactions between primary minerals in the sediment may produce or consume dissolved species, as can the growth of authigenic minerals. In previous chapters, for example, we have considered dissolution-precipitation reactions. Redox reactions, as well, may affect the concentration of transition metal ions, like Mn^{2+} or Fe^{2+}, in solution. Biological processes associated with decay or respiration can act in a number of ways to change the abundance of organic compounds in the sediment column, and to modify the composition and redox state of pore waters.

As in our discussions of transport processes, it is important to recognize that chemical reactions at low temperature are commonly arrested in midflight. That is, the phase assemblages and compositions we observe in a sediment column frequently represent some intermediate step along the way toward equilibrium, rather than equilibrium itself. Because this is so, thermodynamics is often inadequate to interpret the chemical variability observed in sediments. On the other hand, it may be possible to apply principles of chemical kinetics to minerals and fluids in the sediment, and thus to deduce the *reaction pathways* that dominate during diagenesis, as well as the physical or transport pathways.

Lasaga (1981) makes a useful distinction between *elementary* chemical reactions, which occur *as written* at the molecular level, and *overall* reactions, which describe a net change that involves several intermediate steps and takes place along several competing, parallel pathways. We have also encountered this distinction briefly in Chapter 1, where thermodynamic and kinetic approaches to geochemistry were contrasted. Let us consider a few short examples to see how this distinction affects the way we treat reactions during diagenesis.

The chemical reactions that occur during diagenesis include some processes like carbonate precipitation for which we can write an elementary reaction:

$$Ca^{2+} + CO_3^{2-} \longrightarrow CaCO_3.$$

The rate at which molecules of $CaCO_3$ are produced is directly proportional to the probability that a Ca^{2+} ion will collide with a CO_3^{2-} ion in solution. The appropriate rate law in this case, therefore, is

$$\frac{dn_{CaCO_3}}{dt} = kn_{Ca^{2+}}n_{CO_3^{2-}},$$

in which $n_{Ca^{2+}}$, $n_{CO_3^{2-}}$, and n_{CaCO_3} are the concentrations of each species, and k is a linear rate constant. In fact, for *any* elementary reaction, the rate is simply proportional to the abundance of reactant species. This statistical concept is the one we applied in Worked Problem 3-11.

In contrast, the rates at which overall reactions occur are commonly nonlinear functions of concentration. They may be influenced by processes which occur at the surface of a growing or dissolving particle, or by the production of transient, metastable species, or by transport of dissolved material toward or away from the interface. As a result, rate laws may be quite complex, reflecting a variety of inhibiting or enhancing processes. Keir (1980), for example, has determined by experiment that the rate of calcite dissolution (R) in seawater follows the nonlinear relationship

$$R = k\left(1 - \frac{a_{Ca^{2+}}a_{CO_3^{2-}}}{K_{cal}}\right)^{4.5},$$

in which k is an empirical constant that varies widely among samples and is most

Kinetic Factors in Diagenesis

strongly a function of grain size (Figure 5.6). The large exponential factor has been hypothesized to arise from the absorption of phosphate ions on crystal surfaces, where they serve as dissolution inhibitors. Similarly, Cody and Hull (1980) have demonstrated in the laboratory that some organic acids are preferentially adsorbed on nuclei of gypsum, inhibiting further growth and leading instead to low-temperature formation of anhydrite from saturated calcium sulfate solutions. In both of these cases, the rate law cannot be predicted without detailed knowledge of the reaction mechanism or, as is more often the case, an exhaustive set of empirical observations.

For many overall reactions that affect diagenesis, we have little theoretical knowledge and few relevant observations. The formation of sulfide in sediments, for example, is associated with bacterial reduction of sulfate in pore waters. We can write a reasonable reaction to express this process,

$$7CH_2O + 4SO_4^{2-} \longrightarrow 2S_2^{2-} + 7CO_2 + 4OH^- + 5H_2O,$$

but it is valuable only as a guide to system mass balance. The species labelled "CH_2O" is a generalized compound with the average stoichiometry of organic matter in sediments. The rate of sulfide formation during diagenesis will depend on a large number of factors that are not indicated in the schematic form of the overall reaction. These may include the identity of specific organic molecules in the fresh sediment and any which may form after burial, surface reactions on growing sulfides, and absorption by clay surfaces, among others. Our understanding of reaction kinetics in such cases is primitive.

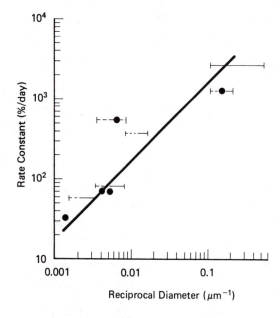

Figure 5.6 The rate of dissolution of calcite (expressed as the rate constant) increases rapidly with decreasing grain size because of the increased surface area in fine-grained materials. (Modified from Keir 1980)

Summary: The Diagenetic Equation

With this brief survey, we have introduced each of the important processes that affect the mass balance in a sediment during diagenesis. Our final task is to combine them, in what has been called a General Diagenetic Equation (Berner, 1980). The rate at which concentration of some interesting chemical species in a particular volume of sediment changes can be expressed by

$$\frac{\partial C}{\partial t} = -\frac{\partial J}{\partial x} + \sum R. \tag{5-14}$$

Here, the first term on the right side describes the total flux of the species through the volume per unit time, and the second term includes the rates (R) of all relevant chemical reactions proceeding in the volume. The distance variable, x, is taken relative to the sediment-water interface. From discussions in previous sections, we know that J must include both the diffusive and the advective flux:

$$J = -D\frac{\partial C}{\partial x} + UC. \tag{5-15}$$

Equation 5-14, then, becomes

$$\frac{\partial C}{\partial t} = \frac{\partial\left(D\frac{\partial C}{\partial x}\right)}{\partial x} - \frac{\partial(UC)}{\partial x} + \sum R. \tag{5-16}$$

This general equation implicitly includes each of the factors we have discussed in previous sections, as well as many we have not, such as the effect of bioturbation.

CEMENTATION

The primary porosity in a sediment may decrease as the result of compaction, as we have seen, or as the result of precipitation of authigenic mineral matter. In this section, we will present examples of two common problems related to cementation during early chemical diagenesis. The first of these concerns the remobilization of transition metal ions under conditions of variable oxidation state below the sediment-water interface. Such processes are commonly diffusion-controlled, and are solved by focusing attention on the first term in equation 5-16. The second examines the development of a concretion during fossilization, where the balance among competing compositional gradients may determine the final appearance of the concretion.

Growth of an Oxidized Surface Layer

Bottom waters throughout most of the ocean are oxygenated, as we discussed in Chapter 4. Within the sediment column below the ocean floor, however, oxygen is rapidly consumed by bacteria that feed on organic debris. As mineral dissolution reactions take place in the sediment, then, iron and manganese remain as mobile Fe^{2+} and Mn^{2+} rather than becoming oxidized and relatively immobile. These ions migrate, primarily by diffusion, toward the sediment surface, where they become oxidized to Fe^{3+} and Mn^{3+} or Mn^{4+} and precipitate to form a stable layer of Fe-Mn-oxyhydroxides. Manganese nodules are formed on the ocean floor by a process closely related to this. A good introductory treatment of the broader problem can be found in Boudreau and Scott (1978).

Manganese remobilization has been modelled in detail by Burdige and Gieskes (1983). Figure 5.7a, derived from their work, illustrates the Mn^{2+} variations in pore water with depth in a sediment core from the east equatorial Atlantic Ocean. The solid curve is their solution to equation 5-16 for the boundary conditions imposed by the geometry of this system, using measured values of porosity, sedimentation rate, and other key parameters in the sample area. The analytical form of their solution, although very

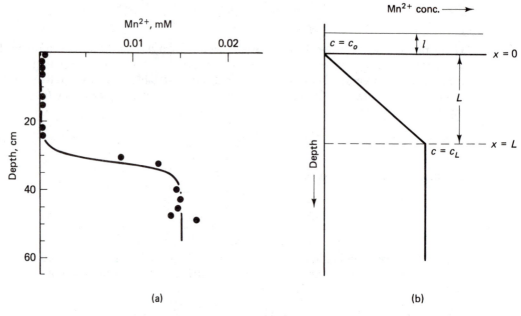

Figure 5.7 (a) Plot of the Mn content of pore water as a function of depth in a sediment core from the eastern equatorial Atlantic Ocean. Solid curve is the model profile calculated by D. J. Burdige and J. M. Gieskes. See their article, A pore water/solid phase diagenetic model for manganese in marine sediments, *Amer. J. Sci.* 283: 29–47. Copyright 1983 by the American Journal of Science. Reproduced with permission. (b) Concentration profile in the simplified redox model for marine Mn redistribution discussed by Berner (1980).

successful in modelling the data in Figure 5-7a, is unfortunately too complex to consider here. Instead, we will briefly consider a simpler model for an almost identical system, proposed by Berner (1980).

Worked Problem 5-6

How does the distribution of manganese in pelagic sediments change with time? Berner's model, illustrated in Figure 5-7b, assumes that the zone of mineral precipitation grows upward from the sediment-water interface ($x = 0$), and consists only of oxyhydroxide minerals and fluid-filled pore spaces. This is only possible if *both* the sedimentation rate and bioturbation are insignificant. Berner further assumes that there are no porosity gradients, and that Mn^{2+} is not adsorbed on clay surfaces, so that advective flow and reaction terms in the general diagenetic equation are negligible. Finally, because the concentration gradient in the transition zone between the precipitation layer and the top of the source region ($x = L$) is nearly linear, he makes the approximation

$$\frac{\partial C}{\partial x} = \frac{(C_L - C_0)}{L}, \qquad (5-17)$$

where C_L and C_0 are the concentrations of Mn^{2+} at $x = L$ and $x = 0$.

The concentration profile in this model consists, then, of an uninterestingly constant segment below depth L, and a linear segment above it described by equation 5-17, the slope of which changes with time. It bears a crude resemblance to the measured curve in Figure 5.7a, offset vertically so that the zone of mineral deposition lies entirely above the sediment. Rather than using this model simply to calculate the concentration profile, however, Berner shows how it is possible to estimate how long it takes to grow a monomineralic surface layer of any given thickness.

The flux of dissolved Mn at $x = 0$ at any given time can be calculated from Fick's first law, adjusted for sediment porosity and tortuosity:

$$J_0 = -D_s \frac{\partial C}{\partial x} = -D_s \frac{(C_L - C_0)}{L}. \tag{5-18}$$

The thickness of the transition zone (that is, the value of L) varies as a function of time that is directly proportional to J_0:

$$\frac{dL}{dt} = \frac{V_d D_s (C_L - C_0)}{F_d L}. \tag{5-19}$$

Here, V_d = the volume of dissolving mineral matter which contains one mole of Mn^{2+}, F_d = the volume fraction of dissolving matter in the bulk sediment.

In a similar way, the thickness of the oxyhydroxide layer (l) grows according to

$$\frac{dl}{dt} = \frac{V_p D_s (C_L - C_0)}{F_p L}, \tag{5-20}$$

in which V_p and F_p are similar to V_d and F_d, but refer to the mineral matter in the precipitation zone.

The thickness of the transition zone can be calculated by integrating equation 5-19 for the boundary conditions $L = 0$ when $t = 0$ and $L = L$ when $t = t$:

$$L = \left[\frac{2V_d D_s}{F_d} (C_L - C_0) t \right]^{1/2} \tag{5-21}$$

If we then insert this result in equation 5-20 and integrate for the boundary conditions $l = 0$ when $t = 0$ and $l = l$ when $t = t$, we have derived an expression for t:

$$t = \frac{l^2 F_p^2 V_d}{2 V_p^2 F_d D_s (C_L - C_0)}. \tag{5-22}$$

We can now determine how long it takes to form an oxidized surface layer of an arbitrary thickness (1 cm) by placing reasonable values for each of the variable parameters in equations 5-21 and 5-22:

$$D_s = 2.4 \times 10^{-6} \text{ cm}^2 \text{ sec}^{-1}$$
$$V_p = V_d = 20 \text{ cm}^3 \text{ mol}^{-1}$$
$$F_p = 0.2$$
$$C_0 = 0.0 \text{ M}$$

For several pairs of F_d and C_L values, we calculate t and L:

F_d	C_L (mol cm^{-3})	L (cm)	t (yr)
0.001	1×10^{-7}	200	133,500
0.002	1×10^{-7}	100	66,770
0.005	1×10^{-7}	40	26,700
0.010	1×10^{-7}	20	13,350
0.010	1×10^{-6}	20	1,335

The geometry of this model is similar in many ways to the more complex example presented by Burdige and Gieskes, and the numerical results may be generally comparable as well. Notice, however, that the predicted time scale for growth of the authigenic layer is long enough to stress some of the initial assumptions of the model. In nature, it is unlikely that a shallow sediment column would remain undisturbed by bioturbation and chemical perturbations for 10^3 to 10^5 years.

Fossilization and the Growth of Concretions

Fossil collectors know that specimens from many localities are found in the cores of concretions. The local concentration of authigenic minerals, evidently related in some geochemical fashion to the preserved organism, serves as a mechanical and chemical armor for it. There can be considerable variety among concretions, however, even from a single collecting area. Some, for example, will contain pyrite or other sulfides, others, only carbonates. In some, fossil remains are well-preserved; in others, there may be only the faintest trace of organic debris, or none at all.

Variations in outward appearance—the size and shape of concretions—can generally be ascribed to local differences in transport properties. We have just examined relationships among time, sediment and fluid composition, and diffusion coefficients that govern the growth rate of an authigenic layer. These same relationships can be recast in spherical coordinates and applied to the diffusion-controlled growth of concretions, with comparable results. Interested readers may wish to see Berner's (1980) treatment of this problem. Also, advection may be important in concretion growth, particularly in sandy aquifers or soils. In those circumstances, the mathematics are altered by the addition or substitution of flow terms for diffusive transport terms, but are otherwise similar. Quantitative discussions of concretion shape are more difficult, because they involve transport properties in more than one dimension. In nonisotropic media, variations in diffusivity or hydraulic conductivity lead to growth rates that depend on direction.

It is difficult to make quantitative models that adequately describe the variations in chemistry from one concretion to another. These commonly depend on interactions among several competing concentration gradients, about which we generally do not have sufficient information for detailed modelling. To illustrate how you may view the problem qualitatively, however, consider the following example.

Worked Problem 5-7

Figure 5.8 is a photograph of a concretion collected from a black shale in the Desmoinesian Series (Pennsylvanian) in Pike County, Indiana, and described by Woodland and Richardson (1975). The prominent feature at its center is a tree branch, now preserved as calcite (96 wt. %) and residual organic matter (4 wt. %), but retaining its cellular structure well enough to be identified at the species level. The matrix of the concretion consists of fine-grained calcite and finely disseminated pyrite, cementing clay- to silt-sized sediments. Proportions of these three constituents vary, but are approximately 83 : 12 : 4, except in a diffuse but distinctive pyrite-rich halo about 0.2 to 2.5 cm inward from the surface of the concretion and in a second, narrow pyritic halo adjacent to the wood. Calcite also occupies large "veins," apparently shrinkage fractures, within and immediately around the wood.

How was this mineralogical structure formed, and what variations might we expect to find in other concretions from the area? As organic matter decays by bacterial and inorganic processes, both CO_2 and NH_3 are released, causing an increase in the alkalinity of pore waters in the vicinity. The overall result is an increase in the carbonate ion concentration, through reactions that can be summed up schematically by

$$2NH_3 + CO_2 + H_2O \longleftrightarrow 2NH_4^+ + CO_3^{2-}.$$

Under anoxic marine conditions, the electron acceptor of choice among bacteria is the sulfate ion, a major constituent in seawater. Oxidation of organic matter by bacterial decay, therefore, also involves a set of reactions which can be generalized as

$$2CH_2O + SO_4^{2-} \longleftrightarrow H_2S + 2HCO_3^-,$$

where CH_2O is a reasonable stoichiometric shorthand for the composition of average or-

Figure 5.8 Concretion around fossilized wood from the Blackfoot #5 mine, Pike County, Indiana. The long axis of the concretion is 37 cm long. (Photo by B. Woodland, Field Museum of Natural History, Chicago, printed with permission.)

ganic debris. These reactions also increase carbonate alkalinity and the concentration of CO_3^{2-}. As a consequence, pore waters gradually reach saturation with respect to calcite. Precipitation of carbonate minerals will continue as long as there is a source of alkalinity.

Anoxia and the availability of NH_4^+ and organic chelating agents contribute to the mobility of reduced metal ions, primarily Fe^{2+}. As H_2S is released from the sulfate reduction process, much of the dissolved iron is reprecipitated as pyrite. The progress of pyrite formation is not uniform, however. It depends on the intensity of H_2S and Fe^{2+} gradients around the decaying organic matter. Where concentrations of both constituents build up sufficiently, their product exceeds the K_{sp} for pyrite. The precipitation process, though, interferes with further diffusion of either H_2S or soluble iron and limits the local extent of pyrite formation. If both constituents continue to be produced, the product of their concentrations may eventually exceed the pyrite K_{sp} at another distance from the nucleus of the concretion. In the case we are describing here, this has led to the formation of two zones of pyrite enrichment. These are analogous to the rhythmic bands of iron oxide observed in many sandstones and attributed to *Liesegang phenomena* (Stern, 1954).

These many processes will continue until interrupted by the exhaustion of one or more reactants, or by mechanisms which block their free movement. If a mass of organic matter is small and the sediment is fairly porous, the decay process is likely to proceed to completion, converting all organic compounds to CO_2. For a larger mass, or if the initial porosity is low, cementation will quickly isolate the organic matter by eliminating advective or diffusive pathways. Decay will then cease, leaving a fraction of the organic matter at the core of the concretion. If the sediment contains very little soluble iron, or if the pore waters have a low sulfate content, pyrite may not precipitate, or may be disseminated rather than forming a discrete band. Finally, dissolved species may be consumed (or, per-

Cementation

haps, produced in excess) by neighboring bacterial communities, working to decompose other pieces of buried organic matter.

THE FATE OF ORGANIC MATTER DURING DIAGENESIS

As sediments accumulate, they commonly include organic as well as inorganic matter. The composition of this organic matter varies greatly. Part of this is because of differences among the populations of organisms from which it is derived. Clearly, organic matter at the bottom of a pond, derived from leaves, grasses, and fresh-water organisms, should not be identical to organic matter in pelagic or shelf sediments. Differences also arise because of reactions that take place prior to burial. In Chapter 4, we reported that an extremely high proportion of the biomass in surface waters is consumed during descent to the ocean floor. Suess and Muller (1979) find empirically that the organic carbon flux at any water depth can be predicted from

$$J_{\text{Org-C}} = 5.9 \, pd^{-0.616},$$

where p is the rate of primary productivity in surface waters and d is the water depth in meters. This trend is confirmed in Table 5-1, in which we also see that the proportions of carbon, nitrogen, and phosphorus change during descent. Nitrogen- and phosphorous-bearing compounds are preferentially destroyed in the water column. In shallow water, this fractionation is negligible.

TABLE 5-1 BIOGENOUS ELEMENT FLUXES IN THE OCEAN*

	Fluxes (g m^{-2} yr^{-1})			Ratios				
	C	N	P	C	:	N	:	P
Open Ocean								
Productivity	50	8.8	1.2	106		16		1
Photic Zone	25	2.1	0.16	106		7.9		0.25
Water Column	0.80	0.030	0.0022	106		3.5		0.11
Accumulation	0.0031	0.00069	0.000069	106		20		0.91
California Margin								
Productivity	250	44	6.1	106		16		1
Photic Zone	158	21	2.15	106		12		0.56
Water Column	42	4.6	0.58	106		10		0.56
Accumulation	2.2	0.23	0.032	106		10		0.60

*Modified from Suess and Muller (1980).

A CRASH COURSE IN ORGANIC NOMENCLATURE

Organic chemistry is a vast subject, largely unexplored by geochemists, who are more at home in the world of minerals. From outside the discipline, perhaps the most unfathomable aspect of organic chemistry is its system of naming compounds and illustrating them on paper. The complexity arises because most compounds are made of only three elements: carbon, hydrogen, and oxygen. Stoichiometry and bulk chemistry, therefore, must play a secondary role to structure if a chemist wants to make sense of the immense number of possible compounds.

Organic compounds consist of a reasonably small number of component parts which cannot exist as independent species, but which are recognizable as structural units from one molecule to the next. These are divided, for convenience, into two large classifications of groups: *saturated hydrocarbon* groups and *functional*, or *reactive*, groups. Saturated groups consist only of carbon and hydrogen atoms, linked by single bonds, which may be assembled to form chain structures (called *aliphatic* compounds, *alkanes,* or *paraffins*) or simple cyclical structures (called *alicyclic* compounds, *cycloalkanes,* or *napthenes*). An example of a saturated hydrocarbon group, shown in Figure 5.9, is the *butyl* group, which may form the aliphatic molecule *butane* with the addition of a single hydrogen atom. The *cyclobutyl* group, in similar fashion, is the basic unit of the alicyclic compound, *cyclobutane*. The tendency toward easy polymerization leads to a great many linear and branched structures based on saturated groups.

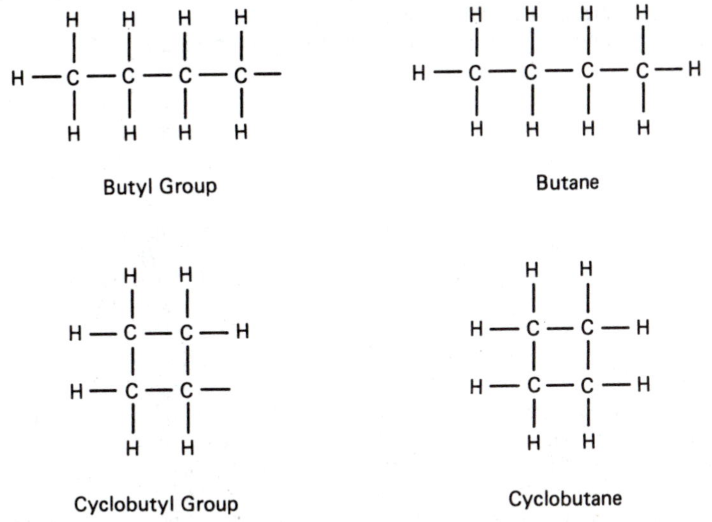

Butyl Group

Butane

Cyclobutyl Group

Cyclobutane

Figure 5.9 Saturated hydrocarbon groups and related aliphatic (butane) or alicyclic (cyclobutane) molecules.

Three styles of representation of saturated hydrocarbon groups are in use, as illustrated by the formula for isobutane shown in Figure 5.10. Detailed formulas indicate separately all of the carbon and hydrogen atoms, and show the bonds between them as straight lines, in the fashion of ball-and-stick models, which you may have used in mineralogy classes. Because this format takes valuable space, condensed formulas commonly omit horizontal lines between adjacent carbon atoms (although vertical ones are retained), and indicate where hydrogen atoms are attached by writing them immediately to the right of carbons to which they are bonded. A yet more economical style uses only straight line segments, representing single carbon-carbon bonds. It is generally understood that a carbon atom sits at each end of any line segment, and is surrounded by as many hydrogen atoms as necessary (2 or 3) to satisfy its net negative charge.

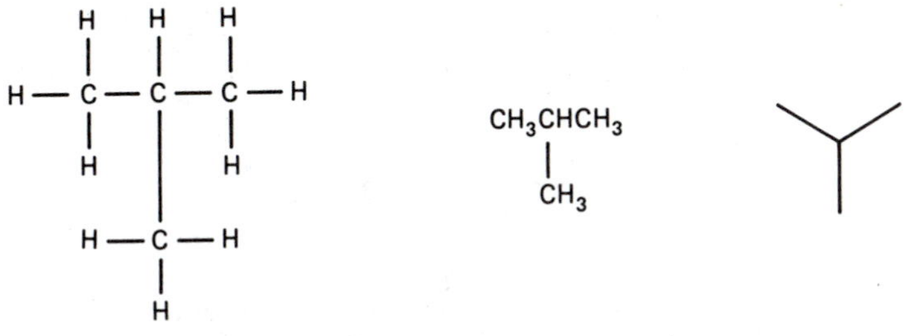

CH_3CHCH_3

CH_3

Figure 5.10 The formula for isobutane, an aliphatic hydrocarbon, can be written in detailed form, in a compressed form, or in a skeletal form. The function of all three styles is to convey information about bonding and molecular structure, although none attempts to indicate the three-dimensional configuration of atoms.

The Fate of Organic Matter During Diagenesis

The simplest functional groups consist of double or triple bonds between carbon atoms. These may join with one another or with saturated groups to form *unsaturated* aliphatic or alicyclic molecules, or may form a third class of compounds known as *aromatic* hydrocarbons (Figure 5.11). In these, the fundamental unit is a ring of six carbon atoms sharing valence electrons. These are commonly represented by *Kekulé* formulas, in which alternating single and double bonds describe a hexagonal unit. Structural studies of aromatic rings indicate that they are symmetrical, however, so that each of the six bonds has an equal probability of being either a single or a double bond. Therefore, it is incorrect to think of the double bond in this structure as a distinct functional group. For that reason, aromatic rings are sometimes drawn as a circle inside a hexagon, as indicated in Figure 5.11.

Figure 5.11 Aromatic structures are based on six-membered rings of carbon atoms. Kekulé formulas (left and center) give the erroneous impression that single and double bonds alternate around a ring. Because all bonds in an aromatic ring are statistically equivalent, many organic chemists now prefer representations like the one on the right.

Some functional groups involve combinations of hydrogen atoms with one or more atoms of another element, most commonly oxygen, nitrogen, or sulfur. Some of these, like hydroxyl (—OH), are familiar from inorganic chemistry. *Alcohols* and *phenols,* both common in nature, are based on combinations of —OH with aliphatic or aromatic saturated groups, respectively. *Sugars* and their polymerized cousins, *starches,* are all aliphatic hydrocarbons bearing a large number of hydroxyl groups in place of hydrogen atoms. Collectively, these compounds constitute the major class known as *carbohydrates*. The sulfur-bearing functional group analogous to hydroxyl (—SH) appears in *thioalcohols* (or *mercaptans*) and *thiophenols*. These also occur in proteins and other complex molecules, where they are responsi-

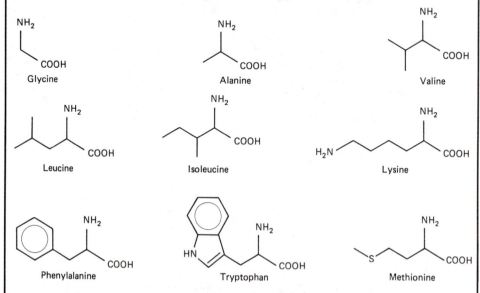

Figure 5.12 All amino acids contain both an amino (NH_2) group and an acid (COOH) group. These 9 amino acids, plus 11 others, are found in living organisms, where they constitute the building blocks of proteins.

ble for pungent odors and flavors, as in garlic and onion. The functional groups

$$-NH_2, -NH, \text{ and } -N-$$

are derivatives of ammonia, and serve as the basic units in aliphatic or aromatic *amines*. These are among the most important compounds for living organisms, because they at the heart of many metabolic processes and are common constituents of *amino acids* and *proteins*, discussed elsewhere in this section. Twenty amino acids occur in living organisms; some of these are shown in Figure 5.12.

Still other functional groups contain carbon atoms linked to some other atom by double or triple bonds. Most of these we will not mention here. The carbon-oxygen double bond by itself, however, is an important functional unit known as the *carbonyl* group. It may occur as a simple functional group or in combination with other groups to form major classes of organic compounds. The *carboxyl* group, for example, consists of a combination of a carbonyl and a hydroxyl unit,

$$\begin{matrix} O \\ \parallel \\ -C-OH. \end{matrix}$$

The hydrogen atom is easily removed by hydrolysis, yielding an acid. Many *carboxylic acids* play important roles in animal and plant metabolism. Acetic acid is a good example, used by living organisms to synthesize long-chain fatty acids and steroid hormones.

Interested readers are urged to expand on this brief foray into organic nomenclature by consulting any introductory organic chemistry text.

Chemically, the material that settles toward the bottom consists of *biopolymers*, most of which fall into four large classes: *carbohydrates, proteins, lignin,* and *lipids*. Plants are composed predominantly of carbohydrates, which are polymers of simple sugars like glucose and fructose. Starch and cellulose are common long-chain carbohydrates, in some cases consisting of thousands of simple sugar units. Higher plants also contain lignin, which is a more rigid substance with a high proportion of its carbon atoms in aromatic rings. This polyphenol provides strength to support the weight of trunks and branches. Figure 5.13 shows generalized structures of carbohydrates and lignin.

Animals do not synthesize lignin. They do contain carbohydrates, some of which, like chitin, provide skeletal strength; these are unknown in plants. The dominant class of molecules in animal tissue, however, is protein. Proteins are polymers of amino acids and contain most of the nitrogen bound in tissues. In some organisms, like corals, proteins serve as skeletal material. They also combine with other organic compounds to form substances like nucleic acids, enzymes, and antibodies. Proteins comprise less than 10% of the mass of most plants.

Lipids, the last major group of biogenous compounds, consist of straight or branched long-chain hydrocarbons (paraffins) formed by the combination of fatty acids and alcohols (Figure 5.14). Molecules with odd numbers of carbon atoms are much more common in living organisms than those with even numbers. Organisms on land tend to produce paraffins containing 27, 29, or 31 carbon atoms (referred to in shorthand as C_{27}, C_{29}, and C_{31} paraffins). These are used as insulation in animals, and fill an important role in regulating the transport of proteins and other substances in body fluids. In plants, they are concentrated in seeds and spores. Lipids of higher weight are waxy, and commonly appear as a protective coating on leaves or spores. Marine organisms, by contrast, synthesize predominantly C_{15}, C_{17}, and C_{19} paraffins. These are generally liquids.

The Fate of Organic Matter During Diagenesis

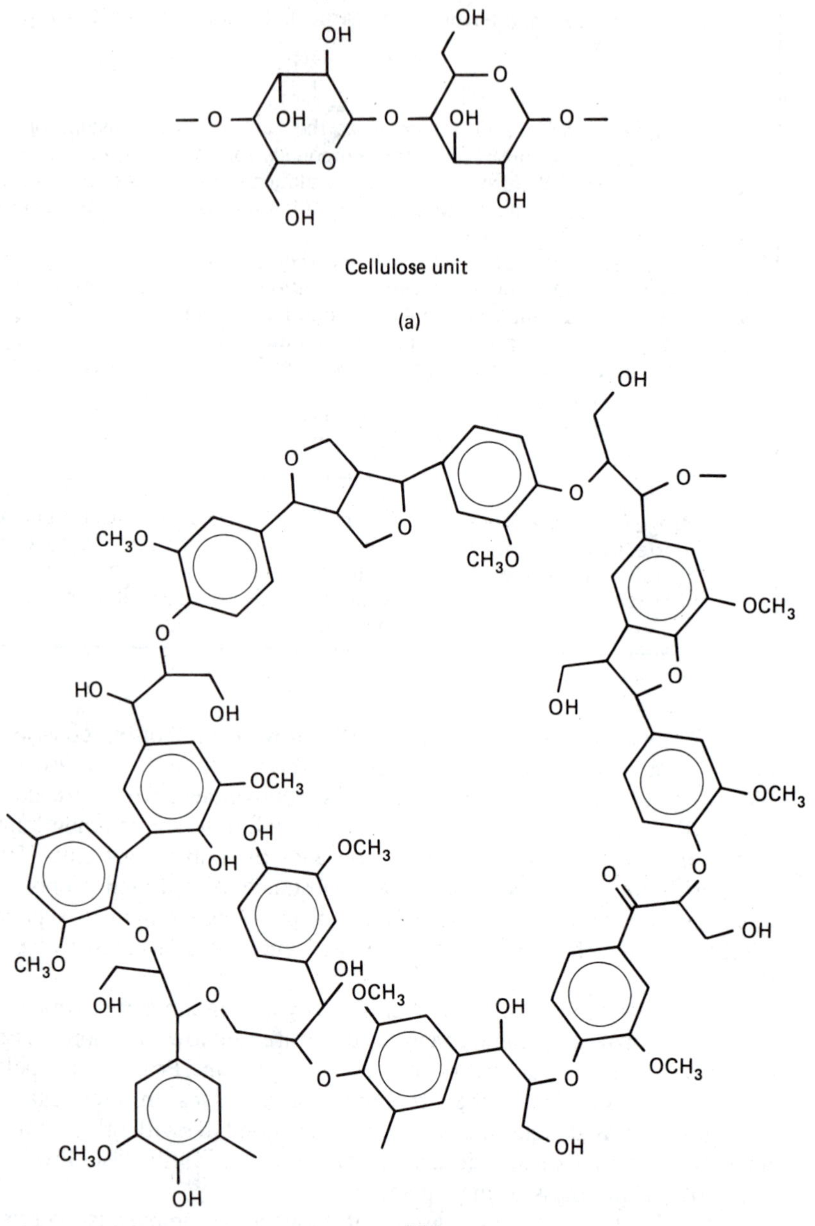

Cellulose unit

(a)

(b)

Figure 5.13 (a) Cellulose, a complex carbohydrate, is composed of a large number of units like this one. (b) Lignin consists of a complicated network of aromatic and alicyclic ring structures.

Among the other natural organic compounds that constitute a small but important part of the inventory are *essential oils* like camphor and menthol, which yield the fragrances of many flowers and leaves. *Resins* are composed of complex alicyclic compounds, which polymerize easily when exposed to air and are highly resistant to decay. They are common in trees and may persist through diagenesis and be found in coal. *Pigments* like chlorophyll and various carotenoids serve both to protect tissues and to

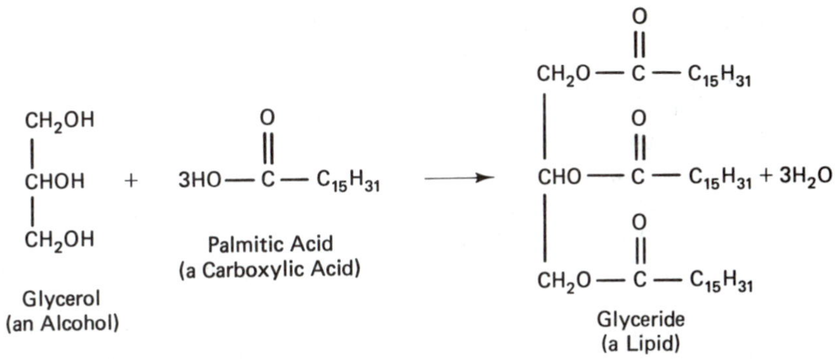

Glycerol (an Alcohol)

CH₂OH / CHOH / CH₂OH → shown as chemical structures

$$CH_2OH$$
$$|$$
$$CHOH \quad + \quad 3HO-\overset{\overset{\displaystyle O}{\|}}{C}-C_{15}H_{31} \quad \longrightarrow$$
$$|$$
$$CH_2OH$$

Glycerol (an Alcohol)

Palmitic Acid (a Carboxylic Acid)

$$CH_2O-\overset{\overset{\displaystyle O}{\|}}{C}-C_{15}H_{31}$$
$$|$$
$$CHO-\overset{\overset{\displaystyle O}{\|}}{C}-C_{15}H_{31} \quad + \quad 3H_2O$$
$$|$$
$$CH_2O-\overset{\overset{\displaystyle O}{\|}}{C}-C_{15}H_{31}$$

Glyceride (a Lipid)

Figure 5.14 Lipids consist of alcohol units bonded to fatty acids. The example shown here is a branched C_{31} paraffin.

act as vital parts of the photosynthetic process. Many essential oils, resins, and pigments are characteristic of specialized groups of organisms and, if they survive diagenesis, can be used as tracers, or *biomarkers*.

Most buried organic matter, however, does not survive diagenesis. Burrowing animals feed on some of it, converting it temporarily into a benthic biomass. In the process, they trap some dissolved nutrients. In marine sediments, this accounts for some of the increase shown on the rows labelled "Accumulation" in Table 5-1. The increase in nitrogen is also due to adsorption of some dissolved amino acids and humic[3] complexes on clay mineral surfaces. The greatest portion of it, however, is eventually degraded by micro-organisms, which use enzymes to convert it to simple *biomonomers*. These, then, can be assimilated as food. Among them are the amino acids and fatty acids that once formed the building blocks of proteins and fats; sugars from the degradation of carbohydrates; and other basic compounds such as phenols. Much of this material is resynthesized into the cell walls of the bacteria or converted to gases and lost from the sediment, but some of it is reassembled to form complex compounds of high molecular weight, known as *geopolymers*.

Many different reaction mechanisms are involved in the formation of geopolymers. One is the Maillard-Amadori reaction, sometimes called the "browning" reaction. To observe it, you do not necessarily have to examine a sediment column, but can watch the surface of a recently cut apple. In this scheme, sugars and amino acids undergo rearrangements and dehydration to form melanoidins, which are yellow-brown aliphatic humic substances. Structurally, these are more complex than the sugars and other biomonomers. To a geochemist, they are among the first indications that diagenesis has begun.

Some bacteria also decompose the lignin found in terrestrial plants by oxidation or demethylation, although they cannot assimilate the products of this reaction. The oxidation products include syringic and vanillic acids, commonly derived from angiosperm lignin, and parahydrobenzoic acid, which may be derived from both angiosperms and gymnosperms. To a degree, at least, the type of humic substance preserved in sediments is an indication of the living population from which it was derived. In general, humic substances formed by oxidation of lignin all have more con-

[3] The class of geopolymers known as humic substances, or humus, is very large. It is divided into three fractions: *humin*, which is alkali-soluble, *humic acids*, which are alkali-soluble but acid-insoluble, and *fulvic acids*, which are soluble in both acids and alkalis. As a group, humic substances are the precursors of *kerogen*, an extremely high molecular weight geopolymer that, at higher temperatures, may be decomposed to form the geomonomers common in both coal and petroleum.

The Fate of Organic Matter During Diagenesis

densed, aromatic structures than melanoidins, because lignin itself is more aromatic. Humic substances have been used, therefore, as a way to distinguish between organic material of terrestrial affinity and that derived from marine sources.

The reaction paths along which organic compounds are changed depend on environmental conditions. Because this is so, some reaction products can be used as indicators of temperature, oxidation state, or compositional variables during diagenesis. In a sense, this procedure is analogous to the way petrologists use mineral assemblages to infer a grade of metamorphism.

Consider, as an example, the reactions that commonly affect the pigment chlorophyll once it is buried. As shown in Figure 5.15, the chlorophyll molecule consists of a complex cycloalkane structure (a porphyrin) called chlorin, attached to a relatively simple aliphatic structure (known as an *isoprenoid*). The bond between these two is easily broken by hydrolysis, yielding a modified porphyrin and an isoprenoid hydrocarbon known as phytol.[4] Under reducing conditions, phytol is readily hydrogenated to form dihydrophytol, which in turn undergoes dehydration and further hydrogenation to produce phytane and release a methane molecule. Under oxidizing conditions, phytol is subjected first to dehydrogenation and oxidation to produce phytanic acid and then to decarbonation to form pristane. The structures of all compounds in these parallel reaction paths are shown in Figure 5.15. Both phytane (Ph) and pristane (Pr) are relatively stable and can be detected in many sediments. By determining their relative abundances, an organic geochemist can therefore estimate the oxidation state during diagenesis. If $Pr/Ph < 1.0$, then the environment was reducing; if $Pr/Ph > 1.0$, then it was more oxidizing.

As diagenesis proceeds, several major changes can be discerned in organic matter. Unlike the compounds in freshly buried biopolymers, for example, a large fraction of the geopolymers is insoluble in nonoxidizing acids (like HCl), alkalis, or common organic solvents like benzene. This fraction increases with temperature and time. Baedecker et al. (1977) sampled organic matter in fresh near-shore sediments off the coast of California, and found that 44% of it was insoluble. Upon heating to 65°C for 30 days, however, the insoluble fraction increased to almost 90%. In common parlance, this material is referred to as *kerogen;* the soluble fraction is known as *bitumen.*

Because a significant amount of rearrangement takes place, the predominance of odd-numbered chain lengths in paraffins, which we remarked earlier is usually observed in living organisms, gradually disappears during diagenesis. Two different quantities have been used to report the extent of this change. One is the *Carbon Preference Index* (CPI), which is the ratio of odd- to even-chain length paraffins in a selected range of molecular weights. The other is a more complex measure, the *Odd-Even Predominance* (OEP), defined as

$$OEP = \left[\frac{C_i + 6C_{i+2} + C_{i+4}}{4C_{i+1} + 4C_{i+3}} \right]^{(-1)^{i+1}},$$

in which C_i is the relative weight percentage of a paraffin containing i carbon atoms per molecule. Neither index usually approaches 1.0 until temperatures well above the limits of diagenesis, where thermal degradation of kerogen generates a new suite of paraffins of random chain length. There are some indications from samples of shallow pelagic sediments, however, that *CPI* and *OEP* decrease even at temperatures as low as 50°C, given sufficient time.

---

[4] The porphyrin unit is an excellent chelating agent. In living organisms, a Mg^{2+} ion is usually bonded to the nitrogens in the center of the ring. In the sediment column, however, Mg^{2+} is often exchanged for Ni^{2+} or other metal ions. This is one of several ways in which organic compounds may affect the solubility and transport of mineral matter during diagenesis.

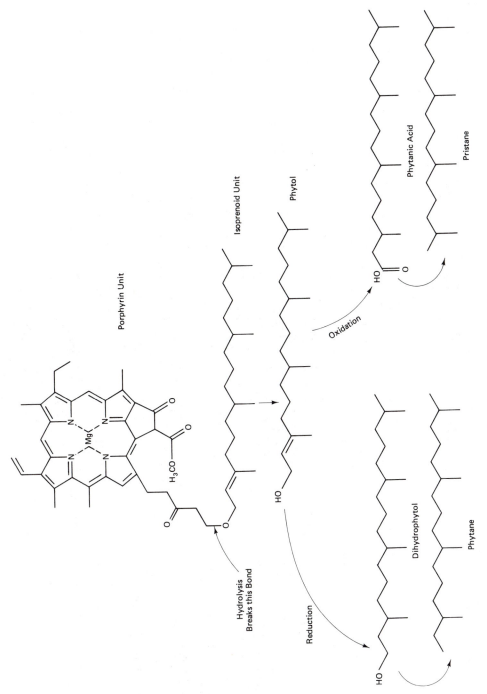

Figure 5.15 The porphyrin unit and the phytol unit in chlorophyll are easily separated by hydrolysis during diagenesis. Depending on oxidation state of the surrounding sediments, the phytol unit may then react to form phytane or pristane. Chelated Mg^{2+} in the porphyrin structure is commonly exchanged for Ni^{2+} or vanadyl ions.

The extent of diagenesis can also be estimated by examining structural variations in organic molecules. The tetrahedral arrangement of bonds around a carbon atom, for example, often makes it theoretically possible for several symmetrically distinct configurations, or *stereoisomers,* of a given molecule to exist. Because living organisms prefer to produce only one of the possible forms, however, recent sediments accumulate a specialized collection. Reactions during diagenesis tend to reorganize this collection and to generate the missing stereoisomers. The progress of these reactions, as gauged by the relative abundance of primary and secondary stereoisomers, can be used as a guide to diagenesis. In other molecules, alicyclic structural units may be dehydrogenated and converted to aromatic units, again yielding an assemblage of primary and diagenetic compounds that can be used to infer a time-temperature history. An extremely large number of molecules have been examined with this purpose in mind, with varying degrees of success. Most thermal history models based on organic reactions are semiquantitative at best, but many have opened a window on a range of diagenetic conditions that could never be studied adequately with inorganic reactions alone.

As a simple example, the amino acid isoleucine is illustrated in Figure 5.16 in its two possible forms. L-isoleucine and D-alloisoleucine are nonsuperimposable mirror images of each other. The prefixes "L-" and "D-" stand for "Levo-" and "Dextro-," which mean "left-handed" and "right-handed." These two forms are identical from a thermodynamic point of view: they have the same free energies of formation, the same melting points, and so forth. The metabolic pathways in living organisms, however, synthesize only L-isoleucine, so it is the only stereoisomer in freshly buried organic matter. This situation gradually changes during diagenesis. As functional groups are detached and reassembled, they do so in a random fashion, no longer constrained by the metabolism of any organism. What began as a sample of L-isoleucine, therefore, is transformed into a mixture of L- and D-molecules. Such a mixture is called *racemic.* Experiments have been performed to determine the rate of racemization of amino acids found in many organic materials, such as shells and bone (Wehmiller, 1984; Kimber, 1986). In general, the kinetics can be described by a first order rate law, but the rate constants depend strongly on both temperature and the presence of water, as well as the degree to which amino acids are "free" rather than bound in proteins or peptide fragments. Despite intense interest among geochemists, the results of these studies, therefore, can only be applied in a qualitative sense to natural materials.

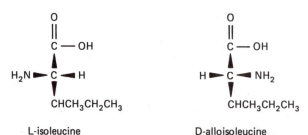

L-isoleucine   D-alloisoleucine

Figure 5.16 The D– and L– stereoisomers of isoleucine are asymmetrical molecules, and therefore will rotate a polarized light beam. This property, called *optical activity,* can be used to distinguish "right-handed" from "left-handed" amino acids. A racemic mixture of D– and L– molecules is not optically active. These formulas, unlike others in this section, attempt to show the three-dimensional nature of part of the amino acid molecule. In each, the carboxyl (—COOH) and amine (—NH$_2$) functional groups, the saturated hydrocarbon group (—CHCH$_3$CH$_2$CH$_3$), and a hydrogen atom occupy the four corners of a tetrahedron centered on a carbon atom. The bond to each group is fattened on one end to indicate that the atom on that end is closest to the viewer.

Impressive progress has been made in the development of kinetic models that exploit knowledge of several isomerization or aromatization reactions simultaneously. Mackenzie and McKenzie (1983) and Mackenzie et al. (1984) have studied the thermal history of sediments by evaluating the progress of three elementary reactions shown in Figure 5.17. First-order rate constants and expressions of temperature dependence for these reactions have been derived from geochemical analyses of samples from well-characterized sedimentary basins. These, then, can be included in diagenetic equations like equation 5-16 and applied to the geologic history of the area. The result is an empirical model that can be used to study the rate and extent of diagenesis in other similar basins.

Other models have used variations in physical properties of organic matter as a measure of the level of organic maturation, and attempted to correlate these with time-temperature histories. A commonly used property is the reflectance of vitrinite, the dense material that is visible as bright bands in coal and is also disseminated as part of the kerogen in sedimentary rocks. One model based on reflectivity (Bustin et al., 1977) is illustrated in graphical form in Figure 5.18. The maturation level, as indicated by reflectance, is an exponential function of temperature, but is approximately linear with time. Another successful model (Lopatin, 1971) makes use of this observation to define a Time Temperature Index (TTI):

$$TTI = \sum_{n(\text{min})}^{n(\text{max})} (\Delta t_n)(r^n), \tag{5-23}$$

in which

> $r = 2$ (assuming that maturation level doubles for each $10°C$ temperature rise)
> $\Delta t_n =$ the time (in 10^6 years) spent by sediments in time interval n
> $n(\text{max}) =$ the highest $10°C$ temperature interval experienced by the sediment
> $n(\text{min}) =$ the lowest $10°C$ interval.

The value of n is set to 1 for the interval from $100°$ to $110°C$, and is incremented (or decreased) by one unit for each $10°C$ change. Higher values of TTI, calculated by equation 5-23, indicate higher levels of diagenesis. The TTI scale can be correlated with vitrinite reflectance by

$$R_0 = -0.10528 \log TTI + 0.20647 (\log TTI)^2 + 0.5011 \tag{5-24}$$

(Bustin et al., 1985), provided that TTI is greater than 3.0.

Worked Problem 5-8

Suppose that organic matter is buried in basinal sediments, and that it takes 40 m.y. to reach a depth of 2000 m. Following this, the area is uplifted and eroded, so that the sediment reaches the surface in another 2.0 m.y.. If we assume a geothermal gradient of $30°C$ km^{-1} and a surface temperature of $20°C$, and assume further that the rates of burial and uplift-erosion were linear, what should be the TTI and vitrinite reflectance for the organic matter?

The burial rate can easily be converted to a heating rate by

$$\frac{\Delta T}{\Delta t} = 30°C \text{ km}^{-1} \frac{2.0 \text{ km}}{40 \text{ m.y.}} = 1.5°C \text{ m.y.}^{-1}.$$

In a similar way, the rate of uplift and erosion is equivalent to a cooling rate of

$$\frac{\Delta T}{\Delta t} = 30°C \text{ km}^{-1} \frac{2.0 \text{ km}}{2.0 \text{ m.y.}} = 30°C \text{ m.y.}^{-1}.$$

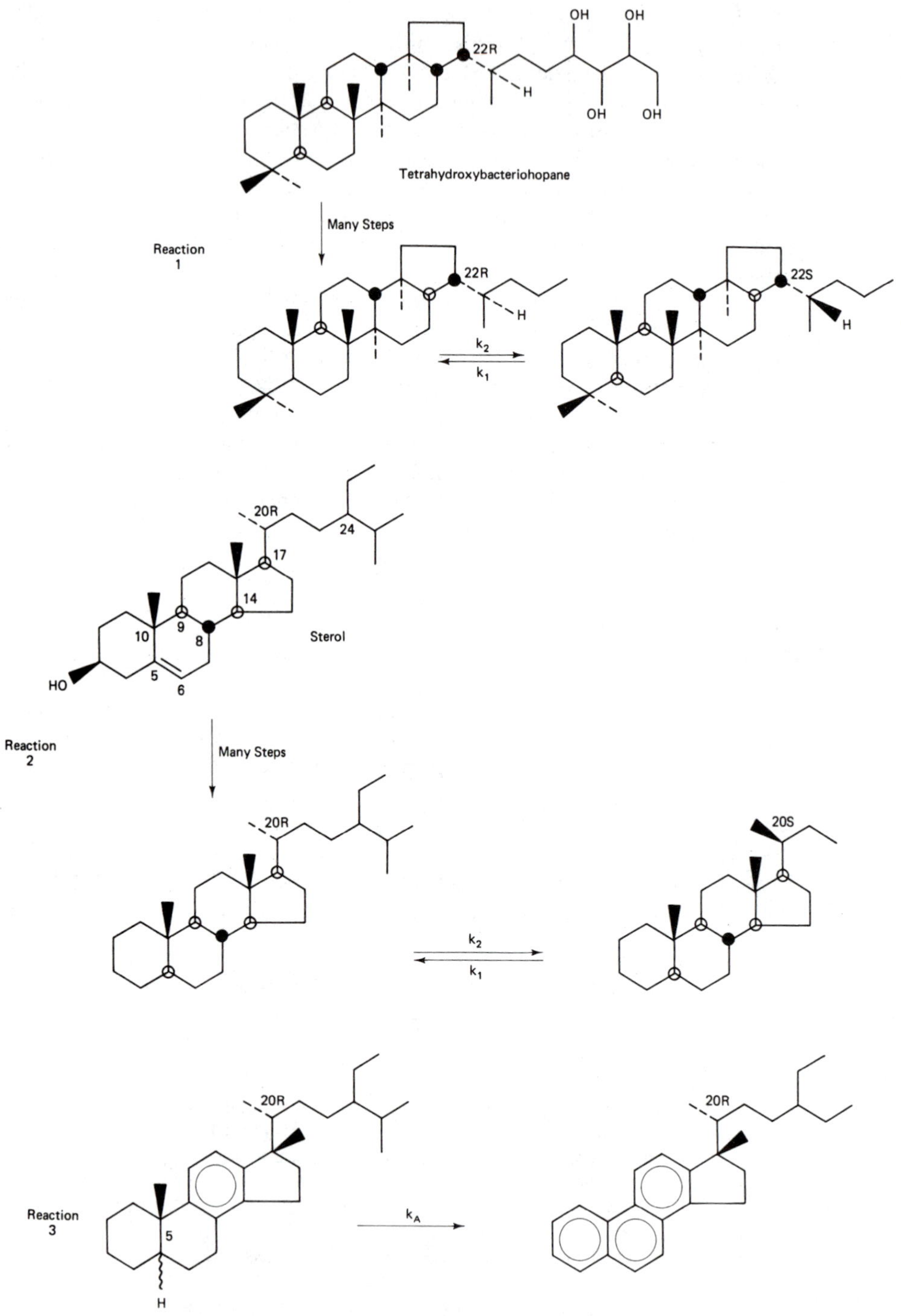

Figure 5.17 These three isomerization and aromatization reactions have been used to model time-temperature paths during sedimentary basin evolution. They involve subtle changes in bonding at specific sites in organic molecules commonly found in petroleum. The rates of these reactions have been studied in the laboratory and can be used to estimate rates in nature. (Modified from Mackenzie et al. 1984.)

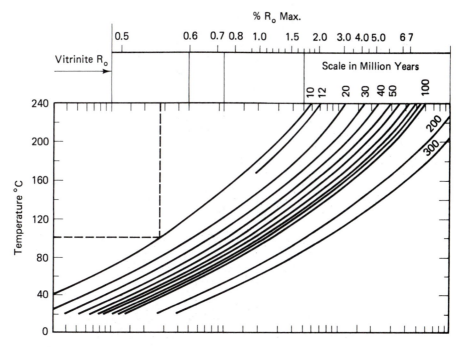

Figure 5.18 Vitrinite reflectance is a function of both temperature and time. This diagram can be used to estimate possible combinations of temperature and duration of diagenesis. The dashed lines indicate, for example, that a coal could have attained a measured reflectance of 0.55 after 10 m.y. at 100°C. (After R. M. Bustin, M. A. Barnes, and W. C. Barnes, Diagenesis 10.: Quantification and modelling of organic diagenesis. *Geoscience Canada* 12: 4–21. Copyright 1977 by the National Research Council of Canada. Reproduced with permission.)

Immediately after burial, the organic matter is at 20°C. It remains in the temperature interval between 20° and 30°C ($n = -7$) for 6.7 m.y., then spends 6.7 m.y. in each of the intervals from 30° to 40° ($n = -6$), 40° to 50° ($n = -5$), 50° to 60° ($n = -4$), 60° to 70° ($n = -3$), and 70° to 80° ($n = -2$). After 40 m.y., the sediment reaches a maximum of 80°C. During uplift, the sediment spends an additional 0.33 m.y. in each of the six intervals back to 20°C. We calculate *TTI* from equation 5-23, using the total times (burial + uplift) spent in each interval:

$$TTI = 7.0(2^{-7}) + 7.0(2^{-6}) + 7.0(2^{-5}) + 7.0(2^{-4}) + 7.0(2^{-3}) + 7.0(2^{-2})$$
$$= 3.45.$$

Vitrinite reflectance can be calculated from equation 5-24:

$$R_0 = -0.10528 \log (3.45) + 0.20647 (\log (3.45))^2 + 0.5011 = 0.50\%$$

It has also proven useful to examine the evolutionary path of kerogens during diagenesis by studying changes in major element ratios. A common tool for this purpose is the Van Krevelen diagram, which compares the atomic ratios H/C and O/C for a sample. These ratios have characteristic values for each of the classes of organic compounds we have discussed in this chapter. Aliphatic hydrocarbons, for example, contain mostly single carbon-carbon bonds, so the net charge on their chain structures is balanced by a large number of hydrogen atoms. Their H/C value, therefore, is high, reaching a maximum for methane (CH_4). Alicyclic compounds have a slightly lower ratio. Compounds with a highly aromatic nature, like the phenols that comprise lignin, have values of

The Fate of Organic Matter During Diagenesis **161**

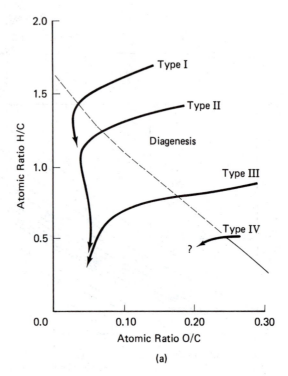

(a)

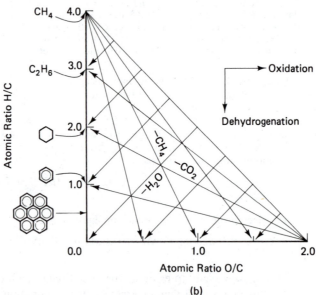

(b)

Figure 5.19 (a) Three major types of fresh organic matter can be characterized in sediments on the basis of H/C and O/C ratios. During diagenesis, these evolve along tracks indicated by the arrows on this Van Krevelen diagram. (b) H/C and O/C values can be changed by several processes, as indicated by the arrows on this diagram. (After A. Gize, Analytical approaches to organic matter in ore deposits. In *Organics and Ore Deposits,* ed. W. E. Dean. Copyright 1986 by the Denver Region Exploration Geologists Society. Reproduced with permission.)

H/C less than 1.0. In a similar fashion, hydroxyl-rich compounds have relatively high O/C ratios. These include carbohydrates, for example.

On the basis of these broad generalizations, three major types of fresh organic matter, shown in Figure 5.19a, are commonly recognized in sediments. Type I consists of normal and branched paraffins, with some napthenes and aromatics, characteristic of marine, algal-derived material often associated with oil shales and carbonates. Type II comprises largely napthenes and aromatic molecules associated with spore walls, cuticles, and the herbaceous cellular structure of plants. Type III is composed primarily of polycyclic aromatic hydrocarbons, plus some waxy paraffins and carbohydrates typical of continental material, which originally contained a high proportion of lignin and cellulose. Some studies have also defined a Type IV material that has been severely weathered or oxidized prior to burial and therefore has a very low H/C value and relatively high O/C.

With increasing degrees of diagenesis, changes take place in all three types of organic matter. For Type I material, the transition largely means a loss of hydrogen. This may take place by the gradual restructuring of saturated paraffins to form aromatic compounds, a process which is called *dehydrogenation*. As indicated by the arrows emanating from the upper left corner in Figure 5.19b, it may also involve the net loss of methane (*demethanation*). For Types II and III material, which from the beginning is less hydrogen-rich, oxygen loss is more characteristic. This can take place by *dehydration*, so that H/C and O/C decrease in a fixed proportion, or may involve the evolution of CO_2 (*decarbonation*), which will cause a decrease in O/C but an increase in H/C. Common evolutionary tracks, along which freshly buried organic matter is converted to kerogen during diagenesis, are shown in Figure 5.19a. At much higher grades, all kerogen compositions converge on the low-H, low-O corner of the Van Krevelen diagram as the geopolymers produced during diagenesis break down to form the geomonomers found in coal and petroleum.

The Van Krevelen diagram has an obvious value in the petroleum industry, where it can indicate both the provenance of a kerogen and its degree of maturation. The most dramatic changes in H/C and O/C take place during diagenesis, however, so a Van Krevelen diagram can be even more useful in the hands of a low-temperature geochemist.

SUMMARY

We have spent much of our effort in this chapter discussing the kinetics of transport and reaction. This is appropriate because so many of the changes we associate with diagenesis are incomplete, limited by the rates at which reactants move from one place to another or by how slowly they combine. We have introduced and contrasted diffusion and advection as transport mechanisms, and have experimented with mathematical methods for evaluating both. These will be used in later chapters as well, particularly in Chapter 10. By choosing to look more closely at oxidized layers in marine sediments and at concretions around fossil organisms, we have tried to show how these tools can be applied, in the form of a general diagenetic equation, to problems of practical interest in sedimentary geochemistry.

Our other emphasis has been on the role of organic matter in the diagenetic environment. In this brief survey, we have examined the diversity of natural organic compounds in sediments, and have seen how many of the changes that take place in organic matter can be used to trace the progress of diagenesis. The pathways from biopolymers to geopolymers are complicated but, if studied carefully, can yield valuable information about temperature, time, depth of burial, oxidation state, and other variables.

SUGGESTED READINGS

There have been few books written on the geochemical side of diagenesis, although there is a great deal of interest among geologists. Most of the informative writing is disseminated in journal articles. For general background, you may wish to consult the following sources.

BERNER, R. A. 1980. *Early diagenesis: A theoretical approach.* Princeton, N.J.: Princeton Univ. Press. (This is the best single reference in the field. Much of what is written today about the chemistry of diagenesis, including what we have said in this chapter, was first considered by Berner. The first five chapters establish fundamental principles we have introduced here and are particularly useful.)

GAUTIER, D. L., ed. 1986. *Roles of organic matter in sediment diagenesis.* Tulsa, Okla: Special Pub. 38, Society of Econ. Paleontologists and Mineralogists. (This symposium volume contains a number of papers that can introduce the practicing geochemist to the world of organic matter.)

SCHOLLE, P. A., and SCHLUGER, P. R., eds. 1979. *Aspects of diagenesis.* Tulsa, Okla: Special Pub. 26, Society of Econ. Paleontologists and Mineralogists. (This volume contains widely quoted papers from two symposia on clastic diagenesis.)

WAPLES, D. 1981. *Organic geochemistry for exploration geologists.* Minneapolis, Minn.: Burgess Publishing Co.. (A straightforward text that can easily be understood by those without specialized training in organic chemistry.)

Readers who are interested in organic geochemistry should also consult issues of the journal *Organic Geochemistry,* paying particular attention to annual volumes entitled *Advances in Organic Geochemistry.*

The following sources were referenced in this chapter. The reader may wish to examine them for further details.

BAEDECKER, M. J.; IKAN, R. I.; ISHIWATARI, R.; and KAPLAN, I. R. 1977. Thermal alteration experiments on organic matter in recent marine sediments as a model for petroleum genesis. In *Chemistry of marine sediments,* ed. T. F. Yen, 55–72. Ann Arbor, Mich: Science Publishers.

BOUDREAU, B. P., and SCOTT, M. R. 1978. A model for the diffusion-controlled growth of deep-sea manganese nodules. *Amer. J. Sci.* 283: 903–29.

BURDIGE, D. J., and GIESKES, J. M. 1983. A pore water/solid phase diagenetic model for manganese in marine sediments. *Amer. J. Sci.* 283: 29–47.

BUSTIN, R. M.; HILLS, L. V.; and GUNTHER, P. R. 1987. Implication of coalification levels, Eureka Sound Formation, northeastern Arctic Canada. *Can. J. Earth Sci.* 14: 1588–97.

BUSTIN, R. M.; BARNES, M. A.; and BARNES, W. C.

1985. Diagenesis 10.: Quantification and modelling of organic diagenesis. *Geoscience Canada* 12: 4–21.

CODY, R. D., and HULL, A. 1980. Experimental growth of primary anhydrite at low temperatures and water salinities. *Geology* 8, 505–09.

CRANK, J. 1975. *The mathematics of diffusion.* London: Oxford University Press.

KEIR, R. S. 1980. The dissolution of biogenic calcium carbonate in seawater. *Geochim. Cosmochim. Acta* 44: 241–54.

KIMBER, R. W. L. 1986. The use of amino acid racemization in establishing time frameworks and correlations for geological and archaeological materials. In *2nd Australian Archaeometry Conf. Proc.,* Canberra, Australia.

LASAGA, A. C. (1979) The treatment of multi-component diffusion and ion pairs in diagenetic fluxes. *Amer. J. Sci.* 279: 324–46.

LI, Y-H., and GREGORY, S. 1975. Diffusion of ions in sea water and in deep-sea sediments. *Geochim. Cosmochim. Acta* 38: 703–14.

LOPATIN, N. V. 1971. Temperature and geologic time as factors in coalification. *Akad. Nauk. SSSR, Ser. Geol. Izvestiya* 3: 95–108. (In Russian)

MACKENZIE, A. S.; BEAUMONT, C.; and McKENZIE, D. P. 1984. Estimation of the kinetics of geochemical reactions with geophysical models of sedimentary basins and applications. *Org. Geochem.* 6: 875–84.

MACKENZIE, A. S., and McKENZIE, D. P. 1983. Isomerization and aromatization of hydrocarbons in sedimentary basins formed by extension. *Geol. Mag.* 120: 417–70.

MANHEIM, F. T., and WATERMAN, L. S. 1974. Diffusimitry (diffusion coefficient estimation) on sediment cores by resistivity probe. *Initial Reports of the Deep Sea Drilling Project* 22: 663–70.

STERN, K. H. 1954. The Liesegang phenomenon. *Chem. Rev.* 54: 79–99.

SUESS, E., and MULLER, P. J. 1979. Productivity, sedimentation rate and sedimentary organic matter in the oceans, II.—Elemental fractionation. In *Biogeochimie de la Matière Organique a l'Interface Eau-Sediment Marin,* ed. R. Dumas, 293: 17–26. Paris: Colloques Internat. du C.N.R.S..

WEHMILLER, J. F. 1984. Relative and absolute dating of Quaternary mollusks with amino acid racemization: Evaluation, applications and questions. In *Quaternary dating methods,* ed. W. C. Mahaney, 171–93. New York: Elsevier.

WOODLAND, B. G., and RICHARDSON, C. K. 1975. Time factors of differentially preserved wood in two calcitic concretions in Pennsylvanian black shale from Indiana. *Fieldiana Geology* 33: 179–92.

Diagenesis: A Study in Kinetics   Chap. 5

PROBLEMS

1. During a catastrophic flood, a barrel of liquid nuclear waste containing radium is carried away by a major stream and dropped on the flood plain, where it bursts and is spread in a thin layer. Assume that overbank deposits cover the material rapidly, and that no process affects the distribution of radium other than diffusion. Using the diffusion coefficient for Ra^{2+} from Figure 5.4, calculate the effective vertical distance over which radium would be distributed in 6 months.

2. A zinc smelter is required by local statute to dispose of slurry from its pickling operation, containing 50 ppm Cd^{2+}, into a "safe" environment. The management wants to pump it into a clay pit abandoned by a neighboring brick factory. Use equation 5-6 or the computer program SEMIINF on your program diskette to calculate the one-dimensional distribution of cadmium in the clay after the pit has been used for three years. Use a value of D from Figure 5.4 and assume that porosity is 0.1 and tortuosity is 1.25. Assume that no process other than diffusion takes place.

3. It is determined that porosity in sediments at the bottom of a lake follows the relationship

$$\phi = 0.8 \exp(-0.07\,x) + 0.08.$$

Assume that sediment accumulates at the rate of 0.15 cm yr^{-1} and that compaction is in steady state. Calculate and graph the porosity, rate of sediment burial (ω), and vertical fluid advection rate as functions of depth.

Chapter 6

Chemical Weathering: Dissolution and Redox Processes

OVERVIEW

The geochemical processes that most profoundly affect the surface of continents–the environment with which we have the closest personal familiarity–can be classed together as *chemical weathering* processes. These include a variety of reactions involving water and either acids or oxygen in the decomposition of rocks. In the two previous chapters, we have referred to these implicitly as the source of sediments and ionic species for oceanic and diagenetic processes. In this chapter, we examine them explicitly. Our focus is on the weathering of silicates, which constitute the bulk of continental rocks.

We begin the chapter by considering simple reactions that dominate the low-temperature SiO_2-MgO-Al_2O_3-(Na_2O, K_2O)-water system, to show how these control the stability of key minerals like quartz, feldspars, and clays on the earth's surface. We then look more closely at the roles of organic compounds and CO_2 in the weathering process, and finish by examining oxidation-reduction reactions. Throughout the chapter, we emphasize the graphical representation of equilibria, using activity-activity diagrams.

The results of this theoretical treatment can be applied to two closely related geochemical processes: the formation of residual minerals and the chemical evolution of surficial and ground waters. We introduce both of these by example in this chapter and end the chapter by showing how our knowledge can also be applied to problems of economic importance, thus completing the survey of low-temperature pathways that we began in Chapter 3.

Silica Solubility

Equilibria between water and silica are important in two contexts. First, a great many sedimentary rocks and most igneous and metamorphic rocks contain quartz. Chemical weathering of these involves the removal of SiO_2 from a discrete phase. In a broader sense, because virtually all common crustal minerals contain SiO_2, our appreciation of the simple silica-water system is important to understanding more general reactions with silicates.

For any silica mineral, the primary reaction with water is

$$SiO_2 + 2H_2O \longleftrightarrow H_4SiO_4. \tag{6-1}$$

If the silica phase is quartz, the equilibrium constant K_q at 25°C is 1×10^{-4}. If it is opaline or *amorphous* quartz, the equilibrium constant K_{qa} at 25°C is 2×10^{-3}. Assuming that both water and the silica phase have unit activity, K_q (or K_{qa}) is also numerically equal to the activity of H_4SiO_4 that enters the solution as a result of reaction 6-1.

Like H_2CO_3, H_4SiO_4 is a weak acid. At *pH* values above 9, it dissociates first to form $H_3SiO_4^-$ and then $H_2SiO_4^{2-}$. The appropriate reactions are

$$H_4SiO_4 \longleftrightarrow H^+ + H_3SiO_4^-, \tag{6-2}$$

for which

$$K_1 = \frac{a_{H^+} a_{H_3SiO_4^-}}{a_{H_4SiO_4}} = 1.26 \times 10^{-10} \text{ at } 25°C,$$

and

$$H_3SiO_4^- \longleftrightarrow H^+ + H_2SiO_4^{2-}, \tag{6-3}$$

for which

$$K_2 = \frac{a_{H^+} a_{H_2SiO_4^{2-}}}{a_{H_3SiO_4^-}} = 5.01 \times 10^{-11} \text{ at } 25°C.$$

No further dissociation takes place. With the exception of some polymerization reactions which occur only in very alkaline solutions and have a negligible effect on total solubility, the concentration of all silica species is given by

$$\sum Si = m_{H_4SiO_4} + m_{H_3SiO_4^-} + m_{H_2SiO_4^{2-}}. \tag{6-4}$$

Notice the similarity between this expression and equation 4-7, which described the sum of dissolved carbon species, ΣCO_2. Just as we manipulated that equation to produce equation 4-12, we can now combine equations 6-2 through 6-4 to write

$$\sum Si = m_{H_4SiO_4}\left[1 + \frac{K_1}{a_{H^+}} + \frac{K_1 K_2}{a_{H^+}^2}\right], \tag{6-5}$$

in which activity coefficients have been omitted for simplicity.

This result is shown graphically in Figure 6.1. At *pH* values below about 9, only H_4SiO_4 contributes significantly to ΣSi. At *pH* 9.9, however, the first and second

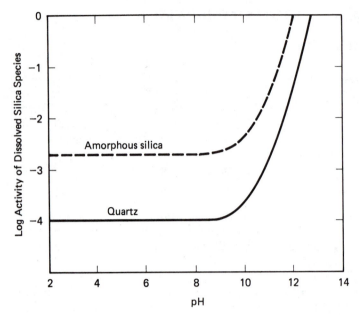

Figure 6.1 Activities of aqueous silica species at 25°C in equilibrium with quartz and with amorphous silica. The lines are Σ Si. Modified from Drever (1982).

terms in equation 6-5 are numerically equal, so $m_{H_4SiO_4} = m_{H_3SiO_4^-}$. This is analogous to the crossover point between H_2CO_3 and HCO_3^- at *pH* 6.35 in the carbonate system. In increasingly alkaline solutions, $m_{H_4SiO_4}$ becomes even less important. At *pH* 11.7, according to equation 6-5, $m_{H_3SiO_4^-} = m_{H_2SiO_4^{2-}}$, analogous to the crossover point between bicarbonate and carbonate ions. The silica-water system, then, has a *pH* buffering capacity in natural waters. It is clear, however, that since both of its crossover points lie outside the range of normal *pH* values, it is generally less important than CO_2-water.

Solubility of Magnesian Silicates

Following the example of silica solubility, it is convenient to write reactions between magnesian silicates and water in such a way that they eventually lead to solubility expressions in terms of *pH*. For now, we will consider only *congruent* reactions (that is, those that produce only dissolved species) rather than those that produce an intermediate hydrated phase. We disregard, for example, the reaction

$$3Mg_2SiO_4 + 2H_2O + H_4SiO_4 \longleftrightarrow 2Mg_3Si_2O_5(OH)_4$$

forsterite   serpentine

and consider only

$$Mg_2SiO_4 + 4H^+ \longleftrightarrow 2Mg^{2+} + H_4SiO_4,$$

for which

$$K_{fo} = \frac{a_{H_4SiO_4} a_{Mg^{2+}}^2}{a_{H^+}^4} = a_{H_4SiO_4}\left[\frac{a_{Mg^{2+}}}{a_{H^+}^2}\right]^2$$

Chemical Weathering: Dissolution and Redox Processes   Chap. 6

and

$$\log a_{H_4SiO_4} = \log K_{fo} - 2 \log \left[\frac{a_{Mg^{2+}}}{a_{H^+}^2} \right]. \tag{6-6}$$

One justification for doing this is that reactions between magnesian silicates and water in nature strongly favor total dissolution, so that comparatively little magnesium is retained in residual phases after chemical weathering has progressed to completion. Direct observations to confirm the dissolution mechanism, however, have been difficult to make. Petit et al. (1987) have shown, by means of a highly sensitive resonant nuclear imaging technique, that dissolution of diopside ($CaMgSi_2O_6$) takes place as water diffuses slowly into the silicate to produce a hydrated zone about 1000Å thick. This zone is very porous near the grain surface, but shows no evidence of secondary precipitates. This observation strongly suggests that diopside dissolves congruently, and lends support to the notion that other magnesian silicates may also.

A large number of reactions like 6-6 can be written. For example,

$$Mg_3Si_2O_5(OH)_4 + 6H^+ \longleftrightarrow 3Mg^{2+} + 2H_4SiO_4 + H_2O,$$

serpentine

for which (assuming that $a_{H_2O} = 1$)

$$K_{serp} = \frac{a_{H_4SiO_4}^2 a_{Mg^{2+}}^3}{a_{H^+}^6} = a_{H_4SiO_4}^2 \left[\frac{a_{Mg^{2+}}}{a_{H^+}^2} \right]^3$$

and

$$\log a_{H_4SiO_4} = \tfrac{1}{2} \log K_{serp} - \tfrac{3}{2} \log \left[\frac{a_{Mg^{2+}}}{a_{H^+}^2} \right]. \tag{6-7}$$

For talc,

$$Mg_3Si_4O_{10}(OH)_2 + 6H^+ + 4H_2O \longleftrightarrow 3Mg^{2+} + 4H_4SiO_4,$$

talc

and

$$K_{tc} = \frac{a_{H_4SiO_4}^4 a_{Mg^{2+}}^3}{a_{H^+}^6} = a_{H_4SiO_4}^4 \left[\frac{a_{Mg^{2+}}}{a_{H^+}^2} \right]^3,$$

or

$$\log a_{H_4SiO_4} = \tfrac{1}{4} \log K_{tc} - \tfrac{3}{4} \log \left[\frac{a_{Mg^{2+}}}{a_{H^+}^2} \right]. \tag{6-8}$$

For sepiolite,

$$Mg_4Si_6O_{15}(OH)_2 \cdot 6H_2O + 8H^+ + H_2O \longleftrightarrow 4Mg^{2+} + 6H_4SiO_4,$$

sepiolite

and

$$K_{sep} = \frac{a_{Mg^{2+}}^4 a_{H_4SiO_4}^6}{a_{H^+}^8} = a_{H_4SiO_4}^6 \left[\frac{a_{Mg^{2+}}}{a_{H^+}^2} \right]^4,$$

or

$$\log a_{H_4SiO_4} = \tfrac{1}{6} \log K_{sep} - \tfrac{2}{3} \log \left[\frac{a_{Mg^{2+}}}{a_{H^+}^2} \right]. \tag{6-9}$$

Fundamental Solubility Equilibria **169**

Finally, we consider the solubility of brucite, although it is not a silicate, and find that

$$Mg(OH)_2 + 2H^+ \longleftrightarrow Mg^{2+} + 2H_2O,$$

and

$$K_{br} = a_{Mg^{2+}}/a_{H^+}^2, \tag{6-10}$$

so $\log a_{H_4SiO_4}$ is independent of $\log \left[\dfrac{a_{Mg^{2+}}}{a_{H^+}^2}\right]$.

Equations 6-6 through 6-10 are linear, first-order equations and can easily be shown in graphical form, as in Figure 6.2. The solubility of quartz, which is independent of $\log a_{H_4SiO_4}$, is also shown. Because we have already seen that H_4SiO_4 dissociates appreciably above pH 9, this diagram is not reliable for highly alkaline environments. Each of the magnesium silicate equilibria is a straight line whose y-intercept (at infinitesimal H_4SiO_4 activity) is equal to a reaction K, calculated from standard free energies of formation. Their slopes in each case depend on the stoichiometry of the solubility reaction (that is, on the Mg/Si ratio in the silicate mineral). The serpentine reaction, for example, has a slope of $-3/2$, and talc has a slope of $-3/4$.

Notice that the dissolved species are stable on the lower left side of each equilibrium line, so that each divides the diagram into a region in which fluid is supersaturated with respect to a silicate mineral (to the upper right) and one in which it is undersaturated (to the lower left). The solid line combining segments of the brucite, serpentine, talc, and quartz equilibria, therefore, represents the only set of $\log a_{H_4SiO_4} - \log(a_{Mg^{2+}}/a_{H^+}^2)$ values for which fluid and a mineral are stable. Sepiolite and forsterite are unstable (or metastable) in contact with water at 25°C.

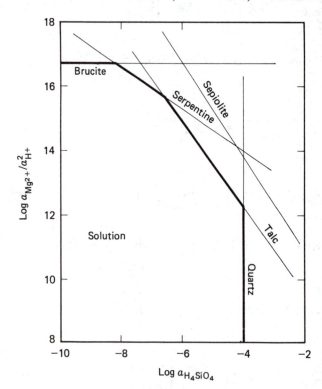

Figure 6.2 Solubility relationships among magnesian silicates in equilibrium with water at 25°C. Lines represent equations 6-7 through 6-10 in the text. The dissolution of forsterite (Eq. 6-11) lies too far to the right to appear on this diagram (see Problem 6.7). Modified from Drever (1982).

Solubility of Gibbsite

Dissolved aluminum in acid solutions exists as the free ion, Al^{3+}. With increasing alkalinity, however, Al^{3+} reacts with water to form ion pairs, which dominate at *pH* greater than 5. When we considered the solubility of magnesian silicates, there was no mention of complex species, because they have little effect in that system. The ion pair $Mg(OH)^+$, for example, is abundant only if *pH* is above 11, outside the range of acidity in most natural waters. By contrast, the solubility relations of $Al(OH)_3$ and aluminosilicate minerals are influenced greatly by the balance among complex species, and therefore are more complicated than those of $Mg(OH)_2$ and magnesium silicates.

Beginning in highly acid solutions, dissolution of gibbsite is described by

$$Al(OH)_3 + 3H^+ \longleftrightarrow Al^{3+} + 3H_2O,$$

for which

$$K_1 = a_{Al^{3+}}/a_{H^+}^3. \tag{6-11}$$

As *pH* increases beyond 5, $Al(OH)_2^+$ dominates instead. The dissolution equilibrium is described by

$$Al(OH)_3 + H^+ \longleftrightarrow Al(OH)_2^+ + H_2O$$

and the constant

$$K_2 = a_{Al(OH)_2^+}/a_{H^+}. \tag{6-12}$$

Finally, above *pH* = 7 the major species in solution is $Al(OH)_4^-$, which forms by

$$Al(OH)_3 + H_2O \longleftrightarrow Al(OH)_4^- + H^+$$

and for which

$$K_3 = a_{Al(OH)_4^-} a_{H^+}. \tag{6-13}$$

Thus, to calculate the total concentration of dissolved aluminum (Σ Al) in waters in equilibrium with gibbsite, we write

$$\sum Al = a_{Al^{3+}} + a_{Al(OH)_2^+} + a_{Al(OH)_4^-} \tag{6-14}$$

and substitute appropriately from equations 6-11 through 6-13 to get

$$\sum Al = K_1 a_{H^+}^3 + K_2 a_{H^+} + K_3/a_{H^+}. \tag{6-15}$$

Notice that equation 6-15 differs from both equation 4-12 (for Σ CO_2) and 6-5 (for Σ Si) in that it predicts high solubility in both very alkaline and very acid solutions, but low solubility throughout most of the range of *pH* observed in streams and most ground waters.

Worked Problem 6-1

Decomposition of organic matter in the uppermost centimeters of a soil may decrease the *pH* of percolating water to 4 or lower. As weathering reactions consume H^+, we expect the dominant form of dissolved aluminum and the total solubility, Σ Al, to change. Ignoring any possible reactions involving the organic compounds themselves, by how much should Σ Al vary between, say, *pH* 4 and *pH* 6?

The three equilibrium constants in equation 6-15 can be calculated from tabulated values of $\Delta \bar{G}_f^0$. We use free energies for water from Helgeson (1978) and for aluminum species at 25°C from Couturier et al. (1984), except for $Al(OH)_2^+$, which is from Reesman (1969):

Species	$\Delta \bar{G}_f^0$ (kcal mol^{-1})
$Al(OH)_3$	-276.02
Al^{3+}	-117.0
$Al(OH)_2^+$	-216.1
$Al(OH)_4^-$	-313.4
H_2O	-56.69

For equilibrium reaction 6-11,

$$\Delta G_{r1} = \Delta \bar{G}_f^0(Al^{3+}) + 3\Delta \bar{G}_f^0(H_2O) - \Delta \bar{G}_f^0(Al(OH)_3)$$

$$= -117.0 + 3(-56.69) - (-276.02) = -11.05$$

$$K_1 = \exp(-\Delta G_{r1}/RT) = 1.27 \times 10^8.$$

For reaction 6-12,

$$\Delta G_{r2} = \Delta \bar{G}_f^0(Al(OH)_2^+) + \Delta \bar{G}_f^0(H_2O) - \Delta \bar{G}_f^0(Al(OH)_3)$$

$$= -216.1 + (-56.69) - (-276.02) = 3.23$$

$$K_2 = \exp(-\Delta G_{r2}/RT) = 4.28 \times 10^{-3}.$$

For reaction 6-13,

$$\Delta G_{r3} = \Delta \bar{G}_f^0(Al(OH)_4^-) - \Delta \bar{G}_f^0(Al(OH)_3) - \Delta \bar{G}_f^0(H_2O)$$

$$= -311.4 - (-276.02) - (-56.69) = 21.31$$

$$K_3 = \exp(-\Delta G_{r3}/RT) = 2.34 \times 10^{-16}.$$

At pH 4, therefore, we calculate from equation 6-15 that

$$\Sigma\, Al = 1.27 \times 10^{-4} + 4.28 \times 10^{-7} + 2.34 \times 10^{-12}$$

$$= 1.27 \times 10^{-4},$$

so

$$a_{Al^{3+}}/a_{Al(OH)_2^+} = 297.$$

At pH 6,

$$\Sigma\, Al = 1.27 \times 10^{-10} + 4.28 \times 10^{-9} + 2.34 \times 10^{-10}$$

$$= 4.64 \times 10^{-9},$$

so

$$a_{Al^{3+}}/a_{Al(OH)_2^+} = 2.97 \times 10^{-2}.$$

With a shift of two pH units, total aluminum concentration drops by four orders of magnitude and, as expected, $Al(OH)_2^+$ eclipses free Al^{3+} as the major dissolved species. Figure 6.3 shows the solubility of gibbsite, calculated from the data in this Worked Problem, over the range from pH 3 to 12.

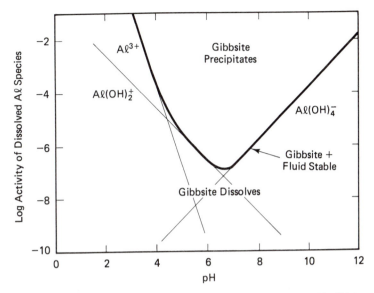

Figure 6.3 Activities of aqueous aluminum species in equilibrium with gibbsite at 25° C. The solid line is Σ Al. Modified from Drever (1982).

Solubility of Aluminosilicate Minerals

In the same way that we constructed Figure 6.2 to illustrate the solubility of magnesium silicates, it would be convenient to draw a diagram to display the solubility behavior of aluminum silicates. From the previous section, however, we now see that this cannot be done easily. Consider, for example, the dissolution of kaolinite. Over the range of natural pH values, each of the following three reactions contributes to Σ Al:

$$0.5Al_2Si_2O_5(OH)_4 + 3H^+ \longleftrightarrow Al^{3+} + H_4SiO_4 + 0.5H_2O,$$
kaolinite

$$0.5Al_2Si_2O_5(OH)_4 + H^+ + 1.5H_2O \longleftrightarrow Al(OH)_2^+ + H_4SiO_4,$$

$$0.5Al_2Si_2O_5(OH)_4 + 3.5H_2O \longleftrightarrow Al(OH)_4^- + H^+ + H_4SiO_4.$$

Equation 6-15 now takes the form

$$\Sigma\,Al = \frac{1}{a_{H_4SiO_4}}\left[K_1 a_{H^+}^3 + K_2 a_{H^+} + \frac{K_3}{a_{H^+}}\right].$$

This result is shown graphically in Figure 6.4, where we see that the solubility surface for kaolinite has cross-sections perpendicular to the $a_{H_4SiO_4}$ axis that are qualitatively similar to Figure 6-3. With increasing silica activity, however, kaolinite solubility at all *pH* values decreases. For minerals like kaolinite or pyrophyllite, a three-dimensional diagram like Figure 6.4 is instructive, or contours at constant $a_{H_4SiO_4}$ can be drawn on diagrams like Figure 6.3. More complicated minerals like feldspars and micas, however, cannot be considered in this way. Furthermore, because the aluminum concentration in most natural waters is too low to measure reliably with ease, diagrams like Figures 6.3 and 6.4 are less practical than Figure 6.2 was for magnesian silicates. For this reason, geochemists commonly write dissolution reactions for aluminum silicates which produce secondary solid aluminous phases instead of dissolved aluminum. Such reactions are said to be *incongruent*.

Fundamental Solubility Equilibria **173**

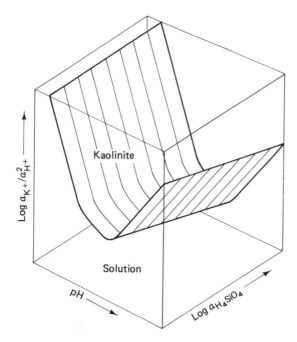

Figure 6.4 Schematic diagram illustrating the kaolinite solubility surface as a function of log $(a_{K^+}/a_{H^+}^2)$, pH, and log $(a_{H_4SiO_4})$. Notice the similarity in form between cross-sections at constant $a_{H_4SiO_4}$ and the gibbsite solubility curve in Figure 6.3. Modified from Garrels and Christ (1965).

One example of such a reaction is

$$0.5Al_2Si_2O_5(OH)_4 + 2.5H_2O \longleftrightarrow Al(OH)_3 + H_4SiO_4, \qquad (6\text{-}16)$$

which can be viewed loosely as an intermediate to any of the kaolinite reactions we considered a moment ago. Notice now that the relative stabilities of kaolinite and gibbsite depend only on the activity of H_4SiO_4 (that is, $K_{kao\text{-}gib} = a_{H_4SiO_4}$). *We have no information about the stability of either with respect to a fluid, unless we also know ΣAl.* This point is emphasized here, before we proceed to develop equations and diagrams commonly used in aluminosilicate systems, because, unfortunately, it is easy to overlook and can lead to misinterpretation of those equations and diagrams.

With that warning, let us now consider the broader set of common minerals which include not only silica and alumina, but other cations as well. In the system $K_2O\text{-}Al_2O_3\text{-}SiO_2\text{-}H_2O$, for example, we can write the following set of reactions between minerals:

$$\underset{\text{K-feldspar}}{KAlSi_3O_8} + H^+ + 4.5H_2O \longleftrightarrow \underset{\text{kaolinite}}{0.5Al_2Si_2O_5(OH)_4} + K^+ + 2H_4SiO_4$$

$$K_{ksp\text{-}kao} = \frac{a_{K^+}}{a_{H^+}} a_{H_4SiO_4}^2 \qquad (6\text{-}17)$$

$$\underset{\text{K-feldspar}}{KAlSi_3O_8} + H^+ + 2H_2O \longleftrightarrow \underset{\text{pyrophyllite}}{0.5Al_2Si_4O_{10}(OH)_2} + K^+ + H_4SiO_4$$

$$K_{ksp\text{-}pyp} = \frac{a_{K^+}}{a_{H^+}} a_{H_4SiO_4} \qquad (6\text{-}18)$$

$$\underset{\text{K-feldspar}}{KAlSi_3O_8} + \tfrac{2}{3}H^+ + 4H_2O \longleftrightarrow \underset{\text{muscovite}}{\tfrac{1}{3}KAl_3Si_3O_{10}(OH)_2} + \tfrac{2}{3}K^+ + 2H_4SiO_4$$

$$K_{\text{ksp-mus}} = \frac{a_{K^+}}{a_{H^+}} a_{H_4SiO_4}^3 \qquad (6\text{-}19)$$

$$\underset{\text{muscovite}}{KAl_3Si_3O_{10}(OH)_2} + H^+ + 1.5H_2O \;\longleftrightarrow\; \underset{\text{kaolinite}}{1.5Al_2Si_2O_5(OH)_4} + K^+$$

$$K_{\text{mus-kao}} = \frac{a_{K^+}}{a_{H^+}} \qquad (6\text{-}20)$$

$$\underset{\text{pyrophyllite}}{0.5Al_2Si_4O_{10}(OH)_2} + 2.5H_2O \;\longleftrightarrow\; \underset{\text{kaolinite}}{0.5Al_2Si_2O_5(OH)_4} + H_4SiO_4$$

$$K_{\text{pyp-kao}} = a_{H_4SiO_4} \qquad (6\text{-}21)$$

Equations 6-16 through 6-21 plot as straight lines on a graph of log $a_{H_4SiO_4}$ vs log (a_{K^+}/a_{H^+}), as we have done in Figure 6.5. This diagram looks superficially like the one we constructed for magnesium silicates. There is one major difference, however, because it has been constructed on the basis of incongruent reactions. Instead of marking the conditions under which a mineral is in equilibrium with a fluid, each line segment in Figure 6.5 indicates that two minerals coexist stably. Reactions 6-16, 6-17, 6-20, and 6-21, consequently, bound a region inside which kaolinite is stable relative to any of the other minerals we have considered. Each of the other fields between reaction lines represents in a similar way a region in which one aluminosilicate mineral is stable.

Several times in earlier chapters we have warned that thermodynamic calculations are valid only for the system as it has been defined, and we raise the same caution flag

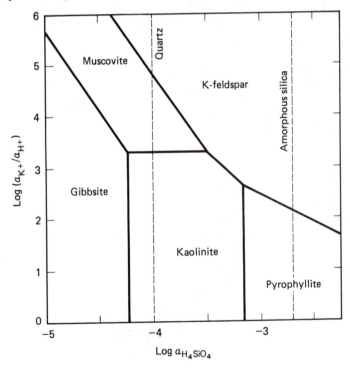

Figure 6.5 Stability relationships among common aluminosilicate minerals at 25° C. Precise locations of field boundaries may vary because of uncertainties in thermodynamic data.

again. Just as Figure 6.2 showed both forsterite and sepiolite to be unstable in the presence of water, we would find that reactions between muscovite and pyrophyllite, K-feldspar and gibbsite, or pyrophyllite and gibbsite are all metastable with respect to the ones we have plotted. In order to verify this, however, we need to write expressions for *all* potential equilibrium constants in the system and insert free energy data to determine which reactions are favored. Particularly as other components like CaO and Na₂O are added and graphical representation once again is difficult, it is no simple matter to identify each of the reactions that bounds a mineral stability field. Figure 6.6 shows how unwieldy graphical methods become with the addition of even one more compositional variable (a_{Na^+}/a_{H^+}). An increasing number of computer programs are now available which do the job laboriously but accurately, so that no significant reaction is inadvertently ignored. The most likely sources of error, then, are ones which arise because the thermodynamic data we give to the program have experimental uncertainties or because we accidentally omit phases from consideration.

Both of these sources of error are more likely than you might think. The crystal chemistry of phyllosilicates is complex: cationic and anionic substitutions are common, as are mixed-layer architectures in which two or more mineral structures alternate randomly. Among (Ca, Na)-aluminosilicates, this is particularly true. Natural smectites,

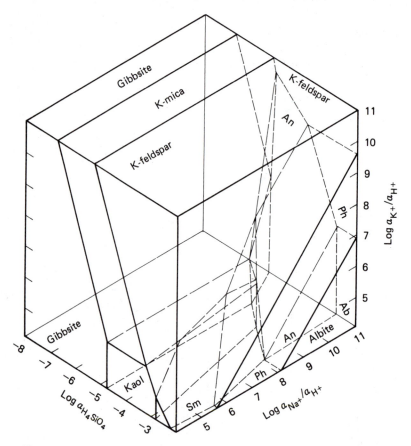

Figure 6.6 Stability relationships among common sodium and potassium aluminosilicate minerals at 25°C. Ab = albite, An = anorthite, Kaol = kaolinite, Ph = phillipsite, and Sm = smectite. (After J. I. Drever, *The geochemistry of natural waters*. Copyright 1982 by Prentice-Hall, Inc. Reproduced with permission.)

for example, include both dioctahedral (pyrophyllite-like) clays like montmorillonite and trioctahedral (talc-like) clays such as saponite. All show high ion-exchange capacity and readily accept Ca^{2+}, Na^+, K^+, Fe^{2+}, Mg^{2+}, and a host of minor cations in structural sites or interlayer positions. Their variable degrees of crystallinity and their generally high surface areas also contribute to large uncertainties in their thermodynamic properties. As a result, thermodynamic data for many aluminosilicates are gathered on phases which bear only a generic resemblance to minerals we encounter in the world at large. The exact placement of field boundaries in diagrams like Figure 6.5 and 6.6 is subject to considerable debate. Complicating the picture are a number of common zeolites, like analcime and phillipsite, which are stable at room temperature and should therefore appear on diagrams of this sort, but are remarkably easy to overlook.

These comments are not intended to be discouraging, but should point out the necessity of choosing data carefully when you try to model a natural aluminosilicate system rather than a hypothetical one. One moderately successful approach to verifying diagrams like Figure 6.5 involves the assumption that subsurface and stream waters are in chemical equilibrium with the weathered rocks over which they flow. If this is so, it should be possible to place field boundaries by analyzing water from wells and streams for *pH*, dissolved silica, and major cations and plotting them and the identity of coexisting minerals. Norton (1974) followed this method in his analysis of springs and rivers in the Rio Tanama basin, Puerto Rico. As indicated in Figure 6.7, the theoretical boundary between stability fields for kaolinite and Ca-montmorillonite corresponds closely to the measured composition of natural waters in equilibrium with both minerals. Drever (1982) has questioned the assumption of equilibrium in studies of this type, but the fact that such studies often show agreement between theory and nature justifies hesitant optimism.

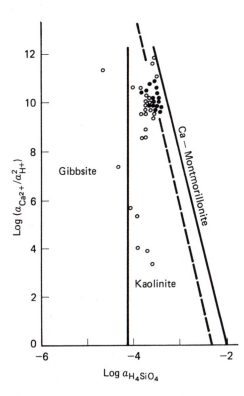

Figure 6.7 Activity diagram illustrating the compositions of water samples from the Rio Tanama, Puerto Rico. Solid circles indicate that sediment samples associated with the water contain both kaolinite and calcium montmorillonite. Open circles identify samples containing only kaolinite. The dashed boundary line, based on these observations, compares well with the solid boundary line, which was calculated from thermodynamic data. (After D. Norton, Chemical mass transfer in the Rio Tanama system, west-central Puerto Rico. *Geochim. Cosmochim. Acta* 38: 267–77. Copyright 1974 by Pergamon Press. Reproduced with permission.)

Fundamental Solubility Equilibria

Worked Problem 6-2

Assuming that reliable stability diagrams can be drawn to illustrate solubility relationships among aluminosilicate minerals, how can we use this information to predict the course of chemical weathering? As an example, consider the reactions that take place as microcline reacts with acidified ground water in a closed system.

In principle, a pathway can be predicted by treating the problem as a titration of acid by microcline, in the same way that we considered the titration of acid by Na_2CO_3 in Chapter 4. The difference here is that incongruent mineral reactions occur along the path, so that species other than aqueous ions have to be taken into account. This is a cumbersome exercise, commonly performed today by computer. These programs, unfortunately, are generally too complex to run efficiently on most small computers. The results we have shown in Figure 6.8 and will discuss here were performed on a mainframe system by the program EQ6 (Wolery, 1979).

Initially, as microcline dissolves in a fixed volume of water, both $a_{H_4SiO_4}$ and the ratio a_{K^+}/a_{H^+} increase from values in the lower left corner of the diagram. Notice, as pointed out earlier, that no secondary phase appears until Σ Al is high enough to saturate the fluid, despite the fact that the water composition lies in the field marked "gibbsite." If pH is below about 4, the dominant reaction at this stage is

$$KAlSi_3O_8 + 4H_2O + 4H^+ \longleftrightarrow K^+ + Al^{3+} + 3H_4SiO_4.$$

Once sufficient aluminum is in solution, gibbsite forms while microcline continues to dissolve. The net reaction releases potassium ions, OH^-, and silica to solution and causes the fluid composition to move toward the upper right:

$$KAlSi_3O_8 + 8H_2O \longleftrightarrow Al(OH)_3 + K^+ + OH^- + 3H_4SiO_4.$$

Eventually, the fluid reaches the gibbsite-kaolinite boundary. At this point, microcline dissolution continues to release H_4SiO_4, but the fluid composition cannot move further to the right because both gibbsite and kaolinite are present. Potassium ions continue to accumulate in solution, however, so the fluid composition follows the gibbsite-kaolinite boundary upwards. Gibbsite is gradually consumed. The appropriate reaction is

$$2KAlSi_3O_8 + 4Al(OH)_3 + H_2O \longleftrightarrow 3Al_2Si_2O_5(OH)_4 + 2K^+ + 2OH^-.$$

When all of the gibbsite is finally converted to kaolinite, the fluid composition is no longer fixed along the field boundary, and both $a_{H_4SiO_4}$ and a_{K^+}/a_{H^+} can increase as kaolinite is precipitated:

$$2KAlSi_3O_8 + 11H_2O \longleftrightarrow Al_2Si_2O_5(OH)_4 + 2K^+ + 2OH^- + 4H_4SiO_4.$$

Once the activity of H_4SiO_4 reaches 1×10^{-4}, it cannot increase further because the fluid is saturated with respect to quartz, which begins to precipitate as microcline dissolution continues. This reaction does not change the identity of the stable aluminosilicate. Kaolinite remains in equilibrium with the fluid as a_{K^+}/a_{H^+} rises.

At the muscovite-kaolinite boundary, the situation is similar to that at the gibbsite-kaolinite boundary earlier, except that neither $a_{H_4SiO_4}$ nor a_{K^+}/a_{H^+} can change until all kaolinite is converted to muscovite by the reaction

$$KAlSi_3O_8 + Al_2Si_2O_5(OH)_4 + H_2O \longleftrightarrow KAl_3Si_3O_{10}(OH)_2 + SiO_2.$$

Still prevented from becoming more silica-rich by the saturation limit for quartz, the fluid continues to follow a vertical pathway across the muscovite field:

$$3KAlSi_3O_8 + 2H_2O \longleftrightarrow KAl_3Si_3O_{10}(OH)_2 + 2K^+ + 2OH^- + 6SiO_2,$$

until it finally reaches the edge of the microcline stability field. No further dissolution takes place, because the fluid is now saturated with respect to microcline.

It is important, once again, to emphasize that the preferred pathway in nature may be distinctly different from what we have calculated in Worked Problem 6-2. Quartz, for example, is generally more difficult to nucleate than amorphous silica, which forms only at higher $a_{H_4SiO_4}$ values (see Figure 6.1). Without quartz, the reaction path is not constrained

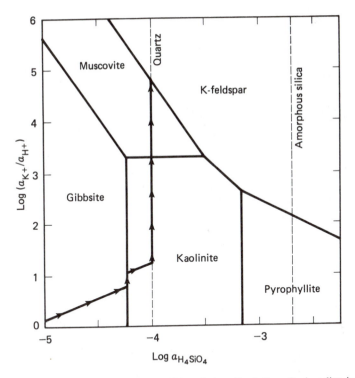

Figure 6.8 Evolution of fluid composition during dissolution of microcline in a closed system at 25°C, calculated by the program EQ6 (Wolery, 1979). Reactions along each segment of the path indicated by arrows are discussed in Worked Problem 6-2.

to move vertically through the kaolinite and muscovite fields. Also, chemical weathering rarely occurs in a closed system. If water flows through a feldspathic parent material instead of stagnating in it, gibbsite and kaolinite concentrate in successive zones rather than back-reacting. The result is a layered weathering sequence.

RIVERS AS WEATHERING INDICATORS: FOUR CASE STUDIES

As we have tried to show in several contexts so far, the type and intensity of weathering reactions will be reflected in an assemblage of residual minerals and in the composition of associated waters. Sometimes it is appropriate to examine the pathways of chemical weathering by comparing soil profiles with mineral sequences that we derive from model calculations, as in Worked Problem 6-2. In other cases, it is more reasonable to study the waters draining a region undergoing weathering. This might be true if soils were poorly developed or mechanically disturbed, or if we were interested in alteration in a subsurface environment from which residual weathering products are not easily sampled. It also might be true if we were concerned about weathering on a regional scale and wanted to integrate the effects of several types of parent rock, or if we were interested in seasonal variations in weathering. We will focus briefly on a few studies to illustrate the strengths of this approach.

Toorongo Granodiorite, Victoria, Australia

In preceding sections, we have focused on the SiO_2-Al_2O_3 framework of silicate minerals and its stability in a weathering environment. We have seen, however, that mineral stabilities during chemical weathering are functions of activity ratios such as a_{K^+}/a_{H^+} or $a_{Mg^{2+}}/a_{H^+}^2$, so this has also been a study of alkali and alkaline earth element behavior. What happens, however, when several cations are released along a reaction pathway? How are they partitioned, relative to each other, among the clay products and the fluid?

Nesbitt et al. (1980) have analyzed a progressively weathered sequence of samples from the Toorongo Granodiorite in Australia, which provide an interesting perspective. The parent rock consists largely of plagioclase, quartz, K-feldspar, and biotite, with some minor hornblende and traces of ilmenite. During initial stages of weathering, sodium, calcium, and strontium in the rock decrease rapidly in comparison with potassium. The primary reason for this decline is the early dissolution of plagioclase, which is much less stable than K-feldspar or biotite, the major potassic minerals. The clay mineral which appears in the least altered samples is kaolinite.

As weathering progresses, vermiculite and then illite join the clay mineral assemblage. These further influence the loss or retention of cations. As the initially low *pH* of migrating fluids begins to increase, charge balance on surfaces and in interlayer sites in clays is satisfied by large cations (K^+, Rb^+, Cs^+, and Ba^{2+}) rather than by H^+. The smaller cations (Ca^{2+}, Na^+, and Sr^{2+}) are less readily accepted in interlayer sites and so are quantitatively removed in ground water and surface runoff. The sole exception is Mg^{2+}, which is largely retained in the soil as biotite is replaced by vermiculite, which has a small structural site to accommodate it. As a result, progressive weathering of the Toorongo Granodiorite produces a residual which is greatly depleted in calcium and sodium and only slightly depleted in potassium and magnesium. Figure 6.9 illustrates this trend, and shows that the composition of local stream waters is roughly complementary (that is, they are relatively rich in Na^+ and Ca^{2+} and poor in K^+ and Mg^{2+}).

These observations are generally consistent with the thermodynamic predictions made in the last section and with observed clay mineral abundances on a global scale (see, for example, Figure 6.10). Gibbsite only appears in weathered sections that are intensively leached (that is, ones in which the associated waters are extremely dilute). Kaolinite dominates where waters are less dilute, and smectites where waters have higher concentrations of dissolved cations. Vermiculite and illite are common as intermediate alteration products where soil waters have concentrations between these extremes, as in most regions of temperate climate and moderate annual rainfall.

Hubbard Brook Watershed, New Hampshire

The Hubbard Brook Watershed lies entirely within a USDA Forest Service experimental forest in the White Mountains of New Hampshire, in the northeastern United States. Because the area is almost unpopulated, and yet is readily accessible, it has served as a valuable source of chemical and hydrologic information in continuous studies since 1963 (see, for example, Johnson et al., 1967; Likens et al., 1977).

Bedrock in most of the area is a sillimanite-grade gneiss. Various lines of evidence, however, suggest that very little weathering takes place on fresh rock. Infiltrating meteoric waters react instead with a layer of glacial till averaging 1.4 m thick. The till was derived largely from the sillimanite gneiss and from a nearby quartz monzonite.

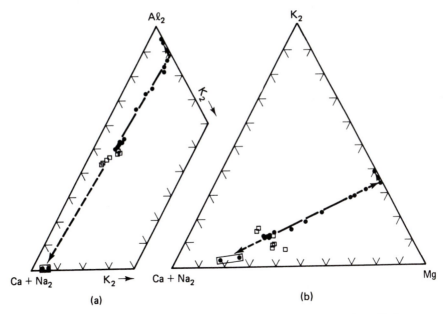

Figure 6.9 Chemical changes during weathering of the Toorongo Granodiorite. Measured compositions of rock samples are indicated by solid circles. Average unweathered rock compositions are indicated by open boxes, and average stream water composition by the solid box. With progressive weathering, rock compositions move away from the Ca + Na₂ corner on both diagrams, becoming more aluminous and relatively enriched in potassium and magnesium. Water compositions show complementary trends. (After H. W. Nesbitt; G. Markovics; and R. C. Price, Chemical processes affecting alkalis and alkaline earths during continental weathering. *Geochim. Cosmochim. Acta* 44: 1659-66. Copyright 1974 by Pergamon Press. Reproduced with permission.)

The region receives about 123 cm of precipitation per year, about 25 to 30% of which is snow. The low *pH* (4.1) and high concentrations of sulfate and nitrate in rainfall reflect major pollution from industrial sources and urban exhaust, typical of much of the eastern United States and Canada. Slightly less than 60% of the precipitation leaves eventually as runoff. The remainder is lost through evapotranspiration.

The residual soil that has developed in this watershed is a typical *spodosol,* or acid forest soil. It has a very organic-rich upper horizon derived from the accumulation of leaves and other plant debris. This is underlain by an A2 horizon which has been severely leached of iron, alkali, and alkaline earth elements and now consists predominantly of quartz. The B horizon is enriched in colloidal organo-metallic chelates and clays derived from the upper zones, and is deeply colored by hydrated iron oxides.

By comparing the bulk compositions of parent material and weathered residue (the A2 horizon), Johnson and co-workers (1967) were able to calculate the amount of leaching that had taken place during chemical weathering. As shown in Table 6-1, the net loss of calcium and magnesium between parent and residuum is roughly twice the net loss of sodium and potassium. If no other changes took place, these ratios should also be observable in analyses of stream water in Hubbard Brook. In fact, however, the actual concentrations of potassium and magnesium in runoff shown in Table 6-1 are much lower than we would predict from the values for Ca^{2+} and Na^+.

We have already seen, in our earlier presentation of data from the Toorongo Granodiorite, a probable explanation for the low Mg^{2+} and at least some of the low K^+ concentration in runoff. The dominant clay minerals in the B horizon are illite and ver-

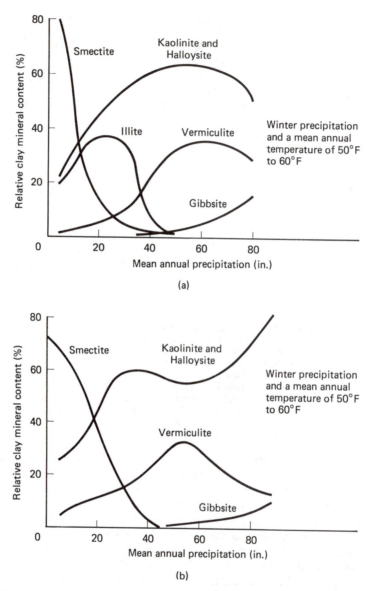

Figure 6.10 Clay minerals in residual soils in California formed on (a) acidic and (b) basic igneous rocks. (After I. Barshad, The effect of variation in precipitation on the nature of clay mineral formation in soils from acid and basic igneous rocks. In *Proc. Internat. Clay Conf. Jerus.* 1, 167–73. Copyright 1966 by the Israel Program for Scientific Translation. Reproduced with permission.)

miculite, which probably trap much of the potassium and magnesium before they can leave the area in ground water. In addition, potassium is readily consumed by growing plants. The area was logged heavily during the first half of the century, and the forest biomass increased steadily during the study period. Likens et al. (1977) estimated that about 85% of the K^+ released by weathering in the Hubbard Brook watershed is incorporated into new plant matter. This fraction may be an overestimate, because it implies an inconsistency among the measured potassium abundances in source material, soil, and runoff. Because they also reported that 45% of the newly released Ca^{2+} is also in-

TABLE 6-1 CHEMICAL CHANGES INDUCED BY WEATHERING IN THE HUBBARD BROOK WATERSHED[*]

Element	Wt % in bedrock	Wt % in residue	Wt % Change due to weathering	Rate of removal (kg hectare^{-1}yr^{-1})
Si	30.7	37.7	+7.0	
Al	8.3	5.1	−3.2	
Fe^{3+}	1.3	0.1	−1.2	
Fe^{2+}	3.1	0.5	−2.6	
Ca	1.4	0.4	−1.0	8.0 ± 0.7
Na	1.6	1.0	−0.6	4.6 ± 0.4
K	2.9	2.4	−0.5	0.1 ± 0.5
Mg	1.1	0.1	−1.0	1.8 ± 0.1

[*]After Johnson et al. (1967).

corporated in plants, however, it is possible that these data indicate that other parent materials (calcite in tills? fresh bedrock?) are important as well as till.

The Amazon River Watershed, South America

Like the Hubbard Brook watershed, the Amazon basin is largely free of urban and industrial pollution. This remarkable situation makes it possible to use the composition of water in this, the world's largest river system, as a clue to weathering processes in the tropics. This was done by Gibbs (1967, 1972), who sampled the river at places where each of its 16 major tributaries joins the main stream. He then used a statistical approach to look for significant correlations among the chemical data and a large number of environmental factors, such as relief, mean temperature, annual rainfall, type of vegetation, and bedrock.

The strongest correlation Gibbs found was between total dissolved solids and relief. Roughly 86% of the dissolved material that reaches the mouth of the Amazon is derived from mountainous slopes in the Andes, which occupy 12% of the drainage basin area. One implication is that chemical weathering proceeds most efficiently where fresh rock is always exposed by mechanical weathering and erosion. In the lower reaches of the Amazon basin, surface waters are isolated from fresh rock by a thick blanket of tropical soil. This hypothesis is bolstered by the secondary observation that the Amazon carries 11 ppm of dissolved silica—very close to the world average of 13.1 ppm (Livingstone, 1963)—but contains only a third of the average amount of bicarbonate. This, coupled with the low concentration of dissolved cations, suggests that dissolved material added in the lower basin is derived from kaolinite and other aluminosilicate minerals. These continue to supply dissolved silica, but not cations balanced by bicarbonate.

Simple dilution may also contribute to the slow weathering rate. Where rains are heavy, a higher proportion of water is diverted directly into runoff without having time to infiltrate and react with soil and rocks. When water does pass through the soil column, it dilutes many of the acids that promote mineral dissolution.

Weathering of Ultramafic Rocks, Western United States

Serpentinization of ultramafic bodies generally takes place in hydrothermal environments at temperatures of 200°–300°C. By studying waters from several springs adjacent to Alpine-type ultramafic rocks in the western United States, however, Barnes and

O'Neil (1969) determined that serpentinization can also result from interactions with shallow groundwater. The net process is nearly isochemical with respect to magnesium and silica, but apparently proceeds in steps and leads to two ranges of composition in associated waters (Table 6-2). By examining these waters, we can deduce the reaction path that produced them.

TABLE 6-2 AVERAGE COMPOSITIONS (IN MG LITER^{-1}) OF SPRINGS IN ALPINE PERIDOTITES AND DUNITES IN THE WESTERN U.S.*

	$Mg^{2+} - HCO_3^-$ waters	$Ca^{2+} - OH^-$ waters
pH	8.6	11.7
Ca^{2+}	7.9	43
Mg^{2+}	126	0.23
Na^+	7.2	33
K^+	0.7	1.3
Cl^-	12	39
SO_4^{2-}	11	0.5
HCO_3^-	656	0
CO_3^{2-}	33	0
OH^-	0.1	53
SiO_2 (total)	19	3.0
B (total)	0.5	0.06

*Calculated from Barnes and O'Neill (1969).

The waters most commonly encountered have large amounts of Mg^{2+} and HCO_3^-, as might be expected from the high solubilities of forsterite and enstatite, which are the dominant silicates in ultramafic rocks. Dissolution causes both $a_{Mg^{2+}}$ and *pH* to increase, by the following reactions:

$$Mg_2SiO_4 + 4H^+ \longleftrightarrow 2Mg^{2+} + H_4SiO_4$$
forsterite

and

$$MgSiO_3 + 2H^+ + H_2O \longleftrightarrow Mg^{2+} + H_4SiO_4.$$
enstatite

Minor amounts of Ca^{2+} are probably also contributed by decomposition of pyroxenes. The *pH* of these waters ranges from 7.84 to 8.97.

In contrast, water from a small number of springs contains virtually no Mg^{2+} or bicarbonate, but is very rich in both Ca^{2+} and OH^-. These waters are also extremely alkaline, with *pH* values close to 12. Barnes and O'Neil calculated, by means similar to those we used at the beginning of this chapter, that waters of this composition are supersaturated with respect to both brucite and serpentine. They hypothesized that as the weathering of primary mafic minerals drives up the *pH*, HCO_3^- is systematically converted to CO_3^{2-} and precipitated from $Mg^{2+} - HCO_3^-$ waters as calcium and magnesium carbonates. Once bicarbonate is exhausted, OH^- accumulates in solution, driving the *pH* higher. Eventually, the saturation limits for brucite and serpentine are exceeded, and the remaining Mg^{2+} and much of the dissolved silica are removed by reactions that we presented earlier as 6-7 and 6-10. Ca^{2+} does not fit readily into serpentine, and is therefore the major surviving cation.

Carbon Dioxide

In Chapter 4, the solubility of calcite was examined in some detail, and it was shown (in Worked Problem 4-3) that $m_{Ca^{2+}}$ depends on the atmospheric partial pressure of CO_2:

$$m_{Ca^{2+}} = \frac{K_{cal} K_1 K_{CO_2} P_{CO_2}}{K_2 m_{HCO_3^-}^2},$$

in which we are assuming that all species behave ideally. It is tempting to conclude that we can simply insert the atmospheric partial pressure of CO_2 into this expression and use it to predict the calcium content of natural waters in limestone terrain. This possibility is evaluated in the following problem.

Worked Problem 6-3

Assuming a reasonable value for atmospheric P_{CO_2}, how well does the expression above predict actual water composition, and is the result consistent with geomorphic observations of regions underlain by limestone?

Charge balance in solution requires that

$$2m_{Ca^{2+}} + m_{H^+} = m_{HCO_3^-} + 2m_{CO_3^{2-}} + m_{OH^-},$$

but under most geologic conditions this can be simplified to

$$2m_{Ca^{2+}} \simeq m_{HCO_3^-}$$

without sacrificing much accuracy. From this, we conclude that

$$m_{Ca^{2+}}^3 \simeq \frac{K_{cal} K_1 K_{CO_2} P_{CO_2}}{4K_2}.$$

If we substitute appropriate values for each of the constants at 25°C, and use 3.2×10^{-4} as the present value of atmospheric P_{CO_2}, this equation predicts that the concentration of Ca^{2+} in streams draining limestone terrain should be 4.7×10^{-4} mol kg^{-1} or 19 mg kg^{-1}. Data in Holland (1978), however, suggest that water in wells and springs from such regions actually contains between 50 and 90 mg Ca^{2+} kg^{-1}. The answer to the first part of our question, then, is that the equilibrium expression does not appear to predict Ca^{2+} concentrations in river water very well at all.

Runoff from an average continental surface is about 30 cm yr^{-1}. Holland (1978) has calculated that if $m_{Ca^{2+}}$ in streams draining limestones were, in fact, 19 mg kg^{-1}, then calcite should be dissolved from the rocks at about 1.4 g cm^{-2} per 1000 years. This is roughly one-tenth of the mean continental erosion rate today. If the weathering of limestone were controlled by the atmospheric CO_2 pressure, carbonate rocks should be much more resistant than average crustal rocks, and should consistently crop out on ridges and mountain tops. Clearly, this is not the case. We must be overlooking something.

The apparent error in this calculation can quickly be traced to the value of P_{CO_2}. Furthermore, because $m_{Ca^{2+}}$ increases only in proportion to the cube root of P_{CO_2}, it is evident that small, seasonal variations in CO_2 are not sufficient to account for the discrepancy we have just uncovered. The measured calcium concentrations reported above suggest that the partial pressure of CO_2 in equilibrium with limestones in nature is 10 to 100 times higher than atmospheric P_{CO_2}. Surprisingly, these high values are common in air held in the pore spaces of soil. Measurements made in arable midlatitude soils indicate that P_{CO_2} is between 1.5×10^{-3} and 6.5×10^{-3} atm—roughly 5 to 20 times that

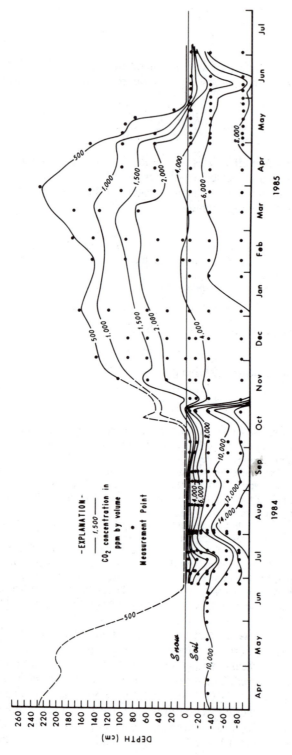

Figure 6.11 Soil CO_2 as a function of depth and time from April 1984 to July 1985 at Brighton, Utah. In general, CO_2 levels are highest during the warm months, when biological activity in the soil is highest. Also, P_{CO_2} decreases toward the surface due to diffusive loss to the atmosphere. Notice, however, that P_{CO_2} in shallow soils is higher during the winter months than during the summer, even though biological CO_2 production is lower. The snow may act as a cap to prevent CO_2 from escaping to the atmosphere. (After D. K. Solomon and T. E. Cerling, The annual carbon dioxide cycle in a montane soil: observations, modeling, and implications for weathering. *Water Resour. Res.* 23: 2257–2265. Copyright 1987 by the American Geophysical Union. Reproduced with permission.)

in the atmosphere. Analyses of a tropical soil reported in Russell (1961) indicate P_{CO_2} as high as 1×10^{-1} atm—330 times the atmospheric value!

These findings indicate that if we want to study the effectiveness of CO_2 as a chemical weathering agent, we must look primarily to those processes which control its abundance in soils. A small amount of CO_2 is generated when ancient organic compounds in sediments are oxidized during weathering. This amount, however, is almost negligible. The greatest proportion is produced by respiration from plant roots and by microbial degradation of buried plant matter (see, for example, Wood and Petraitis, 1984; Witkamp and Frank, 1969). Cawley et al. (1969) provided dramatic confirmation of this CO_2 source by sampling streams in Iceland. They found that those draining vegetated regions had bicarbonate concentrations two to three times higher than streams in areas with the same volcanic rock but without plant cover.

CO_2 generation follows an annual cycle in response to variations in both temperature and soil moisture. Values rise rapidly at the onset of a growing season and fall again as lower air and soil temperatures mark the end of the growing season. The rate of chemical weathering, therefore, should be greatest in most regions during the summer. Even in midwinter, however, soil P_{CO_2} may sometimes be several times the atmospheric value. Solomon and Cerling (1987) have found that snow cover can retard the diffusive loss of CO_2 from some intermontane soils, so that P_{CO_2} in winter can actually exceed the value in summer at shallow depths (see Figure 6.11). Because carbon dioxide is more soluble in cold than warm water, this seasonal pattern may lead to weathering rates that are significantly enhanced during the winter.

CO_2, CHEMICAL WEATHERING, AND THE EARLY CLIMATE ON MARS

Mars today is a cold, dry planet with a very thin CO_2 atmosphere whose total pressure is about 7×10^{-3} atm. Water ice evaporates readily rather than melting, even on those rare occasions when equatorial surface temperatures rise toward its melting point, so that there can be no seas and rivers. Photographs by the Mariner 9 and Viking spacecraft, however, suggest strongly that this was not always the case. Fluvial channels and valley networks have been observed at many places on Mars and have been found, by indirect methods, to be at least 3.5×10^9 years old. The early climate, possibly for as much as 10^9 years, was therefore apparently warm and moist.

Pollack et al. (1987) have used a numerical model of the CO_2 greenhouse effect to infer that P_{CO_2} would need to have been at least 0.75 atm, and maybe as much as 5 atm, in order to maintain a surface temperature above 273 K by solar heating. These estimates involve a number of assumptions regarding the orbit and reflectivity of the planet, as well as the style of atmospheric circulation, but the uncertainties introduced by them are probably small. Their results, however, raise an interesting question: "Where has all of the CO_2 gone?."

We have indicated in this section of the chapter that chemical weathering rate is proportional to P_{CO_2}. On Mars, where there is no vegetation, soil P_{CO_2} can be presumed to be similar to atmospheric. If it were possible to estimate the rate of surficial weathering on early Mars, therefore, we would also have an indirect measure of the rate of CO_2 loss from the primitive atmosphere. By parameterizing the weathering process on Earth (that is, by writing model equations to examine the independent effects of temperature, rock type, surface area, and so forth) and scaling their parameters so that they apply to Mars, Pollack and his coinvestigators have done just that. They estimate that the dense CO_2 atmosphere would have been consumed by weathering crustal silicates in 1.5×10^7 to 5.7×10^7 years. Once again, the result depends on a number of assumptions, but Pollack and his colleagues present arguments which suggest that even the slowest conceivably realistic weathering rate would have removed all CO_2 from the atmosphere in much less than 10^9 years. Because the fluvial channels suggest that CO_2 cannot have been depleted that fast, weathering must have been counterbalanced by some resupply process. The question now becomes "What mechanism can have supplied CO_2 at a rate comparable to loss by weathering for as much as 10^9 years, and why did it stop?"

The likely mechanisms involve global tectonics and volcanism, which draw carbonate-rich sediments into hot portions of the lower crust and devolatilize them. We can only speculate about what kind of tectonic processes prevailed in Mars' early history. Such global-scale processes, however, must operate at rates proportional to integrated heat flow through the crust. To test their concept, therefore, Pollack and his colleagues chose a particularly slow recycling model and determined the rate of CO_2 production from estimates of early Martian heat flow. They found that it was possible, even in this slow limit, to maintain a steady-state P_{CO_2}.

The final step in this inquiry depends on the assumption, common in all modern planetary models, that heat flow was originally high and has decreased with time. The steady state, then, should have fallen apart when planetary heat flow became too small for volcanism to keep the rate of weathering balanced by outgassing. According to this hypothesis, the warm, moist climate on ancient Mars finally succumbed to chemical weathering, and most of the dense CO_2 atmosphere is now buried in carbonate rocks. The evidence for these conclusions is largely circumstantial (see Figure 6.12), but carbonates will be the object of intensive field studies during future missions to the Martian surface.

Figure 6.12 Nedell and McKay (1987) have speculated that these layered rocks, exposed on the walls of Hebes Chasma, a canyon northwest of Valles Marinaris on Mars, are carbonates. If so, they may be an ancient product of weathering by CO_2. (Photo by NASA Viking I spacecraft.)

Organic Acids

Despite considerable study, the role of organic matter in chemical weathering is poorly understood. The previous section suggested that some of the elevated CO_2 concentrations in soil are the result of biodegradation, so organic compounds affect weathering rates indirectly by serving as a source for CO_2. Laboratory and field studies, however, indicate that they have a much stronger direct influence.

Upon decomposition, the organic remains of plants and animals yield a complex mixture of acids. A number of simple compounds, such as acetic and oxalic acid, are commonly found around active root systems and fungi. These and others like malic and formic acids serve the same role as HCl or H_2SO_4 and act as H^+ donors. Depending on the acidity of ground waters, any of these acids can increase mineral solubility.

The role of high molecular weight acids is less well understood. Humic and fulvic acids are composed of a bewildering array of aliphatic and phenolic structural units,[1] with a high proportion of —COOH and phenolic —OH radicals, as suggested schematically in Figure 6.13. At least some of these acids are soluble or dispersible in water, where limited dissociation of these radicals can occur, particularly if soil *pH* is low. The effect is twofold. First, dissociation releases hydrogen ions into solution, where they influence solubility equilibria in the same way that inorganic acids do. Second, the organic acid molecule, following dissociation, develops a surface charge that helps to repel other similarly charged particles. It therefore adopts a dispersed or colloidal character, which increases its mobility in soil waters.

This second role is particularly important because several studies indicate that organic colloids may be important in controlling both solubility and transport of mineral matter (Schnitzer, 1986). Provided that free cations are not so abundant that they completely neutralize the surface charge, humic and fulvic acids can attach significant amounts of Fe^{3+} and Al^{3+}, and lesser amounts of alkali and alkaline earth cations, without losing their mobility as colloids. In this way, the stability of colloidal organic matter may, for example, be a controlling factor in the downward transport of iron and aluminum in acidic forest soils (spodosols and alfisols) (DeConick, 1980).

Finally, organic acids in soils commonly act as chelating agents. Chelates of fulvic acids and of oxalic, tartaric, and citric acid are thought to be responsible for most of the dissolved cations in forest soils (Schnitzer, 1969). As we first saw in Chapter 3, these can contribute significantly to the mobility of metals. Fulvic and humic acids also both form mixed ligand complexes with phosphate and other chemically active groups.

The quantitative effect of organic acids can be striking. Baker (1986) has performed a series of experiments in which minerals were held in various solutions for 24 hours at room temperature, to examine the rate and amount of dissolution. Samples were placed in water saturated with CO_2 at its atmospheric partial pressure and in a 500 ppm solution of humic acids extracted from bracken fern. Because of the possibility that it is CO_2 derived from organic acids, rather than the organic acids themselves, that affects mineral solubility in nature, Baker also performed dissolution experiments in water charged with enough CO_2 to simulate the effect of converting all of the organic matter in the humic acid solution to carbon dioxide. A portion of his results are shown in Table 6-3. The data indicate that organic acids can greatly enhance the dissolution

---

[1] See Chapter 5 for a brief introduction to the terminology used to describe organic compounds of geologic interest.

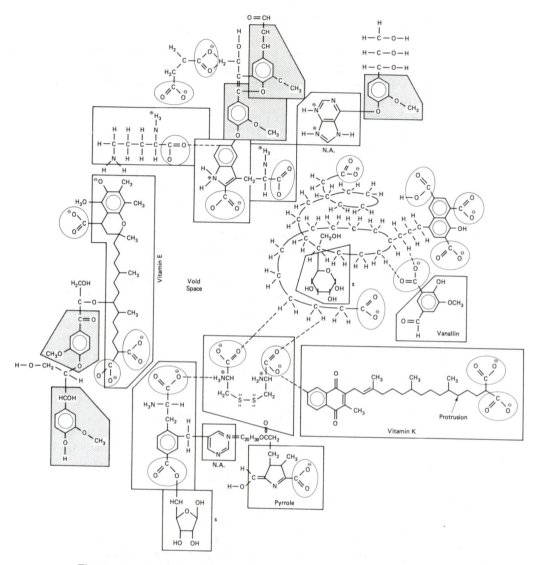

Figure 6.13 Chemical-structural model of a humic acid with a molecular weight of about 5000. Boxes outline various biological precursor units incorporated into the humic acid. Dark shaded boxes are lignin degradation products. Other boxes are amino acids, sugars (labelled *S*), nucleic acids (*N.A.*), or other compounds as labelled. Small circles around plus (+) and minus (−) symbols, electron pairs (..) on nitrogen, oxygen, and sulfur, and large circles around carboxyl ions (O-C-O⁻) show sites where the humic acid can interact with metals through ion exchange and chelation. (After J. S. Leventhal, Roles of organic matter in ore deposits. In *Organics and Ore Deposits*, ed. Dean, W. E., 7–20. Copyright 1986 by the Denver Region Exploration Geologists Society. Reproduced with permission.)

process, especially where reactions take place in an open, flow-through system. Unfortunately, stability constants for most relevant organic complexes are not known, so measurements of this sort cannot generally be used to constrain thermodynamic models for chemical weathering.

TABLE 6-3 DISSOLUTION OF SELECTED MINERALS AFTER 24 HOURS AT 25°C*

Mineral	Ion	Atmospheric H_2O/CO_2	Humic acid	Humic acid (H_2O/CO_2)
Calcite	Ca	0.65	17.40	2.75
Dolomite	Ca	0.25	7.50	0.45
Dolomite	Mg	0.14	5.65	0.33
Hematite	Fe	0.06	0.58	
Pyrite	Fe	0.04	0.24	
Galena	Pb	0.06	2.62	0.24
Gold	Au	10 ng/ml	190 ng/ml	

*All solution concentrations are in mg l^{-1} except as noted. Based on data selected from Baker (1986).

OXIDATION-REDUCTION PROCESSES

Thermodynamic Conventions for Redox Systems

In previous sections and in preceding chapters, we have confined discussion to reactions that involve atoms with only one common oxidation state. Solubility equilibria in these systems are characterized either by hydrolysis or acid-base reactions. Most natural systems, however, also contain transition metal elements, which may form ions in two or more different oxidation states.

Reactions in which the oxidation states of component atoms change are called *redox* reactions. These are not restricted to reactions involving transition metals, although it is most often those systems that attract attention.

In many cases, redox reactions involve the actual exchange of oxygen between phases. For example:

$$2 Fe_3O_4 + \tfrac{1}{2}O_2 \longleftrightarrow 3Fe_2O_3.$$

In a great many cases, however, oxygen is not a reactant in redox systems. In fact, it is misleading to emphasize the role of oxygen, because it is the change in the transition metal ion that characterizes these systems. It is more practical in many cases, therefore, to write redox reactions to emphasize the exchange of *electrons* between reactant and product assemblages. The example above can be rewritten as

$$2 Fe_3O_4 + H_2O \longrightarrow 3 Fe_2O_3 + 2 H^+ + 2 e^-, \qquad (6\text{-}22)$$

in which the fundamental change involves the oxidation of ferrous iron:

$$Fe^{2+} \longrightarrow Fe^{3+} + e^-.$$

Geochemists may write reactions according to either convention. At low temperatures, or when a fluid phase is present, it is usually convenient to write redox reactions in terms of electron transfer. As we will see shortly, reaction potentials in aqueous systems are easily measured as voltages. At high temperatures, where these measurements are much more difficult to perform, it is generally easier to determine oxygen fugacity and to write redox equilibria in terms of oxygen exchange. In this chapter, we will devote most of our attention to aqueous systems.

A reaction like 6-22, although conceptually useful, cannot take place in isolation in nature because an electron cannot exist as a free species. Such a reaction should be viewed in tandem with another similar reaction which consumes electrons. Together, the pair of *half-reactions* represents the exchange of electrons between atoms in one phase, which become oxidized by loss of electrons, and those in another, which gain electrons and are therefore reduced. A companion to reaction 6-22, for example, might be

$$\tfrac{1}{2}H_2 \longrightarrow H^+ + e^-.\qquad(6\text{-}23)$$

A natural reaction involving both 6-22 and 6-23, but resulting in no free electrons, would be

$$2Fe_3O_4 + H_2O \longleftrightarrow 3Fe_2O_3 + H_2.$$

By writing these illustrative reactions, we introduce two standard conventions. First, geochemists usually write half-reactions with free electrons on the right side; that is, they are written as oxidation reactions. It should be clear that the progress of a pair of half-reactions depends on their *relative* free energies of reaction, so that there is no special significance in this convention. Reaction 6-22 may proceed to the right (that is, Fe_3O_4 may actually be oxidized to form Fe_2O_3) if it is coupled with a half-reaction with a higher free energy of reaction, but may be reversed if its companion half-reaction has a lower ΔG_r.

The second standard convention involves reaction 6-23, which is arbitrarily chosen to serve as an analytical reference for all other half-reactions. To explain this standard and its value in solution chemistry, we have drawn a schematic illustration (Figure 6.14) of an apparatus known as the *Standard Hydrogen Electrode* or *SHE*. It consists of a piece of platinum metal immersed in a 25°C solution with $pH = 1$, through which H_2 gas is bubbled at 1 atm pressure. The platinum serves only as a reaction catalyst and as a means of making electrical contact with the world outside the apparatus. If the SHE had a means of exchanging electrons with some external system, it would be the physical embodiment of half-reaction 6-23. By agreement, chemists declare that $\Delta \overline{G}_f$ of H^+ and of e^- is zero. Hence, reaction 6-23 has its free energy of reaction defined arbitrarily as zero. Because $a_{H^+} = 1$ in the SHE, $a_{e^-} = 1$ as well.[2] By agreeing, in addition,

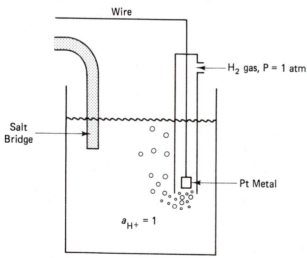

Figure 6.14 Schematic diagram of the Standard Hydrogen Electrode (SHE). The half-cell reaction is equation 6-23. Platinum metal serves as a catalyst on the electrode. The salt bridge is a tube filled with a KCl solution and closed with a porous plug.

[2] The "activity of electrons" is a slippery concept, because it is a measure of a quantity which has no physical existence. In this sense, a_{e^-} is not equivalent to ion activities we have considered before. Instead, it is best to treat it as a measure of the solution's tendency to release or attract electrons from a companion half-cell.

that there is no electrical potential between the platinum electrode and the solution, we have created a laboratory device against which the reaction potential of any other half-reaction cell can be measured.

Figure 6.15 illustrates an experiment in which the SHE is paired with another system containing both Fe^{3+} and Fe^{2+} in solution. The platinum electrode of the SHE is connected by a wire with a similar electrode in the iron cell, and electrical contact between the solutions is maintained through a KCl solution in a U-tube between them. When the circuit is completed, a current will flow as electrons are transferred from the solution with the higher activity of electrons to the one with the lower activity. If a voltage meter is placed in the line, it will measure the difference in electrical potential (E) between the SHE and the $Fe^{2+} \mid Fe^{3+}$ cell.

In this example and most others of interest to aqueous geochemists, the voltage measurement is given the special symbol Eh, where the h informs us that the reported value is relative to the standard hydrogen electrode. It is also common to express the redox potential of a cell in terms of pe ($= -\log_{10}a_{e^-}$). Eh and pe are easily related by

$$pe = \frac{F}{2.303RT} Eh, \qquad (6\text{-}24)$$

in which F is Faraday's constant (23.062 kcal volt^{-1} eq^{-1}) and 2.303 is the constant for conversion between natural (base e) and common (base 10) logarithms.

The implication of the last few paragraphs is that chemical potential and electrical potential are interrelated, so that it is fair to speak either of the Eh of a half-cell or of its free energy. A standard relationship from electrochemistry states this equivalence:

$$\Delta G = -nFEh, \qquad (6\text{-}25)$$

where n is the number of electrons exchanged in a reaction and F, again, is Faraday's constant. For the two half-cells in this example, we can write equilibrium constants:

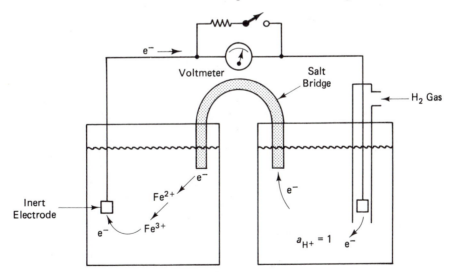

Figure 6.15 A pair of half-cells described by equations 6-22 and 6-23. When the switch is closed, electrons can be transferred through the salt bridge and wires between the two cells. When it is opened, the voltage meter measures the potential difference between them. The direction of current flow depends on the Fe^{2+}/Fe^{3+} ratio in the left-hand cell.

$$K_{eq} = \frac{a_{Fe^{3+}} a_{e^-}}{a_{Fe^{2+}}} \quad \text{and} \quad K_{SHE} = \frac{a_{H^+} a_{e^-}}{P_{H_2}^{1/2}} = 1.$$

The value of K_{eq} is calculated in now-familiar fashion from tabulated $\overline{\Delta G}_f$ values for Fe^{3+} and Fe^{2+}, and a_{e^-} is thus seen to be proportional to the ratio $a_{Fe^{2+}}/a_{Fe^{3+}}$. For the net reaction,

$$Fe^{2+} + H^+ \longleftrightarrow Fe^{3+} + \tfrac{1}{2}H_2,$$

the free energy of reaction is given by

$$\Delta G_r = \Delta G_r^0 + RT \ln (K_{eq}/K_{SHE}),$$

which is simply

$$\Delta G_r = \Delta G_r^0 + RT \ln \frac{a_{Fe^{3+}}}{a_{Fe^{2+}}}.$$

From equation 6-25, we see that this result is equivalent to

$$Eh = E^0 - \frac{RT}{nF} \log \frac{a_{Fe^{3+}}}{a_{Fe^{2+}}}. \tag{6-26}$$

The quantity E^0, called the *standard electrode potential*, is analogous to the standard free energy of reaction. It is the Eh that the system would have if all of the chemical species in both half-cells were in their standard states (that is, if each had unit activity). The logarithmic term is simply the ratio of the activity product of oxidized species (here, $a_{Fe^{3+}}$) to the activity product of reduced species (here, $a_{Fe^{2+}}$). This equation, in its general form, is called the *Nernst equation*.

From this general discussion, it should now be clear that the thermodynamic treatment of redox reactions in terms of Eh is fully compatible with the procedure we have observed since Chapter 3. In the same way that the direction of a net reaction depends on whether the products or the reactants have the lower free energy, a half-cell reaction may be viewed as "oxidizing" or "reducing" relative to the SHE, depending on the arithmetic sign on Eh. The advantage of using Eh rather than ΔG_r is that electrochemical measurements are relatively easier than calorimetry to perform.

Worked Problem 6-4

Iron and manganese are commonly carried together in river waters, but are separated quite efficiently upon entering the ocean. Ferric oxides and hydroxides precipitate in near-shore environments or estuaries, but manganese remains as soluble Mn^{2+} even in the open ocean, except where it is finally bound in nodules on the abyssal plain. Some of this behavior is because of colloid formation. Can we also explain it in terms of electrochemistry?

Two half-reactions are of interest in this problem. Iron may be converted from fairly soluble Fe^{2+} to relatively insoluble Fe^{3+} by the reaction

$$Fe^{2+} \longrightarrow Fe^{3+} + e^-,$$

for which $E^0 = +0.77$ volt. The manganese half-reaction of interest is

$$Mn^{2+} + 2 H_2O \longrightarrow MnO_2 + 4H^+ + 2e^-,$$

for which $E^0 = +1.23$ volt. We construct the net reaction by doubling the stoichiometric coefficients in the $Fe^{2+} \mid Fe^{3+}$ reaction and subtracting it from the manganese reaction to eliminate free electrons:

$$MnO_2 + 4H^+ + 2Fe^{2+} \longleftrightarrow Mn^{2+} + 2H_2O + 2Fe^{3+}.$$

The standard electrode potential for the net reaction is not calculated in the way you might expect. Unlike enthalpies and free energies, *voltages are not multiplied by the coefficients we used to generate the net reaction*. The reason is apparent from equation 6-25, from which we see indirectly that

$$E^0 = \frac{\Delta G_r^0}{nF}.$$

As the stoichiometric coefficients of the $Fe^{3+} \mid Fe^{2+}$ reaction are doubled, so is its value of ΔG_r^0 *and* the number of electrons in the reaction, n. The half-cell E^0, therefore, is unaffected. The standard potential for the net reaction is $+0.77 - (+1.23) = -0.46$ volt. This corresponds to a net ΔG_r^0 of -21.2 kcal mol^{-1}. By analogy with equation 6-26, *Eh* for this problem is calculated from

$$Eh = -0.46 + \frac{2.303R\,(298)}{2F} \log \frac{a_{Mn^{2+}}a_{Fe^{3+}}^2}{a_{H^+}^4 a_{Fe^{2+}}^2}.$$

Sundby et al. (1981) found that waters entering the St. Lawrence estuary have a dissolved manganese content of 10 $\mu g/l^{-1}$, or approximately 1.8×10^{-7} mol l^{-1}. On average, the ratio of total iron to manganese in the world's rivers is about $50:1$, approximately the ratio in average crustal rocks. We estimate, therefore, that $a_{Fe^{2+}} + a_{Fe^{3+}} = 9 \times 10^{-6}$. Unfortunately, river concentrations of Fe^{3+} are rarely reported, so we must choose a value for the activity ratio of Fe^{3+} to Fe^{2+} arbitrarily. Divalent iron is almost certainly the more abundant species but, for reasons that soon become apparent, $a_{Fe^{3+}}$ cannot be much below 1×10^{-6}. For purposes of this problem, we will assume that $a_{Fe^{3+}}/a_{Fe^{2+}}$ is $1:6$, so $a_{Fe^{3+}} = 1.3 \times 10^{-6}$ and $a_{Fe^{2+}} = 7.7. \times 10^{-6}$. We will also assume that the *pH* of water entering the St. Lawrence system is 6.0, which is typical of streams draining forest-covered terrain in this climate.

By making these assumptions and substituting appropriate values for R and F in our earlier expression for *Eh*, we calculate that

$$Eh = 0.46 - \frac{0.059}{2} \log \frac{1.8 \times 10^{-7}(1.3 \times 10^{-6})^2}{a_{H^+}^4(9 \times 10^{-6})^2}.$$

At *pH* $= 6$, the value of *Eh* predicted by this expression is very close to zero. This indicates, therefore, that the net reaction we have written is very nearly in equilibrium. As river water flows further into the estuary system and is mixed with more alkaline seawater, however, calculated values of *Eh* become progressively more negative. At *pH* $= 7$, *Eh* is -0.121 volt; at *pH* $= 8$, it is -0.239 volt. As with values of ΔG_r, negative electrical potentials indicate a tendency for the reaction to proceed to the right, in this case producing Fe^{3+} at the expense of Fe^{2+}. Ferric ions cannot accumulate much before the solubility product for a number of ferric oxides and hydroxides is exceeded, and iron is lost to sediments. The same reaction, however, favors the production of Mn^{2+} ions, which remain in solution and are washed out to sea.

Eh-pH Diagrams

As we have shown in the previous section, many redox reactions are functions of both *Eh* and *pH*. The net reaction in Worked Problem 6-4, for example, consumes four hydrogen ions and involves the transfer of two electrons. It can be easily shown that reactions in a great many systems of interest to aqueous geochemists can be illustrated on diagrams of the type we have already seen in Figures 6.5 and 6.8, but using *Eh* and *pH* as variables instead of the activities of H_4SiO_4 and aqueous cations. We will do this by discussing Figure 6.16, which represents a number of equilibria among iron oxides and water.

The uppermost and lowest lines in Figure 6.16 establish the stability limits of water in an atmosphere with a total pressure of 1 atm (that is, at the surface of the earth). These limits are derived from the net reaction

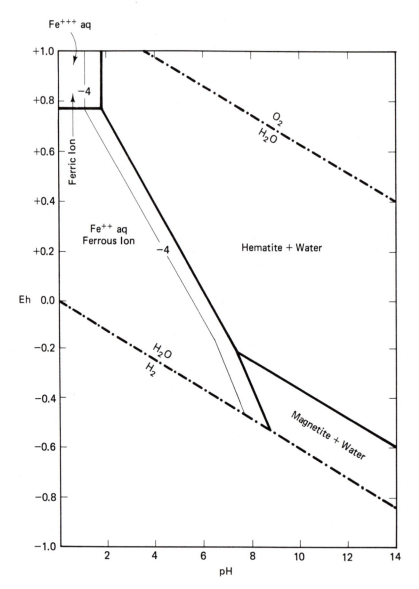

Figure 6.16 Stability fields for hematite, magnetite, and aqueous ions in water at 25°C and 1 atm total pressure. Boundaries involving aqueous species are calculated by assuming that the total activity of dissolved species is 10^{-6}. Lighter-weight lines indicate the field boundaries when the sum of dissolved species activities is 10^{-4}.

$$2H_2O \longleftrightarrow 2H_2 + O_2,$$

which is the sum of the half-reaction

$$2H_2O \longrightarrow O_2 + 4H^+ + 4e^- \qquad (6\text{-}27)$$

and reaction 6-23, which defines the SHE. *Eh* for this reaction at 25°C, therefore, is given by

$$Eh = E^0 + \frac{2.303RT}{4F} \log \frac{K_{eq}}{K_{SHE}}$$

$$= E^0 + \frac{0.059}{4} \log \frac{P_{O_2} a_{H^+}^4}{a_{H_2O}}.$$

Under the conditions we have stated, P_{O_2} is at its maximum, 1 atm. The activity of water, as a pure phase, is unity. The expression for Eh, then, simplifies to

$$Eh = E^0 + \frac{0.059}{4} \log a_{H^+}^4$$

$$= E^0 - 0.059 \, pH.$$

This is the equation for a straight line with slope -0.059 and y-intercept (E^0) equal to $+1.23$ volts, as shown in Figure 6.16. The lower stability limit for water is derived from the SHE half-reaction itself, because it is that half of the net reaction that defines the oxidation of H_2. Therefore,

$$Eh = E^0 + \frac{0.059}{2} \log \frac{a_{H^+}^2}{P_{H_2}}$$

and (because E^0 for the SHE is defined to be zero, and P_{H_2} cannot be higher than the total pressure of 1 atm),

$$Eh = -0.059 \, pH.$$

The boundaries between stability fields of iron oxides are constructed in similar fashion. The reaction forming hematite by oxidation of magnetite, for example, can be written

$$2Fe_3O_4 + \tfrac{1}{2}O_2 \longleftrightarrow 3Fe_2O_3,$$

or can be written in terms of electron exchange by combining this equation with the half-reaction for the breakdown of water (equation 6-27). We saw the result earlier as equation 6-22:

$$2Fe_3O_4 + H_2O_2 \longrightarrow 3Fe_2O_3 + 2H^+ + 2e^-.$$

Following the procedure we just used for the stability of water, we find that

$$Eh = E^0 + \frac{0.059}{2} \log \frac{a_{Fe_2O_3}^3 a_{H^+}^2}{a_{Fe_3O_4}^2 a_{H_2O}}$$

$$= E^0 + \frac{0.059}{2} \log a_{H^+}^2$$

$$= E^0 - 0.059 \, pH.$$

The value of E^0, calculated from standard free energies, is $+0.221$ volt.

For reactions involving aqueous species, we need additional information. Consider, for example, the reaction forming hematite from dissolved Fe^{2+}:

$$2Fe^{2+} + 3H_2O \longrightarrow Fe_2O_3 + 6H^+ + 2e^-.$$

Here, the equilibrium constant involves not only hydrogen ions and electrons, but ferrous ions as well, so that

$$Eh = E^0 + \frac{0.059}{2} \log \frac{a_{H^+}^6}{a_{Fe^{2+}}^2}$$

$$= E^0 - 0.059 \log a_{Fe^{2+}} - 0.177 \, pH,$$

where $E^0 = +0.728$ volt. In order to draw a line to represent this reaction in *Eh-pH* coordinates, therefore, we will need to specify an activity of ferrous ions. In drawing the Fe^{2+}-hematite boundary in Figure 6.16, we have assumed that $a_{Fe^{2+}} = 10^{-6}$, and have indicated by use of a lighter line the way that the boundary would shift if $a_{Fe^{2+}}$ were 10^{-4}. The boundary between Fe^{2+} and magnetite can be calculated in the same way.

The boundary between aqueous Fe^{3+} and hematite is defined by the reaction

$$Fe_2O_3 + 6H^+ \longleftrightarrow 2Fe^{3+} + 3H_2O,$$

which involves no free electrons. The standard free energy of reaction, ΔG_r^0, is equal to 2 kcal mol^{-1}, so

$$\log K_{eq} = \frac{-\Delta G_r^0}{2.303\ RT} = \log \frac{a_{Fe^{3+}}^2}{a_{H^+}^6} = -1.45.$$

Rearrange this result to solve for *pH*, and we conclude that

$$pH = -0.18 - \tfrac{1}{3} \log a_{Fe^{3+}},$$

which plots as a vertical line with a position that depends, again, on the activity of dissolved iron.

Finally, the line between the aqueous Fe^{3+} and Fe^{2+} fields indicates the conditions under which activities of the two species are identical. The redox reaction is

$$Fe^{2+} \longrightarrow Fe^{3+} + e^-,$$

which involves no hydrogen ions and is therefore independent of *pH*. We see easily that

$$Eh = E^0 + 0.059 \log \frac{a_{Fe^{3+}}}{a_{Fe^{2+}}}.$$

Because $a_{Fe^{2+}} = a_{Fe^{3+}}$, however, this reduces simply to $Eh = E^0$, which in this case is $+0.771$ volt.

HOW IS *pH* MEASURED?

Electrodes are commercially available in a variety of physical configurations, and can be designed for sensitivity to a particular ion or group of ions, to the exclusion of others. Regardless of the details of design, however, all rely on ion exchange to do their job. It is the transfer of ions between the electrode and a test solution that results in a current, which can then be registered in a metering circuit.

One standard design is known as the *glass-membrane* electrode; the basic operating principles of this design are illustrated in Figure 6.17. It consists of a thin, hollow bulb of glass, filled with a solution with a known activity of a cation to be measured. The glass of the bulb contains the same cation, introduced during manufacture. A wire, immersed in the filling solution, completes the electrode.

In operation, the bulb is placed in a solution to be analyzed. Because the solution inside the bulb and the solution outside generally have different activities of the cation to be measured, a chemical potential gradient is immediately established between them. Ions on the more concentrated side minimize the difference by attaching themselves to the glass, while those on the other side leave the glass to join the less concentrated solution. The glass, thus, serves as a semi-permeable membrane. Because charged particles are passing from one side to the other, an electrical imbalance is created. By connecting the wire inside the bulb to an inert or reference electrode immersed in the test solution, we allow electrons to flow and neutralize the charge difference. A voltage measured in this system is the electrode potential.

A standard *pH* electrode, found in almost any chemistry lab, is a glass electrode of this type, filled with an HCl solution of known a_{H^+}. The outer bulb is made of a Na-Ca-silicate glass, from whose surfaces sodium ions are readily leached and replaced by H$^+$. When it is

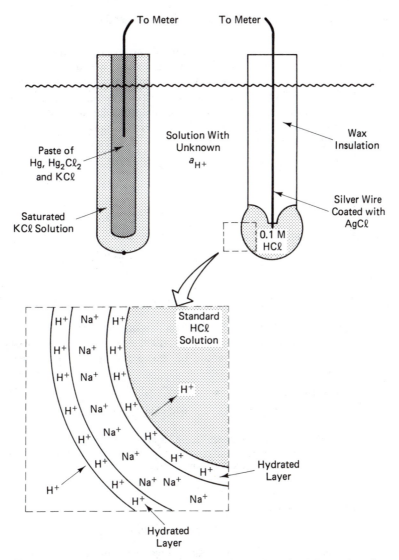

Figure 6.17 A standard glass-membrane *pH* electrode (on the right) consists of a Na-glass bulb filled with 0.1 M HCl, at the tip of a Ag-AgCl reference electrode. Inset shows the ion exchange path between the electrode and a solution with unknown *pH*. This electrode must be paired with a second one, which serves as a reference half-cell to complete the electrical circuit. The one shown on the left in this figure is a *calomel* electrode, and is based on the redox pair Hg^0 | Hg$^+$.

placed in a test solution, the glass gains extra hydrogen ions on one side and yields them on the other. Instead of a simple wire, its internal electrical connection is actually a second electrode, which consists of a silver wire coated with silver chloride. This serves as an internal reference electrode. The charge imbalance that develops within the filling solution in the glass bulb is resolved by the silver wire in the inner electrode, which reacts with the filling solution to release or gain an electron in the half-reaction: Ag + Cl$^-$ → AgCl + e$^-$. The charge imbalance in the test solution is ultimately resolved by transferring electrons between the silver wire and a reference electrode (like the SHE or a more portable equivalent) which is also immersed in the test solution. In this rather roundabout way, the a_{H^+} difference between the test solution and the filling solution is expressed as a measurable voltage.

Redox Systems Containing Carbon Dioxide

Redox reactions in weathering environments always occur in an atmosphere whose P_{CO_2} is at least as high as the atmospheric value and, as we saw earlier in this chapter, may be several hundred times greater. Under some conditions, therefore, we might expect to find siderite ($FeCO_3$) stable in the weathered assemblage. Including carbonate equilibria in the redox calculations we have just considered is not difficult. It merely requires us to write half-cell reactions to include the new phases. For the equilibrium between siderite and magnetite, we write

$$3FeCO_3 + H_2O \longrightarrow Fe_3O_4 + 3CO_2 + 2H^+ + 2e^-$$

and

$$Eh = 0.319 + 0.089 \log P_{CO_2} - 0.059\, pH. \tag{6-28}$$

For that between siderite and hematite, we write

$$2FeCO_3 + H_2O \longrightarrow Fe_2O_3 + 2CO_2 + 2H^+ + 2e^-$$

and

$$Eh = 0.286 + 0.059 \log P_{CO_2} - 0.059\, pH. \tag{6-29}$$

Notice that each of these would plot as a diagonal line with slope of -0.059 on an Eh-pH diagram, if we were to choose a fixed value of P_{CO_2}. For successively higher values of P_{CO_2}, the boundary lines would be drawn through higher values of Eh at any selected pH.

Worked Problem 6-5

What values of Eh, pH, and P_{CO_2} define the stability limits for siderite formed in weathering environments?

The answer, in graphical form, is shown by the shaded area in Figure 6.18. To see how it is derived, we rewrite equations 6-27 and 6-28 by grouping together the E^0 and log P_{CO_2} terms in each. For siderite-magnetite,

$$Eh = (0.319 + 0.089 \log P_{CO_2}) - 0.059\, pH.$$

For siderite-hematite,

$$Eh = (0.286 + 0.059 \log P_{CO_2}) - 0.059\, pH.$$

In order to form siderite in a weathering environment by either of these reactions, water must be present. Consequently, its lower stability limit is imposed by the lower stability limit of water, defined by equation 6-23, namely:

$$Eh = 0.0 - 0.059\, pH.$$

The siderite-hematite line coincides with this limiting condition when

$$0.0 = 0.286 + 0.059 \log P_{CO_2},$$

which occurs when $P_{CO_2} = 1.4 \times 10^{-5}$ atm. The siderite-magnetite line, however, coincides with the breakdown of water when

$$0.0 = 0.319 + 0.089 \log P_{CO_2},$$

which is true when P_{CO_2} is almost 20 times higher, at 2.6×10^{-4} atm. We conclude, therefore, that if the partial pressure of CO_2 in a weathering environment were less than 2.6×10^{-4} atm, no siderite could be formed. The stable phase, instead, would be magnetite.

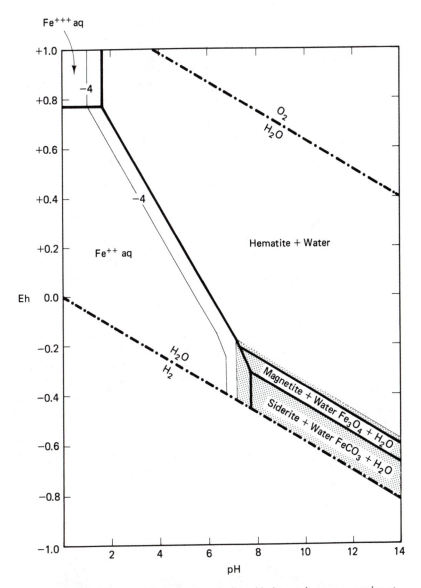

Figure 6.18 Stability of hematite, magnetite, siderite, and aqueous species at 25°C and 1 atm total pressure. The limits of the siderite field (shaded) are drawn to show its probable maximum stability in surficial weathering environments. Boundaries involving aqueous species have been drawn by assuming that the total activity of dissolved species is 10^{-6}.

The upper stability limit of siderite produced during weathering is set by the natural availability of CO_2. An absolute upper limit on P_{CO_2} is 1 atm, which we also adopted earlier as the maximum possible value for either P_{O_2} or P_{H_2}. A more realistic limit, perhaps, is 10^{-1} atm, measured by Russell (1961) in a tropical soil. At this P_{CO_2} value, the siderite-magnetite boundary is defined by

$$Eh = 0.23 - 0.059 \, pH.$$

The siderite-hematite-boundary is given by

$$Eh = 0.227 - 0.059 \, pH.$$

These two lines are almost coincidental. As a fluid initially in equilibrium with siderite is oxidized, however, it encounters the siderite-hematite boundary and passes out of the siderite stability field just before it reaches the siderite-magnetite line. In fact, at any P_{CO_2} higher than 8×10^{-2} atm, at which point equations 6–22 and 6–27 coincide, it cannot co-exist stably with magnetite at all.

In very alkaline solutions, siderite is stable over the range of Eh and P_{CO_2} values we have just found. In more acidic solutions, however, siderite breaks down according to the reaction

$$FeCO_3 + 2_{H^+} \longleftrightarrow Fe^{2+} + CO_2 + H_2O.$$

No free electrons are involved, so this boundary is a vertical line. For the maximum P_{CO_2} we chose above, assuming that $a_{Fe^{2+}} = 10^{-6}$, this line lies at $pH = 6.79$. This condition is almost the same as the magnetite-Fe^{2+} boundary shown in Figure 6.16.

We conclude from these various constraining equations that siderite may be formed only sparingly in weathering environments where P_{CO_2} is similar to the atmospheric value. In more CO₂-rich environments like tropical soils, however, siderite may be formed under roughly the same conditions in which magnetite is produced in low-CO₂ surroundings.

Activity-Activity Relationships: The Broader View

Figure 6.18 illustrates a simplified portion of the three-dimensional diagram which recognizes P_{CO_2} as a variable as well as Eh and pH. Similar diagrams for systems containing sulfur species or phosphate or nitrate are also common in the geochemical literature. Two-dimensional sections through any of these can be drawn and interpreted by analogy with the Eh-pH diagram for the system iron-water. The finished product may be a P_{CO_2}-pH diagram, a P_{CO_2}-P_{S_2} diagram, an Eh-P_{S_2} diagram, or any number of other possibilities. It is worth noting, in fact, that all of the redox relationships we have discussed are similar in form to the solubility expressions we considered in the opening sections of this chapter. You may have been struck by now with the similar appearance

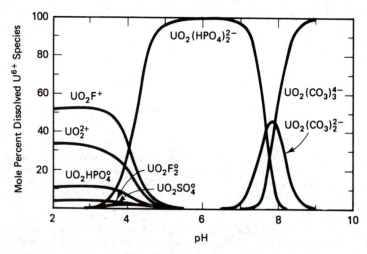

Figure 6.19 Distribution of uranyl complexes vs pH for some typical ligand concentrations in ground waters of the Wind River Formation, Wyoming, at 25°C. $P_{CO_2} = 10^{-2.5}$ atm, Σ F = 0.3 ppm, Σ Cl = 10 ppm, Σ SO_4^{2-} = 100 ppm, Σ PO_4^{3-} = 0.1 ppm, Σ Si = 30 ppm. (After D. Langmuir, Uranium solution-mineral equilibria at low temperatures with applications to sedimentary ore deposits. *Geochim. Cosmochim. Acta* 42: 547–69. Copyright 1978 by Pergamon Press. Reproduced with permission.)

of *Eh-pH* diagrams and earlier figures like 6.5. In each case, the stabilities of minerals or fluids have been expressed in terms of relative activities of key species. Generically, these may be called *activity-activity* diagrams. We will introduce an increasing number of these in later chapters.

At this juncture, we also want to emphasize that activity-activity diagrams are not the exclusive property of geochemists who study chemical weathering, despite the fact that we have presented them in that context. In fact, Figure 6.5 first gained broad use among geochemists through studies of wall-rock alteration by hydrothermal fluids (Hemley and Jones, 1964; Meyer and Hemley, 1967), and is in wide use by economic geologists. *Eh-pH* diagrams also are applied to a variety of problems in mining and mineral extraction.

Worked Problem 6-6

Much of the world's mineable uranium occurs as uraninite (UO_2) in sandstone-hosted deposits, in association with both vanadium and copper minerals. What redox conditions are necessary to form such a deposit, and how can that information be used as a guide to mining uranium?

Before we address the question directly, let us review a few features of solution chemistry relevant to the geologic occurrence of uranium. Uranium can exist in several oxidation states, of which U^{4+} and U^{6+} are most important in nature. In very acid solutions, U^{4+} can occur as a free ion under reducing conditions or can take part in fluoride complexes. More commonly, however, tetravalent uranium combines with water to form soluble hydroxide complexes. Under reducing conditions, uranium may also be present in solution in its 5+ state as UO_2^+. The tendency for uranium to combine with oxygen is even more pronounced when it is in the hexavalent state. Throughout most of the natural range of *pH*, U^{6+} forms strong complexes with oxygen-bearing ions like CO_3^{2-}, HCO_3^-, SO_4^{2-}, PO_4^{3-}, and AsO_4^{3-}, which are present in most oxidized stream and subsurface waters. At 25°C and with a typical ground water P_{CO_2} of about 10^{-2} atm, the most abundant of these are the uranyl carbonate species, which are stable down to a *pH* of about 5. Langmuir (1978), however, has shown that phosphate and fluoride complexes can be at least as abundant in some localities (see Figure 6.19). Below *pH* = 5, U^{6+} is generally in the form of UO_2^{2+}. In addition to these inorganic species, a large number of poorly understood organic chelates also contribute to uranium solubility.

In most near-surface environments, therefore, uranium is easily transported in natural waters. Very little uranium can be deposited in the 6+ state. In some localities where P_{CO_2} is probably close to the atmospheric value, uranyl complexes combine with vanadate to precipitate carnotite, $K_2(UO_2)_2(VO_4)_2$, or tyuyamunite, $Ca(UO_2)_2(VO_4)_2$, or with phosphate or silica to form autunite or uranophane. These comprise only a small fraction of mineable uranium, however. Most uranium is deposited as U^{4+} in the major ore minerals uraninite, UO_2, and coffinite, $USiO_4$.

Some of the field observations of soluble and precipitated uranium discussed in this short preamble are summarized in Figure 6.20, in which the stability fields of abundant species have been calculated in terms of *Eh* and *pH* from experimentally determined stability constants. This diagram, though quantitative, is general in nature because it has been constructed to show only those species in the U-O_2-CO_2-H_2O system that are stable at 25°C with $P_{CO_2} = 10^{-2}$ atm. With it, we can deduce qualitative fluid paths that may characterize the formation of sandstone-hosted uranium deposits.

The primary source of sedimentary uranium is probably granitic rocks, which may have as much as 2 to 15 ppm uranium. Surface waters, as we have found, have *pH* values which typically range from about 5 to 6.5. *Eh* values of these waters are usually in excess of 0.4 volt. Therefore, during weathering of the igneous source rocks, uranium is readily stabilized in solution as $UO_2CO_3^0$ or $UO_2(CO_3)_2^{2-}$. In earlier discussions (for example, in Worked Problem 6-2), we showed that progressive weathering of aluminosilicate minerals causes an increase in *pH*. This, then, may lead $UO_2(CO_3)_3^{4-}$ to play a more important role in solution. Mobile uranyl complexes may be reduced to precipitate uraninite in a number of ways, often involving the simultaneous oxidation of iron, carbon, or sulfur. Langmuir

Oxidation-Reduction Processes

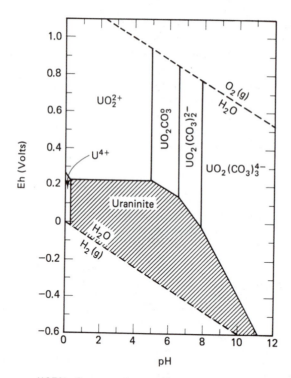

Figure 6.20 Stability of uraninite and aqueous uranium complexes at 25°C for $P_{CO_2} = 10^{-2}$ atm. Boundaries involving aqueous species are drawn by assuming the total activity of dissolved species is 10^{-6}. (Modified from Langmuir 1978.)

(1978), for example, suggests that at $pH = 8$, HS^- generated by bacterial processes may be oxidized to SO_4^{2-} by the reaction

$$4[UO_2(CO_2)_3^{4-}] + HS^- + 15H^+ \longleftrightarrow 4UO_2 + SO_4^{2-} + 12CO_2 + 8H_2O,$$

or Fe^{2+} may be oxidized to form an insoluble hydroxide mineral by

$$UO_2(CO_3)_3^{4-} + 2Fe^{2+} + 3H_2O \longleftrightarrow UO_2 + 2Fe(OH)_3 + 3CO_2.$$

Another reduction mechanism involves the adsorption of uranyl complexes on hydrocarbons, leading to their gradual oxidation and the formation of uraninite. As a result, the fluid path follows a trend toward lower Eh at roughly constant pH to enter the stability field for uraninite.

One mining technique, used at several prospects in the southwestern United States, involves reversing this reaction pathway. Typically, a system of wells is drilled into a uranium-bearing sandstone unit. Through some of these, fluorine-rich acidic solutions with a moderately high Eh are injected into the unit. Fluids are then withdrawn by pumping from the remaining wells. The extracted solution contains hexavalent uranium as a uranyl fluoride complex or as UO_2^+. In this way, uranium can be mined efficiently without a significant amount of excavation.

SUMMARY

In this chapter, we have explored ways to apply our knowledge of solubility equilibria to problems in chemical weathering. We have shown that, although silica minerals and magnesian silicates dissolve congruently, most major rock-forming minerals decompose to produce secondary minerals. Prominent among these are clay minerals and various oxides and hydroxides.

We can examine the chemical pathways along which these changes occur by studying either the residual phases or associated waters. To appreciate the driving forces for weathering, however, we need to understand processes that control the abundance of major weathering agents: CO_2, organic acids, and oxygen. We have seen that it is often desirable to express the last of these in redox reactions, written to show the exchange of electrons between oxidized and reduced species.

Finally, we have shown that the concepts and methods used in this chapter can be used for problems other than those in chemical weathering. Economic geologists, in particular, are fond of using *Eh-pH* diagrams and other activity-activity diagrams to interpret the environments of ore transport and deposition. We will explore other types of activity-activity diagrams and their uses in later chapters.

SUGGESTED READINGS

The literature on chemical weathering is voluminous, and many good references are available in applied fields of soil science, water chemistry, and economic geology. This list is intended to be representative, not exhaustive:

BARNES, H. L., ed. 1979. *Geochemistry of hydrothermal ore deposits*, 2d ed., New York: Wiley Interscience. (This massive volume contains basic papers on a variety of geochemical topics of interest to the economic geologist. Chapter 5, by A. W. Rose and D. M. Burt, contains good examples and discussions of hydrothermal alteration.)

BOWERS, T. S.; JACKSON, K. J.; and HELGESON, H. C. 1984. *Equilibrium activity diagrams*. New York: Springer Verlag. (This book contains only 8 pages of text, but is the most extensive compilation of activity-activity diagrams available.)

COLMAN, S. P., and DETHIER, D. P., eds. 1986. *Rates of chemical weathering of rocks and minerals*, New York: Academic Press. (This is an excellent compendium of papers at the introductory and advanced level on mechanisms and kinetics of chemical weathering.)

DREVER, J. I. 1982. *The geochemistry of natural waters*. Englewood Cliffs, N.J.: Prentice-Hall. (This text provides a good introduction to the geochemistry of natural waters. Chapters 5, 8, and 11 are particularly relevant to topics in this chapter.)

GARRELS, R. M., and CHRIST, C. L. 1965. *Solutions, minerals, and equilibria*. New York: Harper and Row. (This book has been the standard in aqueous geochemistry for a generation of geologists. There is more to be learned from this single text than from almost any other one you might choose.)

The following papers were referenced in this chapter. You may wish to consult them for further information.

BAKER, W. E. 1986. Humic substances and their role in the solubilization and transport of metals. In *Mineral exploration: Biological systems and organic matter*, ed. Carlisle, D.; Berry, W. L.; Kaplan, I. R.; and Watterson, J. R., 377–407. Englewood Cliffs, N. J.: Prentice-Hall.

BARNES, I. and O'NEIL, J. R. 1969. The relationship between fluids in some fresh Alpine-type ultramafics and possible modern serpentinization, western United States. *Geol. Soc. Amer. Bull.* 80: 1947–60.

BARSHAD, I., 1966. The effect of variation in precipita-

tion on the nature of clay mineral formation in soils from acid and basic igneous rocks. In *Proc. Internat. Clay Conf. Jerus.* 1, 167–73.

CAWLEY, J. L.; BURRUSS, R. C.; and HOLLAND, H. D. 1969. Chemical weathering in central Iceland: An analog of pre-Silurian weathering. *Science* 165: 391–92.

COUTURIER, Y.; MICHARD, G.; and SARAZIN, G. 1984. Constantes de formation des complexes hydroxydés de l'aluminum en solution aqueuse de 20 à 70°C. *Geochim. Cosmochim. Acta* 48: 649–60.

DECONICK, F. 1980. Major mechanisms in formation of spodic horizons. *Geoderma* 24: 101–28.

GIBBS, R. J. 1967. The geochemistry of the Amazon River system: I. The factors that control the salinity and the composition and concentration of the suspended solids. *Geol. Soc. Amer. Bull.* 78: 1203–32.

GIBBS, R. J. 1972. Water chemistry of the Amazon River. *Geochim. Cosmochim. Acta* 36: 1061–66.

HELGESON, H. C.; DELANY, J. M.; NESBITT, H. W.; and BIRD, D. K. 1978. Summary and critique of the thermodynamic properties of rock-forming minerals. *Amer. J. Sci.* 278A: 1–229.

HEMLEY, J. J., and JONES, W. R. 1964. Chemical aspects of hydrothermal alteration with emphasis on hydrogen metasomatism. *Econ. Geol.* 59: 538–569.

HOLLAND, H. D. 1978. *The chemistry of the atmosphere and oceans*. New York: Wiley Interscience.

JOHNSON, N. M.; LIKENS, G. E.; BORMANN, F. H.; and PIERCE, R. S. 1968. Rate of chemical weathering of silicate minerals in New Hampshire. *Geochim. Cosmochim. Acta* 32: 531–45.

LANGMUIR, D. 1978. Uranium solution-mineral equilibria at low temperatures with applications to sedimentary ore deposits. *Geochim. Cosmochim. Acta* 42: 547–69.

LEVENTHAL, J. S. 1986. Roles of organic matter in ore deposits. In *Organics and ore deposits*, ed. Dean, W. E., 7–20. Wheat Ridge, Colo: Denver Region Exploration Geologists Society.

LIKENS, G. E.; BORMANN, F. H.; PIERCE, R. S.; EATON, J. S.; and JOHNSON, N. M. 1977. *Biogeochemistry of a forested ecosystem*. New York: Springer Verlag.

LIVINGSTONE, D. A. 1963. Chemical composition of rivers and lakes. *U.S. Geol. Survey Prof. Paper 440G*.

MEYER, C., and HEMLEY, J. J. 1967. Wall rock alteration. In *Geochemistry of hydrothermal ore deposits,* 1st ed., ed. Barnes, H. L., 166–235. New York: Holt, Rinehart, and Winston.

NEDELL, S. S., and McKAY, C. P. 1987. Possible formation of carbonates in ancient lakes in the Valles Marineris, Mars: A search of the Mariner 6/7 IRS dataset. In *Abstr. XVIII Lunar Planet. Sci. Conf.,* 712–13. Houston, TX: Lunar and Planetary Institute.

NESBITT, H. W.; MARKOVICS, G.; and PRICE, R. C. 1980. Chemical processes affecting alkalis and alkaline earths during continental weathering. *Geochim. Cosmochim. Acta* 44: 1659–66.

NORTON, D. 1974. Chemical mass transfer in the Rio Tanama system, west-central Puerto Rico. *Geochim. Cosmochim. Acta* 38: 267–77.

PETIT, J-C.; DELLAMEA, G.; DRAN, J-C.; SCHOTT, J.; and BERNER, R. A. 1987. Mechanism of diopside dissolution from hydrogen depth profiling. *Nature* 325: 705–07.

POLLACK, J. B.; KASTING, J. F.; RICHARDSON, S. M.; and POLIAKOFF, K. 1987. The case for a wet, warm climate on early Mars. *Icarus* 71: 203–224.

REESMAN, A. L.; PICKETT, E. E.; and KELLER, W. D. 1969. Aluminum ions in aqueous solutions. *Amer. J. Sci.* 267: 99–113.

RUSSELL, E. W. 1961. *Soil conditions and plant growth.* New York: John Wiley and Sons.

SCHNITZER, M. 1986. Reactions of humic substances with metals and minerals. In *Mineral exploration: Biological systems and organic matter,* ed. Carlisle, D.; Berry, W. L.; Kaplan, I. R.; and Watterson, J. R., 408–27. Englewood Cliffs, N.J.: Prentice-Hall.

SOLOMON, D. K., and CERLING, T. E. 1987. The annual carbon dioxide cycle in a montane soil: Observations, modeling, and implications for weathering. *Water Resour. Res.* 23: 2257–2265.

SUNDBY, B.; SILVERBERG, N.; and CHESSELET, R. 1981. Pathways of manganese in an open estuarine system. *Geochim. Cosmochim. Acta* 45: 293–307.

WITKAMP, M., and FRANK, M. L. 1969. Evolution of CO_2 from litter, humus, and subsoil of a pine stand. *Pedobiologia* 9: 358–65.

WOLERY, T. J. 1979. Calculation of chemical equilibrium between aqueous solution and minerals: the EQ3/6 software package. (NTIS Document UCRL-52658.) Livermore, Calif: Lawrence Livermore Laboratory.

WOOD, W. W., and PETRAITIS, M. J. 1984. Origin and distribution of carbon dioxide in the unsaturated zone of the southern high plains of Texas. *Water Resour. Res.* 20: 1193–1208.

PROBLEMS

1. Using appropriate data from this chapter or from literature sources, calculate the value of $a_{H_4SiO_4}$ for which gibbsite and kaolinite are in equilibrium with an aqueous fluid at 25°C.

2. Following the approach used in Worked Problem 6-2, give a qualitative description of the fluid reaction path for pure water during dissolution of potassium feldspar, assuming that amorphous silica rather than quartz is precipitated. Write all relevant reactions.

3. How much total silica (Σ Si in ppm) is in equilibrium with quartz at $pH = 9$? At $pH = 10$? At $pH = 11$?

4. A typical bottled beer has a H^+ activity of about 2×10^{-5}. Assuming that it remains in the bottle long enough to reach equilibrium with the glass and that the effect of organic complexes is negligible, what concentrations of H_4SiO_4, $H_3SiO_4^-$, and $H_2SiO_4^{2-}$ should we expect to find in the beer?

5. Assume that two species, A and B, can each exist in either an oxidized or a reduced form. For the redox reaction

$$A_{ox} + B_{red} \longleftrightarrow A_{red} + B_{ox},$$

in which only one electron is transferred, what must be the value of E^0 at 25°C if the equilibrium ratios A_{red}/A_{ox} and B_{ox}/B_{red} are found to be 1000 : 1?

6. What is the oxidation potential (Eh) for water in equilibrium with atmospheric oxygen ($P_{O_2} = 0.21$ atm) if its pH is 6?

7. Using data from Helgeson (1969), calculate the value of log K_{fo} in equation 6-6 and plot a line representing the congruent dissolution of forsterite on Figure 6.2. Explain why forsterite is metastable with respect to the other magnesian silicates shown.

8. Using data in Worked Problem 6-1, estimate the total dissolved aluminum content (Σ Al in ppm) of water near the mouth of the Amazon River. Assume that runoff is in equilibrium with kaolinite, that Σ Si is 11 ppm, and the pH is 6.3.

9. By writing reactions in the system $Fe-O_2-H_2O$ in terms of transfer of oxygen rather than transfer of electrons, redraw Figure 6.15 as an f_{O_2}-pH diagram.

10. E^0 for the reaction $Ce^{3+} \rightarrow Ce^{4+} + e^-$ is -1.61 volt. The ratio Ce^{3+}/Ce^{4+} in surface waters in the ocean has been estimated to be 10^{17}. If the redox half-reaction for cerium is in equilibrium with the half-reaction $Fe^{2+} \rightarrow Fe^{3+} + e^-$ at 25°C, what should be the Fe^{2+}/Fe^{3+} ratio in these waters? How does this value compare with the ratio you would calculate from the reaction

$$4Fe^{3+} + 2H_2O \longleftrightarrow 4Fe^{2+} + 4H^+ + O_2,$$

assuming that the pH of seawater is 8.2?

11. Write the half-reaction for the reference electrode shown on the left side of Figure 6.17. What is its standard potential at 25°C? What is the net reaction for this pH measurement system?

Chapter 7

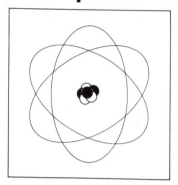

Using Stable Isotopes

OVERVIEW

This is the first of two chapters on isotope geochemistry. In it, we introduce the language and some of the basic concepts used by geochemists who use *stable* isotopes of hydrogen, carbon, nitrogen, oxygen, and sulfur to interpret geologic processes and environments. Then, with the aid of several case studies, we show how these concepts can be applied to problems of practical interest. In each case study, the common principle is mass fractionation. We explain how the different stable isotopes of a single element can be separated from each other in a variety of thermal or biochemical processes, and how we can trace the chemical history of a system by measuring the abundance ratios of these isotopes in coexisting phases. Some fractionation processes can be interpreted by the familiar tools of equilibrium thermodynamics. Among these are simple isotopic exchange reactions that have been used successfully as geothermometers. Other processes are kinetically controlled. In many cases, abundance patterns produced by these can yield insights into the pathways along which isotopic changes have occurred.

THE LANGUAGE OF ISOTOPE GEOCHEMISTRY

Some of the most dramatic advances of this century in the geological sciences have been made in the field of isotope geochemistry. It is a discipline with a surprisingly short history, having attracted talented geochemists in large numbers since only about 1950. Because of this intense interest, however, and because methods of isotope geochemistry have proven useful in a wide range of problems, we cannot say all that we would like in a single chapter. So we begin by examining many of the fascinating types of information that can be gleaned by studying *stable* isotopes in nature. The basis for this area of geochemistry lies in Harold C. Urey's discovery of deuterium in 1932, and his recognition that the isotopes of hydrogen vary in relative abundance from one geologic environment to another. Most of the major advances, however, have been made since World War II. Beginning in the late 1940s, Urey and his students were pioneers in the use of oxygen isotopes for geothermometry and in an increasing number of provenance studies. Geochemists have also developed ways of using isotopes of carbon, sulfur, and nitrogen to trace chemical pathways in the crust.

More specifically, in this chapter we consider equilibrium and kinetic effects that result in the fractionation of stable isotopes, and show how they may be used to interpret petrologic, hydrologic, and ore-forming processes. The study of *radioactive* isotopes has also offered us a number of tools for unraveling the record of crust-mantle interactions and for understanding part of the thermal engine that drives plate tectonics, as well as providing geologists with their most reliable means for dating the earth. That, however, is another story, which we will address in Chapter 12.

Terminology and Symbols

Most of the topics we discuss in this text deal with the interactions between electrons on the outer fringes of atoms. It is these low-mass, negatively-charged particles that are engaged in forming bonds between atoms and that are exchanged or rearranged during chemical reactions. In this chapter and its later companion (Chapter 12), however, we discuss the nuclei of atoms instead. The nuclei constitute most of the mass of atoms but contribute only in subtle ways to interactions among them.

Nuclear physics has now revealed an extremely large list of particles known to make up the nucleus of an atom. For our purposes, however, we recognize only two: *protons,* which carry a unit positive charge, and *neutrons,* which are only slightly more massive and bear no electrical charge at all. Taken together, protons and neutrons are referred to as *nucleons.*

Three numbers, describing the abundance and type of nucleons in an atom, are in common use. The first of these, symbolized by the letter Z, counts the number of protons in a nucleus and is known as the *atomic number*. It is this number which identifies the element that the atom represents (any atom with $Z = 8$, for example, is an atom of oxygen, one with $Z = 16$ is sulfur, and so forth). Because an electrically neutral atom must have an equal number of protons and electrons, Z also tells us the number of electrons in an uncharged atom. The second useful number is the *neutron number,* symbolized by the letter N, which indicates the number of neutrons in the nucleus. Within limits imposed by the attractive and repulsive forces among nucleons, atoms of any element may have different numbers of neutrons. Atoms for which Z is the same but N is different are called *isotopes.* The third number in common use is the sum of Z and N, given the symbol A. Because neutrons and protons differ in mass by only a factor of 1 part in 1836, it is common practice to assign each the same arbitrary unit mass. The quantity A, then, represents the total mass of the nucleus or (because electrons have an almost negligible mass) the entire atom. From a practical perspective, it is easier to measure the total mass of an atom than to count its neutrons, so A is a more useful quantity to identify than N. In shorthand form, therefore, we convey our knowledge about a particular atom (or *nuclide,* as it is generally called when the focus is on nuclear properties) by writing its chemical symbol, a familiar alternative for Z, preceded by a superscript that specifies A. For the nuclide consisting of 8 protons and 8 neutrons, for example, we write ^{16}O. For the one with 26 protons and 30 neutrons, we write ^{56}Fe.

Roughly 1700 nuclides are known. For reasons we will explore more fully in

The Language of Isotope Geochemistry

Chapter 12, however, only about 260 of these are stable. The remainder have nuclei that may spontaneously disintegrate to produce subatomic particles and leave a nuclide in which Z or N has changed. In this chapter, we will investigate only a small group of light stable isotopes of hydrogen, carbon, nitrogen, oxygen, and sulfur which have proven useful in geochemistry. Isotopes of silicon and boron have also been used by geochemists, but have drawn less attention and will not be discussed here.

What Makes Stable Isotopes Useful?

Interpretations based on studies of these stable isotopes all depend on the fact that various kinetic and equilibrium processes separate or *fractionate* light from heavy isotopes. By understanding the mechanisms involved, we can measure the relative abundances of light and heavy isotopes of an element in natural materials and interpret the pathways and environments they have seen. Isotopes of the five elements listed above share a set of characteristics that make them particularly well-suited for this sort of work:

1. They all have low atomic mass. Nuclides with Z greater than 16 do not fractionate efficiently in nature.

2. The relative difference in mass between heavy and light isotopes of each of these nuclides is fairly large. The difference is largest for hydrogen, whose heavy isotope, ^2H (commonly known as deuterium and given the symbol D) is twice as massive as the light isotope, ^1H. At the other extreme on our short list of stable nuclides, ^{34}S is still 6.25% heavier than ^{32}S. Because isotopes fractionate on the basis of relative mass, this is an important criterion.

3. All five of the elements are abundant in nature and constitute a major portion of common earth materials. Oxygen, in particular, makes up almost 47% of the crust by weight (nearly 92% by volume).

4. Carbon, nitrogen, and sulfur exist in more than one oxidation state and may therefore participate in processes over a wide range of redox conditions.

5. Each of the five elements forms bonds with neighboring atoms that may range from ionic to highly covalent. All other factors being equal, fractionation of heavy and light isotopes will be greatest between phases that have markedly different bond types or bond strengths. For this reason, most other light elements like magnesium or aluminum, which form the same type of bonds in almost all common earth materials, do not show marked isotopic fractionation.

6. For each of the five elements, the abundance of the least common isotope still ranges from a few tenths of a percent to a few percent of the most common isotope. From an analytical point of view, this means that the precision of measurements is quite high.

Three different types of notation to express the degree of isotopic fractionation are in common use. Chemists generally describe the distribution of stable isotopes between coexisting phases A and B in terms of a *fractionation factor*, $\alpha_{A\text{-}B}$, defined by

$$\alpha_{A\text{-}B} = R_A/R_B, \tag{7-1}$$

in which R is the ratio of the heavy to the light isotope in the phase indicated by the subscript. The fractionation of ^{18}O and ^{16}O between quartz and magnetite, for example, would be indicated by the magnitude of

$$\alpha_{\text{qz-mt}} = \frac{(^{18}O/^{16}O)_{\text{qz}}}{(^{18}O/^{16}O)_{\text{mt}}}.$$

Concentrations, rather than activities, are used in calculating fractionation factors because activity coefficients for isotopes of the same element are nearly identical and would cancel each other. Values for α are generally between 1.0000 and 1.0040.

Because it is commonly inconvenient to use absolute isotopic ratios (R_A/R_B) and because α values commonly differ only in the third or fourth decimal place, geochemists commonly express their data in terms of delta (δ) notation instead. Using this convention, the isotopic ratio in a sample is compared to the same ratio in a standard with the formula

$$\delta = \frac{R_{\text{sample}} - R_{\text{standard}}}{R_{\text{standard}}} \times 1000. \qquad (7\text{-}2)$$

The numerical value that results from this procedure is a measure of the deviation of R in parts per thousand, or *permil* (‰, by analogy with percent) between the sample and the standard. Samples with positive values of δ are said to be isotopically heavy (that is, they are enriched in the heavy isotope relative to the standard); those with negative values are isotopically light.

For each isotopic system, laboratories have agreed on one or two readily available substances to serve as standards. For hydrogen and oxygen, the universal standard is *Standard Mean Ocean Water* (SMOW). In older studies, $\delta^{18}O$ values were commonly referred to the isotopic ratio in a Cretaceous marine belemnite from the Pee Dee formation in South Carolina. Today, the PDB scale for ^{18}O is only used in paleoclimatology studies. Carbon isotopic measurements, however, are almost always compared to $^{13}C/^{12}C$ in the Pee Dee Belemnite, so most people think of PDB as a standard for carbon rather than oxygen.[1] Nitrogen measurements are all reported relative to the $^{15}N/^{14}N$ ratio in air, and sulfur measurements are reported relative to the $^{34}S/^{32}S$ ratio in troilite (FeS) from the Canyon Diablo meteorite (whose impact produced Meteor Crater in Arizona). In practice, each laboratory develops its own standards, which are then calibrated against these universal standards.

Geochemists also use the symbol Δ to compare δ values for coexisting substances. We can derive this third quantity from either of the other systems of notation. From equation 7-1, we see that

$$\alpha_{A\text{-}B} - 1 = \frac{R_A - R_B}{R_B}.$$

From equation 7-2, we also see that

$$\delta_A = \frac{R_A - R_{\text{standard}}}{R_{\text{standard}}} \times 1000$$

and

$$\delta_B = \frac{R_B - R_{\text{standard}}}{R_{\text{standard}}} \times 1000,$$

[1] In more recent literature, SMOW and PDB are called V-SMOW and V-PDB, referring to reference standards available from the International Atomic Energy Agency (I.A.E.A.) in Vienna. Comparisons between these standards, to earlier samples of SMOW and PDB, and to other commonly used oxygen standards, can be found in Coplen et al. (1983).

so

$$\alpha_{A\text{-}B} - 1 = \frac{\delta_A - \delta_B}{1000 + \delta_B}.$$

Because δ_B is small compared to 1000, we can write this last result as

$$1000(\alpha_{A\text{-}B} - 1) \simeq \delta_A - \delta_B.$$

For each of the light stable isotope pairs except D-H, fractionation factors for exchange between geologic materials at equilibrium are on the order of 1.000 to 1.004. For values in this range, it is a very good approximation to write

$$\ln \alpha_{A\text{-}B} \simeq \alpha_{A\text{-}B} - 1.$$

(The natural logarithm of 1.004, for example, is 0.00399). With this approximation, we finally see that

$$1000 \ln \alpha_{A\text{-}B} \simeq \delta_A - \delta_B = \Delta_{A\text{-}B}. \tag{7-3}$$

In this way, fractionation factors between coexisting minerals can be calculated from their respective δ values, measured relative to a universal laboratory standard.

Worked Problem 7-1

If atmospheric CO_2 has a $\delta^{13}C$ value of $-7\permil$ and HCO_3^- in a sample of river water has a measured value of $+1.24\permil$, what are the equivalent values of Δ and the fractionation factor $\alpha_{HCO_3^-\text{-}CO_2}$?

We can easily substitute these values into equation 7-3 to see that

$$\delta_{HCO_3^-} - \delta_{CO_2} = 1.24\permil - (-7\permil) = 8.24\permil$$

and

$$\alpha_{HCO_3^-\text{-}CO_2} = \exp\left(\Delta_{HCO_3^-\text{-}CO_2}/1000\right) = 1.00827.$$

Fractionation factors and delta values can be derived in several different ways, which generally yield compatible results. Experiments involving isotopic exchange can be performed in a laboratory setting, and fractionation can be determined by analyzing the run products. In some cases, pairs of natural materials can also be analyzed if independent information about the conditions of isotopic exchange between them is available. For a great many situations, however, values of α are calculated from principles of statistical mechanics, using spectroscopic data. Because our approach to thermodynamics in this textbook has been macroscopic rather than statistical, we have not laid the foundation for a quantitative appraisal of this third technique. In a brief qualitative fashion, however, it is possible to appreciate the thermodynamic driving force for isotopic fractionation.

Mass Fractionation and Bond Strength

Fractionation among isotopes occurs because some thermodynamic properties of materials depend on the masses of the atoms of which they are composed. The internal energy of a molecule in a gas, for example, gains contributions from translational, rotational, and vibrational components in its movement relative to other atoms in the gas. In a solid or liquid, the internal energy is related largely to the stretching or vibrational frequency of bonds between the atom and adjacent ligands. According to principles of statistical mechanics, the vibrational frequency of a molecule is inversely proportional to its mass. Therefore, if two stable isotopes of a light element are distributed randomly into molecules of the same substance, the vibrational frequency associated with bonds to the lighter isotope will be greater than that with bonds to the heavier one. As a result, bonds formed with the lighter isotope will be somewhat weaker and more easily broken than bonds with the heavier one. The effect of isotopic composition on the physical properties of a phase can be surprisingly large, as we have shown by example in Table 7-1. Molecules containing the lighter isotope are more readily extracted from a material during processes like melting or evaporation. In general, then, the lighter isotope of a pair like D-H or ^{18}O-^{16}O tends to fractionate into a vapor phase rather than a liquid or into a liquid rather than a solid. This is the basis for many kinetic mass fractionation processes.

TABLE 7-1 PHYSICAL PROPERTIES OF WATER AS A FUNCTION OF ISOTOPIC COMPOSITION

	$H_2^{16}O$	$D_2^{16}O$
Melting point	0.0°C	3.81°C
Boiling point	100.0°C	101.42°C
Max. density	at 3.98°C	at 11.23°C
Viscosity	8.9 mpoise	10 mpoise

By inverting this logic, we can see that an isotope, given a choice between coexisting phases, will fractionate between them on the basis of bonding characteristics. Bonds to small, highly charged ions have higher vibrational frequencies than those to larger ions with lower charge. Substances with such bonds tend to accept heavy isotopes preferentially, because by doing so they can reduce the vibrational component of their internal energy. This, in turn, reduces the energy available for chemical reactions—the free energy of the system—and makes the phase more stable.

The effect of substituting heavy isotopes into a substance whose bonds already have lower vibrational frequencies is generally less pronounced. There is a tendency, consequently, for substances like quartz or calcite, in which oxygen is bound to small Si^{4+} or C^{4+} ions by covalent bonds, to incorporate a high proportion of ^{18}O. These minerals, therefore, have characteristically large positive values of $\delta^{18}O$. In feldspars, this tendency is less dramatic, largely because a great proportion of the Si^{4+} ions in the silicate framework have been replaced by Al^{3+}. The contrast is even greater in magnetite, where oxygen forms ionic bonds to Fe^{2+} and Fe^{3+}, which are much larger and less highly charged than Si^{4+}. These bonds, therefore, have a lower vibrational frequency. Magnetite, therefore, tends to prefer the lighter isotope, ^{16}O. At the extreme, uraninite (UO_2) is commonly among the most ^{18}O-depleted minerals found in nature, as we might predict from the large radius and mass of the U^{4+} ion.

GEOLOGIC INTERPRETATIONS BASED ON ISOTOPE FRACTIONATION

The distribution of stable isotopes among minerals and fluids has proven to be a highly fruitful area of research in geochemistry. In this section, we will examine some of the more prominent problems to which isotopic interpretations have been applied. This survey is not intended to be comprehensive, but rather to suggest the wide variety of geologic questions that can be addressed from the perspective of stable isotope fractionation.

Thermometry

Fractionation of stable isotopes between minerals can be described by writing simple chemical reactions in which the only differences between the reactants and the products are their isotopic compositions. Water and calcite, for example, may each contain both ^{16}O and ^{18}O, so the isotopic exchange between them is given by

$$CaC^{16}O_3 + H_2^{18}O \longleftrightarrow CaC^{18}O_3 + H_2^{16}O. \qquad (7-4)$$

The equilibrium constant, K_{eq}, for this reaction is equal to

$$K_{eq} = \frac{a_{CaC^{18}O_3} a_{H_2^{16}O}}{a_{CaC^{16}O_3} a_{H_2^{18}O}},$$
$$= \frac{(^{18}O/^{16}O)_{CaCO_3}}{(^{18}O/^{16}O)_{H_2O}}$$
$$= \alpha_{CaCO_3-H_2O}.$$

The free energy change for this reaction is due only to the very slight difference between the vibrational energy states of a ^{16}O-C bond and a ^{18}O-C bond in calcite and between a ^{18}O-H bond and a ^{16}O-H bond in water. Unlike mineral reactions we have discussed so far, in which ΔG_r^0 is on the order of tens of kilocalories per mole, the free energy associated with this exchange is therefore only a few calories per mole. As written, the exchange reaction has a small negative ΔG_r^0, so $\alpha_{CaCO_3-H_2O}$ has a positive value.

In one of the first significant geochemical studies involving stable isotope fractionation, Nobel laureate Harold Urey (1947) recognized that the isotopic exchange between ^{16}O and ^{18}O (presented earlier in this chapter as equation 7-4) could be used as the basis of a paleothermometer to estimate the temperature of the ancient ocean. At 25°C, the fractionation factor for the calcite-water system has a value of 1.0286, which means that $\delta^{18}O$ is equal to +28.6‰, if the water we have been talking about is SMOW. The value of α is a function of temperature, such that with decreasing temperature, calcite becomes more ^{18}O-enriched relative to seawater. If accurate determinations of $\delta^{18}O$ could be made on marine limestones, it should be possible to calculate the ocean temperature at their time of formation.

Despite early promise, this particular paleothermometer has not proven reliable except in sediments deposited since about the Miocene epoch. Its most serious limitation is that the oxygen isotopic composition of seawater has varied considerably through time, so that the isotopic fluctuations in carbonate sediments do not reflect changes in temperature alone. Later in this chapter, for example, we will discuss kinetic effects that cause polar ice to be isotopically lighter than mean seawater. As glacial epochs have come and gone and the balance of ^{18}O and ^{16}O has shifted between ice and the oceans, the $\delta^{18}O$ of the oceans may have varied by as much as 1.4‰. Other practical problems related to postburial changes in the isotopic composition of limestone and to the biochemical fractionation of oxygen by shell-forming organisms have contributed further complications to carbonate paleothermometry.

Although Urey's original idea for a marine carbonate thermometer has turned out to be of limited value in practice, the basic concept is a good one. Especially at metamorphic or igneous temperatures, kinetic complications are rarer than at 25°C, and a large number of silicate and oxide mineral pairs have been shown to yield valid temperature estimates. Oxygen isotope thermometry is particularly attractive in studies of deep crustal processes, because isotopic fractionation takes place almost independently of pressure. This is a substantial advantage over conventional geothermometers, which we will examine in Chapter 8.

Figure 7.1 shows how a number of fractionation factors for oxygen isotopes between quartz and other rock-forming minerals vary with temperature. These have been calculated from spectroscopic data, using a quantum mechanical model for lattice dynamics. More commonly, plots of this type are constructed from experimental data. At temperatures above 1000 K it has been shown that the natural log of α is proportional to $1/T^2$, so it is convenient to construct plots like Figure 7.1 on which equations of the form

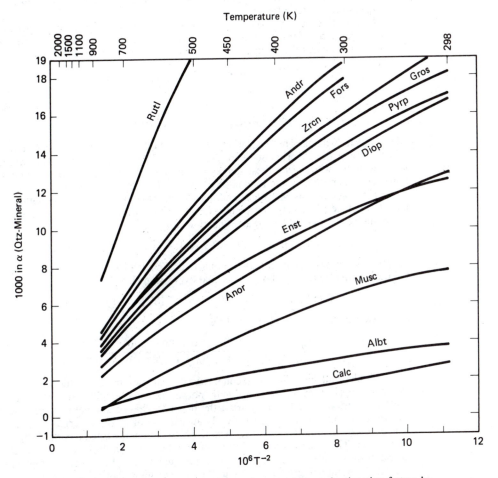

Figure 7.1 Temperature dependence of oxygen isotope fractionation factors between quartz and common rock-forming minerals, calculated from spectroscopic data. Abbreviations: *Rutl* = Rutile; *Andr* = Andradite; *Fors* = Forsterite; *Zrcn* = Zircon; *Gros* = Grossular; *Pyrp* = Pyrope; *Diop* = Diopside; *Enst* = Enstatite; *Anor* = Anorthite; *Musc* = Muscovite; *Albt* = Albite; *Calc* = Calcite. (After O'Neil, J. R., Theoretical and experimental aspects of isotopic fractionation. In *Stable isotopes in high temperature geological processes*, ed. Valley, J. W.; Taylor, H. P., and O'Neill, J. R., 1–40. Reviews in Mineralogy 16. Copyright 1986 by the Mineralogical Society of America. Reproduced with permission.)

$$1000 \ln \alpha = A(10^6 \, T^{-2}) + B \qquad (7\text{-}5)$$

plot as straight lines. A and B are empirical constants. For many pairs of geologic materials, this proportionality can be used to extrapolate to temperatures much lower than 1000 K. Deviations are most obvious below 500 K but are often small enough to make extrapolation reasonable. The following problem is typical of many in isotope geothermometry.

Worked Problem 7-2

The Crandon Zn-Cu massive sulfide deposit in northeastern Wisconsin is a volcanogenic ore body within a series of intermediate to felsic volcanic rocks. Munha et al. (1986) have sampled andesitic flows and tuffs adjacent to the ore body and determined the oxygen isotope compositions of quartz and plagioclase in them. How can these be used to estimate the temperature at which the ore body formed?

Matsuhisa et al. (1979) performed a series of isotopic exchange experiments between quartz and water, and other experiments between plagioclase and water. By performing least-squares fits of equation 7-5 to their data, they determined that

$$1000 \ln \alpha_{\text{qz-H}_2\text{O}} = 3.13 \times 10^6 \, T^{-2} - 2.94,$$
$$1000 \ln \alpha_{\text{Ab-H}_2\text{O}} = 2.39 \times 10^6 \, T^{-2} - 2.51,$$

and

$$1000 \ln \alpha_{\text{An-H}_2\text{O}} = 1.49 \times 10^6 \, T^{-2} - 2.81,$$

where T is in Kelvins. These equations are strictly valid only over the range from 673 K to 773 K, where the experiments were performed. Considering the likely uncertainties we will encounter when using these curves to determine temperatures from natural samples, however, it is probably safe to apply them somewhat outside this range.

If we assume, as Matsuhisa and his co-workers did, that the fractionation factor for an intermediate plagioclase varies in a linear fashion with its anorthite content, then we can combine the two feldspar-water equations to yield

$$1000 \ln \alpha_{\text{pl-H}_2\text{O}} = (2.39 - 0.9 \, \text{An}) \times 10^6 \, T^{-2} - (2.51 + 0.3 \, \text{An}).$$

To describe the fractionation between quartz and plagioclase, we subtract this result from the quartz-water equation above. The final expression is

$$1000 \ln \alpha_{\text{qz-pl}} = (0.74 - 0.9 \, \text{An}) \times 10^6 \, T^{-2} - (0.43 - 0.3 \, \text{An}),$$

which we can solve for temperature to get

$$T = \left[\frac{(0.74 - 0.9 \, \text{An}) \times 10^6}{(\delta^{18}\text{O}_{\text{qz}} - \delta^{18}\text{O}_{\text{pl}}) + 0.43 - 0.3 \, \text{An}} \right]^{1/2}.$$

Munha and his colleagues collected two samples from the footwall rocks at the Crandon deposit. One was from a slightly porphyritic andesite flow, the other from a chlorite lapilli tuff about 2 km away from the deposit. These show petrographic evidence of *propylitic* alteration (characterized by the appearance of chlorite, serpentine, and epidote in hydrothermally altered andesitic rocks), which commonly causes marked albitization of plagioclase. The mole fraction of anorthite in feldspars from these rocks is roughly 0.15. In the flow sample, the measured $\delta^{18}\text{O}$ for quartz is $+9.38‰$ and for plagioclase is $+6.71‰$. By inserting these values in the last equation above, we find that $T = 533$ K (260°C). In the tuff, $\delta^{18}\text{O}_{\text{qz}} = +9.53‰$ and $\delta^{18}\text{O}_{\text{pl}} = +6.44‰$, from which we calculate that $T = 503$ K (230°C). These are compatible with temperatures from 240° to 310°C measured by independent techniques.

Isotope geothermometry is perhaps most heavily used by economic geologists, for problems like the one shown above. Sulfur as well as oxygen isotopes are commonly measured to determine the temperature of ore formation. A few of the most popular sulfur isotope thermometers are indicated in Figure 7.2 and Table 7-2. Where sulfide minerals can be shown to have formed in isotopic equilibrium with a hydrothermal fluid, these can provide valuable insights into the conditions of mineralization. Unfortunately, however, sulfur isotopic systems are easily upset by kinetic factors. The oxidation state and *pH* of ore-forming fluids, for example, can seriously affect the rates of isotopic exchange between aqueous species and ore minerals, particularly at low temperature.

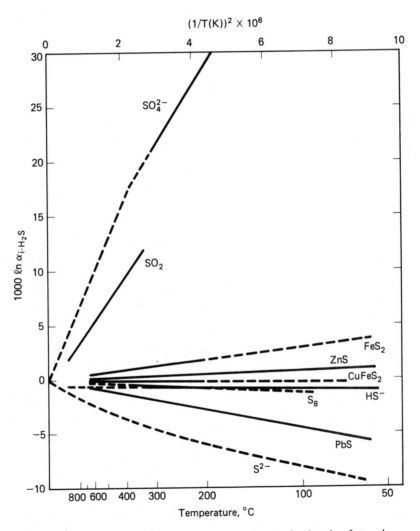

Figure 7.2 Temperature dependence of sulfur isotope fractionation factors between minerals or fluid species and H_2S. Solid curves are experimentally determined. Dashed curves are estimated or extrapolated. (After Ohmoto, H. and Rye, R. O., Isotopes of sulfur and carbon. In *Geochemistry of hydrothermal ore deposits.* 2d ed, ed. H. L. Barnes, 509–67. Copyright 1979 by John Wiley and Sons. Reproduced with permission.)

Geologic Interpretations Based on Isotope Fractionation

TABLE 7-2 SELECTED SULFUR ISOTOPE THERMOMETERS*
(T IN KELVINS; $\Delta = \delta^{34}S_A - \delta^{34}S_B = 1000 \ln \alpha_{A\text{-}B}$)

Pyrite-Galena	$T = \dfrac{(1.01 \pm 0.04) \times 10^3}{\Delta^{1/2}}$
Sphalerite-Galena or Pyrrhotite-Galena	$T = \dfrac{(0.85 \pm 0.03) \times 10^3}{\Delta^{1/2}}$
Pyrite-Chalcopyrite	$T = \dfrac{(0.67 \pm 0.04) \times 10^3}{\Delta^{1/2}}$
Pyrite-Pyrrhotite or Pyrite-Sphalerite	$T = \dfrac{(0.55 \pm 0.04) \times 10^3}{\Delta^{1/2}}$

*Values from Ohmoto and Rye (1979).

Isotopic Evolution of the Oceans

In Chapter 4, we found that SO_4^{2-}, at an average concentration of 2700 ppm, is the second most abundant anion in seawater. The total mass of sulfur in the oceans is approximately 3.7×10^{21} g. These values have probably not changed significantly since the Proterozoic Era. The isotopic composition of oceanic sulfate, however, has varied considerably. Today, marine SO_4^{2-} is very nearly homogeneous, with a $\delta^{34}S$ value of almost $+20‰$. In the past 600 million years, that value has ranged from about $+10$ to almost $+30‰$. As illustrated in Figure 7.3, these systematic variations, reflected in the isotopic composition of sulfate evaporites, offer promise as a stratigraphic tool.

The specific causes of variation are not well understood, but they are widely interpreted to represent a dynamic balance between redox processes that supply and remove sulfur from the ocean. Sulfur enters the ocean primarily as sulfate derived from a variety of continental sources and dissolved in river water. Among these sources are sulfidic shales and carbonate rocks, volcanics and other igneous rocks, and evaporites. Sulfur leaves the ocean as sulfate evaporates or in reduced form as pyrite and other iron sulfides in anoxic muds.

In general, sulfides are isotopically light compared to sulfates. This is due, in part, to equilibrium fractionation (see Figure 7.2), but is also a product of kinetic fractionation by bacteria. Metabolic pathways in sulfate-reducing bacteria involve enzyme-catalyzed reactions that favor ^{32}S over ^{34}S, because ^{32}S-O bonds are more easily broken than ^{34}S-O bonds. The degree of fractionation depends on sulfate concentration as well as temperature. It is not uncommon to find sedimentary sulfide that is more than $50‰$ lighter than contemporaneous gypsum in marine evaporites.

The isotopic composition of sulfate entering the oceans, therefore, may vary as a function of time because of fluctuations in global volcanic activity, or because different proportions of evaporites and sulfidic shales are exposed to continental weathering and erosion. In the same way, changes in the rate of deposition of pyritic sulfur in marine sediments may affect the $\delta^{34}S$ value in seawater sulfate. Berner (1987) has shown that such changes may be the indirect result of fluctuations in the mass and diversity of plant life during its evolution on the continents, or may reflect changes in the relative dominance of euxinic and "normal" marine depositional environments. By studies of this type, the grand-scale processes that have moderated the global cycle for sedimentary sulfur (and carbon and oxygen, which mimic or mirror the history of sulfur) are be-

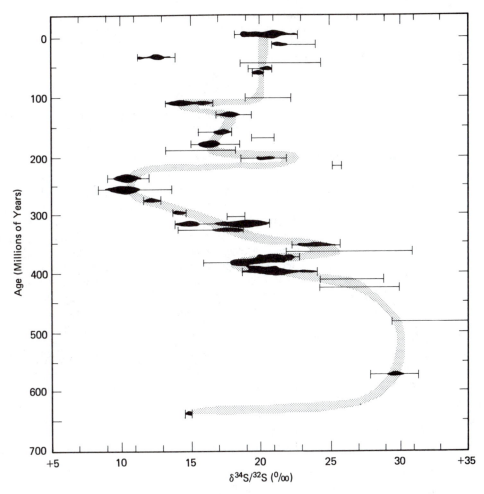

Figure 7.3 The variation of the sulfur isotopic composition of sulfate evaporites through the Phanerozoic Era has been used to estimate the isotopic composition of the oceans through time. Most measured values lie within the dark shaded areas. Dashed lines indicate the range of extreme values. (Modified from Holser and Kaplan, 1966)

coming well understood. The events that have given rise to specific fluctuations in the marine $\delta^{34}S$ record, however, remain largely unknown.

Worked Problem 7-3

What is the maximum change in $\delta^{18}O$ that could have occurred in ocean water due to the weathering of igneous and metamorphic rocks to form sediments during the history of the earth?

To answer, we need a few basic pieces of information. The average $\delta^{18}O$ value for igneous rocks is about $+8‰$, and that for sedimentary and metamorphic rocks is roughly $+14‰$. These numbers are not very well constrained, but they are good enough for a rough calculation. The mass of the oceans (Appendix C) is 1.4×10^{24} g, and the mass of sedimentary and metamorphic rocks in the crust is about 2.1×10^{24} g.

Assume that all of the sedimentary and metamorphic rocks were once igneous, and therefore once had a $\delta^{18}O$ value of $+8\%o$, or $8/1000$ heavier than the present ocean. Their $\delta^{18}O$ value, then, has increased by $+6\%o$. The complementary change in the composition of the hydrosphere must have been

$$+6\%o\frac{2.1 \times 10^{24} \text{ g}}{1.4 \times 10^{24} \text{ g}} = 9\%o.$$

The oxygen isotopic mix in the oceans, in other words, has been becoming lighter gradually through geologic time as a result of crustal weathering reactions.

Fractionation in the Hydrologic Cycle

Because there are two stable isotopes of hydrogen (^1H and ^2D) and three of oxygen (^{16}O, ^{17}O, and ^{18}O), there are nine different ways to build isotopically distinct water molecules. Masses of these molecules range from a low of 18 for $^1H_2\,^{16}O$ to a high of 22 for $^2D_2\,^{18}O$. In an earlier section, we found that the vibrational frequency associated with bonds to a light isotope is greater than that for bonds to a heavy isotope. Consequently, "light" water molecules escape more readily from a body of water into the atmosphere than do molecules of "heavy" water. When water evaporates from the ocean surface, it has a δD value of about $-80\%o$ and a $\delta^{18}O$ value around $-9\%o$. The degree of fractionation is, of course, a function of temperature. The result at any place on the earth, however, is that atmospheric water vapor always has negative δD and $\delta^{18}O$ values relative to SMOW.

The reverse process occurs when water condenses in the atmosphere. Condensation is nearly an equilibrium process, favoring heavy water molecules in the rain or snow. The first rain to fall from a "new" cloud over the ocean, therefore, has values for both δD and $\delta^{18}O$ that are close to $0\%o$. Water vapor remaining in the atmosphere, however, is systematically depleted in deuterium and ^{18}O by this process. Subsequent precipitation is derived from a vapor reservoir that has delta values even more negative than freshly evaporated seawater. The isotopic ratio (R) in the remaining vapor is given by

$$R = R_0 f^{(\alpha-1)}, \tag{7-6}$$

where R_0 is the initial $^{18}O/^{16}O$ value in the vapor, f is the fraction of vapor remaining, and α is the fractionation factor. Equation 7-6, which is known as the *Rayleigh distillation equation*, will also appear in a later chapter as we consider element fractionation during crystallization from a melt.

Worked Problem 7-4

Suppose that rain begins to fall from an air mass whose initial $\delta^{18}O$ value is $-9.0\%o$. If we assume that the fractionation factor ($\alpha_{\text{liq-vap}}$) is 1.0092 at the condensation temperature, what should we expect the isotopic composition of the air mass to be after 50% of the vapor has recondensed? What happens when 75% or 90% has been recondensed? If we were to collect all of the rain that fell as the air mass approached total dryness, what would its bulk $\delta^{18}O$ value be?

Using Equation 7-2, we can convert the R values to equivalent delta notation:

$$(\delta + 1000) = 1000(R/R_{\text{SMOW}}). \tag{7-7}$$

Equation 7-6, therefore, can be written in the form

$$\frac{R}{R_0} = \frac{\delta^{18}O + 1000}{(\delta^{18}O)_0 + 1000} = f^{(\alpha-1)}$$

or

$$\delta^{18}O = [(\delta^{18}O)_0 + 1000]f^{(\alpha-1)} - 1000. \tag{7-8}$$

For the problem at hand, we substitute appropriate values into this expression. After 50% recondensation,

$$\delta^{18}O = [(-9.0‰) + 1000](0.5)^{(0.0092)} - 1000$$
$$= -15.3‰.$$

After 75%,

$$\delta^{18}O = [(-9.0‰) + 1000](0.25)^{(0.0092)} - 1000$$
$$= -21.6‰.$$

After 90%,

$$\delta^{18}O = [(-9.0‰) + 1000](0.1)^{(0.0092)} - 1000$$
$$= -29.8‰.$$

What happens to the isotopic composition of *rainwater* during this evolution? To find out, we write

$$\alpha = \frac{R_{rain}}{R_{vapor}} = \frac{\delta^{18}O_{rain} + 1000}{\delta^{18}O_{vapor} + 1000}.$$

By rearranging terms, we find that

$$\delta^{18}O_{rain} = \alpha(\delta^{18}O_{vapor} + 1000) - 1000. \tag{7-9}$$

We use the results of the first calculation to find that after 50% condensation, $\delta^{18}O$ of rain is

$$1.0092(-15.3 + 1000) - 1000 = -6.2‰.$$

After 75%, $\delta^{18}O_{rain} = -12.6‰$ after 90%, it reaches $-20.9‰$. With continued precipitation, therefore, rainwater becomes lighter. Notice, however, that if we collected *all* of the rain that falls, conservation of mass would require that its bulk $\delta^{18}O$ value be the same as the initial $\delta^{18}O$ value of the water vapor.

Because of this fractionation process, $\delta^{18}O$ values for fresh water are always negative compared to seawater. A similar fractionation affects hydrogen isotopes. For both isotopic systems, the separation between rainwater and seawater becomes more pronounced as air masses move farther inland or are lifted to higher elevations. Also, because the fractionation factors for both oxygen and hydrogen isotopes become larger with decreasing temperature, the difference between seawater and precipitation increases toward the poles. The lightest natural waters on Earth are in snow and ice at the South Pole, where $\delta^{18}O$ values less than $-50‰$ and δD less than $-450‰$ have been measured. These various trends are illustrated for the North American continent in Figure 7.4.

These observations make it possible to recognize the isotopic signature of meteoric waters in a particular region and to use that information to study the evolution of surface and subsurface waters. Analyses of $\delta^{18}O$ and δD are commonly displayed on a two-isotope plot, like Figure 7.5. Meteoric waters on such a plot lie along a straight line given by

$$\delta D = 8 \, \delta^{18}O + 10.$$

Two-isotope plots have been used for a number of interesting purposes. Shallow ground waters tend to retain the isotopic pattern they inherit from rain. Water in deep

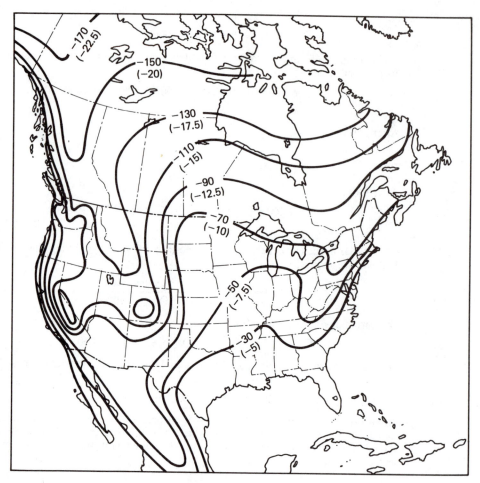

Figure 7.4 Values of δD and δ^{18}O (in parentheses) in rainwater become increasing negative as moist air moves inland, to higher elevations, or to higher latitudes. (After Sheppard, S. F. M.; Nielsen, R. L.; and Taylor, H. P., Oxygen and hydrogen isotope ratios of clay minerals from porphyry copper deposits. *Econ. Geol.* 64: 755-77. Copyright 1969 by the Economic Geology Publishing Co. Reproduced with permission.)

aquifers, however, can deviate from local rainwater in many ways. Because of climatic changes, for example, the isotopic composition of modern rainwater may be significantly different from that of water which fell thousands of years ago and is now buried in deep aquifers, even though both lie on the meteoric water line. Siegel and Mandle (1984) have sampled well waters drawn from the Cambrian-Ordovician aquifer system in southern Minnesota, Iowa, and northern Missouri. They find that δ^{18}O values in the north, where the aquifer is exposed, are close to modern rainwater values. To the south, however, δ^{18}O becomes steadily more negative. In northern Missouri, well water is as much as 11‰ lighter than present precipitation. A contour map of δ^{18}O values (Figure 7.6), therefore, can serve in this case as a qualitative guide to rates of groundwater movement during the past 12,000 years.

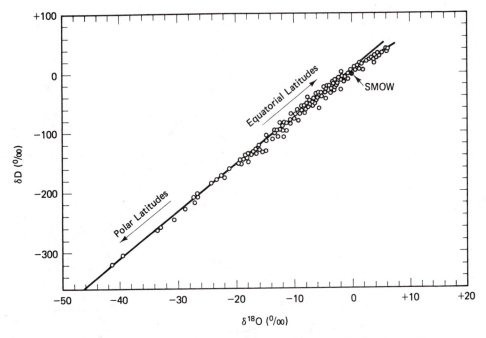

Figure 7.5 Oxygen and hydrogen isotopic values in meteoric water vary with temperature, but always lie along the meteoric water line in this two-isotope plot. (Modified from Craig, 1961)

Fractionation in Geothermal and Hydrothermal Systems

Where water and rocks combine chemically, shifts can be much larger and more obvious. Exchange reactions between ground water and rocks generally cause isotopic values to leave the meteoric water line. This tendency is perhaps most easily seen in analyses of geothermal brines. The amount of hydrogen in most minerals is small, so most of the hydrogen in a water-saturated rock is in the water. Exchange reactions, therefore, produce only negligible shifts in the deuterium content of geothermal brines. Particularly at elevated temperatures, however, large changes in $\delta^{18}O$ are possible. More than half of the oxygen in a typical water-rock system is in the rock. Reactions like 7-4 or those involving silicates commonly enrich geothermal waters in ^{18}O relative to rainwater. Four horizontal arrows on Figure 7.7 indicate the trends measured at Steamboat Springs (Colorado), and Lassen Park, the Salton Sea, and The Geysers (all in California) by Ellis and Mahon (1977). These data have helped to resolve a long-standing question about the relative importance of juvenile (mantle-derived) waters and recycled meteoric water in geothermal systems. Trends like the four shown here should converge on the composition of juvenile water if any appreciable mixing had taken place. Because they do not, geochemists conclude that water in geothermal systems is derived almost entirely from local precipitation.

This conclusion appears reasonable for any water-dominated hydrothermal system, but what about those where water must circulate through plutonic rocks or through country rock with a very low permeability? In these, the water-to-rock ratio should be

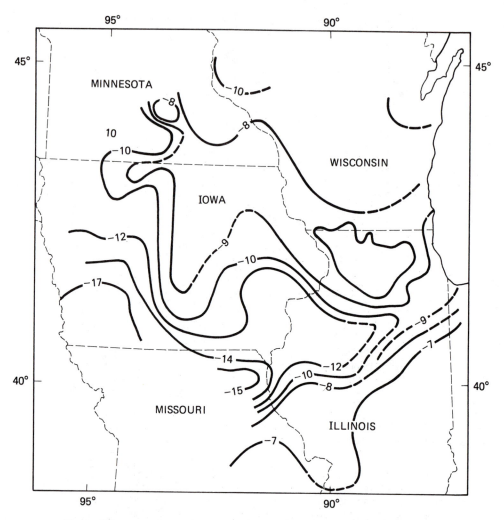

Figure 7.6 Variation of $\delta^{18}O$ in well waters drawn from the Cambrian-Ordovician aquifer system in the north-central midwestern United States. Values are more negative to the southwest, where the aquifer is deepest. These waters may have been introduced from glacial melt-water sources during the late Pleistocene epoch. (Modified from Seigel and Mandle, 1984.)

very low, and there may be comparable amounts of exchangeable hydrogen in the rock and in circulating fluid. In rocks with low permeabilities, therefore, we should observe that both $\delta^{18}O$ and δD shift in the fluid phase.

This is an exciting possibility, because it offers a way to solve a problem that has interested economic geologists for a long time. Particularly where circulating meteoric waters have mobilized and concentrated economically important metals, economic geologists want to know *how much* water has been involved in exchange reactions. Without making some assumptions about the nature of the subsurface plumbing in an area, it is difficult to calculate the ratio of water to rock during a period of hydrothermal alteration. Suppose, though, that we observed shifts in δD and $\delta^{18}O$ for the mineralizing fluid in a given deposit. Maybe we could use those observations to estimate how much water had passed through. The following problem illustrates one attempt of this type.

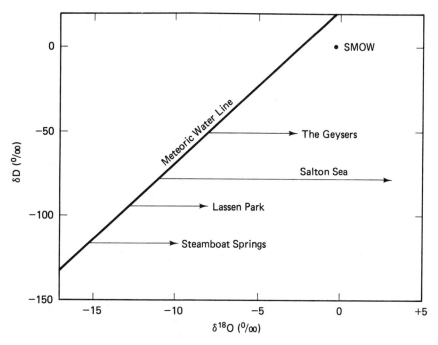

Figure 7.7 Isotopic composition of geothermal waters. (Data from Ellis and Mahon, 1977.)

Worked Problem 7-5

The San Cristobal ore body in the Peruvian Andes is a wolframite-quartz vein deposit associated with the intrusion of Tertiary quartz monzonite into Paleozoic and lower Permian metamorphic rocks. Oxygen and hydrogen isotopic data have been gathered in the area by Campbell et al. (1984). How can these data be used to estimate the water to rock ratio during hydrothermal mineralization?

Taylor (1979) has pointed out that isotopic mass balance for either oxygen or hydrogen must be expressed by

$$W \, \delta_{H_2O}(i) + R \, \delta_{rock}(i) = W \, \delta_{H_2O}(f) + R \, \delta_{rock}(f),$$

in which W and R are the total atomic percent of exchangeable oxygen or hydrogen in water and rock, respectively, and i and f stand for initial and final states in the system. The delta values for fresh meteoric water and unaltered rock are easily obtained by direct measurement. The values for the fluid or rock in the final state, however, cannot be determined without assuming that they were in isotopic equilibrium. If this assumption is reasonable, then we can use an equation in the form of Equation 7-5 to calculate $1000 \ln \alpha_{rock-H_2O}$ ($= \Delta = \delta_{rock} - \delta_{H_2O}$) at the temperature of mineralization. If $\delta_{rock}(f)$ is known, we can then rearrange the first equation and use the definition of Δ to eliminate $\delta_{H_2O}(f)$. This yields the expression

$$\frac{W}{R} = \frac{\delta_{rock}(f) - \delta_{rock}(i)}{\delta_{H_2O}(i) - (\delta_{rock}(f) - \Delta)}.$$

In this problem, we are actually more interested in finding out how final values of δD_{H_2O} and $\delta^{18}O_{H_2O}$ are affected by relative masses of water and rock, so it is best to recast this expression again. With minor manipulation to eliminate $\delta_{rock}(f)$ instead of $\delta_{H_2O}(f)$, we find the isotopic composition of water in equilibrium with granite to be given by

Geologic Interpretations Based on Isotope Fractionation

$$\delta_{H_2O}(f) = \frac{w}{w + zr} \delta_{H_2O}(i) + \frac{zr}{w + zr}(\delta_{rock}(i) - \Delta).$$

The quantities w and r now refer to mass units of water and rock instead of atomic proportions of oxygen or hydrogen in each, and a new quantity z has been added to correct for the different amounts of hydrogen or oxygen in water and rock. Water is 89 weight % oxygen, and granite contains roughly 45 weight % oxygen, so z for oxygen is 45/89—about 0.5. Bulk rock analyses of granites typically contain about 0.6 weight percent water, so z for hydrogen is approximately 0.006.

At San Cristobal, Campbell and his co-workers determined that fresh granite has a $\delta^{18}O$ value of 7‰ and $\delta D = -70$‰. The initial composition of meteoric water can be estimated by analyzing fluids trapped at low temperature in late-stage barite. These values are $\delta D = -140$‰ and $\delta^{18}O = -20$‰. The value of Δ for hydrogen was calculated from a fractionation equation for biotite and water, since biotite is assumed to contain most of the OH in granite:

$$\Delta = \delta D_{bio} - \delta D_{H_2O} = -21.3 \times 10^6 \, T^{-2} - 2.8.$$

The value of Δ for oxygen was calculated in a similar way from the fractionation equation for plagioclase (An_{30}) and water:

$$\Delta = \delta^{18}O_{pl} - \delta^{18}O_{H_2O} = 2.68 \times 10^6 \, T^{-2} - 3.53.$$

The temperature for both of these calculations was taken to be 400°C (673K).

By inserting these various parameter values into the expressions for $\delta_{H_2O}(f)$ and varying the relative magnitudes of w and r, Campbell and colleagues generated the plot shown in Figure 7.8. From it, we see that hydrothermal fluids that are produced by exchange between meteoric water and rock at high w/r ratios (> 0.2) show only the $\delta^{18}O$ shift observed in geothermal systems. With decreasing water/rock ratios, however, changes in $\delta^{18}O$ are minimal while δD begins to become more positive. The actual water to rock ratio at San Cristobal can be estimated by plotting delta values measured in trapped fluids, or values calculated from equilibrium fractionation equations between quartz and water or wolframite and water. These values, also indicated on Figure 7.8, tell us that w/r was probably between 0.1 and 0.003.

As this worked example indicates, the isotopic composition of hydrothermal fluids, and of coexisting rocks, depends greatly on the relative amounts of water and rock in a system. Because this is so, it is not always clear that an interpretation like the one by Campbell et al. (1984) is appropriate. How could we distinguish between fluids that are meteoric waters modified by exchange with rock and fluids that are simply mixtures of meteoric and juvenile water? In the San Cristobal study, Campbell and company used the equations for Δ_{pl-H_2O} and Δ_{bio-H_2O} to calculate the composition of water in equilibrium with granite at 800°C. This "magmatic" water, plotted on Figure 7.8, is colinear with fresh meteoric water and the calculated hydrothermal fluid in the box labeled "wolframite." The fluid in equilibrium with quartz, however, has a much lower δD value. Campbell concludes from this that wolframite at San Cristobal may have been deposited by a mixed meteoric-magmatic fluid, but that quartz probably was not.

Another approach has been used to distinguish between meteoric and magmatic fluids in large-scale porphyry metal deposits. Mineralization in porphyry copper and molybdenum deposits characteristically occurs at the border between a porphyritic granite or granodiorite intrusion and country rock. Economically extractable sulfide minerals are disseminated in a broad, highly altered zone and in veins and fractures that cut across both the intrusion and its host. It has long been recognized that this mineralization was promoted by the migration of heated water. Before isotopic data began to

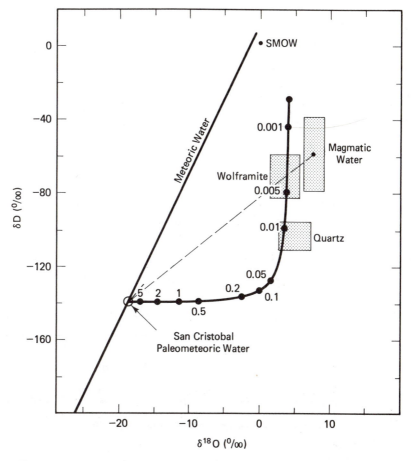

Figure 7.8 Isotopic compositional values of waters in the San Cristobal, Peru, hydrothermal system. Values along the solid curve are calculated for various ratios of water to rock. By comparing these to values for water in isotopic equilibrium with quartz and wolframite in the deposit, Campbell et al. (1984) estimated that the water-to-rock ratio during ore-formation was between 0.01 and 0.003. Wolframite could also have been deposited from a mixture of meteoric and magmatic waters, as indicated by the dashed line.

accumulate on these systems, however, the source of this water was widely believed to be magmatic.

Several research teams during the 1970s (summarized by Taylor, 1979) investigated the oxygen and hydrogen systematics in sericite, pyrophyllite, and hypogene clay minerals from North American porphyry metal deposits. As shown in Figure 7.9, delta values for each of these lie in a belt parallel to the meteoric water line. As we compare the deposits to each other, we see also that there is a clearly recognizable latitude trend among them. Both δD and $\delta^{18}O$ decrease consistently as we sample northward from Arizona and New Mexico (Santa Rita, Safford, Copper Creek, Mineral Park) through Utah, Colorado, and Nevada (Gilman, Bingham, Ely, Climax) to Montana, Washington, and Idaho (Butte, San Poil, Ima, Wickes). This trend looks so much like what we

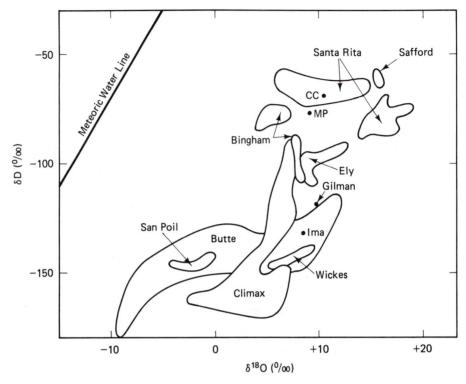

Figure 7.9 Isotopic compositions of OH-bearing minerals from a selection of North American porphyry metal deposits. Values are more negative for northern deposits, indicating that the hydrothermal fluid was probably circulating meteoric water. (Data from Sheppard et al., 1969 and Taylor, 1979.)

see along the meteoric water line itself that it is now commonly believed that porphyry metal deposits are generated largely by recycling local meteoric water.

Fractionation in Sedimentary Basins

Basin brines present yet another class of trends on a two-isotope plot for oxygen and hydrogen. It was once widely believed that pore waters in deep sedimentary basins were *connate* fluids (that is, samples of seawater trapped at the time of sediment accumulation). Clayton et al. (1966) and others more recently have shown that waters within a given basin are generally distinct. Values for both $\delta^{18}O$ and δD show considerable scatter, but the least saline brines in any basin tend to lie close to the meteoric water line while more saline fluids generally contain heavier oxygen and hydrogen. Furthermore, as with porphyry copper and molybdenum deposits, isotopic values display a latitudinal variation similar to that for surface waters. As indicated on Figure 7.10, Gulf Coast basin brines are much heavier than those in the Alberta Basin. The primary source for these brines, therefore, seems to be local rainwater. The degree of scatter and the disparate slopes of isotopic trends from basin to basin, however, suggest that other sources (silicate or carbonate rocks, true connate water, clays, or hydrocarbons) are also important.

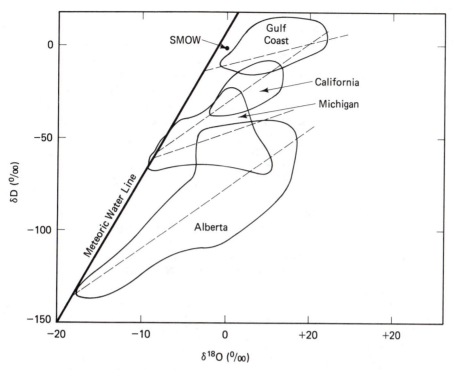

Figure 7.10 Isotopic compositions of basin brines are increasingly negative from south to north, indicating the influence of meteoric waters. The scatter in values, and the variable trends within basins (indicated by dashed lines) are due to the influence of other sources such as true connate waters or magmatic fluids, and to isotopic exchange with host rocks. (Summarized from data in Taylor, 1979.)

Fractionation Among Carbon Compounds

Two stable isotopes of carbon, ^{12}C and ^{13}C, exist in nature. On a global scale, $\delta^{13}C$ values generally range from about $-30‰$ to $+25‰$ relative to PDB, although methane in some natural gas fields has been found with values as low as $-70‰$, and some carbonates in meteorites have measured $\delta^{13}C$ values in excess of $+60‰$. In general, carbon in the reduced compounds found in living and fossil organisms is isotopically light, while carbonate minerals (particularly in marine environments) are isotopically heavy. This twofold distinction itself is useful in characterizing the sources of secondary carbonate in sediments. The carbonate cap rocks on salt domes, for example, have $\delta^{13}C$ values that are very negative (about $-36‰$) compared to marine carbonates (near $0‰$). This has been taken as evidence that they are produced by oxidation of methane in adjacent hydrocarbon-rich strata, rather than from recrystallized marine limestones. In a similar way, fresh-water carbonates and dripstone in caves are characteristically light, because they are formed from waters that are charged with soil CO_2. As we saw in Chapter 6, soil CO_2 is largely the product of plant respiration and decay, so it should be ^{12}C-rich relative to atmospheric CO_2.

The complex biochemical pathways in living organisms offer a large number of opportunities to fractionate carbon isotopes. Some of these have been investigated, with limited success, for use as paleoenvironmental indicators or as guides in provenance studies. Green plants fractionate carbon isotopes in at least three metabolic steps during photosynthesis, each of which favors the retention of ^{12}C rather than ^{13}C. In broad terms, these involve (1) the physical assimilation of CO_2 across cell walls, (2) the conversion of that CO_2 into intermediate compounds by enzymes, and (3) the synthesis of large organic molecules. The degree of fractionation in each step is kinetically controlled by such factors as the atmospheric or intracellular partial pressure of CO_2.

In addition, terrestrial plants can be divided into two large groups on the basis of the enzyme that dominates in step two and the types of large organic molecules that form in step three. Those in one group, which includes trees, bushes, and grains like wheat and rice, are known as C_3 plants because the initial product of CO_2 fixation is a molecule with three carbon atoms. In these, lignin, cellulose, and waxy lipids constitute a large portion of the total organic matter. By contrast, C_4 plants contain relatively high proportions of carbohydrate and protein. These include corn, sugar cane, many aquatic plants, and some grasses. The two groups are roughly the same ones we identified with Type III and Type I hydrocarbons, respectively, at the end of Chapter 5.

WHAT DID EARLY MAN EAT?

Many of the early social and economic systems that marked the rise of civilized mankind were attempts to guarantee a steady and nutritious diet. With this in mind, archaeologists try to determine the diets of early societies in order to gain insights into their daily life and community structure. Unfortunately, there are few reliable ways to ascertain what early man ate. Operating on the theory that "you are what you eat," however, a small group of geochemists working with stable carbon and nitrogen isotopes have begun working on the problem.

Just as $\delta^{13}C$ values can be used to distinguish between C_3 and C_4 plants, or between terrestrial and marine vegetation, stable nitrogen isotope ratios can be used to separate plants into two groups. Those which can assimilate N_2 directly from the atmosphere, called *legumes*, have $\delta^{15}N$ values near 0‰. Beans and peas belong in this group. Nonleguminous plants, like wheat and corn, cannot use nitrogen directly from the air, but rely instead on ammonia and nitrates released by bacteria in the soil. Before the advent of modern chemical fertilizers, these had $\delta^{15}N$ values that averaged around +9‰.

In the early 1980s, Michael DeNiro and Christine Hastorf, a geochemist and an archaeologist at UCLA, became curious about cooking pots unearthed at an archaeological site in the central highlands of Peru. These typically contained a layer of dark organic material which, upon close inspection, turned out to be food that had burned during cooking. Laboratory simulations convinced DeNiro and Hastorf (1985) that the food had been vegetables, rather than meat, and that it had retained its original isotopic composition. By studying the isotopic ratios in these encrustations, they were able to determine that early inhabitants of the area (from 200 B.C. to 1000 A.D.) ate nonleguminous C_3 plants, like tubers. Later inhabitants, however, began to cook C_4 plants (almost certainly corn). This shift in diet probably reflects advances in agricultural methods in the region.

Other observations can help isotope geochemists distinguish between marine and terrestrial foods. In laboratory experiments in which animals are fed diets of known isotopic composition, no significant shift in carbon isotope ratios between plant and animal tissue has been observed. These results are consistent with the observation that $\delta^{13}C$ values in marine animals are generally 5‰ less negative than in terrestrial animals. The same experiments, however, have shown that with each step up a food chain, organisms tend to concentrate more ^{15}N. Carnivores, therefore, have $\delta^{15}N$ values about 3‰ more positive than herbivores. In the oceans, where food chains are longer, animal tissues have $\delta^{15}N$ values as much as 10‰ more positive than terrestrial herbivores.

These trends can be shown on a two-isotope plot of carbon and nitrogen, as we have done in Figure 7.11. We have also indicated average delta values from bone samples collected at three archaeological sites, summarized by DeNiro (1987). From these data, he concludes

---

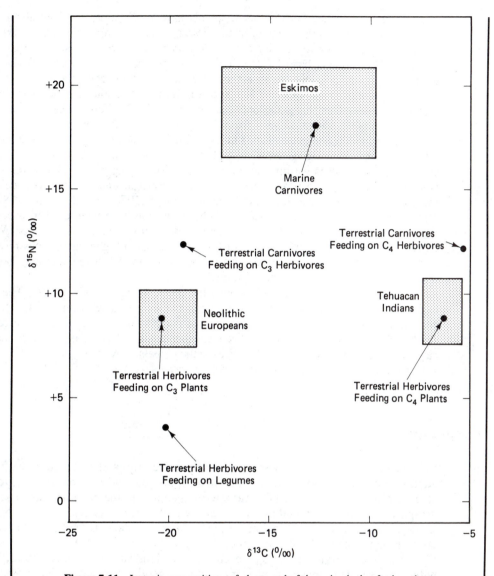

Figure 7.11 Isotopic compositions of plants and of the animals that feed on them can be used to determine what the diets of prehistoric men may have been. Carbon and nitrogen isotopic values have been measured in collagen from human bones at several archaeological sites (DeNiro, 1987; DeNiro and Hastorf, 1985). Values for each of the three societies indicated by shaded rectangles in this figure suggest diets that are compatible with what we know about each of the societies from other sources.

that Neolithic man in Europe fed largely on C_3 plants or on herbivores (cattle, deer) that ate C_3 vegetation. Tehuacan indians in Mexico, on the other hand, apparently had a diet based on C_4 plants like corn. Before European settlers arrived, Eskimos lived primarily on fish. These conclusions agree with what we know about these societies from other sources, and give us some assurance that the same technique can be applied to less familiar societies.

Their significance here is that C_3 plants have $\delta^{13}C$ values between about $-22‰$ and $-33‰$, while C_4 plants range from $-9‰$ to $-16‰$. Only a few plants like algae, lichens, and cacti occupy an isotopic middle ground between these two major plant groups. Marine plants have $^{13}C/^{12}C$ ratios that are about $7.5‰$ less negative than C_3 terrestrial plants, possibly because they assimilate carbon from marine HCO_3^- $(0‰)$ rather than atmospheric $CO_2(-7‰)$. Too little is known about how marine plants fix carbon to be sure that this is the only reason for the $\delta^{13}C$ difference between terrestrial and marine plants. Plankton vary from $-17‰$ to about $-27‰$.

Sackett and Thompson (1963) measured $\delta^{13}C$ values in Gulf Coast sediment and found that they decrease systematically shoreward from about $-21‰$ to $-26‰$. Using as a guide the patterns we introduced in the previous paragraph, they inferred that the isotopic trend reflects a near-shore mixing of terrestrial and marine plant debris. In more recent studies (for example, Schultz and Calder, 1976; Gearing et al., 1984), isotopic signatures of individual species of phytoplankton have been used to account for the spatial variability of ^{13}C in estuarine and shallow marine sediments. Without painstakingly detailed work, studies of this sort can yield highly ambiguous results because of the large number of hidden source materials and processes that may have contributed to present isotopic compositions.

Ultimately, diagenesis renders even the most obvious trends useless. The transition from primary organic debris to kerogen causes an enrichment in ^{12}C which becomes more severe as kerogen is converted to petroleum or coal. Kerogen ranges from $-17‰$ to $-34‰$. In this, and in some Cenozoic coals and crude oils, source materials can still be recognized isotopically. Pre-Tertiary coals, however, all have $\delta^{13}C$ values very near to $-25‰$, and values in petroleum cluster very close to $-28‰$.

SUMMARY

In this chapter, we have discussed some of the basic processes that lead to the fractionation of light stable isotopes from one another in geologic environments. Hydrogen, carbon, nitrogen, oxygen, and sulfur are all abundant in the crust and atmosphere/hydrosphere, and may be found in a wide variety of different chemical compounds. Variations in bond strength and site energy among these compounds lead them to favor different proportions of the stable isotopes.

The free energies of isotopic exchange reactions vary with temperature, but not with pressure. This makes them very useful as geothermometers. The most popular ones involve equilibrium fractionation of oxygen or sulfur isotopes.

Kinetic fractionation effects are also commonly exploited by geochemists. The small difference in vapor pressure between "light" water and "heavy" water leads to a significant fractionation of oxygen and hydrogen isotopes with increasing latitude, altitude, and distance from the ocean. These systematic trends in meteoric water have been useful in recognizing the source of water in aquifers, sedimentary basins, and hydrothermal systems.

The role of living organisms in fractionating light stable isotopes has made it possible to determine the provenance and, to some degree, the environmental history of organic matter. Isotopic biomarkers in kerogen and young fossil hydrocarbons can be helpful stratigraphic guides.

SUGGESTED READINGS

The following books discuss principles of stable isotope geochemistry at a level appropriate for the beginning student. In most cases, they emphasize the application of methods to specific field problems.

BARNES, H. L., ed. 1979. *Geochemistry of hydrothermal ore deposits.* 2d ed. New York: Wiley Interscience. (This collection of basic articles by geochemists at the leading edge of research in ore-forming systems has two long chapters, chapters 6 and 10, devoted to light stable isotopes.)

CLAYTON, R. N. 1981. Isotopic thermometry. In *Thermodynamics of minerals and melts,* ed. Newton, R. C.; Navrotsky, A.; and Wood, B. J., 85–109. New York: Springer Verlag. (This concise review paper is a very good way to learn the first principles of stable isotope geothermometry.)

FAURE, G. 1986. *Principles of isotope geology.* 2d ed. New York: John Wiley and Sons. (This is an easy-to-read text dealing largely with radioactive isotopes. Chapters 18 through 21, however, give a good introduction to light stable isotopes.)

HOEFS, J. 1980. *Stable isotope geochemistry.* 2d ed. New York: Springer Verlag. (This is an excellent text, covering all the basics, with a focus on field examples.)

JAGER, E., and HUNZIKER, J. C. 1979. *Lectures in isotope geology.* Berlin: Springer Verlag. (The lectures in this book were given in 1977 as an introductory short course for geologists. Most deal with radioactive isotopes, but the four final lectures are a useful overview of hydrogen, oxygen, carbon, and sulfur.)

VALLEY, J. W.; TAYLOR, H. P.; and O'NEIL, J. R., eds. 1986. *Stable isotopes in high temperature geological processes,* Reviews in Mineralogy 16, Washington, D.C.: Mineralogical Society of America. (Papers in this volume are of particular interest to igneous and metamorphic petrologists who use stable isotopes to study processes in the deep crust.)

We have referred to the following papers in Chapter 7. The interested student may wish to consult them further.

BERNER, R. A. 1987. Models for carbon and sulfur cycles and atmospheric oxygen: Application to Paleozoic geologic history. *Amer. J. Sci.* 287: 177–96.

CAMPBELL, A.; RYE, D.; and PETERSEN, U. 1984. A hydrogen and oxygen isotope study of the San Cristobal mine, Peru: Implications of the role of water to rock ratio for the genesis of wolframite deposits. *Econ. Geol.* 79: 1818–32.

CLAYTON, R. N.; FRIEDMAN, I.; GRAF, D. L.; MAYEDA, T. K.; MEENTS, W. F.; and SCHIMP, N. F. 1966. The origin of saline formation waters, I: Isotopic composition. *J. Geophys. Res.* 71: 3869–82.

COPLEN, T. B.; KENDALL, C; and HOPPLE, J. 1983. Comparison of stable isotope reference samples. *Nature* 302: 236–38.

CRAIG, H. 1961. Isotopic variations in meteoric waters. *Science* 133: 1702–03.

DEINES, P.; LANGMUIR, D.; and HARMON, R. S. 1974. Stable carbon isotope ratios and the existence of a gas phase in the evolution of carbonate ground water. *Geochim. Cosmochim. Acta* 38: 1147–64.

DENIRO, M. J. 1987. Stable isotopy and archaeology. *Amer. Scientist* 75: 182–91.

DENIRO, M. J., and HASTORF, C. A. 1985. Alteration of $^{15}N/^{14}N$ and $^{13}C/^{12}C$ ratios of plant matter during the initial stages of diagenesis: Studies utilizing archaeological specimens from Peru. *Geochim. Cosmochim. Acta* 49: 97–115.

ELLIS, A. J., and MAHON, W. A. J. 1977. *Geochemistry and geothermal systems.* New York: Academic Press.

GEARING, J. N.; GEARING, P. J.; RUDNICK, P. T.; REQUEJO, A. G.; and HUTCHINS, M. J. 1984. Isotopic variability of organic carbon in a phytoplankton-based temperate estuary. *Geochim. Cosmochim. Acta* 48: 1089–98.

HATTORI, K., and HALAS, S. 1982. Calculation of oxygen isotope fractionation between uranium dioxide, uranium trioxide, and water. *Geochim. Cosmochim. Acta* 46: 1863–68.

HOLSER, W. T., and KAPLAN, I. R. 1966. Isotope geochemistry of sedimentary sulfates. *Chem. Geol.* 1: 93–135.

MATSUHISA, Y., GOLDSMITH, J. R., and CLAYTON, R. N. 1979. Oxygen isotopic fractionation in the system quartz-albite-anorthite-water. *Geochim. Cosmochim. Acta* 43: 1131–40.

MUNHA, J.; BARRIGA, F. J. A. S.; and KERRICH, R. 1986. High $\delta^{18}O$ ore-forming fluids in volcanic-hosted base metal sulfide deposits: Geologic, $^{18}O/^{16}O$, and D/H evidence from the Iberian Pyrite belt; Crandon, Wisconsin; and Blue Hill, Maine. *Econ. Geol.* 81: 530–52.

OHMOTO, H., and RYE, R. O. 1979. Isotopes of sulfur and carbon. In *Geochemistry of hydrothermal ore deposits.* 2d ed., ed. Barnes, H. L., 509–67. New York: Wiley Interscience.

O'NEIL, J. R. 1986. Theoretical and experimental aspects of isotopic fractionation. In *Stable isotopes in high temperature geological processes,* ed. Valley, J. W.; Taylor, H. P.; and O'Neill, J. R., 1–40. Reviews in Mineralogy 16. Washington, D.C.: Mineralogical Society of America.

RYE, R. O. 1974. A comparison of sphalerite-galena sulfur isotope temperatures with filling temperatures of fluid inclusions. *Econ. Geol.* 69: 26–32.

SACKETT, W. M., and THOMPSON, R. R. 1963. Isotopic organic carbon composition of recent continental derived clastic sediments of the eastern Gulf Coast, Gulf of Mexico. *Bull. Amer. Assoc. Petrol. Geol.* 47: 525–31.

SHEPPARD, S. F. M.; NIELSEN, R. L.; and TAYLOR, H. P. 1969. Oxygen and hydrogen isotope ratios of clay minerals from porphyry copper deposits. *Econ. Geol.* 64: 755–77.

SHULTZ, D. J. and CALDER, J. A. 1976. Organic carbon $^{13}C/^{12}C$ variations in estuarine sediments. *Geochim. Cosmochim. Acta* 40: 381–86.

SIEGEL, D. I., and MANDLE, R. J. 1984. Isotopic evidence for glacial melt-water recharge to the Cambrian-Ordovician aquifer, North-Central United States. *Quaternary Res.* 22: 328–35.

TAYLOR, H. P. 1979. Oxygen and hydrogen isotope relationships in hydrothermal mineral deposits. In *Geochemistry of hydrothermal ore deposits*. 2d ed, ed. Barnes, H. L., 236–77. New York: Wiley Interscience.

UREY, H. C. 1947. The thermodynamics of isotopic substances. *J. Chem. Soc. London* 562–81.

PROBLEMS

1. The relative atomic abundances of oxygen and carbon isotopes are:

$$^{16}O : {}^{17}O : {}^{18}O = 99.759 : 0.0374 : 0.2039$$
$$^{12}C : {}^{13}C = 98.9 : 1.1.$$

   Calculate the relative abundances of (a) CO_2 molecules with masses 44, 45, and 46; (b) CO molecules with masses 28, 29, and 30.

2. The fractionation factor for hydrogen isotopes between liquid water and vapor at 20°C is 1.0800. Make a graph to indicate how the value δD changes in rainwater during continued precipitation from an air mass that initially has a value of −80‰. Show how the δD value of the remaining water vapor changes during this process.

3. The following ^{34}S data were collected on samples from Providencia, Mexico by Rye (1974). On the basis of these data, what is your estimate for the temperature at which these ores formed?

Sample #	Galena	Sphalerite
60-H-67	−1.15‰	+0.95‰
60-H-36-57	−1.11	+0.87
62-S-250	−1.42	+1.03
63-R-22	−1.71	+0.01

4. Hattori and Halas (1982) have determined that the fractionation factor for oxygen exchange between UO_2 and water varies with temperature according to the relationship

$$1000 \ln \alpha_{UO_2\text{-}H_2O} = 3.63 \times 10^6\, T^{-2} - 13.29 \times 10^3\, T + 4.42.$$

   Using the similar expression for quartz-water fractionation from Worked Problem 7-2, devise an oxygen isotope geothermometer to estimate temperatures from coexisting quartz and uraninite.

5. The fractionation of carbon isotopes between calcite and CO_2 gas can be calculated from $1000 \ln \alpha_{\text{cal-CO}_2} = 1.194 \times 10^6\, T^{-2} - 3.63$ (Deines et al., 1974). If the $\delta^{13}C$ value of atmosphere CO_2 is −7.0‰, what should be the $\delta^{13}C$ value of calcite precipitated from water in equilibrium with the atmosphere at 20°C?

6. Rain and snow from a large number of sampling sites have been analyzed and found to have the following $\delta^{18}O$ values:

$\delta^{18}O$	Mean annual air temp. (°C)
− 3.18‰	15
− 8.04	8
−11.52	3
−20.55	−10
−30.98	−25

Write an equation, based on these data, that could be used to predict the oxygen isotopic composition of precipitation at any locality, given its mean annual temperature. Write a second equation based on these data and the meteoric water line, relating δD to mean annual temperature.

7. Use the procedure by Campbell et al. (1984), described in Worked Problem 7-5, to determine what the final values of $\delta^{18}O$ and δD would be in water which initially had values of $-30‰$ and $-230‰$, but which had equilibrated in a closed system with granite at 400°C. Assume the same isotopic values for unaltered granite that were used for the study at San Cristobal, Peru, and a water to rock ratio (w/r) of 0.05.

Chapter 8

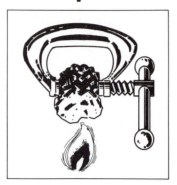

Temperature and Pressure Changes: Thermodynamics Again

OVERVIEW

This chapter expands upon the thermodynamic principles introduced in Chapter 2 and provides the preparation for the next chapter, which is about phase diagrams. We first give a thermodynamic definition for the equilibrium state. We also derive the phase rule and explore its use as a test for equilibrium. The effects of changing conditions of temperature and pressure on the free energy of a system at equilibrium are analyzed, and a general equation for *dG* at *P* and *T* is formulated. *P-T* and *G-T* diagrams for one-component systems are introduced. The utility of the Clapeyron equation in constructing or interpreting phase diagrams is explored.

We then introduce concepts that are necessary for understanding systems with more than one component. Raoult's law and Henry's law are defined, and expressions for the chemical potentials of components that obey these laws are derived. The concept of standard states is introduced, and we examine how activity coefficients are defined in terms of mixing models. Geothermometers and geobarometers that utilize many of the concepts introduced in this chapter are presented.

WHAT DOES EQUILIBRIUM REALLY MEAN?

In previous chapters, we have introduced and used the concept of thermodynamic equilibrium. By this time you should be comfortable with the idea that the observable properties of a system at equilibrium will undergo no change with time as long as physical conditions imposed on the system remain constant. Although this statement is commonly used to define equilibrium, it is actually a consequence of equilibrium rather than a strict definition of it. A more rigorous definition of the equilibrium state that we have not seen before is the following: equilibrium is a condition of minimum energy for that portion of the universe under consideration.

We have also learned that Gibbs Free Energy is a convenient way of describing the energy of a system. The relationship of this thermodynamic quantity to temperature, pressure, and the amounts of the constituent components of a system is given by the following:

$$dG = (\partial G/\partial T)_{P,n_i}dT + (\partial G/\partial P)_{T,n_i}dP + \sum_i (\partial G/\partial n_i)_{P,T,n_{j\neq i}}dn_i, \qquad (8\text{-}1)$$

where n_i represents the number of moles of component i. The quantities $(\partial G/\partial n_i)_{P,T,n_{j\neq i}}$ are chemical potentials, defined earlier in Chapter 2. The goal of this chapter is to learn how to evaluate all of the quantities in this equation. At equilibrium, the chemical potential of any component will be the same in all of the phases that contain it. From equation 8-1, we can surmise that if temperature and pressure are specified, a system at equilibrium will have the lowest possible free energy. This is an extremely important conclusion, and we explore its ramifications in later sections.

DESCRIBING THE EQUILIBRIUM CONFIGURATION OF A SYSTEM

The Phase Rule

How can we determine whether a geologic system is in a state of equilibrium? The answer, of course, is that we cannot be completely sure. However, we can at least argue for equilibrium on the basis of our experience with reactions in the laboratory and of the absence of any observational evidence for disequilibrium. Consider, for example, a metamorphic rock consisting of biotite, garnet, plagioclase, sillimanite, and quartz. We know from experiments that these minerals can constitute a stable assemblage under conditions of high temperature and pressure. Let's assume that textural observations of this rock provide no indication of arrested reactions between any of the constituent minerals. Furthermore, grains of all five minerals are observed to be in direct contact, and no zoning or other kind of chemical heterogeneity within grains is present. In this case we might infer that the rock system is in a state of equilibrium, but this indirect evidence may be less than convincing. Is there another way to test for equilibrium?

Let us describe the equilibrium configuration of a system in more quantitative terms. The general relationships among phases, components, and physical conditions are given by the *phase rule,* first formulated by J. Willard Gibbs more than a century ago. The phase rule describes the maximum number of phases that can exist in an equilibrium system of any complexity. As we will see, this provides an additional test for equilibrium.

The *variance* (f), also called *degrees of freedom,* of a system at equilibrium is the number of independent parameters that must be fixed or determined in order to specify the state of the system. The number of independent parameters equals the number of unknown variables minus the number of known (dependent) relationships between them. Assume that there are p phases and every phase contains one or more of the c components that constitute the system. For each phase $(A, B, \ldots p)$ in the system, we can recast equation 8-1 to write expressions of the following form:

For ϕ_A,

$$dG_A = -SdT_A + VdP_A + \mu_{1A}dn_{1A} + \mu_{2A}dn_{2A} + \cdots \mu_{cA}dn_{cA}.$$

For ϕ_B,

$$dG_B = -SdT_B + VdP_B + \mu_{1B}dn_{1B} + \mu_{2B}dn_{2B} + \cdots \mu_{cB}dn_{cB}.$$

For ϕ_p,

$$dG_p = -SdT_p + VdP_p + \mu_{1p}dn_{1p} + \mu_{2p}dn_{2p} + \cdots \mu_{cp} dn_{cp}.$$

Recall also a major conclusion of Chapter 2, that the temperatures and pressures in all phases at equilibrium must be identical, and that μ_c must be the same in all phases. This is the set of relationships expressed in equations 2–45a–i. We have one of these quantities associated with each term in any of the equations for dG above. The total number of potentially unknown variables, therefore, is c (the number of chemical potential terms per equation) plus 2 (for temperature and pressure). The number of known relationships among them is p (the number of dG equations). The variance, therefore, is the difference:

$$f = (c + 2) - p. \tag{8-2}$$

Algebra aficionados will recognize that we have approached this statement of the phase rule by solving a set of p simultaneous equations in $(c + 2)$ unknowns.

Equation 8-2 is a mathematical statement of the phase rule. To illustrate its use, consider an ordinary glass of water at room temperature and atmospheric pressure. The number of components necessary to specify this system is 1 (H_2O), and the number of phases is also 1 (liquid water). Thus, from equation 8-2, the variance of the system is 2. This means that two variables (in this case, temperature and pressure) can be changed independently with no resulting effect on the system. Our everyday experience tells us that this is true—it would still be a glass of water if transported from sea level to a mountaintop where the temperature is cooler and atmospheric pressure is less—but there is some finite limit to this variation. For example, the phase in this system (liquid water) would certainly change if we were to lower the temperature drastically by putting the glass of water in a freezer. However, in the case of a frozen glass of water, the variance would still be 2, because there is still only 1 component (H_2O) and 1 phase (ice in this case). Are there situations in which the variance of this system is other than 2?

The *phase diagram* for water, illustrated in Figure 8.1, summarizes how this system reacts to changing conditions of temperature and pressure. This is actually only a part of the H_2O phase diagram, because various ice polymorphs occur at higher pressures, but this part of the diagram will suffice for our purposes. The diagram consists of fields in which only one phase is stable, separated by boundary curves along which combinations of phases coexist in equilibrium. The three boundary lines intersect at a point, called the *triple point,* at which three phases occur. The boundary between the liquid water and water vapor fields terminates in a *critical point,* above which there is no distinction between liquid and vapor. From inspection of this figure, it is obvious that a single phase is stable under most combinations of temperature and pressure. Within these one-phase fields the variance is 2, and the fields are said to be *divariant.*

However, there are certain special combinations of temperature and pressure at which several phases coexist at equilibrium. For example, if we calculate the variance at some point along any of the boundary curves, the presence of two coexisting phases results in a variance of 1. (Prove this yourself.) These boundary curves are thus *univari-*

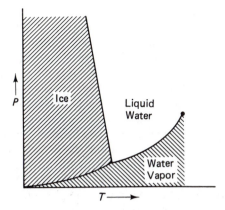

Figure 8.1 Schematic phase diagram for the system H_2O. Most of the area of the diagram consists of regions in which only one phase is stable at equilibrium. Boundary curves are combinations of pressure and temperature along which two phases are stable, and three phases occur at the triple point. The boundary between liquid water and water vapor terminates in a critical point, above which the distinction between these phases is meaningless.

ant. Coexistence of three phases at the triple point produces a variance of 0, forming an *invariant* point.

For a more complex system like a rock, of course, more components are required to describe the system, and there will probably be more phases. In an assemblage of minerals, it is unlikely that every component will occur in every phase. However, the phase rule is a general result and is valid for systems in which some components do not occur in all phases.

Worked Problem 8-1

What is the variance for the metamorphic rock described at the beginning of this section? There are five phases, and components that describe the compositional variation of each phase are listed beside each mineral in the list that follows. From our discussion of components in Chapter 2, we should recognize that the components listed are not unique; you might try devising your own set of components to replace the ones illustrated here. (Note: It would be necessary to choose additional components if some of these minerals were more complex solid solutions.)

> *biotite* – $KMg_3AlSi_3O_{10}(OH)_2$, $KFe_3AlSi_3O_{10}(OH)_2$
> *garnet* – $Fe_3Al_2Si_3O_{12}$, $Ca_3Al_2Si_3O_{12}$
> *sillimanite* – Al_2SiO_5
> *plagioclase* – $NaAlSi_3O_8$, $CaAl_2Si_2O_8$
> *quartz* – SiO_2

As we learned in Chapter 2, not all of these components are necessary to define the rock system, because they are not all independent. A mathematical relationship exists between the grossular component of garnet, sillimanite, quartz, and the anorthite component of plagioclase. We can write a reaction between these components, even though such a reaction might never take place in nature:

$$Ca_3Al_2Si_3O_{12} + 2Al_2SiO_5 + SiO_2 = 3CaAl_2Si_2O_8.$$

Thus we can express anorthite in terms of the other three components. In fact, any one of these four components can be expressed in terms of the other three by rearranging the reaction, and so we can delete any one of these as a system component. We then are left with five phases and seven components, which give a variance of four. This means that temperature and pressure can be varied independently (within some reasonable limits), as well as the compositions of two phases (for example, biotite, garnet, or plagioclase), without changing the number of phases in the system at equilibrium.

Open Versus Closed Systems

The discussion so far assumes that the system is closed. However, there are examples of geologic systems in which some components are free to migrate in and out of the system. A common example of such a *mobile* component is the fluid phase in some metamorphic rocks. We may describe this situation using the same kind of free energy equations that we formulated before, but in this case we must distinguish between mobile and immobile components. In the closed system case, the chemical potentials of all components were fixed, but in an open system the chemical potentials of mobile components are controlled by phases or other parameters outside of the system. Consequently, we have an additional constraining equation for each mobile component. These additional chemical restrictions, like those of constant temperature and pressure, are reflected in the phase rule as additional known relationships. In the case of m mobile components, the number of known relationships is therefore p plus m. Substitution of this value into equation 8-2 produces the relation

$$f = (c + 2) - (p + m). \qquad (8\text{-}3)$$

This open-system modification provides an alternative form of the phase rule for systems in which mobile components can be recognized.

Now let us return to our discussion of the utility of the phase rule as a test for equilibrium. As with other kinds of observational evidence, we will not be able to demonstrate equilibrium with certainty, but we can now cite an additional factor that is consistent with equilibrium. In the phase diagram for water (Figure 8.1), it is apparent that most of the area of the diagram is occupied by divariant fields, and phase assemblages with lower variance comprise only a tiny fraction of the total area. The same

Figure 8.2 Photomicrograph of coexisting kyanite (ky) and sillimanite (sil) in a metamorphic rock from the Panamint Mountains, California. The low variance of this rock, as well as the textural evidence for replacement of kyanite by sillimanite, suggest that this is a disequilibrium assemblage. (Courtesy of T. C. Labotka.)

holds true for complex systems like rocks. Although mineral assemblages with low variance surely are formed under just the right combinations of temperature and pressure, the probability of finding these in the field is slight compared to the likelihood of sampling divariant assemblages. For example, locating precisely the kyanite ↔ sillimanite transformation (the sillimanite isograd) in the field may be difficult because most outcrops will contain only one of these polymorphs. We can conclude that most rocks should have variances of 2 or higher at equilibrium. Assemblages that appear to be univariant or invariant (such as the kyanite + sillimanite rock in Figure 8.2) probably represent disequilibrium in most cases, unless it can be shown by field mapping that these rocks lie between appropriate divariant assemblages.

CHANGING TEMPERATURE AND PRESSURE

You are probably aware that water boils at a different temperature at the seashore than it does in the mountains. The boundary curve between liquid water and water vapor in Figure 8.1 defines the boiling point of water at any given pressure, and its slope explains this phenomenon. Changing conditions of temperature and pressure obviously exert influences on more complex geological systems as well.

In Chapter 2, we learned how temperature and pressure are related to other thermodynamic functions, but it may be worthwhile to review these relationships in a slightly different way. Molar internal energy, entropy, and volume for a system under various conditions can be represented by a plane in an $\bar{E}$-$\bar{S}$-$\bar{V}$ plot, as shown in Figure 8.3. We have already discovered that

$$(\partial\bar{E}/\partial\bar{V})_S = -P,$$

so the slope of this surface in the $\bar{E}$-$\bar{V}$ plane of Figure 8.3 is negative pressure. And be-

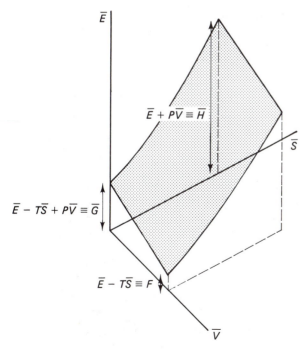

Figure 8.3 This surface represents molar internal energy, entropy, and volume for a hypothetical system. The second derivatives of E are positive, so this surface is concave upward at all points. The slope of the intersection of the surface with the $\bar{E}$-$\bar{S}$ plane is temperature, and the slope of the intersection with the $\bar{E}$-$\bar{V}$ plane is negative pressure. Free energy, enthalpy, and work function can be derived geometrically in this diagram, and their values are illustrated by vertical arrows.

cause

$$(\partial \overline{E}/\partial \overline{S})_V = T,$$

its slope in the $\overline{E}$-$\overline{S}$ plane is temperature. Using these slopes, the relationships of the other thermodynamic functions can be derived geometrically, and their values are shown in Figure 8.3.

From this kind of geometric argument, we can readily see how all of these thermodynamic functions are related to T and P. However, it is most convenient to describe modifications in the state of a geochemical system in terms of changes in free energy ($\Delta \overline{G}$), because differentials of $\Delta \overline{G}$ with respect to T and P are easily evaluated. You may recall that the concept of free energy was developed in Chapter 2 for fixed T and P conditions. In this section, we will see how to evaluate $\Delta \overline{G}$ under different sets of temperature and pressure conditions.

Temperature Changes and Heat Capacity

First we examine the effect of changing only temperature on a system at equilibrium. Differentiating the expression $d\Delta \overline{G} = \Delta \overline{V}\, dP - \Delta \overline{S}\, dT$ with respect to temperature at constant pressure gives

$$(\partial \Delta \overline{G}/\partial T)_P = -\Delta \overline{S}. \qquad (8\text{-}4)$$

It is often not easy to evaluate entropy directly, and we must resort to using another more readily measureable quantitity, in this case enthalpy.

Equation 8-4 is equivalent to

$$(\partial \Delta \overline{G}/\partial T)_P = (\Delta \overline{G} - \Delta \overline{H})/T.$$

Because pressure is constant, we can rewrite this as

$$d\Delta \overline{G} = (\Delta \overline{G} - \Delta \overline{H})dT/T.$$

Dividing both sides of this expression by T and rearranging yields

$$d\Delta \overline{G}/T - (\Delta \overline{G})dT/T^2 = -(\Delta \overline{H})dT/T^2.$$

The left side of this equation is equivalent to $d(\Delta \overline{G}/T)$, so

$$d(\Delta \overline{G}/T) = -(\Delta \overline{H})dT/T^2.$$

If we consider the expression above at standard conditions, then $\Delta \overline{G}$ and $\Delta \overline{H}$ become $\Delta \overline{G}_r^\circ$ and $\Delta \overline{H}_r^\circ$, respectively. For a system at equilibrium, $\Delta \overline{G}_r^\circ = -RT \ln K$, so that

$$d \ln K/dT = \Delta \overline{H}_r^\circ/RT^2 \qquad (8\text{-}5a)$$

or

$$-d \ln K = [\Delta \overline{H}_r^\circ/R]d(1/T). \qquad (8\text{-}5b)$$

These expressions, known in either form as the *van't Hoff equation*, provide a useful way to determine enthalpy changes.

Worked Problem 8-2

Determine $\Delta \overline{H}_r^\circ$ for the breakdown of annite (an end-member component of biotite) by the following reaction:

$$KFe_3AlSi_3O_{10}(OH)_2 = KAlSi_3O_8 + Fe_3O_4 + H_2.$$
$$\text{annite} \qquad\qquad \text{sanidine} \quad \text{magnetite}$$

From equation 8-5b, we can see that $\Delta \overline{H}_r^\circ$ can be obtained from a plot of $\ln K$ versus $1/T$.

The slope of this line is equivalent to $-\Delta \overline{H}_r^o/R$.

Eugster and Wones (1962) found experimentally that the equilibrium constant for this reaction could be expressed by

$$\log K = -(9215/T) + 10.99.$$

In other words, a plot of $\log K$ versus $1/T$ gave a straight line of slope -9215 deg. Therefore,

$$-(\Delta \overline{H}_r^o/2.303\ R) = -9215,$$

and

$$\Delta \overline{H}_r^o = 2.303(1.987\ \text{cal deg}^{-1}\ \text{mol}^{-1})(9215\ \text{deg})$$

$$= 42.2\ \text{kcal mol}^{-1}.$$

How do we then determine enthalpy differences under nonstandard conditions? First let us define a new variable, the *heat capacity at constant pressure*, $\overline{C}_P$:

$$\overline{C}_P = (dq/dT)_P = (\partial \overline{H}/\partial T)_P.$$

If a material is heated at constant pressure, then its molar enthalpy increases with temperature, according to the relationship

$$d\overline{H} = \overline{C}_P\ dT, \tag{8-6}$$

where $d\overline{H}$ is the increase in enthalpy through a temperature interval dT. Tabulations of thermodynamic data for minerals normally give enthalpy at 298 K, and it is necessary to integrate equation 8-6 in order to obtain enthalpy at any other temperature. Thus we have

$$\int_{298}^{T} d\overline{H} = \int_{298}^{T} \overline{C}_P\ dT$$

or

$$\overline{H}_T - \overline{H}_{298}^o = \overline{C}_P\ \Delta T, \tag{8-7}$$

where $\Delta T = T - 298$. However, heat capacities are not generally independent of temperature, so the integrated form of the expression for $\overline{C}_P$ is somewhat complex. The experimental data from which empirically integrated $\overline{C}_P$ values are determined are commonly given in the form of a power series:

$$\overline{C}_P = a + bT + c/T^2,$$

where a, b, and c are experimentally determined constants.

Coincident with a change in enthalpy with temperature (equation 8-6), the system will also experience a change in entropy, because

$$d\overline{H} = T\ d\overline{S}$$

at constant pressure. Consequently, entropy can also be expressed in terms of the heat capacity:

$$d\overline{S} = (\overline{C}_P dT)/T.$$

We can obtain the entropy at any temperature from the integrated form of the following equation:

$$\overline{S}_T - \overline{S}_{298}^o = (\overline{C}_P/T)\Delta T \tag{8-8}$$

where $\Delta T = T - 298$.

Changing Temperature and Pressure

We now have the necessary tools to determine the effect of a change in temperature on the free energy of a system. An expression for $\Delta \bar{G}$ with which we are already familiar (equation 2-34) is

$$\Delta \bar{G} = \Delta \bar{H} - T\Delta \bar{S}.$$

$\Delta \bar{G}$ at any temperature can be evaluated by substituting enthalpy (equation 8-7) and entropy (equation 8-8) values at the appropriate temperature into the equation above:

$$\Delta \bar{G}_T = \Delta \bar{H}^o_{298} + \int_{298}^T \Delta \bar{C}_P \, dT - T\left(\Delta \bar{S}^o_{298} + \int_{298}^T (\Delta \bar{C}_P/T) \, dT\right). \qquad (8\text{-}9)$$

Worked Problem 8-3

As an example of the use of equation 8-9, determine the free energy change for the following reaction at 800°C and atmospheric pressure:

$$MgSiO_3 + CaCO_3 + SiO_2 \longrightarrow CaMgSi_2O_6 + CO_2.$$

$$\text{enstatite} \quad \text{calcite} \quad \text{quartz} \qquad \qquad \text{diopside}$$

The thermodynamic data for these phases are given in Table 8-1.

After ensuring that the reaction is balanced, we calculate values for products (Di + CO_2) minus reactants (En + Cc + Qz):

$$\Delta \bar{H}^o_{298} = 14720 \text{ cal mol}^{-1}$$
$$\Delta \bar{S}^o_{298} = 37.01 \text{ cal mol}^{-1} \text{ K}^{-1}$$
$$\Delta \bar{C}_P = 2.69 - 8.24 \times 10^{-3}T - 2.62 \times 10^5/T^2 \text{ cal deg}^{-1} \text{ mol}^{-1}$$
$$T = 1073 \text{ K}$$

Substitution of these values into equation 8-9 gives $\Delta G^o_r = -23924.42$ cal mol^{-1}.

An interesting variation on this problem is to specify the value of P_{CO_2}, say at 0.1 bar. To solve this problem now, we must first write an expression for the equilibrium constant K_{eq}, which in this case is

$$K_{eq} = P_{CO_2} = 0.1 \text{ bar}.$$

From the relationship

$$\Delta \bar{G}_r = \Delta \bar{G}^o_r + RT \ln K_{eq},$$

we obtain a value of $\Delta \bar{G}_r = -28833.65$ cal mol^{-1}. Program DELTAG on the diskette which accompanies this text will solve problems such as this, and we encourage you to use the program in future solutions.

TABLE 8-1 THERMODYNAMIC DATA FOR WORKED PROBLEMS 8–3 AND 8–5*

	$\bar{H}^o_{298,1}$(cal mol^{-1})	$\bar{S}^o_{298,1}$(cal mol^{-1}K^{-1})	$\bar{V}_{298,1}$(cal mol^{-1} bar^{-1})
En	−370,100	16.22	0.752
Cc	−288,420	22.15	0.883
Qz	−217,650	9.88	0.542
Di	−767,400	34.20	1.579
CO$_2$	−94,050	51.06	584.73

$$\bar{C}_P(\text{En}) = 24.55 + 4.74 \times 10^{-3}T - 6.28 \times 10^5 T^{-2}$$
$$\bar{C}_P(\text{Cc}) = 24.98 + 5.24 \times 10^{-3}T - 6.20 \times 10^5 T^{-2}$$
$$\bar{C}_P(\text{Qz}) = 11.22 + 8.20 \times 10^{-3}T - 2.70 \times 10^5 T^{-2}$$
$$\bar{C}_P(\text{Di}) = 52.87 + 7.84 \times 10^{-3}T - 15.74 \times 10^5 T^{-2}$$
$$\bar{C}_P(\text{CO}_2) = 10.57 + 2.10 \times 10^{-3}T - 2.06 \times 10^5 T^{-2}$$

*After Helgeson (1978).

Pressure Changes and Compressibility

The effect of changing pressure on a system at equilibrium can be determined by differentiating $d\Delta\overline{G} = \Delta\overline{V}dP - \Delta\overline{S}dT$ with respect to pressure at constant temperature:

$$(\partial\Delta\overline{G}/\partial P)_T = \Delta\overline{V}. \tag{8-10}$$

Consequently, the change in free energy of a reaction with respect to pressure alone is equal to the change in the molar volume of products and reactants. Volume changes can be determined from simple physical measurements of product and reactant volumes. For reactions involving only solid phases under most geological conditions, the effects of temperature and pressure on $\Delta\overline{V}$ are small and can be ignored. In this case, the integration of equation 8-10 gives

$$\Delta\overline{G}_r = \Delta\overline{G}_{T,1}^0 + \Delta\overline{V}\int_1^P dP, \tag{8-11}$$

where $\Delta\overline{G}_{T,1}^0$ is the free energy for the reaction at 1 bar and temperature T.

Worked Problem 8-4

As an example, consider the polymorphic transformation of aragonite to calcite. Use equation 8-11 to calculate the pressure at which these two minerals coexist stably at 298K.
At equilibrium, $\Delta\overline{G} = 0$, so

$$\int_1^P dP = -\Delta\overline{G}_{T,1}^0/\Delta\overline{V},$$

or

$$P - 1 = -\Delta\overline{G}^\circ/\Delta\overline{V}.$$

Appropriate thermodynamic data are $\Delta\overline{G}_{298}^0$ (kJ mol^{-1}) $= -1127.793$ for aragonite, and -1128.842 for calcite; $\overline{V}$ (J bar^{-1}) $= 3.415$ for aragonite, and 3.693 for calcite. For the reaction aragonite $\rightarrow$ calcite, $\Delta\overline{G}_r^\circ = -1049$ J mol^{-1} and $\Delta\overline{V} = 0.278$ J bar^{-1}. By substituting these values into the equation above we calculate that $P = 3767$ bars.
Aragonite has the lower molar volume and is thus the phase that is stable on the high pressure side of this reaction. Therefore, at 298 K, calcite is the stable polymorph to approximately 3.7 kilobars pressure. From this computation, we can readily see why aragonite precipitated by shelled organisms at low pressure is metastable.

Equation 8-11 expresses an extremely useful relationship for determining the effect of pressure on many equilibria of geological interest. For any reactions involving fluids (such as many metamorphic reactions), however, or for solid phase reactions at very high pressures and temperatures (such as those under mantle conditions), we cannot assume that $\Delta\overline{V}$ is independent of temperature and pressure. In these cases, we must take into account the expansion or contraction of a fluid or crystal lattice due to thermal effects and the compression or relaxation because of changing pressure. For gases, it is permissible to use an equation of state, such as the ideal gas law $\overline{V} = RT/P$, to describe the effects of changing temperature or pressure, that is:

$$(\partial\overline{V}/\partial P)_T = -RT/P$$

and

$$(\partial\overline{V}/\partial T)_P = R/P.$$

More complex equations of state for fluids have been introduced in Chapter 3. For solid phases, we can use the *isobaric thermal expansion* α_P, defined as

$$\alpha_P = 1/\bar{V}(\Delta\bar{V}/\Delta T)_P,$$

where $\bar{V}$ is the molar volume of the phase at the temperature at which α_P is measured. The temperature dependence of α_P is large near absolute zero, but becomes small at the temperatures at which most geologic processes operate. If we ignore this small variation, this expression can be rearranged to give

$$\Delta\bar{V} = \bar{V}^0_{298,1}(1 + \alpha_P\Delta T),$$

where $\bar{V}^0$ is the molar volume at 1 bar and 298°C and $\Delta T = T - 298$. The *isothermal compressibility* β_T is defined by

$$\beta_T = -1/V(\Delta\bar{V}/\Delta P)_T,$$

where $\bar{V}$ is the molar volume of the phase at the pressure at which β_T is measured and $\Delta P = P - 1$. The effect of pressure on β_T can as a first approximation be neglected in petrological problems at pressures less than 20 or 30 kilobars, although it is very important in interpreting mantle seismic data. If we regard β_T as constant in the pressure range where most observable geological processes operate, this expression becomes

$$\Delta\bar{V} = \bar{V}^0_{298,1}(1 - \beta_T\,\Delta P),$$

where $\bar{V}^0$ is the molar volume at 1 bar and 298°C. By combining the definitions for α_P and β_T in a total differential expression for $\bar{V}$, we obtain

$$d\bar{V} = (\partial\bar{V}/\partial T)_P\,dT + (\partial\bar{V}/\partial P)_T\,dP$$
$$= \alpha_P\bar{V}\,dT - \beta_T\bar{V}\,dP$$

or

$$\int_{V_{298,1}}^{V_{T,P}} d\bar{V}/\bar{V} = \int_{298}^{T} \alpha_P\,dT - \int_{1}^{P} \beta_T\,dP.$$

This produces the result

$$\bar{V}_{T,P} = \bar{V}_{298,1}\exp\left[\alpha_P\,\Delta T - \beta_T\,\Delta P\right].$$

Because the term in brackets is very small, we can use the approximation $\exp x \approx 1 + x$ to yield

$$\bar{V}_{T,P} = \bar{V}_{298,1}\left[1 + \alpha_P\,\Delta T - \beta_T\,\Delta P\right]. \qquad (8\text{-}12)$$

This last expression allows us to gauge the effect of changing thermal expansion and isothermal compressibility on the volume change for a reaction occurring under mantle conditions. Table 8-2 gives some representative values for α_P and β_T for various mineral and fluid species. From these comparisons, you can see that the molar volumes of some phases are much more affected by temperature and pressure than others.

TABLE 8-2 REPRESENTATIVE VALUES FOR THERMAL
EXPANSION AND ISOTHERMAL COMPRESSIBILITY

	$\alpha_P \times 10^6$ at 400°C (deg^{-1})	$\beta_T \times 10^6$ at 1000 bars (bar^{-1})
microcline	17	14.92
almandine	25	1.27
forsterite	38	1.82
quartz	69	26.71
diopside	28	1.07

The new quantities described in this section, $\bar{C}_P$, α_P, and β_T, can be visualized geometrically by referring again to Figure 8.3. These three quantities are related to the second derivatives of $\bar{E}$, $\bar{S}$, and $\bar{V}$, so they define the curvature of the $\bar{E}$-$\bar{S}$-$\bar{V}$ surface in this figure.

Temperature and Pressure Changes Combined

A general equation for the effect of both temperature and pressure on free energy can be obtained by combining the equations we have already derived for the effects of temperature alone or pressure alone, as follows:

$$\Delta\bar{G}_{P,T} = \Delta\bar{H}_{298}^0 + \int_{298}^{T} \Delta\bar{C}_P \, dT - T\left(\Delta\bar{S}_{298}^0 + \int_{298}^{T} (\Delta\bar{C}_P/T) \, dT\right)$$

$$+ \bar{V}_{298,1}^0\left[1 + \int_{298}^{T} \alpha_P \, dT - \int_{1}^{P} \beta_T \, dP\right]. \tag{8-13}$$

Worked Problem 8-5

Consider again the reaction enstatite + calcite + quartz→diopside + CO_2. In Worked Problem 8-3, we calculated the free energy of this reaction at 800°C and 1 bar. Now calculate the value of $\Delta\bar{G}_r$ at 800°C and 2000 bars pressure, assuming that $P_{CO_2} = 0.1$ bar.

To do this, we need to make some assumption about $(\partial\bar{V}/\partial P)_T$. It is clear that $\Delta\bar{V}$ is not constant, because CO_2 is quite compressible. Let us assume instead that all of the compressibility is due to CO_2, and pretend that it is an ideal gas so that $\Delta\bar{V} = RT/\Delta P$. Under these conditions, is it the reactants or the products that are stable? [Program DELTAG on the disk can be used to solve this problem if you wish.]

From the data in Table 8-1, we can calculate $\Delta\bar{H}_{298,1}^0$, $\Delta\bar{C}_P$, and $\Delta\bar{S}_{298}^0$. (In fact, we have already done this in Worked Problem 8-3). The change in molar volume $\Delta\bar{V}$ of CO_2 between 1 bar and 2000 bars pressure is $(1.987 \text{ cal deg}^{-1} \text{ mol}^{-1})(1073 \text{ deg})/2000 \text{ bar} = 1.066 \text{ cal bar}^{-1}$. Substituting these values into equation 8-12 gives the following results:

$$\Delta\bar{G}_{800,2000} = -30027.05 \text{ cal mol}^{-1}.$$

Because $\Delta\bar{G}$ is negative, the reaction proceeds as written and diopside + CO_2 are stable under these conditions.

A GRAPHICAL LOOK AT CHANGING CONDITIONS: THE CLAPEYRON EQUATION

A phase diagram of temperature versus pressure summarizes a large quantity of data on the stabilities of various phases and combinations of phases in a system. It is important to remember, however, that such a diagram gives information about the thermodynamic properties of these phases as well. The phase diagram for the aluminosilicate polymorphs kyanite, andalusite, and sillimanite is illustrated in Figure 8.4. Ignore, for the moment, the dashed lines. Like the phase diagram for water, this shows a one-component system, because the compositions of all the phases can be described by the single component Al_2SiO_5. At the beginning of this chapter, we noted that equilibrium is defined as a condition of minimum energy for a system. Therefore, the divariant fields in Figure 8.4 represent regions of P-T space where the free energy of the system is minimized by the occurrence of only one phase; the univariant curves and the invariant point define special combinations of pressure and temperature at which two or more coexisting phases provide the lowest free energy.

In order to illustrate this relationship between free energy and phase diagrams graphically, we will construct qualitative $\bar{G}$-T diagrams at constant pressure for the alu-

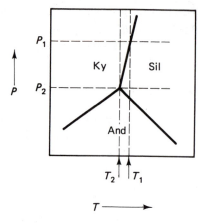

Figure 8.4 Schematic phase diagram for the system Al_2SiO_5. Dashed horizontal lines represent the traces of isobaric sections used to construct $\overline{G}$-T diagrams.

minosilicate system. The traces of such isobaric sections are represented by the horizontal dashed lines in Figure 8.4. At pressure P_1, kyanite has a lower free energy than the other Al_2SiO_5 polymorphs, so it must be the stable phase. This is shown in Figure 8.5a. Because $(\partial \overline{G}/\partial T)_P = -\Delta \overline{S}$ and the entropy of any phase must be greater than zero, free energy decreases in all phases as temperature increases. At temperature T_1, the $\overline{G}$-T lines for kyanite and sillimanite intersect, and the coexistence of both phases provides the lowest free energy configuration. At temperatures above T_1, sillimanite has

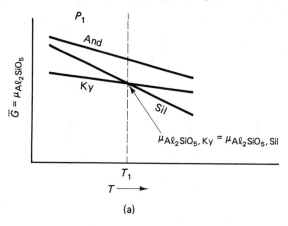

(a)

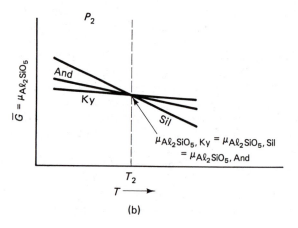

(b)

Figure 8.5 $\overline{G}$-T diagrams for the system Al_2SiO_5 at constant pressure: (a) corresponds to pressure P_1 in Figure 8-4, and (b) corresponds to P_2. At any temperature, the stable phase or combination of phases is that with the lowest free energy.

the lowest free energy and is the stable phase. Andalusite never appears at this pressure because its free energy is higher than those of the other polymorphs. The $\overline{G}$-T diagram at pressure P_2, shown in Figure 8.5b, is similar to the P_1 case except that the $\overline{G}$-T line for andalusite is depressed so that it intersects those of kyanite and sillimanite at temperature T_2, the aluminosilicate triple point. At temperatures other than T_2, however, the other polymorphs still offer lower free energy configurations.

There are other useful relationships between phase diagrams and thermodynamic quantities. In Chapter 2, we derived the following Maxwell relationship from $d\overline{H}$ (equation 2-27):

$$(\partial P/\partial T)_V = (\partial \overline{S}/\partial \overline{V})_T,$$

which is equivalent to

$$dP/dT = \Delta\overline{S}/\Delta\overline{V} = \Delta\overline{H}/T\Delta\overline{V}. \qquad (8\text{-}14)$$

This expression is known as the *Clapeyron equation,* and it is extremely useful in constructing and interpreting phase diagrams. Notice that the left side of the Clapeyron equation is the slope of any line in a *P-T* phase diagram. Thus this relationship enables us to calculate phase boundaries from thermodynamic data, provided that one point on the boundary is known. As an alternative, ratios of thermodynamic quantities can be obtained from experimentally determined reactions.

Worked Problem 8-6

To illustrate how the Clapeyron equation is used, calculate the approximate slope of the kyanite-andalusite boundary curve in Figure 8.4. The data we will need are the following: $\overline{V}$ (J bar^{-1}) = 4.409 for kyanite, 5.153 for andalusite, and 4.990 for sillimanite; $\overline{S}_{800}$ (J mol^{-1} K^{-1}) = 242.42 for kyanite, 251.31 for andalusite, and 252.94 for sillimanite. Inspection of these data indicates the relative positions of the phase fields for the three polymorphs. Kyanite, which has the smallest molar volume, should be stable at the highest pressures, and sillimanite should be stable at the highest temperatures because of its high molar entropy. Substituting values of $\overline{S}$ and $\overline{V}$ for kyanite and andalusite into equation 8-14, we obtain

$$dP/dT = (251.37 - 242.42 \text{ J mol}^{-1} \text{ K}^{-1})/(5.153 - 4.409 \text{ J bar}^{-1}) = 12.03 \text{ bars K}^{-1}$$

for the slope of the kyanite-andalusite boundary curve.

To complete the construction of this line, we must fix its position by determining the location of some point along it. The triple point has been located by experiment, although not without controversy (see the accompanying box), and its position can be used to locate this boundary in *P-T* space. There is a considerable amount of potential error in most calculated *P-T* slopes because of uncertainties in thermodynamic data. For this reason, it is preferable to determine phase diagrams experimentally. However, calculations like these can be valuable in estimating how systems behave under conditions other than those under which experiments have been carried out. Calculations also enable the experimental geochemist to limit runs to the probable location of a line or point, thus saving time and money.

PHASE EQUILIBRIA IN THE ALUMINOSILICATE SYSTEM

The tortuous history of the experimental determinations of phase equilibria in the aluminosilicate system illustrates the difficulties that can be encountered in this kind of work. Zen summarized the state of affairs in 1969, after decades of experiments, in the diagram shown as Figure 8.6a. Because of the constant shifting of the position of the aluminosilicate phase boundaries with each new experimental determination, this situation came to be known popularly as the "flying triple point." Zen discussed a number of causes for this experimental difficulty. One possibility was experimental error, such as incorrect pressure calibration. This might be due to failure to take into account the strength of the material, friction, and other

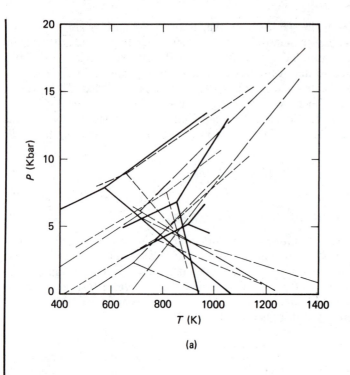

(a)

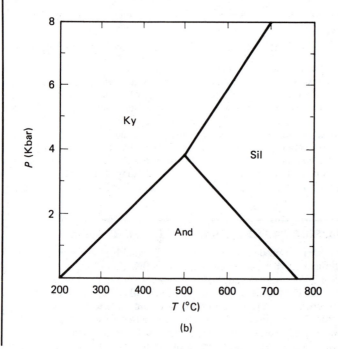

(b)

Figure 8.6 (a) Experimental determinations of phase equilibria in the system Al_2SiO_5. These data, summarized by Zen (1969), illustrate the difficulties in these measurements. (b) The phase relationships determined by Holdaway (1971) are the most commonly accepted values.

factors in converting the gauge pressure to actual pressure in the sample. He also observed that most of the experiments to that time were synthesis experiments; that is, they were based only on the first appearance of a phase and either failed to demonstrate reversibility or represented some synthesis reaction that did not occur in metamorphic rocks. For example, in one set of experiments, sillimanite was synthesized from gels or kaolinite. Another potential source of error was that some of the polymorphs were synthesized in the presence of water. If they contained small quantities of H_2O, they would not be true one-component phases.

Many metamorphic petrologists now accept the experimental determination of Holdaway (1971) as the best available phase equilibria for this system. Discrepancies between previous data and Holdaway's experiments may be explained by intense strain due to grinding that reduced the stability of kyanite, the presence of fibrolite (a rapidly grown, fine-grained "sillimanite" with generally unverified stoichiometry), and Al-Si disorder in sillimanite. This last factor is probably very important, as X-ray-diffraction studies suggest that some disorder is common in natural sillimanites, and that experimentally produced sillimanites may be even more disordered. Al-Si disorder in sillimanite is also suggested by the fact that the slope of the andalusite-sillimanite curve is steeper than predicted from the volume and entropy change of reaction using the Clapeyron equation. Holdaway was able to show that substitution of other components, such as Fe^{3+} for Al, was not really an important factor in this case. (However, such substitutions may be very important in interpreting the P-T conditions under which minerals other than Al_2SiO_5 polymorphs formed.)

Holdaway's phase diagram is presented in Figure 8.6b. The triple point is located at 501°C and 3.76 kbar.

HENRY'S LAW AND RAOULT'S LAW: MIXING OF SEVERAL COMPONENTS

So far in this chapter, we have considered only systems with one component. In systems with several components, it is necessary to understand how the various components interact. For example, if we define our system as a grain of olivine, we must determine the behavior of the components Mg_2SiO_4 and Fe_2SiO_4 when mixed together. The mixing characteristics of these two components may be very different from the mixing behavior of some trace olivine component such as Ni_2SiO_4. The interaction of various components in combinations of coexisting minerals, magmatic liquid, or hydrous fluid is in some cases nonideal, resulting in their selective concentration into one phase or another as explored briefly in Chapter 3. We will now examine the thermodynamic basis for various kinds of mixing behavior.

In an ideal solution, mixing of components with similar volumes and molecular forces occurs between lattice sites without any change in the molecular energy states or total volume of the system. Under these conditions, mixing is neither endothermic nor exothermic, that is, $\Delta \overline{H}_r = 0$. The activities of components (a_i) mixing on ideal sites are thus equal to their concentrations:

$$a_i = X_i, \tag{8-15}$$

which is a statement of *Raoult's law*.

Interactions between molecules or ions, however, may cause the chemical potentials of individual species to increase or decrease on a certain site, resulting in nonideal mixing. In this case, enthalpy of mixing is not equal to zero, and activities will depart from the ideal mixing curve, as illustrated in Figure 8.7. However, as the nonideal component is more and more diluted (that is, X_i becomes much less than 1), the component becomes so dispersed that eventually it is surrounded by a uniform environment of other ions or molecules. Therefore, at high dilution its activity becomes directly pro-

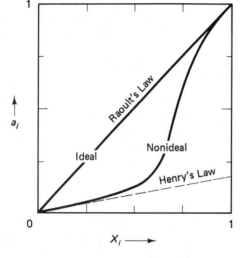

Figure 8.7 Relations between activity and mole fraction for solutions obeying Raoult's law and Henry's law. Raoult's law may be a good approximation for nonideal solutions in which X_i is large, and Henry's law is applicable to highly diluted components.

portional to its concentration, as shown by the straight dashed line in Figure 8.7. In other words,

$$a_i = h_i X_i, \tag{8-16}$$

where h_i is a proportionality constant. This is a statement of *Henry's law*.

These two laws can be used to explain the solution behavior of many geological materials. Raoult's law behavior is a common assumption for many major components of solid solution series, as in the case of the forsterite and fayalite components of olivine. Henry's law is commonly used to describe the behavior of many trace elements, such as nickel in olivine. It is also common to describe trace element distributions in terms of *distribution coefficients,* which are readily measureable quantities related to Henry's law constants. This relationship is explored below.

Worked Problem 8-6

Hakli and Wright (1967) analyzed the nickel contents of olivine phenocrysts and quenched liquid in samples taken at different temperatures from an Hawaiian lava lake. Their results can be expressed in terms of a *Nernst distribution coefficient*[1] defined as

$$K_D = c_{Ni,oliv}/c_{Ni,liq},$$

where c refers to measured concentration. What is the relationship of the distribution coefficient to Henry's law?

At equilibrium, we know that

$$\mu_{Ni_2SiO_4,liq} = \mu_{Ni_2SiO_4,oliv},$$

or

$$\mu^0_{Ni_2SiO_4,liq} + RT \ln a_{Ni_2SiO_4,liq} = \mu^0_{Ni_2SiO_4,oliv} + RT \ln a_{Ni_2SiO_4,oliv}.$$

If we apply Henry's law,

$$a_{Ni_2SiO_4,liq} = h_1 X_{Ni_2SiO_4,liq}$$

and

$$a_{Ni_2SiO_4,oliv} = h_2 X_{Ni_2SiO_4,oliv}.$$

Thus

---

[1] For equilibria in which the same phase or phases appear on both sides of the equation, K_{eq} is commonly called K_D, the Nernst distribution coefficient.

$$\mu^0_{\text{Ni}_2\text{SiO}_4,\text{liq}} - \mu^0_{\text{Ni}_2\text{SiO}_4,\text{oliv}} = RT \ln (h_2 X_{\text{Ni}_2\text{SiO}_4,\text{oliv}}/h_1 X_{\text{Ni}_2\text{SiO}_4,\text{liq}}) = RT \ln K_{\text{eq}}.$$

This simplifies to

$$(h_2 X_{\text{Ni}_2\text{SiO}_4,\text{oliv}}/h_1 X_{\text{Ni}_2\text{SiO}_4,\text{liq}}) = K_{\text{eq}},$$

or

$$K_{\text{eq}}(h_1/h_2) = X_{\text{Ni}_2\text{SiO}_4,\text{oliv}}/X_{\text{Ni}_2\text{SiO}_4,\text{liq}}.$$

This last term equals K_D because the concentration of Ni is equivalent to the mole fraction of nickel in olivine. Thus we see that the Nernst distribution coefficient incorporates the Henry's law constants in its formulation.

STANDARD STATES AND ACTIVITY COEFFICIENTS

Chemical potentials are functions of temperature, pressure, and composition. It is common practice to define the chemical potential in terms of a chemical potential at some reference value of temperature and pressure (called the *standard state*) and another term that corrects this value for deviations in temperature, pressure, and composition:

$$\mu_i = \mu^0_i + RT \ln a_i, \tag{8-17}$$

where μ_i is the chemical potential of component i in the phase of interest at T and P, μ^0_i is its value at some standard set of conditions, and a_i is the activity of the component. Note that when the phase of interest is in its standard state, its activity, a_1, must equal 1, so that its logarithm is zero and the second term drops out.

The standard state can be chosen arbitrarily, because it has no effect on the final result as long as we are consistent in its use. It might seem most logical to choose the standard state as a pure end-member component at 298 K and 1 bar. However, because of the availability of tabulated thermodynamic data, it is often more convenient to select the standard state as a pure component at the temperature and/or pressure of interest.

The relationships between μ_i and the activity term in equation 8-17 can be most readily seen in graphical form. Note that equation 8-17 is in the form of an equation for a straight line ($y = b + mx$) on a y versus x plot. Figure 8.8a shows μ_i versus $\ln a_i$. The line illustrated is not straight, because it is for a real substance that does not mix ideally at all concentrations. Instead, the line has two linear segments (shown extended with dashed lines). In the straight line segment extending from pure component $i(a_i = 1$, so $\ln a_i = 0$) to slight dilution with another component, component i obeys Raoult's law ($a_i = X_i$). The slope of this segment is RT and its intercept on the μ_i axis is μ^0_i. In the Raoult's law region, chemical potential can be expressed as

$$\mu_i = \mu^0_i + RT \ln X_i. \tag{8-18}$$

The Henry's law region extends from infinite dilution of component i to some modest value and, because $a_i = h_i X_i$, it is also represented by a straight line in Figure 8.8a. Its slope is also RT, but its intercept on the μ_i axis is not μ^0_i. Instead, this intercept is now μ^0_i plus the added term $RT \ln h_i$, which is a function of temperature, pressure, and the composition of the component with which i is diluted, but not a function of X_i. The expression for chemical potential in the Henry's law region therefore is

$$\mu_i = \mu^0_i + RT \ln X_i + RT \ln h_i,$$

where h_i is the Henry's law constant for component i in this phase. Combining the two logarithmic terms in this equation gives

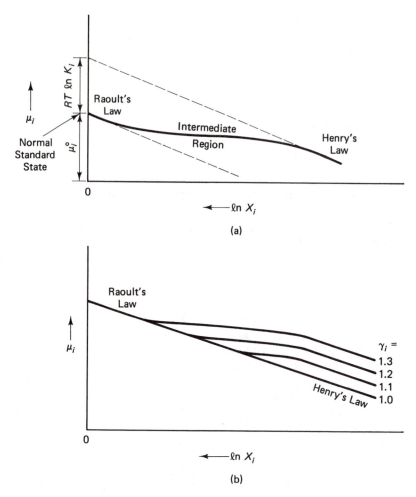

Figure 8.8 (a) Plot of μ_i versus $\ln X_i$ showing the dependence of chemical potential on composition. In the Raoult's law region, $\mu_i = \mu_i^o$; in the Henry's law region, $\mu_i = \mu_i^o + RT \ln h_i$. The slope of the Raoult's and Henry's law regions is RT. The normal standard state for component i in a pure substance at P and T is μ_i^o. (b) Plot of μ_i versus $\ln X_i$ showing how the intermediate region changes with activity coefficient. With ideal mixing ($\gamma_i = 1$), the Raoult's and Henry's law regions lie along the same trend. As γ_i becomes larger, the inflection appears and the intermediate region becomes larger.

$$\mu_i = \mu_i^0 + RT \ln h_i X_i. \tag{8-19}$$

The intermediate region between Henry's and Raoult's law behavior is characterized by a curved segment that connects the two straight line segments (Figure 8.8a). This region is thus expressed by an equation that is intermediate between 8-18 and 8-19. In order to formulate such an equation, we can describe the degree of nonideality by using γ_i, the activity coefficient, introduced in Chapter 3. The activity coefficient is a function of composition, such that in the Raoult's law region,

$$\gamma_i \longrightarrow 1, \text{ so that } a_i \longrightarrow X_i$$

and in the Henry's law region,

$$\gamma_i \longrightarrow h_i, \text{ so that } a_i \longrightarrow h_i X_i.$$

Thus we can formulate a general equation for chemical potential of a pure component i in any phase at any temperature and pressure:

$$\mu_i = \mu_i^0 + RT \ln \gamma_i X_i. \tag{8-20}$$

We can see that as component i becomes more similar to the other component with which it is mixing, $h_i \to 1$. Moreover, similarities in these components will increase the size of the Raoult's and Henry's law regions in this diagram, to the point where the intermediate region disappears and the Raoult's and Henry's law regions connect. In such a case, we would then have one straight mixing line in this diagram (equivalent to the line that extrapolates to μ_i^0), and the mixing would be ideal. The effects of changing activity coefficients on the various mixing regions are illustrated in Figure 8.8b.

Besides the standard state we have just considered, there are others that could be used; in fact, it is advantageous to do so under certain conditions. Obviously, we must

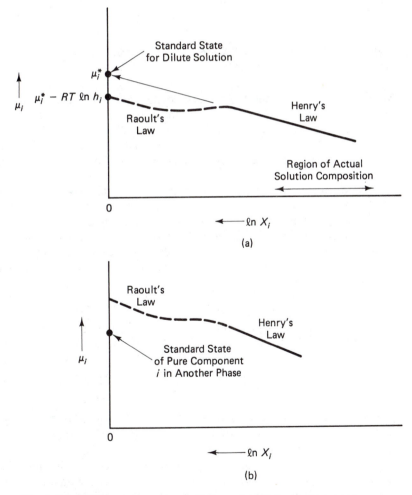

Figure 8.9 Diagrams of μ_i versus $\ln X_i$ illustrating (a) the standard state for component i in a dilute solution extrapolated from the Henry's law region, and (b) the standard state for component i taken as the activity of i in another phase at P and T.

Standard States and Activity Coefficients

use another standard state if it is not possible to make the phase of interest out of pure component i. For example, one cannot synthesize a garnet of lanthanum silicate, so μ_{La}^o in garnet based on the pure phase cannot be determined directly. In this case, the standard state could be a hypothetical garnet with properties obtained by extrapolating the Henry's law region to the pressure and temperature of interest. Such a situation is illustrated in Figure 8.9a. This is the standard state we assumed without any explanation in Chapter 3. In fluids, it is not possible to measure the activity of, for example, pure Na^+ ions, so we must extrapolate from the dilute salt solution.

We might also envision a case where component i could not be studied directly in the phase of interest, but its chemical potential could be analyzed in another phase. This could occur, for example, if we were interested in determining the chemical potential of Al_2O_3 in a magma but found it easier to study glass quenched from the magma. We could use the activity of Al_2O_3 in a glass of pure Al_2O_3 composition to define the standard state. Such a standard state is illustrated in Figure 8.9b. The expressions for chemical potentials using other standard states can be derived from equation 8-19.

SOLUTION MODELS: NUMERICAL EXPRESSIONS FOR ACTIVITY COEFFICIENTS

Many different equations for activity coefficients as a function of composition have been formulated. Such models may be theoretical, or they may be constructed empirically from observed mixing behavior. The only restrictions on them are that they must obey the constraints given above for the Raoult's and Henry's law regions. That is, they must explain mixing at the extreme ends of the compositional spectrum where $X_i \rightarrow 1$ and $X_i \rightarrow 0$.

One situation model that is often used in solving geochemical problems assumes ideal mixing. This is a trivial model, because the expression for $\gamma_i = 1$. From Figure 8.8a, we can see that the assumption will be correct in the Raoult's law region, but will become increasingly less accurate in moving toward the intermediate and Henry's law regions. Thus this solution model works best in cases where X_i is large.

Another solution model that is commonly used is that for a *regular solution*. For this model,

$$RT \ln \gamma_i = WX_j^2 = W(1 - X_i^2), \tag{8-21}$$

where W is an *interaction parameter* that is a function of temperature and pressure but not composition, and X_j is the mole fraction of the other component with which component i is mixing. Substitution of this expression in equation 8-19 gives the chemical potential for component i using a regular solution model:

$$\mu_i = \mu_i^o + RT \ln X_i + WX_j^2. \tag{8-22}$$

Note that this equation obeys the necessary restrictions in the Raoult's and Henry's law regions. In the case of Raoult's law, as $X_i \rightarrow 1$, $WX_j^2 \rightarrow 0$ and $\gamma_i \rightarrow 1$. For Henry's law, as $X_i \rightarrow 0$, $WX_j^2 \rightarrow W$ and $\gamma_i \rightarrow h_i$.

In the regular solution model, the inflection point in the intermediate region on the $\mu_i - \ln X_i$ plot is flanked by symmetrical limbs, as illustrated in Figure 8.10. However, in real systems, mixing behavior need not be this regular. The *asymmetric solution* model is a somewhat more complex formulation that is also very useful in some situations. This model involves two interaction parameters, W_i and W_j, rather than one as

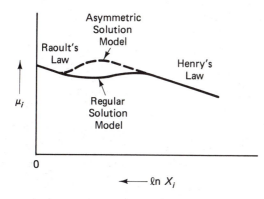

Figure 8.10 The difference in mixing behavior for the regular and asymmetric solution models is illustrated by the shapes of the intermediate regions. The regular solution model produces symmetrical limbs flanking the inflection point, and the asymmetrical model does not.

in the regular solution model. Asymmetric interaction parameters are sometimes called *Margules parameters*.[2] For the asymmetric solution model,

$$RT \ln \gamma_i = X_j^2(W_i + 2X_i(W_j - W_i)). \tag{8-23}$$

This model can be used in situations where mixing is not symmetric, as illustrated in Figure 8.10. Note that when $W_i = W_j$, the asymmetric solution model reduces to the regular solution model. This mixing model also obeys the necessary restriction in the Raoult's and Henry's law regions. However, in the Henry's law region, $RT \ln \gamma_i \to W_i$ and $RT \ln \gamma_j \to W_j$.

GEOTHERMOMETRY AND GEOBAROMETRY

The formulation of geothermometers and geobarometers involves some of the concepts that were just introduced, so we will use one example of each to illustrate how useful these can be in solving geologic problems. For a particular reaction to be used as a geothermometer, it must be a strong function of temperature but be nearly independent of pressure. Conversely, a geobarometer should be sensitive to pressure but not to temperature. We have already seen that at equilibrium

$$\Delta \overline{G}_r^0 = -RT \ln K_D = \Delta \overline{H} - T\Delta \overline{S} + P\Delta \overline{V}. \tag{8-24}$$

This equation assumes that $\Delta \overline{V}$ is independent of temperature and pressure; otherwise we would have to include expressions for thermal expansion and isothermal compressibility. The form of this equation indicates that if K_D is constant, there is only one equilibrium pressure at any given temperature; fixing K_D determines the position of the equilibrium reaction curve in *P-T* space. Examples of equilibria that would make good geothermometers and geobarometers are illustrated in Figure 8.11. The factors that control the slopes of these equilibrium curves in *P-T* space can be evaluated by differentiating equation 8-24 with respect to temperature at constant pressure and with respect to pressure at constant temperature, as follows:

$$(\partial \ln K_D/\partial T)_P = \Delta \overline{H}/RT^2 \tag{8-25}$$

and

$$(\partial \ln K_D/\partial P)_T = -\Delta \overline{V}/RT. \tag{8-26}$$

[2] Rather than use Margules parameters, which have no physical meaning, a few recent papers attempt to derive free energy of mixing expressions from basic theory. These models usually involve configurational entropy terms obtained from the physics of the ordering process. An excellent example of this new and powerful approach is Davidson's (1985) model for quadrilateral pyroxene compositions.

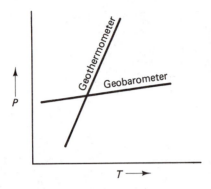

Figure 8.11 Equilibrium curves for re-actions that would make a good geother-mometer and a good geobarometer. The position of each of these curves is fixed at a given K_D, but the geothermometer curve will be translated to the left, paral-lel to its present position, with increas-ing K_D. The geobarometer curve will be translated upward with increasing K_D, but it too will retain the same slope.

If the right side of equation 8-25 is larger than that of 8-26, then the equilibrium de-pends more on temperature than pressure and would be a suitable geothermometer. Thus, we can see that geothermometer reactions should have large values of $\Delta \overline{H}$ and, conversely, that geobarometer reactions should be characterized by large $\Delta \overline{V}$. It will also be necessary to have accurate standard-state thermodynamic data for the reaction available, as well as formulations for activity coefficients.

Worked Problem 8-8

How are geothermometers and geobarometers calibrated? One example of a useful geothermometer is the partitioning of iron and magnesium between coexisting biotite and garnet. The exchange reaction is

$$Fe_3Al_2Si_3O_{12} + KMg_3AlSi_3O_{10}(OH)_2 \longleftrightarrow Mg_3Al_2Si_3O_{12} + KFe_3AlSi_3O_{10}(OH)_2.$$

The enthalpy change for this reaction is large but $\Delta \overline{V}$ is quite small, as is appropriate for a geothermometer. Available data suggest that iron and magnesium mix almost ideally in bi-otite and garnet, so we will assume that they obey Raoult's law and $a_i = X_i$. No formula-tion for their activity coefficients is necessary in this case. The Fe-Mg partitioning be-tween these phases are functions of T (and to a lesser extent P) alone so long as we assume ideal mixing in both garnet and biotite. There are a number of calibrations for this reac-tion. Ferry and Spear (1978) measured the progress of this exchange reaction by experi-ment at a series of temperatures, and obtained the following relationship:

$$\ln K_D = -2109/T + 0.782,$$

where K_D equals $(X_{Mg}/X_{Fe})_{garnet}/(X_{Mg}/X_{Fe})_{biotite}$, and T is the temperature in Kelvins. The coefficients were determined by a linear least-squares fit of the experimental values of $\ln K_D$ versus $1/T$, determined at a pressure of about 2 kilobars. Using the Clapeyron equa-tion, Ferry and Spear were also able to calculate dP/dT and, therefore, to estimate the de-pendence of the reaction on pressure. Their expression for this Fe-Mg exchange reaction at any pressure and temperature is

$$12{,}454 - 4.662T(K) + 0.057P(\text{bars}) + 3RT \ln K_D = 0. \qquad (8\text{-}27)$$

From equation 8-27 it can be seen that the effect of pressure on this exchange equilibrium is small, and that it can be ignored if no pressure estimate is available.

As an example of a reaction that is pressure-dependent, let us consider the follow-ing. Assemblages of plagioclase + garnet + quartz + clinopyroxene can exhibit the re-action

$$3CaAl_2Si_2O_8 + 3CaMgSi_2O_6 \longleftrightarrow 2Ca_3Al_2Si_3O_{12} + Mg_3Al_2Si_3O_{12} + 3SiO_2.$$

This equilibrium involves a significant change in $\Delta \overline{V}$. We will examine a calibration of this geobarometer formulated by Newton and Perkins (1982). As we will see, this reaction is much more complex than the geothermometer we have already examined. Using equation

8-24, and substituting appropriate thermodynamic data, these workers obtained the relationship

$$P(\text{bars}) = 675 + 17.179T(K) + 3.5962\ T \ln K_{eq}, \tag{8-28}$$

where

$$K_{eq} = (a_{Ca}^2 a_{Mg})_{\text{garnet}} / (a_{Ca})_{\text{plagioclase}} (a_{CaMg})_{\text{clinopyroxene}}.$$

The difficulty comes in determining expressions for activities of the various components. Newton and Perkins found that mixing of calcium and magnesium in garnet is non-ideal, and they suggested an interaction parameter of the form

$$W_{\text{Ca-Mg garnet}} = 3300 - 1.5\ T,$$

where T is in Kelvins and W is expressed as calories per mole of $(Ca, Mg)Al_{2/3}SiO_4$. Of course, these are not the only components in garnets, but we will make the simplifying assumption that garnets consist only of Ca, Mg, and Fe aluminosilicate components (grossular, pyrope, and almandine). Interaction parameters for Fe-Mg and Fe-Ca exchange in garnets are apparently much smaller than for Ca-Mg and can be ignored. Assuming a regular solution model, the activity coefficient for calcium in garnet is given by

$$RT \ln \gamma_{\text{Ca garnet}} = W_{\text{Ca-Mg}}(X_{Mg}^2 + X_{Mg}X_{Fe}),$$

and the activity coefficient for magnesium in garnet by

$$RT \ln \gamma_{\text{Mg garnet}} = W_{\text{Ca-Mg}}(X_{Ca}^2 + X_{Ca}X_{Fe}).$$

The activities for calcium and magnesium in garnet to be substituted into the expression for K_{eq} in equation 8-28 are found by multiplying these values by X_{Ca} and X_{Mg}.

The expression for the activity coefficient of anorthite in plagioclase takes the form

$$RT \ln \gamma_{An} = X_{Ab}^2(2025 + 9442X_{An}).$$

As we already discussed in Chapter 3, estimating the activities of various components in minerals can be complicated by the fact that certain components may show different mixing behavior on different crystallographic sites in the same phase. For example, a lattice accommodating a large cation may be less distorted if the cations do not occur in adjacent sites. This is apparently the situation in plagioclase. Here, we use a special mixing model known as aluminum-avoidance, which avoids placing Al-filled tetrahedra next to each other in the feldspar framework structure. The expression for activity of calcium in plagioclase becomes

$$a_{\text{Ca plagioclase}} = \gamma_{An}X_{An}(1 + X_{An})^2/4.$$

The diopside component in clinopyroxene is considered to mix nearly ideally with other pyroxene components, but again we must keep track of the distinct crystallographic sites in this mineral. The expression for the activity of CaMg in clinopyroxene is

$$a_{\text{Ca-Mg clinopyroxene}} = X_{Mg\,M1}X_{Ca\,M2},$$

where $M1$ and $M2$ denote the two structural sites in pyroxenes that contain this cation. Ca, Na, Mn, and Fe^{2+} are assigned to $M2$, and Mg, Fe^{3+}, Ti, Al, and any remaining Fe^{2+} are partitioned into $M1$.

Finally, computed values for the activities of the various components are substituted into equation 8-28 to obtain the equilibrium pressure. An independent determination of temperature should also be made using a suitable geothermometer such as the garnet-biotite equilibrium already discussed, if a precise pressure estimate is to be obtained.

A useful program called PT that calculates temperatures and pressures of equilibration is contained on the diskette available with this book. This program solves simultaneous equations for a geothermometer and a geobarometer for equilibrium values of P and T. Geothermometers that can be employed are garnet-biotite (Ferry and Spear, 1978) or garnet-clinopyroxene (Ellis and Green, 1979); geobarometers include either garnet-plagioclase-aluminosilicate-quartz (Newton and Haselton, 1981), garnet-plagioclase-clinopyroxene-quartz (Newton and Perkins, 1982), or garnet-plagioclase-orthopyroxene-quartz (Newton and Perkins, 1982). A short description of the program with appropriate references is given in Appendix D. Although input parameters for the program are self-explanatory, you are encouraged to read these original references so that you can understand the theory and potential pitfalls in these geothermometers and geobarometers.

SUMMARY

We have reaffirmed that equilibrium is a condition of minimum free energy for a system. The phase rule provides a test for equilibrium, and it is likely that geologic systems at equilibrium will have variances of 2 or higher. The effects of changing temperature and pressure on the free energy of a system can be determined, and the direction of a particular reaction under different conditions can be predicted from appropriate thermodynamic data.

Systems containing more than one component require analysis of the way such components interact in the phases that comprise the system. Components that mix ideally obey Raoult's law; that is, their activities equal their concentrations. Components at high dilution obey Henry's law; that is, their activities are directly proportional to (but not equal to) their concentrations. The chemical potentials of components are generally expressed in the form $\mu_i = \mu_i^o + RT \ln \gamma_i X_i$, which is the chemical potential at some standard state plus an activity term that corrects this value for deviations in conditions and composition. Activity coefficients are equal to 1 for Raoult's law behavior and h_i for Henry's law behavior. Various solution models have been formulated for activity coefficients for components exhibiting intermediate (non-ideal) behavior.

With this thermodynamic background, we are now ready to consider phase diagrams for multi-component systems. These are introduced in the following chapter.

SUGGESTED READINGS

There are a number of excellent textbooks that provide an introduction to thermodynamics, as well as to various geological applications. These references generally contain significantly more detail than has been presented here, and are highly recommended for the serious student.

DENBIGH, K. 1971. *The principles of chemical equilibrium*. London: Cambridge University Press. (A classic thermodynamic text containing rigorous derivations but no examples of geologic interest. A superb reference, but not for the faint-hearted.)

ERNST, W. G. 1976. *Petrologic phase equilibria*. San Francisco: W. H. Freeman and Company. (A very readable introductory text. Chapter 3 illustrates the computational approach to phase diagrams, and Chapter 4 contains descriptions of G-T diagrams.)

ESSENE, E. J. 1982. Geologic thermometry and barometry. In *Characterization of metamorphism through mineral equilibria*, ed. Ferry, J. M., 153–206. Reviews of Mineralogy 10. Washington, D.C.-Mineral. Soc. of America. (An excellent review of the calibration, assumptions, and precautions in determining metamorphic temperatures and pressures.)

POWELL, R. 1978. *Equilibrium thermodynamics in petrology*. London: Harper and Row, Ltd.. (An excellent introductory text that dispenses with proofs. Chapter 2 describes G-T diagrams, and Chapter 3 gives useful formulations for various standard states not presented in this book.)

SAXENA, S. K. 1973. *Thermodynamics of rock-forming crystalline solutions*. New York: Springer-Verlag. (A more advanced book that provides a wealth of information on mixing models and solution behavior for real mineral systems.)

WOOD, B. J., and FRASER, D. G. 1977. *Elementary thermodynamics for geologists*. Oxford: Oxford University Press. (A clear and concise introduction to thermodynamics, with numerous worked examples of geologic interest. Chapter 3 treats regular mixing models, and Chapter 4 provides a much more detailed look at geothermometers and geobarometers than is given in this book.)

The following papers were referenced in this chapter. The interested student may wish to explore them more thoroughly.

DAVIDSON, P. M. 1985. Thermodynamic analysis of quadrilateral pyroxenes. Part I: Derivation of the ternary non-convergent site disorder model. *Contrib. Mineral. Petrol.* 91: 383–89.

ELLIS, D. J., and GREEN, D. H. 1979. An experimental study of the effect of Ca upon garnet-clinopyroxene Fe-Mg exchange equilibria. *Contrib. Mineral. Petrol.* 71: 13–22.

EUGSTER, H. P., and WONES, D. R. 1962. Stability relations of the ferruginous biotite, annite. *J. Petrol.* 3: 82–125.

FERRY, J. M., and SPEAR, F. S. 1978. Experimental calibration of the partitioning of Fe and Mg between biotite and garnet. *Contrib. Mineral. Petrol.* 66: 113–17.

HAKLI, T. A., and WRIGHT, T. L. 1967. The fractionation of nickel between olivine and augite as a geothermometer. *Geochim. Cosmochim. Acta* 31: 877–84.

NEWTON, R. C., and HASELTON, H. T. 1981. Thermodynamics of the garnet-plagioclase-Al$_2$SiO$_5$-quartz geobarometer. In *Thermodynamics of minerals and melts,* ed. Newton, R. C., Navrotsky, A., and Wood, B. J., 131–147. New York: Springer-Verlag.

NEWTON, R. C., and PERKINS, D. III 1982. Thermodynamic calibration of geobarometers based on the assemblages garnet-plagioclase-orthopyroxene (clinopyroxene)-quartz. *Amer. Mineral.* 67: 203–22.

PROBLEMS

1. Using Figure 8.4, construct qualitative $\bar{G}$-P diagrams for the Al$_2$SiO$_5$ system at temperatures T_1 and T_2.

2. From the understanding that $(\partial \Delta \bar{G}/\partial T)_P = -\Delta \bar{S}$ and $(\partial \Delta \bar{G}/\partial P)_T = \Delta \bar{V}$, construct a schematic $\bar{S}$-T diagram at pressure P_1 and a $\bar{V}$-P diagram at temperature T_2 for the Al$_2$SiO$_5$ system (Figure 8.4).

3. Define the system components necessary to describe a rock with the following mineral assemblage: wollastonite CaSiO$_3$, calcite CaCO$_3$, and quartz SiO$_2$. What is the variance of the system? Now assume that the rock is being metamorphosed in equilibrium with a CO$_2$ fluid phase. What are the system components? Would the variance change if CO$_2$ were a mobile component? Explain.

4. Consider an isolated system containing the following phases at equilibrium: (K,Na)Cl, (Na,K)AlSi$_3$O$_8$, and aqueous fluid. (a) Choose a set of components for this system. (b) State the conditions necessary for the system to be in complete equilibrium.

5. Clinoenstatite (MgSiO$_3$) melts incongruently to forsterite (Mg$_2$SiO$_4$) plus liquid at 1557°C at atmospheric pressure. (a) Using thermodynamic data from some reliable source, calculate the chemical potential of SiO$_2$ in the liquid, when these three phases coexist at equilibrium. (b) What will happen to the system if μ_{SiO_2} is increased while T and P are held constant?

6. Using the following data, (a) calculate $\Delta \bar{H}^0_{800}$, $\Delta \bar{S}^0_{800}$, and $\Delta \bar{G}^0_{800}$ for the following reaction:

$$NaAlSi_3O_8 \longrightarrow NaAlSi_2O_6 + SiO_2.$$

$$\text{albite} \qquad\qquad \text{jadeite} \qquad \text{quartz}$$

	H^0_{298} (cal mol^{-1})	S^0_{298} (cal deg^{-1} mol^{-1})	$\bar{C}_P$ (cal deg$_{-1}$ mol$_{-1}$)
albite	−937.146	50.20	$61.7 + 13.9T/10^3 - 15.01 \times 10^5/T^2$
jadeite	−719.871	31.90	$48.2 + 11.4T/10^3 - 11.87 \times 10^5/T^2$
quartz	−217.650	9.88	$11.2 + 8.2 T/10^3 - 2.7 \times 10^5/T^2$

(b) Calculate $\Delta \bar{G}^0_{800}$ for the reaction assuming $\Delta \bar{C}_P = 0$. Compare the two values and comment.

7. In Worked Problem 8-6, we calculated the slope of the andalusite-kyanite boundary in P-T space. Its location was fixed using the experimentally determined triple point of Holdaway discussed in the box on page 249. Using the Clapeyron equation and data in Worked Problem 8-6, construct the remainder of the aluminosilicate phase diagram. Assume that $\Delta \bar{C}_P = 0$ and that $\bar{V}$ is independent of P and T.

8. For the reaction

$$Mg(OH)_2 \rightarrow MgO + H_2O,$$

$$\text{brucite} \quad \text{periclase}$$

(a) calculate the equilibrium temperature for the reaction at 1000 bars pressure using thermodynamic data from some appropriate source. Assume that $\Delta \bar{C}_P = 0$, $\bar{V}_{solids} = $ constant, and that H_2O behaves as an ideal gas. (You may wish to use the DELTAG program on the diskette.) (b) Calculate the equilibrium temperature under the same conditions as before, but this time with $P_{H_2O} = 0.5$.

9. Using the values for thermal expansion and isothermal compressibility of almandine and quartz in Table 8-2, and the molar volumes that follow, calculate $\bar{V}$ for each of these phases at 800°C and 1 kbar.

$$\bar{V}_{298,1\,\text{almandine}} = 118.2 \text{ cm}^3 \text{ mol}^{-1}$$
$$\bar{V}_{298,1\,\text{quartz}} = 22.69 \text{ cm}^3 \text{ mol}^{-1}$$

10. Calculate a_i as X_i increases, using Raoult's Law and a regular solution model with $W = 10$ kJ, and determine how small X_i can become before Raoult's Law is no longer a good approximation.

11. Given the reactions in the following list, discuss what standard states you would choose for the chemical potentials:
(a) leucite + SiO_2 (melt) → orthoclase, at 1 bar and 1160°C;
(b) $EuAl_2Si_2O_8$ (melt) → $EuAl_2Si_2O_8$ (plagioclase), at 1300°C and 1 bar;
(c) gypsum → anhydrite + water, at 100°C and 1 kbar.

12. A metamorphic rock contains garnet, biotite, plagioclase, kyanite, and quartz in apparent equilibrium. Microprobe analyses for the first three phases are presented in the following list (already converted to atomic proportions). Determine the temperature and pressure at which this assemblage equilibrated. (You should use the PT program on the diskette.)

biotite: 1.888 Fe, 2.838 Mg
garnet: 2.219 Fe, 0.588 Mg, 0.150 Ca, 0.183 Mn
plagioclase: $X_{An} = 0.17$

Chapter 9

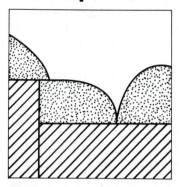

Picturing Equilibria:
Phase Diagrams

OVERVIEW

Many geologists, even after years of experience, find thermodynamics difficult to understand unless they can draw pictures to relate their equations to tangible systems. Phase diagrams provide a convenient and powerful way to picture equilibria. In this chapter, we introduce the principles of $\overline{G}$-X diagrams for binary systems, and illustrate the minimization of free energy graphically. We then construct T-X and P-X diagrams by stacking $\overline{G}$-X sections, using examples of real geochemical systems. Phases terminate in these diagrams by three-phase equilibria, coincidences in composition, limiting values, and critical points to produce eutectics, peritectics, transition loops, and solvi. We examine each of these in turn. We also review equilibrium crystallization paths in binary systems and show how binary phase diagrams can be constructed from thermodynamic data. With this background, we then describe P-T diagrams for binary systems. Finally, we briefly introduce graphical relationships for systems with three components (triangular liquidus and subsolidus diagrams), and provide an example of a diagram based on chemical potentials.

G-X DIAGRAMS

In Chapter 8, we learned that a *phase diagram* is simply a graphical representation of the stability fields of one or more phases, drawn in terms of convenient thermodynamic variables. In a one-component (unary) system, the equilibrium state is determined by the specification of two thermodynamic variables, such as some combination of P, T, $\overline{S}$, or $\overline{V}$. In our earlier discussion of the unary system H_2O, we considered a phase diagram with pressure and temperature as Cartesian axes, which showed single-phase fields for water, ice, and vapor, as well as lines along which two phases coexisted stably. To determine the state of a two-component (binary) system completely, an additional compositional variable must be specified. Thus, a complete phase diagram for a binary system must be drawn in three dimensions. Once again, we are free to choose any three thermodynamic functions as Cartesian axes, but for most geochemical applications it is most convenient to select P, T, and X_2, where X_2 is the mole fraction of component 2. Of course, we could just as easily choose X_1 as our compositional variable, because $X_1 = 1 - X_2$.

Molar Gibbs free energy is a function of all of the variables above [that is, $\overline{G} = \overline{G}(P, T, X_2)$], so let us consider a diagram (Figure 9.1) in which a single phase with a continuous compositional range between $X_2 = 0$ and $X_2 = 1$ is represented. The curved line on this diagram is a graph of the function $\overline{G} = \overline{G}(P, T, X_2)$, considered at an arbitrary but fixed set of values for P and T. It is, in other words, an isothermal, isobaric plot of $\overline{G} = \overline{G}(X_2)$. Given a mixing model from which to calculate the activity coefficients properly, we could write $\overline{G}(X_2)$ explicitly as

$$\overline{G}(X_2) = \mu_1^o X_1 + \mu_2^o X_2 + RT(X_1 \ln X_1 + X_2 \ln X_2 + X_1 \ln \gamma_1 + X_2 \ln \gamma_2)$$
$$= \mu_1^o + X_2(\mu_2^o - \mu_1^o) + RT([1 - X_2][\ln (1 - X_2)\gamma_1] + X_2[\ln X_2 \gamma_2]).$$

Now let's consider some significant features of this diagram.

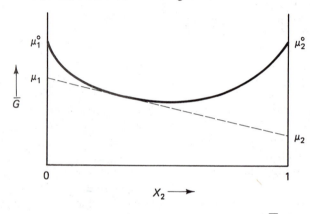

Figure 9.1 $\overline{G}$-X_2 diagram at fixed T and P for a phase with a continuous compositional range from $X_2 = 0$ to 1. At any particular composition, the ends of a tangent to the $\overline{G}(X_2)$ curve define the values for μ_1 and μ_2 in that phase, as illustrated by the dashed line.

We can, of course, define the value of $\overline{G}$ for this phase at any specific composition by reading it directly from the ordinate on the graph. This value will be the molar free energy for the phase at composition X_2. It is more instructive, however, to consider the individual contributions of μ_1 and μ_2 to $\overline{G}$. We do this by constructing a tangent to the $\overline{G}$-X curve at X_2, as illustrated in Figure 9.1 by the dashed line. The point at which this tangent intersects the ordinate at $X_2 = 1$ is the value for μ_2 in the phase; the point at which it intersects the ordinate at $X_2 = 0$ ($X_1 = 1$) is the value of μ_1 in the phase. Notice that despite the fact that, as drawn in Figure 9.1, $\overline{G}$ reaches a minimum value at some point between $X_2 = 0$ and $X_2 = 1$, the value of μ_2 increases monatonically (and the value of μ_1 decreases monotonically) as X_2 increases. Also notice that the value for either chemical potential approaches a maximum as the composition approaches that of the appropriate undiluted chemical component. Therefore, μ_2^o represents the chemical potential of component 2 in the phase at its pure $X_2 = 1$ end member.

In the example above, we have chosen a phase exhibiting solid solution behavior. We can, of course, also encounter pure substances with no compositional variation whatsoever. For such substances, the $\overline{G}(X_2)$ curve sharpens to a point. (This could be represented equally well by a vertical spike, if you prefer.)

If several phases are possible in the system, each will have its own $\overline{G}(X_2)$ curve or point. The equilibrium state for the system is that phase or set of phases which yields a minimum value of $\overline{G}$ for any specified value of X_2 (still, remember, at constant P and T). The equilibrium states for various values of X_2 can best be visualized by imagining a tangent line or set of tangent lines that touch the set of $\overline{G}$ curves or points, forming a sort of floor to the diagram as illustrated by the dashed lines in Figure 9.2a. In this example, we have one pure phase B and two phases (A and C) that exhibit solid solution behavior. Notice that if we start at the left side of the diagram and move gradually toward the right, the tangent to the left limb of the $\overline{G}(X_2)$ curve for phase A changes slope monotonically, which is to say that both μ_{1A} and μ_{2A} vary monotonically. This situation continues until we reach the value of X_2 at which the tangent to the curve for phase A is also tangent to the point for phase B. We now see that $\mu_{1A} = \mu_{1B}$ and $\mu_{2A} = \mu_{2B}$. This is, of course, a condition defined originally in equation 2-45, which we have come to know as chemical equilibrium between phase A and phase B. Beyond this value of X_2, the free energy of the system can be found along the common tangent and is lower than the free energy for phase A alone. The two-phase assemblage A + B continues to be stable until we reach the value of X_2 where the free energy for phase B equals that given by the common tangent at that value of X_2. At this point only phase B is stable, and the tangent pivots instantaneously on B to become the tangent to both B and C. At this point, B + C becomes the stable phase assemblage. A similar pattern occurs as we continue to the right and encounter a region where phase C alone is the stable phase.

We see, therefore, that the chemical potential of either component 1 or component 2 changes monotonically across single-phase fields as we vary X_2, but both chemi-

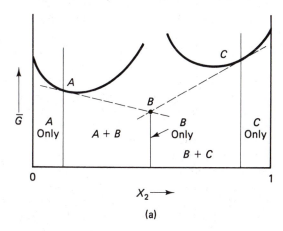

(a)

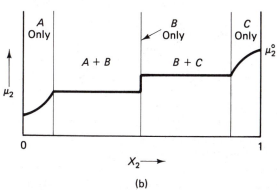

(b)

Figure 9.2 (a) $\overline{G}$-X_2 diagram at fixed T and P for a system containing three phases. Phase B is pure, but phases A and C show limited ranges of composition. The equilibrium states for various values of X_2 are those phases which yield the minimum $\overline{G}$ for the system. These can be readily determined from tangents to the $\overline{G}(X_2)$ curves, as shown by the dashed lines. (b) The chemical potential for component 2 in the stable phase or phase assemblage in (a) increases with X_2, as illustrated in this diagram. μ_2 is fixed by the coexistence of two phases, but changes monotonically if only one phase is present.

cal potentials remain fixed in two-phase fields. This can be seen even more graphically in Figure 9.2b (a schematic construction of the variation of μ_2 in the stable phase or phase assemblage) as X_2 ranges from 0 to 1 in the system in Figure 9.2a. When two phases are stable, a binary system has only two degrees of freedom (T and P), as dictated by the phase rule.

DERIVATION OF T-X AND P-X DIAGRAMS

A $\overline{G}$-X diagram like those we have just considered is valid only for the fixed values of P and T that we specified in drawing it. To construct a phase diagram for the system that takes into account changing temperature and pressure conditions, we must recall the variation of $\overline{G}$ with T and with P at constant X_2:

$$(\partial \overline{G}/\partial T)_{P,X_2} = -\overline{S}, \tag{9-1}$$

and

$$(\partial \overline{G}/\partial P)_{T,X_2} = \overline{V}. \tag{9-2}$$

Because $\overline{S}$ and $\overline{V}$ are inherently positive quantities, all $\overline{G}(X_2)$ curves will shift upward or downward simultaneously with increasing P or T, respectively. However, because the specific $\overline{S}$ and $\overline{V}$ values for each phase are different and because $\overline{S}$ and $\overline{V}$ are themselves functions of composition, *relative* movements of the $\overline{G}$ curves and changes in their shapes will generally occur. Consider, for example, a sequence of $\overline{G}$-X diagrams drawn at T_1, T_2, and T_3, an arbitrarily decreasing set of temperatures, illustrated in Figure 9.3. With increasing temperature, the $\overline{G}(X_2)$ curves for all three phases in the system move upwards, but in this example, the curve for phase B moves faster than the other two. As a result, the curves for all three phases have a common tangent at the unique temperature T_2, and we have the condition $\mu_{1A} = \mu_{1B} = \mu_{1C}$; that is, a three-phase field A + B + C exists at T_2, replacing the fields A + B, B only, and B + C that were stable at temperatures below T_2. A further increase in temperature lifts the curve for phase B above the common tangent to A and C, and thus phase B is no longer stable either alone or in combination with A or C. The diagram for T_1 illustrates this condition. Phase B is said to be *metastable* with respect to A + C.

Now let us visualize $\overline{G}(X_2)$ diagrams (isothermal and isobaric) drawn over a continuous range from T_1 to T_3, and extending to higher and lower temperatures beyond these limits. The one-, two-, and three-phase fields in these diagrams will be seen to change shapes continuously. If we now envision a "stack" of all such diagrams and rotate it so that it can be viewed from the side with temperature as the vertical axis and X_2 as the horizontal one, we have created a T-X_2 (isobaric) diagram for the system. This diagram is illustrated in Figure 9.4.

Notice that, as in the $\overline{G}(X_2)$ diagrams, phase fields for single phases and pairs of phases alternate from left to right across the composition axis. The three-phase field A + B + C exists only at temperature T_2. Because this field is invariant with respect to P and T and only variations in the compositions or proportions of A, B, and C are allowed within the field, it is a univariant field; that is, only one of the three defining variables is not fixed within the field. You may wish to verify that a three-phase assemblage in a binary system is univariant by using the phase rule introduced in Chapter 8.

In Figure 9.4, phase B terminates in a three-phase equilibrium (A + B + C) as temperature increases. Terminations of phases are probably the most complicated and confusing parts of phase diagrams. In addition to three-phase equilibria, phases may also terminate at critical points, at limiting values, and because of coincidences in com-

Picturing Equilibria: Phase Diagrams   Chap. 9

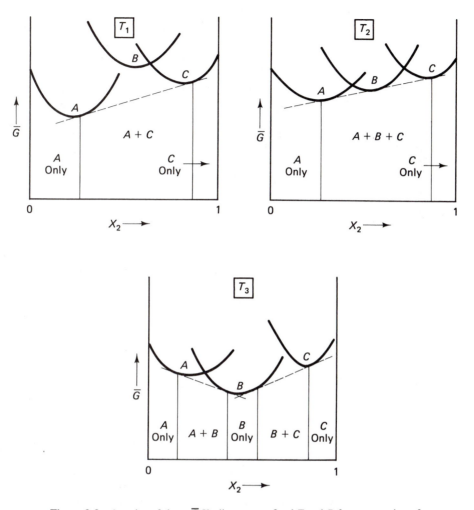

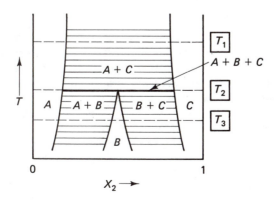

Figure 9.3 A series of three $\overline{G}$-X_2 diagrams at fixed T and P for a succession of temperatures decreasing from T_1 to T_3. The $\overline{G}(X_2)$ curves for phases A, B, and C shift downward at different rates with decreasing temperature, and thus different combinations of phases are stable.

Figure 9.4 A "stack" of $\overline{G}$-X_2 diagrams at various temperatures produces a T-X_2 diagram. This particular diagram results from stacking the three $\overline{G}$-X_2 diagrams shown in Figure 9.3. The compositions of coexisting phases are connected by horizontal lines, because at equilibrium the temperatures of all coexisting phases must be the same.

position. All of these kinds of terminations are illustrated and explained in the following section.

Note also the set of tie-lines parallel to the X_2 axis in Figure 9-4. These connect the compositions of phases A + B, B + C, or A + C that are in equilibrium in the two-phase fields at each temperature. Because it is a necessary condition of equilibrium that the temperatures of all phases be the same, tie-lines must always be isothermal.

It may not be immediately apparent that a point on any tie-line has a significance in terms of the relative proportions of phases in the two-phase field. We can, in fact, define a *lever rule* using such a point. The proportion of each phase in the field at that point varies inversely as the length of the tie-line from the point to the composition of that phase. For example, in the two-phase field A + C in Figure 9.4, the assemblage approaches a limit of single-phase A at its left edge and single-phase C at its right. A point on the tie-line between single-phase limits at a composition one quarter of the way from the left edge (the composition of phase A) represents a mixture of 75% phase A and 25% phase C, and the tie-line segments to the left and right of the point are 25% and 75% of the total length, respectively. This configuration is illustrated in Figure 9.5. Proportions of the phases in a two-phase field can thus be determined by inspection.

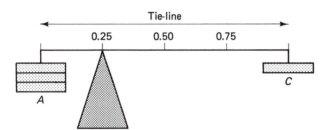

Figure 9.5 This balance schematically illustrates the lever rule. For the two-phase field A + C in Figure 9.4, a point on the tie-line one quarter of the way from the composition of A represents a mixture of 75% phase A and 25% phase C, and the tie-line segments on either side of this point are 25% and 75%, respectively.

It should be obvious that the process by which we developed *T-X* diagrams from $\overline{G}$-X sections (isobarically) can be duplicated to generate *P-X* diagrams (isothermally). In general, however, the appearance of a *P-X* diagram will be upside down from that of a *T-X* diagram because increases in temperature and pressure commonly have opposing effects on most materials; that is, high temperature favors higher $\overline{V}$ and higher $\overline{S}$, but high pressure favors lower $\overline{V}$ and lower $\overline{S}$. *P-X* diagrams are not commonly used in solving geochemical problems. Instead, the effects of pressure are generally gauged from *P-T* diagrams, considered in a later section. For the time being, we will focus on the characteristics of *T-X* diagrams.

T-X DIAGRAMS FOR REAL GEOCHEMICAL SYSTEMS

Let us examine some $\overline{G}$-X diagrams and corresponding *T-X* diagrams for real binary systems, in order to illustrate the geochemically important features commonly encountered in phase diagrams for natural materials. In this exercise, we will be especially interested in univariant conditions (*three-phase equilibria, coincidences in composition, limiting values,* and *critical points*), which is how phases terminate. In most of the following examples, a liquid phase will be stable at high temperatures, and it will solidify to form other phases as temperature is lowered. Each of these systems will be considered only at atmospheric pressure, at least for the moment. We will present a series of $\overline{G}$-X diagrams for various temperatures at constant pressure, and then "stack" these to construct the appropriate *T-X* diagram for that system. We can then explore the characteristics of each *T-X* diagram in detail.

An Example of Three-Phase Equilibria: Diopside-Anorthite Eutectic

The simplest case we will consider is an example of a two-component system that is liquid at high temperature and crystallizes to form two pure solid phases. An example of such a system, diopside-anorthite, is shown in Figure 9.6. First let us consider the $\overline{G}$-X diagrams for this system at T_1 through T_4, representing a progression of decreasing temperatures. At T_1, a tangent to the $\overline{G}(X_{An})$ curve for liquid (L) defines the lowest free en-

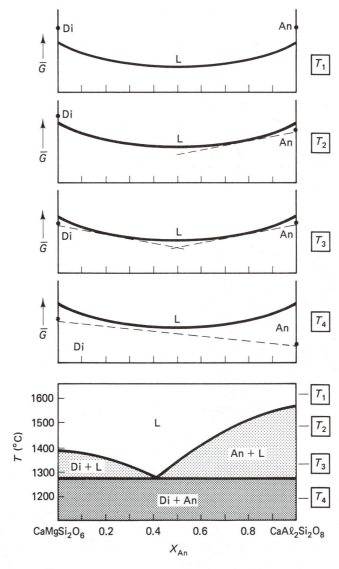

Figure 9.6 $\overline{G}$-X_{An} diagrams for the system diopside-anorthite at $P = 1$ atm and at various temperatures can be stacked to produce the T-X_{An} diagram at the bottom. Phase abbreviations are: diopside (Di), anorthite (An), and liquid (L). This diagram contains a eutectic point, representing the three-phase equilibrium between Di, An, and L. In this and the following phase diagrams, the area between the liquidus and solidus is shown in light shading, and the area below the solidus is in dark shading.

ergy for the system at any X_{An}, so that liquid is the only stable phase. By T_2, the $\overline{G}$ point for anorthite has dropped faster than the curves and points for the other phases, so that a tangent to anorthite and liquid offers the lowest free energy configuration for the system at high X_{An} values. By T_3, the $\overline{G}$ for diopside has lowered to the point where Di + L or An + L are the stable phases over practically the entire compositional range, but at one special composition all three phases are stable. This is an example of the termination of a phase (in this case, liquid) in a three-phase equilibrium. At T_4, liquid is metastable because a tangent to the two solid phases lies below the liquid $\overline{G}$ curve.

By combining the information on phase stability in these $\overline{G}$-X diagrams, we can construct a T-X diagram for the system diopside-anorthite, as illustrated at the bottom of Figure 9.6. Examination of this diagram reveals some interesting features, probably already familiar to petrology students. The lightly shaded area identifies regions where liquid is stable with a solid phase. The upper boundary of this area, called the *liquidus*, thus shows the onset of crystallization in terms of T and X_{An}. The area with dark shading denotes a region where liquid is not a stable phase. Its upper boundary, called the *solidus*, indicates the temperature below which the system is fully condensed (consists only of solid phases). The point at which the liquidus touches the solidus is called a *eutectic*. It represents the composition at which liquid can persist to the lowest temperature, as well as the point where both diopside and anorthite crystallize together with falling temperature (in other words, in *three-phase equilibrium* in which liquid terminates).

Worked Problem 9-1

Describe the crystallization sequence for a liquid in this system with the composition $X_{An} = 0.2$. The final product can be determined from the position of this composition along the horizontal axis of the diagram, using the lever rule. It will be a mixture of 80% diopside and 20% anorthite. But how do we arrive at this final state? The system, initially all liquid, begins crystallization as it intercepts the diopside liquidus on cooling. From Figure 9.6, we can see that at $X_{An} = 0.2$, this will occur at a temperature of approximately 1350°C, at which point diopside begins to form. During subsequent cooling, the composition of the remaining liquid follows the liquidus down slope, changing as diopside is removed. Finally, the liquid reaches the eutectic point, where anorthite joins diopside in the crystallizing assemblage. The system can cool no further until the liquid is fully crystallized. Following solidification of the last drop of liquid, the mixture of diopside and anorthite crystals cools to lower temperature without further change.

An Example of Coincidence of Composition: Leucite-Silica Peritectic

Now we can examine a system in which the $\overline{G}(X_2)$ curves for two phases having the same composition intersect, a situation known as *coincidence of composition*. Figure 9.7 illustrates $\overline{G}$-X diagrams for the system leucite-silica at various temperatures. At T_1, liquid is the stable phase at any composition. First the $\overline{G}$ point for cristobalite, and then the $\overline{G}$ point for leucite, drop with lowering temperature to produce the situation at T_2. So far, this system is no different in its behavior from the diopside-anorthite system already examined. However, all this changes at T_3. The composition of orthoclase can be represented as a mixture of leucite and silica, as shown by its position on the horizontal axis of Figure 9.7. At T_3, the orthoclase $\overline{G}$ point coincides with the $\overline{G}$ curve for liquid, and by T_4 the orthoclase point has pierced the liquid curve to replace liquid as the stable phase over part of the compositional range. Liquid has become metastable by T_5.

The T-X_{SiO_2} diagram produced from this information is presented at the bottom of Figure 9.7. Light and dark shadings again illustrate the positions of the liquidus and

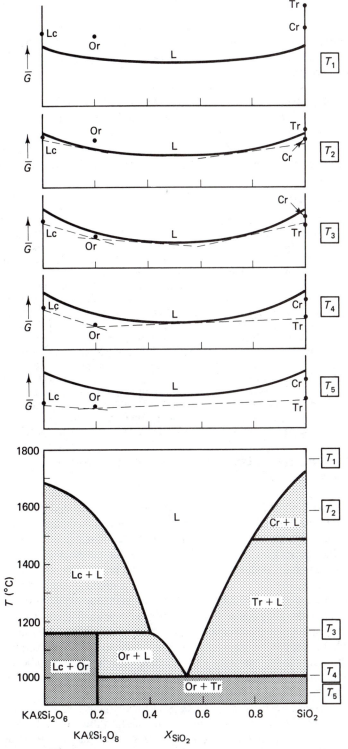

Figure 9.7 $\overline{G}$-X_{SiO_2} diagrams for the system leucite-silica at $P = 1$ atm and at a progression of temperatures are stacked to make the T-X_{SiO_2} diagram at the bottom. Phase abbreviations are the following: leucite (Lc), orthoclase (Or), cristobalite (Cr), tridymite (Tr), and liquid (L). This diagram contains a eutectic between Or, Tr, and L, and a peritectic between Lc, Or, and L. The peritectic represents a coincidence in composition.

solidus for this system. The right side of the diagram contains a eutectic and is rather similar to the system discussed earlier. Notice, however, the polymorphic transformation of cristobalite to tridymite at 1470°C, resulting from the fact that the tridymite $\overline{G}$ point descended more rapidly than that for cristobalite with lowering temperature, overtaking it at the point of transformation. The left side of the T-X diagram contains a new feature, called a *peritectic* (a point of reaction between solid and liquid), produced by the coincidence of orthoclase with liquid. Notice in the $\overline{G}$-X section at T_3, the tangent between leucite and liquid has been replaced by tangents between these two phases and orthoclase. This means that leucite and liquid are no longer stable together at this temperature. Thus these phases react to produce a new phase with intermediate composition, orthoclase.

Worked Problem 9-2

What is the crystallization sequence for a composition $X_{SiO_2} = 0.3$ in the system leucite-silica? The final product can again be ascertained from the position of this composition along the horizontal axis of Figure 9.7. It will be a mixture of tridymite and orthoclase crystals. Application of the lever rule, this time using tridymite and orthoclase (rather than leucite) as the ends of the tie-line, gives about 87% orthoclase and the remainder tridymite. The first phase to crystallize from a liquid of this composition, however, is leucite, as it cools to about 1425°C. With continued cooling, the liquid follows the liquidus down slope until intersecting the peritectic. At this point already crystallized leucite reacts with liquid to produce orthoclase, and the temperature of the system cannot decrease until all leucite is gone. When the reaction is complete, the liquid resumes its descent along the liquidus, all the while crystallizing more orthoclase. When it reaches the eutectic, the crystallization sequence is joined by tridymite, and after final solidification, the mixture of orthoclase and tridymite crystals cools to lower temperature.

Let's now try the crystallization sequence for a composition $X_{SiO_2} = 0.1$ in this system. We will produce the same sequence as before, but in this case the liquid will never be able to get past the peritectic. The final product for a liquid of this composition must be consumed before all of the leucite reacts to form orthoclase.

An Example of Limiting Value: Plagioclase Transition Loop

Another kind of *coincidence in composition* is illustrated by $\overline{G}$-X sections for the system albite-anorthite, as shown in Figure 9.8. In this case, these end-members exhibit complete solid solution, and the composition that is stable at any temperature is determined by the interaction of the $\overline{G}(X_{An})$ curves for plagioclase and liquid. At temperatures above T_1, only liquid is stable. At T_1, the plagioclase $\overline{G}$ curve intersects the liquid $\overline{G}$ curve at $X_{An} = 1$. This is an example of a *limiting value*, as it is obvious that plagioclase cannot contain more X_{An} than pure anorthite. With progressively decreasing temperature, the plagioclase $\overline{G}$ curve sweeps downward across the liquid $\overline{G}$ curve, intersecting it obliquely at lower and lower X_{An} values. Thus at T_2 and T_3, compositions with high X_{An} only have plagioclase stable, intermediate compositions have both plagioclase and liquid, and those with low X_{An} have only liquid. At some value between T_3 and T_4, the plagioclase and liquid $\overline{G}$ curves intersect at $X_{An} = 0$, another *limiting value*. At T_4, liquid is metastable at any composition.

The corresponding T-X diagram for the plagioclase system is shown at the bottom of Figure 9.8. Shaded portions of the diagram illustrate the position of the liquidus and solidus, as in previous diagrams. This kind of diagram illustrates a *transition loop*, a pair of lines along each of which the composition of one phase is specified. The transition loop in this case is bounded on each end by limiting values on the permissible compositional range of plagioclase.

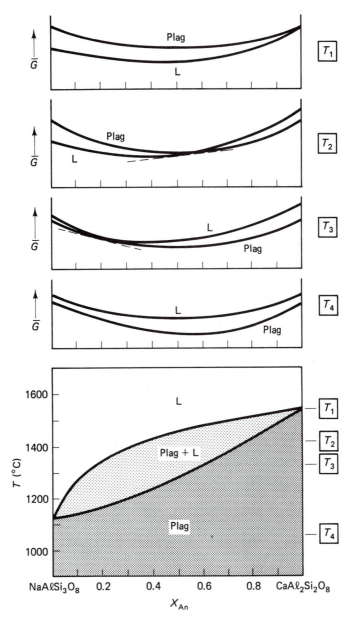

Figure 9.8 $\bar{G}$-X_{An} diagrams for the system albite-anorthite at $P = 1$ atm and at various temperatures are combined to make the T-X_{An} diagram at the bottom. Plagioclase shows a complete solid solution between these end members, forming a transition loop. The ends of the loop represent limiting values.

Worked Problem 9-3

Describe the crystallization path for a composition $X_{An} = 0.65$. The final product will obviously be crystals with composition An_{65}, as this is a solid solution series. The path by which this product is formed is illustrated in Figure 9.9. Crystallization begins when the liquid intersects the liquidus at point a, and the first crystals to form have composition b. As the liquid cools further to point c, the intersections of a horizontal line through this point with the liquidus (point d) and solidus (point e) give the compositions of the liquid

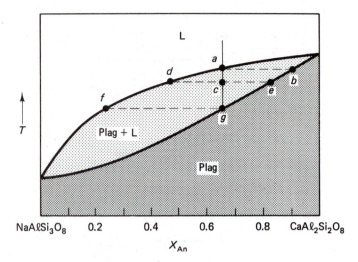

Figure 9.9 An example of an equilibrium crystallization path in the system albite-anorthite. The liquid cools to point *a*, where crystallization begins with crystals of composition *b*. With falling temperature, liquid changes composition progressively from *a* to *f*, and solid from *b* to *g*.

and solid, respectively, that are in equilibrium at this temperature. Recall that two phases in equilibrium must have the same temperature, so their compositions can be connected by a horizontal tie-line in this diagram. With continued cooling, the composition of the liquid follows the liquidus down slope, and that of crystallizing plagioclase follows the solidus. The coincidence in composition seen in the $\overline{G}$-X sections indicates that a reaction takes place between liquid and already precipitated solid, so that calcic plagioclase continually reequilibrates to form more sodic plagioclase that is stable at lower temperatures. Finally, the liquid reaches point *f*, its farthest point of descent along the liquidus. A horizontal line through this point passes through point *g*, the composition of the solid at this temperature. Because this composition is $X_{An} = 0.65$, the required final composition for plagioclase, reaction with liquid must end here and the liquid disappears. Crystals of An_{65} composition then cool to low temperature without further change.[1]

An Example of Critical Point: Alkali Feldspar Solvus

Finally, we will consider a system containing a *critical point*. In a $\overline{G}$-X diagram, this is manifested as an inflection in a $\overline{G}(X_2)$ curve. Inflection points occur wherever the second derivative $(\partial^2 G/\partial X_2)$ equals zero. The trace of all such points on a T-X diagram is called the "spinodal"; we will defer discussion of this until Chapter 10. A natural example of critical behavior is the system albite-orthoclase, a solid solution series at high temperature that unmixes into discrete alkali feldspars at low temperature. For our purposes here, we are only interested in what happens below the solidus in this system, so we will ignore temperatures above the solidus. $\overline{G}$-X sections are illustrated in Figure 9.10. At T_1, the alkali feldspars exhibit complete solution at any composition. At T_2, a small inflection in the alkali feldspar $\overline{G}$ curve occurs. The crest of this inflection is its critical point. Notice that a tangent providing the lowest system free-energy indicates that two stable feldspar phases should exist, at least for a small compositional range near the middle of the diagram. With decreasing temperature, the inflection widens and thereby increases the compositional range affected, as shown at T_3.

---

[1] Actually, there are some subsolidus reactions, similar to those that take place in alkali feldspars discussed in the next example, that may take place in slowly cooled plagioclases, but we will ignore these here.

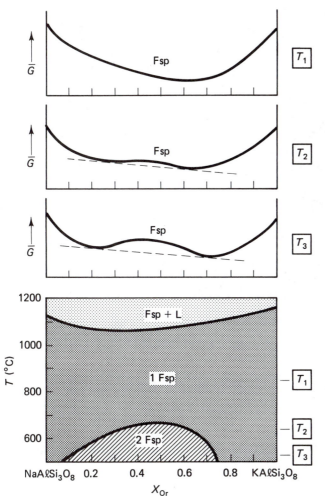

Figure 9.10 $\overline{G}$-X_{Or} diagrams for the system albite-orthoclase at $P = 1$ atm and at various temperatures are stacked to produce the T-X_{Or} diagram at the bottom. An inflection in the $\overline{G}(X_{Or})$ curve with lowering temperature creates a region of unmixing in the solid state, called a solvus. The top of the solvus is a critical point. Abbreviations for phases are feldspar (Fsp) and liquid (L).

The *T-X* diagram appropriate to this kind of behavior is illustrated at the bottom of Figure 9.10. The gradually widening tangent points in the $\overline{G}$-X diagrams define a concave-downward curve, called a *solvus,* that identifies a region of unmixing, or exsolution. Above the solvus, one homogeneous phase is stable, and below it this phase separates into two phases with distinct compositions.

Worked Problem 9-4

What is the subsolidus crystallization path for the composition $X_{Or} = 0.6$ in the alkali feldspar system? With cooling, a homogeneous feldspar of composition $X_{Or} = 0.6$ reaches the solvus at approximately 630°C (Figure 9.10). At this point, the homogeneous feldspar begins to unmix in the solid state, producing two coexisting feldspars. At any temperature below the solvus, intersections of a horizontal (isothermal) line with the two limbs of the solvus give the compositions of the two stable phases. For example, at 550°C the coexisting phases will have compositions of $X_{Or} = 0.14$ and 0.72. Notice that the compositions of these phases become more separated as temperature decreases and the solvus widens. At some low temperature, the diffusion of ions becomes so sluggish that unmixing ceases, and the compositions of the exsolution products are frozen in at that point. A photograph of an exsolved alkali feldspar formed in this way is illustrated in Figure 9.11.

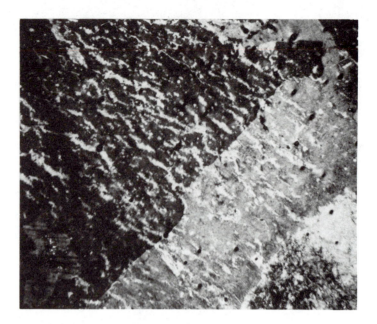

Figure 9.11 Photomicrograph of perthite, an intergrowth of sodic and potassic alkali feldspars, formed by exsolution in the solid state. The light-colored lamellae are Na-feldspar, and the K-feldspar host is twinned. This sample is from the Winns-boro granite, South Carolina. Width of the photomicrograph is 5 mm.

We have now examined the common features of geologic T-X diagrams, and in particular the ways in which phases terminate in binary systems. Termination in three-phase equilibria, the most common situation, produces a eutectic. Coincidences in composition can result in a peritectic or in a transition loop. Limiting values mark the boundaries of the permissible compositional range of a particular phase. Termination in a critical point produces a solvus.

THERMODYNAMIC CALCULATION OF PHASE DIAGRAMS

Up to this point, we have presented without proof a series of $\overline{G}$-X diagrams and their corresponding T-X diagrams. We should confess that in order to do this, we first worked the problems backwards, having used experimentally determined T-X diagrams to construct qualitatively correct $\overline{G}$-X sections. However, if thermodynamic data of sufficient quality are available, it should be possible to calculate rigorous $\overline{G}$-X diagrams, and from them to construct T-X diagrams. We will illustrate how this is done in Worked Problems 9-5 and 9-6. It is often easier to omit the first step (the $\overline{G}$-X diagram) and to calculate the stability of various phases at each temperature directly. This prob-lem is similar to the calculation of phase boundaries in the P-T diagram for the unary system Al_2SiO_5 that was introduced in Chapter 8, but with the added complication that the mixing properties of the two binary components must be modeled.

Worked Problem 9-5

Calculate $\overline{G}$-X diagrams at a series of temperatures for a real binary system that shows un-mixing behavior. The system we will examine is Mg_2SiO_4 (forsterite) = $CaMgSiO_4$ (mon-ticellite). This system was described by Saxena (1973) as a regular solution with a temperature-dependent interaction parameter $W = 35000 - 37247(T \cdot 1000^{-1})$ +

13094(T 1000^{-1})^2 cal mol^{-1} at 5 kbar pressure. Thus W has a value of 10,109 cal mol^{-1} at 1073K, 8804 cal mol^{-1} at 1273K, 8546 cal mol^{-1} at 1473K, and 9335 cal mol^{-1} at 1673 K.

The general expression for the free energy of olivine in this system is

$$\overline{G}_{\text{oliv}} = X_{\text{Fo}}\mu_{\text{Fo}} + X_{\text{Mc}}\mu_{\text{Mc}}$$

$$= X_{\text{Fo}}\overline{G}_{\text{Fo}} + X_{\text{Mc}}G_{\text{Mc}} + X_{\text{Fo}}RT \ln X_{\text{Fo}}\gamma_{\text{Fo}} + X_{\text{Mc}}RT \ln X_{\text{Mc}}\gamma_{\text{Mc}}.$$

The terms containing RT represent the excess free energy over that for mechanical mixing, the so-called *free energy of mixing*. We will be concerned only with these terms. Substituting the expression for $RT \ln \gamma$ for a regular solution (equation 8-21) into the equation above, yields

$$\overline{G}_{\text{mixing}} = RT (X_{\text{Mc}} \ln X_{\text{Mc}} + X_{\text{Fo}} \ln X_{\text{Fo}}) + WX_{\text{Mc}}X_{\text{Fo}}.$$

Using this equation, we can calculate the value for the excess free energy of mixing for various values of X_{Fo} at a given temperature. You may use the program REGULAR on the diskette to solve this problem if you wish. Figure 9.12a was constructed from the results of this calculation.

Now sketch a *T-X* diagram that shows this region of unmixing (solvus) by noting the X_{Fo} of the two tangent points to each of the $\overline{G}_{\text{mixing}}$ curves in Figure 9.12a. Such a *T-X* diagram is shown in Figure 9.12b. It is not possible to complete construction of the crest of the solvus in this system because its temperature is so high that it intersects the solidus.

Worked Problem 9-6

As a second example, let us see how accurately we can construct the plagioclase transition loop. You may use the program called LOOP on the diskette to solve this problem, if you wish. We can write the following four separate equations to express the chemical potentials of albite and anorthite in the plagioclase solid solution and in the liquid:

$$\mu_{\text{Ab}S} = \mu^o_{\text{Ab}S} + RT \ln X_{\text{Ab}S}, \tag{9-3a}$$

$$\mu_{\text{Ab}L} = \mu^o_{\text{Ab}L} + RT \ln X_{\text{Ab}L}, \tag{9-3b}$$

$$\mu_{\text{An}S} = \mu^o_{\text{An}S} + RT \ln X_{\text{An}S}, \tag{9-3c}$$

$$\mu_{\text{An}L} = \mu^o_{\text{An}L} + RT \ln X_{\text{An}L}. \tag{9-3d}$$

The two phases, liquid (L) and solid (S), are indicated by the appropriate superscripts above. In each case, the chemical potentials with the "o" superscript are standard state values (using the conventional standard state defined by pure plagioclase or pure liquid of end-member compositions). For this exercise, we will assume that both the liquid and the solid are ideal solutions, so that $a_i = X_i$.

Within the transition loop (the two-phase field), we know that equations 9-3a and 9-3b must be equal, and that equations 9-3c and 9-3d must also be equal. The consequence of two-phase equilibrium, therefore, is that

$$\mu^o_{\text{Ab}S} - \mu^o_{\text{Ab}L} = RT \ln (X_{\text{Ab}L}/X_{\text{Ab}S}) \tag{9-4a}$$

and

$$\mu^o_{\text{An}S} - \mu^o_{\text{An}L} = RT \ln (X_{\text{An}L}/X_{\text{An}S}). \tag{9-4b}$$

The chemical potentials with the "o" superscripts are simply the free energy values for pure albite, pure anorthite, and the pure liquids of end-member composition. The left sides of equations 9-4a and 9-4b, therefore, correspond to the free energies of melting pure albite and pure anorthite. $\Delta\overline{H}$ and $\Delta\overline{S}$ in this context refer to the enthalpy or entropy changes associated with the melting process. We now have

$$\mu^o_{\text{Ab}S} - \mu^o_{\text{Ab}L} = \Delta\overline{H}_{\text{Ab}} - T\Delta\overline{S}_{\text{Ab}} \tag{9-5a}$$

and

$$\mu^o_{\text{An}S} - \mu^o_{\text{An}L} = \Delta\overline{H}_{\text{An}} - T\Delta\overline{S}_{\text{An}}, \tag{9-5b}$$

which can be substituted into equations 9-4a and 9-4b, respectively, to give

Thermodynamic Calculation of Phase Diagrams

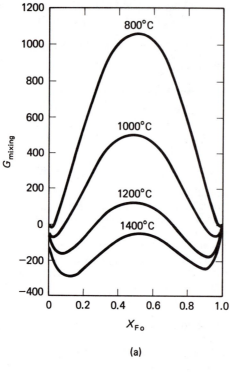

(a)

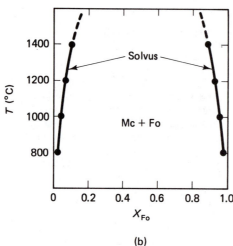

(b)

Figure 9.12 (a) Calculated $\overline{G}_{\text{mixing}}$-$X_{\text{Fo}}$ diagram for the system forsterite-monticellite. (b) T-X_{Fo} diagram for the same system. (After S. K. Saxena, *Thermodynamics of rock-forming crystalline solutions*. Copyright 1973 by Springer-Verlag. Reprinted with permission.)

$$RT \ln (X_{\text{Ab}L}/X_{\text{Ab}S}) = \Delta \overline{H}_{\text{Ab}} - T\Delta \overline{S}_{\text{Ab}} \tag{9-6a}$$

and

$$RT \ln (X_{\text{An}L}/X_{\text{An}S}) = \Delta \overline{H}_{\text{An}} - T\Delta \overline{S}_{\text{An}}. \tag{9-6b}$$

Let us now take the exponential of both sides of equations 9-6a and 9-6b in order to solve for the mole fractions. From 9-6a we get

$$(X_{\text{Ab}L}/X_{\text{Ab}S}) = \exp ((\Delta \overline{H}_{\text{Ab}} - T\Delta \overline{S}_{\text{Ab}})/RT)$$

or

$$X_{\text{Ab}L} = X_{\text{Ab}S}e^{a}, \tag{9-7a}$$

in which we are introducing the quantity a to avoid having to write the exponential out in its full glory from now on. Equation 9-6b similarly yields

$$X_{AnL} = X_{AnS}e^b, \tag{9-7b}$$

in which the new quantity b looks exactly like the exponential that defines a, but refers to melting anorthite rather than albite.

After these first steps, we now recall that the mole fractions X_{Ab} and X_{An} are not independent, but are related by

$$X_{AnL} = 1 - X_{AbL} \tag{9-8a}$$

and

$$X_{AnS} = 1 - X_{AbS}. \tag{9-8b}$$

From equations 9-7a, 9-7b, and 9-8a we can derive

$$X_{AnS}e^b = 1 - X_{AbS}e^a. \tag{9-9}$$

We can eliminate X_{AnS} from this equation by substituting for it from equation 9-8b:

$$(1 - X_{AbS})e^b = 1 - X_{AbS}e^a.$$

If we rearrange this result to solve for X_{AbS},

$$X_{AbS} = (1 - e^b)/(e^a - e^b). \tag{9-10}$$

Now let us go back and look at the quantities a and b more carefully. When the temperature is equal to the melting temperature of albite (that is, when $T = T_{Ab}$), we know that equation 9-5a is equal to zero. In other words, the liquidus and solidus curves coincide for pure albite. Thus

$$\Delta \bar{H}_{Ab} = T_{Ab}\Delta \bar{S}_{Ab}$$

and

$$a = [\Delta \bar{H}_{Ab} - T(\Delta \bar{H}_{Ab}/T_{Ab})]/RT. \tag{9-11}$$

This can be rewritten as

$$a = \Delta \bar{H}_{Ab}/R(1/T - 1/T_{Ab}). \tag{9-12a}$$

By analogy,

$$b = \Delta \bar{H}_{An}/R(1/T - 1/T_{An}). \tag{9-12b}$$

The terms a and b involve enthalpies of melting for albite and anorthite which have been determined calorimetrically. Numerically, the enthalpies are equal to 13.1 kcal/mol and 29.4 kcal/mol, respectively. With these values in hand, we can calculate from equation 9-10 the mole fraction of albite in the solid solution at any temperature above the melting temperature of pure albite. The calculated set of compositions and temperatures defines the solidus. The liquidus curve can be determined from these values with the help of equation 9-8a. These calculated values are tabulated below and plotted graphically in Figure 9.13. This transition loop constructed from thermodynamic data is an almost perfect fit to the plagioclase phase diagram determined in the classic experiments of Bowen (1915).

T (°C)	Solidus (X_{Ab})	Liquidus (X_{Ab})
1100	1.000	1.000
1150	0.983	0.830
1200	0.955	0.689
1250	0.913	0.579
1300	0.852	0.463
1350	0.767	0.366
1400	0.650	0.275
1450	0.491	0.185
1500	0.280	0.095
1550	0.000	0.000

Thermodynamic Calculation of Phase Diagrams

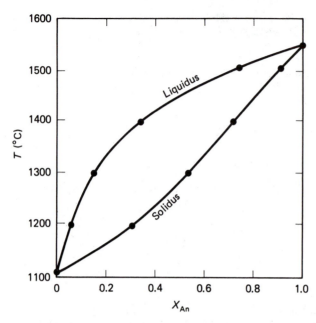

Figure 9.13 The plagioclase transition loop calculated in Worked Problem 9-6.

In the example above, we have calculated a *T-X* phase diagram that was already known from experimental petrology. This is an interesting academic exercise, but it could be considered a waste of time insofar as we have learned nothing new about plagioclase crystallization. However, it illustrates an important point. There are other geochemically important systems for which experimentally derived phase equilibria are difficult to obtain on any reasonable time scale. Calculation of equilibrium relationships in these systems from thermodynamic data may be the most efficient way of attacking such problems.

There is also another cogent reason for calculating phase diagrams. If the phase relationships in a system can be accurately modeled under one set of physical conditions, then these relationships can be calculated under other physicial conditions. In the example above, the phase diagram was determined at atmospheric pressure. Now that we have demonstrated that we can model this system, we could calculate the phase relationships at elevated pressures using appropriate thermodynamic data.

P-T DIAGRAMS

So far, we have been dealing with isobaric slices through *P-T-X₂* space, largely because we are constrained by the two-dimensional nature of a sheet of paper. It is possible to extend our results into three dimensions, however. As an example, let us take the system Fe-O, a sketch of which is illustrated in Figure 9.14a.

What happens to the various univariant equilibria that we have already defined in such a three-dimensional diagram? A three-phase equilibrium traces out a ruled surface in *P-T-X₂* space. The surface is everywhere parallel to the X_2 axis because we have the equilibrium condition that *P* and *T* must be identical in all phases. A coincidence in composition or a critical point traces out a three-dimensional curve in *P-T-X₂* space. An equilibrium involving a limiting value or a pure substance (sometimes this equilibrium is called *degenerate*) defines a two-dimensional curve lying completely in the plane of constant composition (X_2). These univariant curves and surfaces divide *P-T-X₂* space

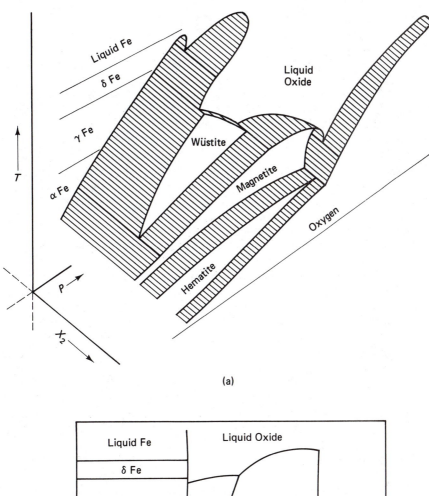

(a)

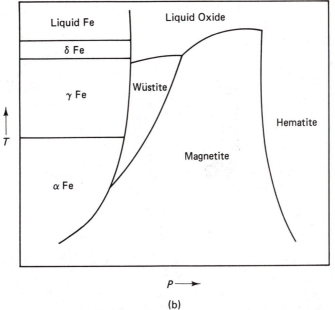

(b)

Figure 9.14 (a) P-T-X_2 diagram for the system Fe-O. (b) P-T diagram for the system Fe-O obtained by projecting the three-dimensional diagram in part (a) onto the P-T plane.

into one- and two-phase volumes (with variance of two or higher) as they intersect at invariant points or lines.

We have already examined T-X_2 slices, and to a lesser extent P-X_2 slices through such a P-T-X_2 diagram, so let us now consider the remaining two-dimensional section, an isocompositional P-T diagram. Of all the diagrams we have discussed, P-T diagrams are perhaps the most commonly used and certainly the most difficult to understand.

Inspection of the Fe-O diagram in Figure 9.14a will suggest the key features of such diagrams. A P-T diagram is obtained by collapsing the three-dimensional diagram onto the P-T plane (that is, projecting parallel to the X_2 axis), as illustrated in Figure 9.14b. On this P-T projection, univariant equilibria appear as curves and invariant equilibria as points. One- and two-phase regions are not shown. A complete diagram includes unary P-T lines for the limiting systems, curves for all critical phases, coincidences in composition, and three-phase equilibria, as well as curves for any degenerate univariant equilibria involving special compositions within the system. Invariant equilibria appear as intersections of these curves, although not all apparent intersections are invariant points, because some curves "intersect" only in projection and not in three-dimensional space.

There are many unique topological ways (14, to be precise) in which univariant curves may intersect on a P-T diagram. Derivation of these, or even presentation of them, is beyond the scope of this book. Here we will consider only one such intersection, just to illustrate how this relates to its corresponding T-X and $\overline{G}$-X sections.

The intersection that we will examine is illustrated in Figure 9.15a. Each of the univariant lines on this diagram describes a three-phase equilibrium, as labeled. Each might also be thought of as a line describing a reaction. To see this, we refer to the T-X diagrams in Figure 9.15b. (It probably is not obvious that these T-X sections are not topologically unique, but that is the case. The topology of the set of T-X sections that can be drawn from this single invariant point will depend on its orientation relative to the P and T axes. In fact, three distinct topologies can be constructed for different arrangements of isotherms and isobars. We will consider only one.)

In the T-X diagram at pressure P_1, there are two univariant three-phase equilibria shown. The first, encountered at low temperature, is the one in which phases B, C, and D coexist stably. The second, at a somewhat higher temperature, is the one in which phases A, B, and D are stable. Inspection of the P-T diagram (Figure 9.15a) indicates the same sequence of equilibria with increasing temperature along the P_1 isobar. Similarly, the T-X diagram drawn at P_3 (Figure 9.15b) can be easily correlated with the sequence of phase equilibria appearing with increasing temperature on the P_3 isobar. The T-X diagram at P_2, however, contains only one line of interest, the one at which phases A, B, C, and D are all stable together. This corresponds to the unique point where all four three-phase lines on the P-T diagram intersect—the invariant point.

Now consider the significance of any of these univariant lines as a reaction. The three-phase line marked ABD, for example, is a line along which A, B, and D coexist. It is also a line separating fields for phases A, B, and D on the low-temperature side from a two-phase field of A + D on the high-temperature side (Figure 9.15b). Therefore, the three-phase line ABD may also be viewed as a description of the reaction

$$B \longleftrightarrow A + D$$

in which B is consumed in the formation of A and D with increasing temperature. Similarly, the three-phase lines ACD, ABC, and BCD may be seen as reactions in which phases C, B, and D are consumed. The invariant (four-phase) point ABCD represents a reaction

$$B + C \longleftrightarrow A + D$$

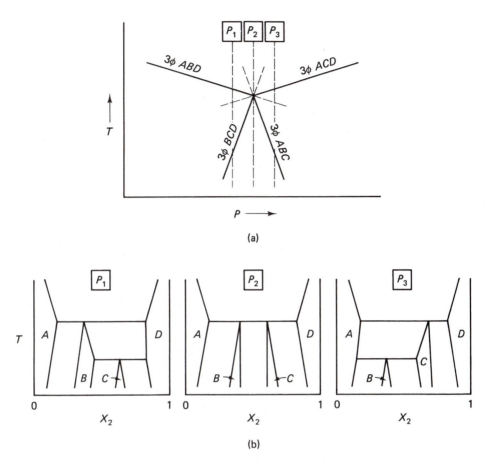

Figure 9.15 (a) An example of one of the 14 ways in which univariant (three-phase) curves can intersect on a P-T diagram. P_1, P_2, and P_3 are three pressures for which representative T-X_2 diagrams are constructed. (b) Isobaric T-X_2 diagrams at the three pressures identified in part (a).

in which both B and C are consumed. These relations should not come as any surprise, of course, because they are what we have already been led to expect from our earlier treatment of $\overline{G}$-X diagrams.

BINARY PHASE DIAGRAMS INVOLVING OTHER INTENSIVE PARAMETERS

In addition to temperature, pressure, and X_2, equilibria may also be influenced by other parameters, the effects of which we might want to show on phase diagrams. For example, fluid pressure is decidedly less than confining (total) pressure in some metamorphic systems because fluids have been progressively driven off. In such a case, P_{fluid} should be considered an independent variable. Employing the same technique we used earlier to derive T-X diagrams, we could envision $\overline{G}$-P_{fluid} diagrams at various values of T and at constant P_{total}, and "stack" these to construct a T-P_{fluid} phase diagram. However, we will spare you the derivation and go directly to a real geochemical example. Phase equilibria in the system MgO-H_2O are illustrated in Figure 9.16. This particular dia-

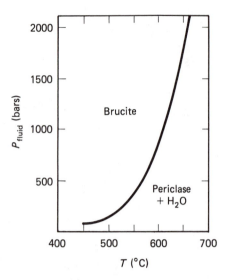

Figure 9.16 P_{fluid}-T diagram for the system MgO-H_2O, showing how the reaction brucite $\leftrightarrow$ periclase + H_2O depends on P_{fluid}, even though confining pressure remains constant.

gram is for a fixed confining pressure of 1000 bars and a fixed system composition such that X_{H_2O} equals X_{H_2O} in brucite. We can readily see that the temperature at which the equilibrium

$$\text{Mg(OH)}_2 \quad \longleftrightarrow \quad \text{MgO} + \text{H}_2\text{O}$$
$$\text{brucite} \qquad\qquad \text{periclase}$$

occurs is controlled by P_{fluid}, even though P_{total} remains constant. The curve describing this reaction, like most involving a fluid phase, is concave to the P_{fluid} axis. This reaction has a large $\Delta \bar{V}$ because of the larger molar volume of the water produced, and the Clapeyron equation dictates that the slope dP/dT should become steeper as $\Delta \bar{V}$ increases markedly at higher fluid pressures.

Another potentially important factor is the redox state of the system. We have ignored this complication so far in this chapter by concentrating on systems that either contain no iron or are at fixed f_{O_2}, but these restrictions need not apply to real geochemical systems.

In order to illustrate how oxygen fugacity can affect equilibria, let us consider the Fe-Ti oxide system. The compositions of coexisting titanomagnetite (Mt) and hemoilmenite (Ilm) solid solutions are governed by the following exchange reaction and oxidation reaction:

$$\text{Fe}_3\text{O}_{4\,\text{Mt}} + \text{FeTiO}_{3\,\text{Ilm}} \quad \longleftrightarrow \quad \text{Fe}_2\text{TiO}_{4\,\text{Mt}} + \text{Fe}_2\text{O}_{3\,\text{Ilm}},$$

and

$$4\text{Fe}_3\text{O}_{4\,\text{Mt}} + \text{O}_2 \quad \longleftrightarrow \quad 6\text{Fe}_2\text{O}_{3\,\text{Ilm}}.$$

The first of these equations is temperature-dependent, and the second is a function of oxygen fugacity. Spencer and Lindsley (1981) devised a solution model for this system based on experimental data. They then predicted the compositions of coexisting titanomagnetite and hemoilmenite phases at various values of temperature and oxygen fugacity and contoured these on a plot of f_{O_2} versus T, as shown in Figure 9.17. This system serves as both a geothermometer and oxygen barometer, as the analyzed compositions of the two coexisting phases uniquely define both parameters. Program FETIOX on the diskette can be used to calculate T and f_{O_2} from the compositions of coexisting Fe-Ti oxides.

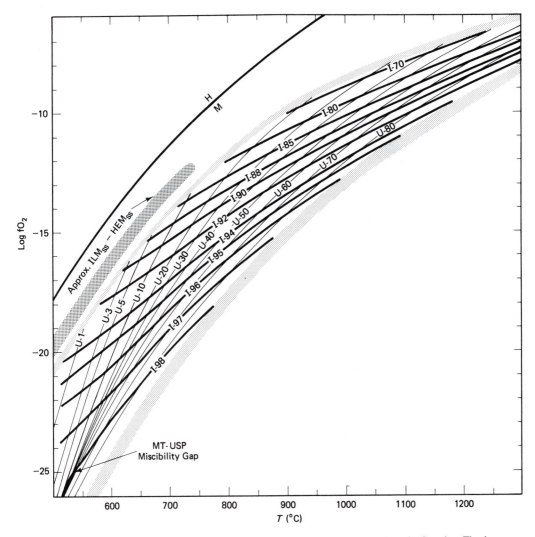

Figure 9.17 T-f_{O_2} grid for coexisting titanomagnetite(u)-hemoilmenite(I) pairs. The intersection of contours for the compositions of these two phases provide a measure of both temperature and oxygen fugacity. H-M is the hematite-ilmenite buffer. (After K. J. Spencer and D. H. Lindsley, A solution model for coexisting iron-titanium oxides. *Amer. Mineral.* 66: 1189–1201. Copyright 1981 by the Mineralogical Society of America. Reprinted with permission.)

SYSTEMS WITH THREE COMPONENTS

It becomes increasingly difficult to treat graphically any systems with three or more components. Although we will not consider this subject in great detail, you should at least be aware of how such systems are presented.

Liquidus Diagrams

For systems containing a liquid phase, it is common to combine three binary T-X sections into one triangular diagram, as illustrated for the system diopside-albite-anorthite in Figure 9.18a. The three binaries are folded up to form a triangular prism whose ver-

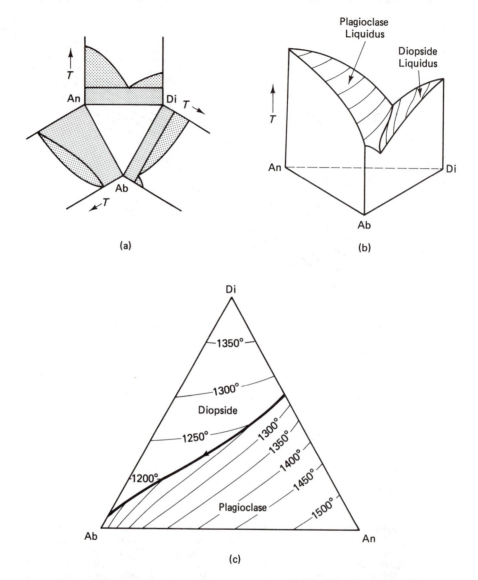

Figure 9.18 Representations for ternary systems. (a) The three binary phase diagrams for the system albite-anorthite-diopside (Ab-An-Di) oriented around a triangle. (b) These binary systems can be stood upright and the liquidus surface within the triangle can be extrapolated from the binary systems and then contoured, as shown here. (c) A projection of the liquidus surface in (b) onto the bottom triangle is shown in this figure. The arrowhead indicates the direction of slope for the boundary between the primary phase fields of diopside and plagioclase.

tical axis represents temperature, as illustrated in Figure 9.18b. The upper lid of the prism is the liquidus, now a curved, isobaric T-X surface. Three-dimensional perspective sketches like Figure 9.18b are difficult to draw and it is almost impossible to derive quantitative information from them, so details of the liquidus surface are commonly projected onto the compositional base of the prism, as shown in Figure 9.18c. The slopes of the liquidus surfaces with temperature are illustrated by isothermal contour lines, and the heavy line delineates the intersection of the liquidus surfaces for different phases, a low-temperature valley called a *boundary curve,* or *cotectic.* Slope on the boundary curve is indicated by the arrowhead. Along such boundary curves the liquid coexists with two solid phases, either crystallizing both simultaneously ("subtraction") or reacting with one solid phase to produce the other ("reaction"). Thus subtraction and reaction boundary curves are equivalent to eutectic and peritectic points, respectively, in binary systems. A more detailed analysis of various kinds of boundary curves and points in ternary phase diagrams is presented in an accompanying box for interested readers.

THE INTERPRETATION OF TERNARY LIQUIDUS PHASE DIAGRAMS

Ternary phase diagrams are divided into *primary phase fields* by *boundary curves.* Each primary phase field represents the first phase to crystallize upon intersecting the liquidus. We will take as an example the phase diagram for the system anorthite-leucite-silica, shown in Figure 9.19a. Such phase diagrams can be analyzed using Alkemade's theorem, which encompasses the following statements:

1. An *Alkemade line* is a straight line connecting the compositions of two phases whose primary phase fields share a common boundary curve. The first step in the analysis of any ternary phase diagram is to draw in all of the Alkemade lines. In Figure 9.19a, most of the Alkemade lines already exist as external boundaries to the triangle. For example, the primary phase fields for leucite and anorthite share a common boundary curve, and the appropriate Alkemade line joining their compositions is the left sideline of the triangle. In this example, the only additional Alkemade line we must draw is a line joining the compositions of anorthite and orthoclase, as illustrated in Figure 9.19b.

2. Alkemade lines divide a triangular diagram into smaller triangles. The ultimate equilibrium crystallization product is defined by the corners of the small Alkemade triangle in which the original liquid composition lies. For example, a liquid of composition X in Figure 9.19b will ultimately crystallize to form leucite + orthoclase + anorthite. If the original liquid lies *on* an Alkemade line, the ultimate assemblage consists of a mixture of the two phases at the ends of the line.

3. The intersection of an Alkemade line with its pertinent boundary curve represents the maximum temperature along that boundary curve. For example, the intersection of the leucite-anorthite boundary curve and Alkemade line occurs along the left side of the triangle. This point represents the highest temperature along that particular boundary curve, and temperature must fall off as one follows the curve to the right.

4. If a boundary curve or its tangent intersects the corresponding Alkemade line, that portion of the boundary curve is a *subtraction curve.* In Figure 9.19b, a tangent to the leucite-anorthite boundary curve at any place along its length will intersect the Alkemade line, so the entire curve is a subtraction curve along which the liquid crystallizes to form two phases, leucite + anorthite. It is conventional to identify subtraction curves with one arrowhead pointing in the direction of lowering temperature. If the boundary curve must be extended to intersect the pertinent Alkemade line, the curve is still a subtraction curve and the point of interesection represents the thermal maximum. The orthoclase-anorthite boundary curve in Figure 9.19b must be extended to intersect its Alkemade line, so it is a subtraction curve and temperature decreases away from the Alkemade line, as shown by the arrowhead.

5. If a boundary curve or its tangent does not intersect the appropriate Alkemade line, but merely intersects an extension of that line, then that portion of the boundary curve is a *reaction curve.* For example, a tangent drawn to any point along the leucite-orthoclase boundary curve in Figure 9.19b does not intersect the leucite-orthoclase Alkemade line, but only intersects an extension of that line. Thus, this boundary is a reaction curve along its entire length, as indicated by the double arrowheads. You might envision a hypothetical boundary curve that changes from subtraction to reaction along its length as its tangent swings outside the limits of its Alkemade line. The intersection

---

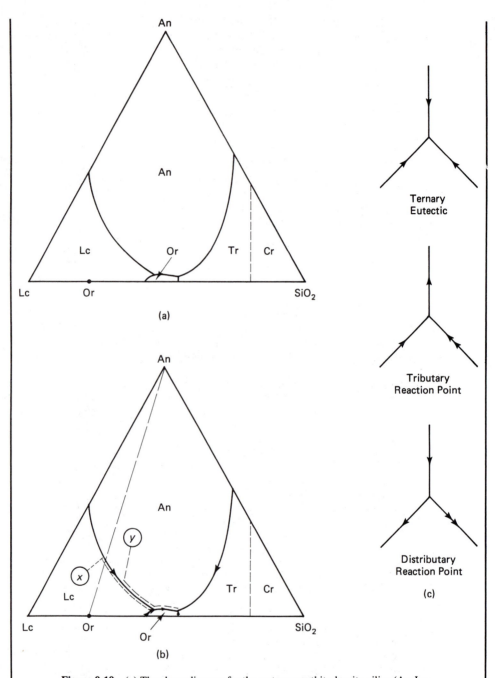

Figure 9.19 (a) The phase diagram for the system anorthite-leucite-silica (An-Lc-SiO₂) at P = 1 atm. Other phase abbreviations are: orthoclase (Or), tridymite (Tr), and cristobalite (Cr). The line between the primary phase fields for Tr and Cr is dashed to indicate that a phase transition between these two polymorphs occurs at a temperature above the eutectic point. (b) The same diagram with the addition of the Alkemade line between An and Lc. Boundaries between primary phase fields are identified as subtraction or reaction curves by single- or double-headed arrows, respectively. (c) These curves may intersect to form three different kinds of invariant points, as illustrated here.

of the curve with the extended Alkemade line is the maximum temperature along the boundary curve.

Once the various kinds of boundary curves have been identified using the rules above, we can recognize several types of invariant points formed by intersections of these curves, illustrated in Figure 9.19c. A *eutectic point* occurs where three boundary curves form a "dead end," and the liquid has no direction of escape with further cooling. An example of a eutectic in Figure 9.19b is the intersection of the orthoclase-anorthite, orthoclase-tridymite, and anorthite-tridymite subtraction curves. A *tributary reaction point* provides one outlet for liquid to follow, once the reaction that occurs at that point is completed. In Figure 9.19b, the intersection of the leucite-anorthite and orthoclase-anorthite subtraction curves with the leucite-orthoclase reaction curve produces this kind of point. A *distributary reaction point*, shown in Figure 9.19c, has two outlets and, after the reaction that occurs at the point is complete, permits the liquid to follow either of two branches, depending on what its final crystallization product must be (determined by the Alkemade triangle in which it resides). There is no example of a distributary reaction point in the anorthite-leucite-silica system.

We can now describe equilibrium crystallization paths for any liquids in this system. We will take two examples, identified by X and Y in Figure 9.19b. Liquid X lies in the Alkemade triangle leucite-anorthite-orthoclase, so those phases must be the final product. These phases only coexist at one point in the diagram, the tributary reaction point, which must be where crystallization will end. Upon intersecting the liquidus, X first begins to precipitate leucite, since it lies in the leucite primary phase field. The liquid composition will move directly away from the leucite corner as leucite forms, finally intersecting the leucite-anorthite subtraction curve. At this point, leucite + anorthite crystallize simultaneously and the liquid follows the curve toward the tributary reaction point. The ratio of leucite to anorthite in the crystallizing assemblage can be found at any point along the subtraction curve by noting the intersection of the tangent to the boundary curve with the Alkemade line and employing the lever rule. When the liquid reaches the tributary reaction point, already crystallized leucite reacts with the liquid to produce orthoclase. We know that the reaction cannot run to completion, because we must have some leucite left in the final product. Therefore, we must run out of liquid before all the leucite can react.

Liquid Y in Figure 9.19b lies in the Alkemade triangle anorthite-orthoclase-silica, so this must be the final product. These phases only coexist at the eutectic point. Anorthite is the first phase to crystallize, and the liquid composition moves from Y directly away from the anorthite corner. Upon intersecting the leucite-anorthite subtraction curve, the crystallization sequence is joined by leucite. Both phases crystallize as the liquid moves down the curve until it reaches the tributary reaction point. Then leucite reacts with the liquid to form orthoclase. In this case, the reaction runs to completion, and the liquid is then free to move down the orthoclase-anorthite subtraction curve, all the while crystallizing these two phases. When it finally reaches the eutectic point, orthoclase + anorthite + tridymite continue to precipitate until the liquid disappears.

Subsolidus Diagrams

For ternary systems without a liquid phase, a similar scheme is devised, but in this case the solid phases are plotted on triangular diagrams. The compositions of coexisting phases are connected with tie-lines. Thus, the diagram consists mostly of two-phase regions that bound smaller triangles that represent areas where three phases coexist. A commonly used example from metamorphic petrology is shown in Figure 9.20. The ACF diagram combines several chemical constituents into components applicable to most metamorphic phases. The ACF components are defined as follows:

$$A = Al_2O_3 + Fe_2O_3 - Na_2O - K_2O$$

$$C = CaO - 3.3P_2O_5$$

$$F = FeO + MgO + MnO$$

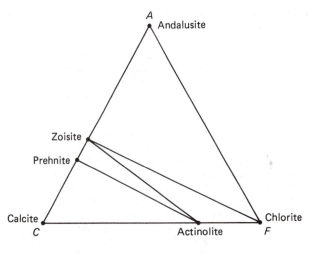

Figure 9.20 An ACF diagram, representing greenschist facies metamorphic conditions, with the compositions of coexisting phases connected by tie-lines. This diagram corresponds to P-T conditions in the lower righthand field of Figure 9.21.

This is an isobaric, isothermal representation, so the stable phases and locations of tie-lines may change with physical conditions.

CHEMICAL POTENTIAL DIAGRAMS

For ternary systems, it is exceedingly difficult to draw $\overline{G}$-X diagrams—in this case, actually $\overline{G}$-X_2-X_3 diagrams. (Notice that we need only two compositional coordinates to define a free energy function in a three-component system, because the third is dependent on the other two.) It is sometimes easier to visualize free energy changes in ternary systems from a consideration of chemical potentials. From the definition

$$\mu_i = (\partial \overline{G}/\partial X_i)_{P,T,X_{j \neq i}},$$

we can recognize that the chemical potential of component i in the phase under study is the slope of a curve in $\overline{G}$-X_i space. It is, by definition, a measure of the change in free energy of the phase as a result of changing the mole fraction of component i in the system, all other parameters remaining unchanged.

In order to compare the relative stabilities of coexisting phases, we define the free energy of each phase by an equation of the form

$$d\overline{G} = -\overline{S}dT + \overline{V}dP + \sum_i \mu_i \, dX_i.$$

If we consider only those changes in the system due to composition (that is, an isothermal, isobaric system) and further evaluate the system only at a set of fixed values of X_i (that is, we are in a closed system), then it is clear that free energy differences between phases are due only to chemical potential differences between them.

The phases prehnite (Pr), pumpellyite (Pm), zoisite (Zo), actinolite (At), and chlorite (Ch) all lie in the system CaO-MgO-FeO-Al_2O_3-SiO_2-H_2O. A P-T diagram for this system is shown in Figure 9.21. The curve marked [Pm] represents those pressure and temperature conditions under which the phases prehnite, chlorite, actinolite, and zoisite coexist in equilibrium. It is true, then, that anywhere on the line [Pm] the chemical potential of CaO is identical in all four phases, as are the chemical potentials of MgO, FeO, Al_2O_3, SiO_2, and H_2O. This is a momentarily disturbing concept unless we

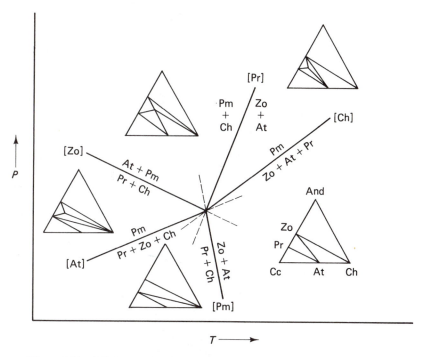

Figure 9.21 *P-T* diagram illustrating reactions between phases in the system CaO-MgO-FeO-Al₂O₃-SiO₂-H₂O. Phase abbreviations are as follows: prehnite (Pr), pumpellyite (Pm), zoisite (Zo), actinolite (At), and chlorite (Ch). Schematic ACF diagrams superimposed in the five regions around the invariant point describe phase relations within those regions of *P-T* space.

recall what a chemical potential is. Our statement regarding chemical potentials along line [Pm] can be rewritten in the following wordy statement: "The slope of the free energy surface with respect to CaO is the same in all phases. The slope with respect to MgO is also the same in all phases, and so forth for the slopes with respect to the other components."

Notice that it is not the *amount* of a component such as CaO that must be the same in all phases at equilibrium, but rather the *slope of the $\overline{G}$-X surface* in each phase. Inspection of Figure 9.21 also illustrates that line [Pm] represents a reaction Zo + At ⟷ Pr + Ch, as indicated in the ACF sketches on both sides of the line (compare with Figure 9.20).

We will now use our example at hand to try a new exercise. This one will be more difficult than the first, because we are going to add a noncompositional dimension to the problem. To simplify, let's compensate by removing most of the compositional dimensions and using only the coordinates ACF, as defined above. The diagram we will consider is a wedge-shaped polyhedron with an ACF diagram at its base and with molar free energy as its vertical axis. To begin, let us use the phase assemblages to the right of line [Pm]. Our polyhedron is illustrated in Figure 9.22a. The compositions of phases and appropriate tie-lines are shown on the base.

When we discussed $\overline{G}$-X diagrams for two-component systems, we concluded that the $\overline{G}(X_2)$ function for a phase with fixed composition was a point. The free energy of the system was a composite of the tangents to all of the individual $\overline{G}(X_2)$ curves in the system. For a familiar view of just such a binary system, consider the $\overline{G}$-C-F face of the

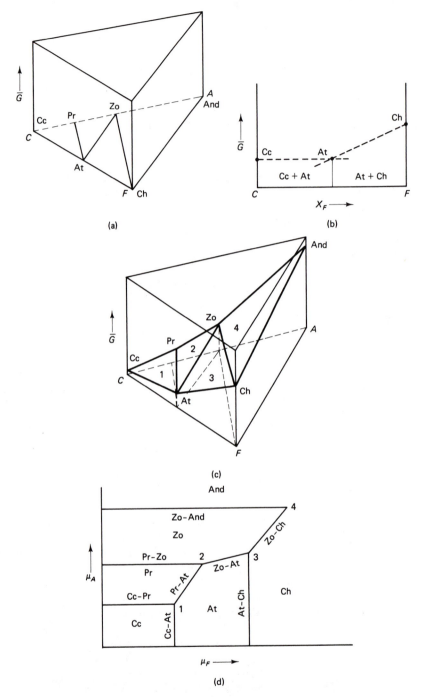

Figure 9.22 (a) ACF projection for phases to the right of line [Pm] in Figure 9.21, with the addition of a vertical $\overline{G}$ axis. (b) Variation in $\overline{G}$ with X_F along the C-F join in (a). Notice that $(\partial\overline{G}/\partial X_F)_{X_A} = \mu_F$ is constant for the assemblage Cc + At, and also for the assemblage At + Ch, although with a higher value. (c) $\overline{G}$-ACF diagram as in (a) with the addition of $\overline{G}(X_F, X_A)$ planes. (d) Plot of μ_A versus μ_F for the system in (c), showing the stability fields for various phases and combinations of phases.

polyhedron (Figure 9.22b). By way of review, notice that as we follow μ_F from left to right across the diagram, it increases. In this case, the increase is instantaneous at actinolite (At). Recall that in general, where phases have variable composition, μ_i increases continuously toward the side of the diagram represented by pure component i.

Now let's go back to the polyhedron (Figure 9.22a). To consider variations in chemical potentials of the three components A, C, and F, we recognize that we are now in a ternary system. The tangents to individual $\overline{G}(X_F, X_A)$ functions are now planes rather than lines. If we try to visualize the tangent planes in this system, we obtain something like Figure 9.22c. The slopes of planes 1, 2, 3, and 4 are defined by

$$(\partial \overline{G}/\partial X_F)_{P,T,X_A} = \mu_F$$

and

$$(\partial \overline{G}/\partial X_A)_{P,T,X_F} = \mu_A.$$

As in the binary example, μ_F and μ_A increase toward the compositional corners F and A, respectively. Also, as in the binary example, the changes in both μ_F and μ_A are abrupt, because once again we are dealing with a set of pure phases rather than solid solutions.

As a final step, let us now construct a new type of diagram, on which we plot changes in μ_F and μ_A (Figure 9-22d). As we start drawing this diagram, recognize that it is oriented so that the lower left corner (minimum values for both μ_A and μ_F) corresponds to the C corner in the previous diagram. The chemical potentials μ_A and μ_F increase toward compositional corners A and F, respectively. The easiest features to plot are the three-phase assemblages Cc-Pr-At, Pr-At-Zo, Zo-At-Ch, and Zo-Ch-And (numbered 1, 2, 3, and 4). At all compositions within these regions, the slope of $\overline{G}(X_F, X_A)$ is constant. Each of these regions, then, can be designated by a point on our μ_A-μ_F diagram. Between each of the three-phase regions in the $\overline{G}$-ACF wedge is a two-phase field. The slope of the $\overline{G}(X_F, X_A)$ surface is not fixed, because it takes three points to define a plane. The slope components in both the A direction and the F direction, however, are bounded by the slopes of the three-phase region on either side. For example, the two-phase field Pr-At is a line in $\mu_A = \mu_F$ space along which both μ_F and μ_A may vary. The upper limit is constrained by Pr-At-Zo. The assemblages Zo-At and Zo-Ch follow the same logic. The remaining two-phase assemblages lie on the binary edges of the $\overline{G}$-ACF wedge, so they will be lines of constant μ_A or μ_F on this diagram. The assemblage Cc-Pr, for example, is a field of constant μ_A but is free to vary in μ_F from our system's lower limit to the value constrained by Cc-Pr-At. Pr-Zo, Zo-And, and Cc-At behave in similar fashion. The remaining regions of interest are single-phase fields, in which both μ_A and μ_F are free to vary. These fields, which were points in the $\overline{G}$-ACF wedge, are now planar regions in μ_A-μ_F space.

Diagrams such as this one are very useful in metamorphic terrains, where changes in environmental parameters may result in a chemical potential gradient, superimposed on an initially homogeneous rock body. A careful study of the phase assemblages produced by mass transfer down the gradient (by diffusion, for example) can help us interpret the driving forces and material source areas for metamorphism. The development of contact aureoles or metasomatic fronts are examples of problems that can be analyzed using μ-μ diagrams. We have also encountered μ-μ diagrams before in Chapter 6, although they were not labelled as such there. *Eh-pH* diagrams can be thought of as plots of chemical potential of electrons versus chemical potential of hydrogen ions, and we have already explored how these diagrams can be used.

Chemical Potential Diagrams

SUMMARY

Phase diagrams provide a useful way of summarizing the effects of intensive and extensive variables on a system. For binary systems, one of the most widely used is the T-X diagram. We have seen how this diagram, as well as the P-X diagram, can be derived from stacking $\overline{G}$-X sections at various temperatures or pressures. We have also explored the ways in which phases terminate in these diagrams as three-phase equilibria, coincidences in composition, limiting values, and critical points to produce binary eutectics, peritectics, transition loops, and solvi. We also discussed how these features appear on P-T diagrams, and we introduced phase diagrams that utilize other parameters such as P_{fluid} or f_{O_2}. The various kinds of boundary curves and invariant points in ternary liquidus phase diagrams and the phase compatibilities in ternary subsolidus phase diagrams have been explored. We have also introduced the concept of μ-μ diagrams for ternary systems.

We have emphasized the relationships between phase diagrams and thermodynamics at every opportunity, and we hope that this treatment will aid you in comprehending these relationships. So far we have considered only systems in equilibrium. The next chapter will explore nonequilibrium conditions.

SUGGESTED READINGS

Phase diagrams are introduced in virtually all igneous-metamorphic petrology textbooks, though few relate these diagrams to thermodynamic functions. Some of the best explanations of phase diagrams are noted below.

BARKER, D. S. 1983. *Igneous rocks.* Englewood Cliffs, N.J.: Prentice-Hall. (Phase diagrams are introduced in Chapter 3 of this comprehensive text.)

COX, K. G.; BELL, J. D.; and PANKHURST, R. J. 1979. *The interpretation of igneous rocks.* London: Allen & Unwin. (Chapter 3 in this introductory text provides an excellent description of binary phase diagrams, and Chapters 4 and 5 discuss ternary diagrams.)

EDGAR, A. D. 1973. *Experimental petrology, basic principles and techniques.* London: Oxford University Press (A techniques book that describes the equipment, procedures, and pitfalls in experimentally determining phase diagrams.)

EHLERS, E. G. 1972. *The interpretation of geological phase diagrams.* San Francisco: Freeman. (A complete and highly readable introduction to the subject that provides much more detail than in this book. Chapter 2 describes binary phase diagrams and Chapter 3 discusses ternary systems.)

ERNST, W. G. 1976. *Petrologic phase equilibria.* San Francisco: Freeman. (An exceptionally fine reference on phase diagrams. Chapter 2 describes the experimental determination of phase equilibria, Chapter 3 provides a look at the computational approach, and Chapter 4 introduces unary, binary, and ternary systems.)

MORSE, S. A. 1980. *Basalts and phase diagrams.* New York: Springer-Verlag. (A thorough and very readable book about phase diagrams. Chapters 5 through 10 discuss binary diagrams, and Chapters 11 and 12 introduce ternary diagrams.)

WINKLER, H. G. F. 1979. *Petrogenesis of metamorphic rocks,* 5th ed. New York: Springer-Verlag. (Chapter 5 gives a good description of subsolidus phase diagrams commonly used in metamorphic petrology.)

YODER, H. S., JR., ed. 1979. *The evolution of the igneous rocks, Fiftieth Anniversary Perspectives.* Princeton: Princeton Univ. Press. (An updated version of Bowen's treatise that uses phase diagrams throughout. Chapter 4, by A. Muan, gives a thorough description of phase diagrams.)

A large number of research papers are devoted to either experimental determination or thermodynamic modeling of phase diagrams. Those cited in this chapter are given below.

BOWEN, N. L. 1913. The melting phenomena of the plagioclase feldspars. *Amer. Jour. Sci.* 35, 4th Series: 577–99.

SAXENA, S. K. 1973. *Thermodynamics of rock-forming crystalline solutions.* New York: Springer-Verlag.

SPENCER, K. J., and LINDSLEY, D. H. 1981. A solution model for coexisting iron-titantium oxides. *Amer. Mineral.* 66: 1189–1201.

THOMPSON, J. B., JR., and WALDBAUM, D. R. 1969. Mixing properties of sanidine crystalline solutions: III. Calculations based on two-phase data. *Amer. Mineral.* 54: 811–38.

PROBLEMS

1. Using Figure 9.6, reconstruct the equilibrium crystallization sequence for a liquid with a composition that corresponds exactly to the eutectic in this system.

2. Equilibrium melting is the reverse process from equilibrium crystallization. From Figure 9.7, determine the melting sequence for a mixture of 90 mole % orthoclase and 10 mole % tridymite.

3. The liquid in Problem 2 is now cooled again from 1200°C to room temperature under equilibrium conditions. In this experiment, you continuously measure the temperature of the system over the four hours it takes to reach room temperature. Sketch a qualitative plot of how temperature in this system changes with elapsed time, and explain why temperature does not fall continuously with time.

4. Using Figure 9.9, describe the melting under equilibrium conditions of plagioclase with composition $X_{An} = 0.2$.

5. Using the experimental results tabulated below, construct a T-X diagram for the system forsterite-fayalite. The glass composition represents the liquid in the experiments.

Bulk composition	T(°C)	Olivine composition	Glass composition
$X_{Fa} = 0.6$	1800	—	$X_{Fa} = 0.60$
	1700	—	0.60
	1600	$X_{Fa} = 0.34$	0.60
	1500	0.42	0.75
	1400	0.58	0.91
	1300	0.60	—
$X_{Fa} = 0.2$	1800	0.08	0.25
	1700	0.17	0.47
	1600	0.20	—

6. (a) Construct $\overline{G}$-X diagrams for a system that is a regular solution with $W = 15$ kJ at 800, 900, 1000, 1100, and 1200K. You may use program REGULAR on the diskette. Then construct the corresponding T-X diagram for this system. (b) At 1000 K what is the effect of changing W from 2 to 5 kcal? Show your work on a $\overline{G}$-X diagram.

7. Test your understanding of μ-μ diagrams by constructing a $\mu_A = \mu_F$ diagram for the mineral assemblages on the left side of the line in Figure 9.21.

8. Draw a series of topologically reasonable isothermal P-X sections that describe the phase relations shown in Figure 9.15a.

10. Calculate T-X_{CO_2} diagram for the reaction in Worked Problem 9-2 at 1 atm for a $CO_2 + H_2O$ fluid that behaves as an ideal gas. You may use program DELTAG on the diskette.

11. Using the T-X diagram on the bottom of Figure 9.7, determine the relative proportions of stable phases at 900, 1100, and 1300°C for a bulk compositon of $X_{SiO_2} = 0.3$.

12. Determine the temperature and oxygen fugacity under which a lava containing the following Fe-Ti oxides equilibrated: titanomagnetite with 43 mol% ulvospinel component, and hemoilmenite containing 89 mol % ilmenite component. You may use program FETIOX on the diskette.

Chapter 10

Kinetics

and Crystallization

OVERVIEW

Not all geochemical processes make the final adjustment to equilibrium conditions. Even those processes that eventually go to completion, such as crystallization of magmas and recrystallization of metamorphic rocks, are sometimes controlled by kinetic factors. In this chapter, we discover what these factors are and how they can be understood. We first examine the effect of temperature on rate processes, which leads to the concept of activation energy. We then consider three specific rate processes: diffusion, nucleation, and growth. Differences between volume diffusion and grain boundary diffusion are explained, and the importance of crystal defects is considered. We introduce nucleation by discussing the extra free energy term related to surface energy, and then consider rates for homogeneous and heterogeneous nucleation. Crystal growth may be dominated by interface- or diffusion-controlled mechanisms, so we discuss each briefly. Finally, we illustrate how the interplay of all three kinetic processes controls crystallization in real geochemical systems.

ARRHENIUS RELATIONS: THE EFFECT OF TEMPERATURE ON KINETIC PROCESSES

It may not have been surprising to learn in Chapter 5 that diagenetic processes are kinetically controlled. Most of us are intuitively aware that reactions are sluggish at low temperature. It may not be so obvious that kinetics also plays a role in many geochemical processes at high temperatures, such as crystallization of magma or recrystallization of metamorphic rocks.

For rate-controlled processes in magmatic or metamorphic systems, it is critical that we take into account the temperature dependence of rate constants. The relationship between any rate constant k and temperature follows the classic equation proposed by Arrhenius in 1889:

$$k = A \exp \frac{-\Delta G^*}{RT}, \tag{10-1}$$

where A is the *frequency factor*, ΔG^* is the *activation energy*, R is the molar gas constant, and T is the absolute temperature. Because a reaction requires the collision of two molecules, the reaction rate must be proportional to the frequency of such collisions. The frequency factor, consequently, indicates the number of times per second that atoms are close enough to react. Not all collisions are comparable, though. Intuition tells us that a very gentle collision between molecules may be insufficient to cause a reaction. The reaction rate, therefore, is assumed to depend on the probability that collisions will have energies greater than some threshold value. The activation energy is an expression of this free energy barrier that must be overcome for the reaction to proceed at an appreciable rate. The term $\exp(-\Delta G^*/RT)$ expresses the fraction of reacting atoms that have energy higher than the average energy of atoms in the system. This probability term is reminiscent of the Boltzmann distribution law of statistics. The exponential dependence of rate constants on temperature explains why this must be considered in high-temperature systems; depending on the size of ΔG^*, rate constants may change by several orders of magnitude over a temperature range of several hundred degrees. An extended discussion of rate constants and the Arrhenius Law can be found in Lasaga (1981).

Worked Problem 10-1

How could we determine activation energy and frequency factor for a real geochemical system? As an example, let us consider the reaction analcite + quartz → albite + water. The following experimental data for the rate of this reaction in NaCl were reported by Mathews (1980):

$k\,(\text{kcal}/RT)$	$T\,(^\circ\text{C})$
9.02×10^{-4}	419
2.55×10^{-3}	435
5.95×10^{-3}	454
9.80×10^{-3}	474

Rearranging equation 10-1 gives

$$\ln k = \ln A - [\Delta G^* R(1/T)]. \tag{10-2}$$

This expression has the form of a straight line ($y = mx + b$), in which the slope is $-\Delta G^* R$ and the y-intercept is $\ln A$. Therefore, by plotting $\ln k$ versus reciprocal temperature, we can graphically obtain values for activation energy and frequency factor.[1] A plot of Mathews' data is shown in Figure 10.1. The intercept of the least squares regression line with the $\ln k$ axis corresponds to a value for A of 1.2×10^{11}. The slope of the line corresponds to a ΔG^* value of 44.5 kcal mol^{-1}.

Unfortunately, there are few experimental data on reaction rates of geochemical interest, like those in Worked Problem 10-1. One notable exception is a study of the

---

[1] A cautionary note is in order here. Values for k in some rate equations may not be independent of time. In such cases, we obtain only an *apparent* activation energy.

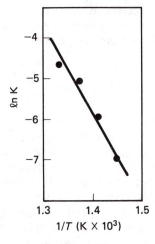

Figure 10.1 Arrhenius plot of ln k versus $1/T$ for the reaction analcite + quartz → albite + water. Activation energy for the reaction can be obtained from the slope of the regression line. (Data are from experiments by Mathews 1980.)

system MgO-SiO_2-H_2O (Greenwood, 1963). Experiments show that at 830°C and 1 kbar, talc dehydrates to form enstatite plus quartz. However, a multistep process involving the breakdown of talc to form anthophyllite as a reactive intermediate is kinetically more favorable than the direct transformation of talc to enstatite plus quartz. From Greenwood's measured rate constants, Mueller and Saxena (1977) calculated the volumes of talc, anthophyllite, enstatite, and quartz produced as functions of time. Their results are summarized in Figure 10.2. Even though the stable assemblage under these conditions is enstatite plus quartz, anthophyllite is the most abundant phase during the time interval from 1000 to 1500 minutes. Thus, experiments of short duration would give the false impression that anthophyllite is the stable phase under these conditions. The rate constants for the formation of anthophyllite from talc and of enstatite from anthophyllite are approximately the same in this case, so neither reaction functions as the rate-limiting step in this sequence.

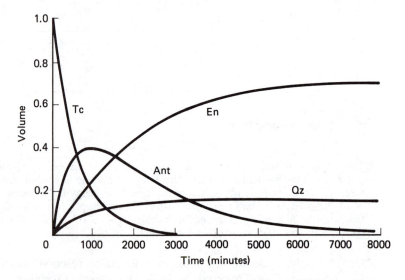

Figure 10.2 Volumes of talc (Tc), anthophyllite (Ant), enstatite (En), and quartz (Qz) as a function of time produced by reaction at 830° C and 1 kbar. (After R. F. Mueller and S. K. Saxena, *Chemical Petrology*. Copyright 1977 by Springer-Verlag. Reprinted with permission.)

Migration of components can occur over free surfaces of grains (*surface diffusion*), along the boundaries between grains (*grain-boundary diffusion*), and through the body of grains or liquids (*volume diffusion*). The latter two styles of diffusion are of most interest in high-temperature systems. When we consider volume diffusion, it is also useful to distinguish between *self-diffusion,* by which we mean the movement of an atom or ion through a system composed, at least in part, of the same element (for example, when oxygen diffuses through a feldspar), and *interdiffusion,* in which the diffusion of one component is dependent on the opposing diffusion of another (for example, Mg-Fe exchange in olivine). Generally, the requirement of local charge balance implied by interdiffusion results in a decrease in the rate of diffusion.

In Chapter 5, when we first discussed diffusion, we presented a macroscopic or empirical view, in which Fick's laws were used to describe the bulk transport of material through a medium that was considered to be continuous in all its processes. That treatment involved the definition of diffusion coefficients (D), which come in a bewildering array of forms. Self-diffusion is commonly described by *tracer* diffusion coefficients (D^*), which are measured experimentally by following the progress of isotopically tagged ions (for example, ^{18}O moving through the oxygen framework of a silicate). Interdiffusion coefficients ($D_{1\text{-}2}$), describing the interaction of ions of species 1 and species 2, are measured empirically or can be calculated from the tracer coefficients of each species, weighted by their relative abundance in the system, as follows:

$$D_{1\text{-}2} = \frac{n_1 D_1^* + n_2 D_2^*}{n_1 + n_2}.$$

In geologic systems of even moderate complexity, this approach can be difficult to employ, because it involves collecting a large amount of experimental data in systems that are sluggish and hard to analyze.

An alternate approach, which considers diffusion from a mechanistic perspective at the atomic level, has been used with some success by metallurgists and ceramists, and is slowly being adopted by geochemists. It offers the advantage of describing diffusive transport in terms of forces between ions in a mineral lattice or other medium. These, in theory, can be predicted from conventional bonding models, given detailed information about crystal structures. This approach holds great promise, but has been applied to very few materials of geologic significance. It is beyond the scope of this chapter, although we can use the atomistic perspective to make some broad generalizations to help interpret the behavior or diffusing ions.

Volume diffusion in solids usually involves migration of atoms or ions through imperfections in the periodic structures of crystals. *Point defects* can arise from the absence of atoms at lattice sites (vacancies). Two kinds of *intrinsic* vacancies in ionic crystals maintain electrical neutrality: Frenkel defects are caused by an ion abandoning a lattice site and occupying an interstitial position, and Schottky defects involve equal numbers of vacancies in cation and anion positions. These imperfections are illustrated in Figure 10.3. Impure crystals may also contain *extrinsic* defects, which arise because foreign ions may have a different charge from the native ions they replace. The requirement for overall charge balance leads to vacancies in the crystalline structure. For example, the incorporation of Fe^{3+} into olivine requires a cation vacancy somewhere in the crystal.

Nonstoichiometric crystals, by definition, also have vacancies. Real crystals commonly contain both intrinsic and extrinsic defects, each of which dominate the overall

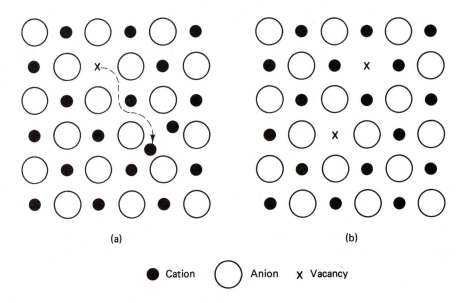

(a)   (b)

● Cation   ◯ Anion   ✗ Vacancy

Figure 10.3 Volume diffusion is facilitated by defects in crystals. (a) Frenkel defects result when ions leave normal crystal sites for interstitial positions, leaving vacancies. (b) Schottky defects occur when cation and anion vacancies preserve charge balance.

diffusion at different temperatures. Intrinsic mechanisms tend to be most effective at high temperatures, where thermal vibrations of atoms are high. As temperatures are lowered, the number of vacancies induced by impurities becomes greater than those generated intrinsically. This can be seen in Figure 10.4, a plot of experimentally determined diffusion coefficients for Fe-Mg along the c axis in olivine as functions of temperature (Buening and Buseck, 1973). In the high-temperature region (left side of Figure 10.4), diffusion occurs mostly by an intrinsic mechanism, which has a steeper Arrhenius slope (larger ΔG^*) than for the extrinsic region. The transition between diffusion mechanisms occurs at approximately 1125° C. The difference in slope tells us, qualitatively, that it takes more energy to move ions between intrinsic defects than be-

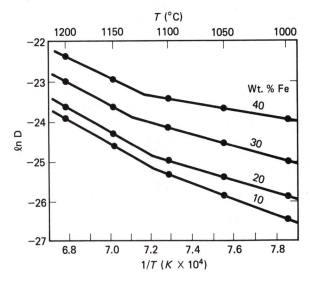

Figure 10.4 Arrhenius plot of ln D versus $1/T$ summarizing data on interdiffusion of Fe and Mg in olivines of different composition (Buening and Buseck, 1973). The kink in these lines at 1125°C probably corresponds to a change from an intrinsic to an extrinsic diffusion mechanism with different activation energy (hence, slope).

tween extrinsic ones (62 kcal mol^{-1} in the intrinsic region, 30 kcal mol^{-1} in the extrinsic region, when extrapolated to pure forsterite). Extrinsic diffusion mechanisms probably predominate in natural systems, because most minerals contain appreciable quantities of impurities, and temperatures for most geologic processes are less than required for intrinsic mechanisms.

Point defects can be treated thermodynamically by considering them to be quasi-chemical species. A crystal containing defects can then be visualized as a solution between the host mineral and vacuum, and a chemical potential can be assigned to each defect. All defect chemical potentials are referred to the perfect crystal as the standard state, and each defect is assigned an effective charge (based on the charge the crystal acquires as a result of the defect). Chemical reactions involving defects can then be written, each of which has an equilibrium constant. Such a treatment can be useful for unraveling the thermodynamic properties of real crystalline systems.

Extended defects also play important roles in volume diffusion. These include dislocations, which are linear displacements of lattice planes, and various kinds of planar defects, such as twin boundaries and stacking faults. Irregular microfractures also fall into this category. Extended defects provide very effective conduits (sometimes called "short circuits") for diffusion, and where they are abundant they may control this process. This accounts in part for the reactivity of strained versus unstrained crystals. However, extended defects are difficult to treat quantitatively.

Grain boundaries are sometimes considered to be extended defects, but this blurs the distinction between volume diffusion and grain-boundary diffusion. It is a commonly held belief that the most efficient path for diffusive mass transfer in metamorphic rocks is along grain boundaries. This idea is grounded in the notion that metamorphic fluids along such boundaries provide faster diffusion paths than volume diffusion through solid grians. Although this idea is probably correct in many cases, the situation is not as straightforward as it may seem. The diffusion cross section offered by grain boundaries is quite small, the tortuosity of these paths may be appreciable, and the amount of fluid may decrease considerably at high metamorphic grades. Freer (1981) suggests that volume diffusion may actually dominate in many systems at high temperatures. Attempts have been made to place limits on grain-boundary diffusion rates in metamorphic rocks, but there are few data with which to test numerical models of this process.

Worked Problem 10-2

Can diffusion be used to constrain the rate at which a magma cools? Taylor et al. (1977) developed a "speedometer" applicable to basalts, using frozen diffusion profiles of Fe/Mg in olivines. Let us examine how this system works.

Buening and Buseck (1973) demonstrated that interdiffusion of Fe^{2+} and Mg in olivine is a function of temperature, composition, f_{O_2}, and crystallographic direction. From their data (illustrated in Figure 10.4), we can derive the following expression for the diffusion coefficient along the c axis, D_c, for temperatures below 1125°C:[2]

$$D_c = 10^2 f_{O_2}^{1/6} \exp\left(-0.0501 X_{Fa} - 14.03\right) \exp \frac{-31.66 + 0.2191 X_{Fa}}{RT}, \quad (10\text{-}3)$$

where X_{Fa} is the mole fraction of Fe_2SiO_4 in olivine. Buening and Buseck also showed that olivine diffusion is highly anisotropic, with transport being fastest in the c direction. The measured profiles of olivine grains, however, reflect reactions limited by the slowest diffusive rate, which is along a. Taylor and his coworkers estimated that D_c was greater than D_a by a factor of about 4 at 1050° C, so that

---

[2] Taylor et al. (1977) actually defined D_i on an atomic basis. Equation 10-4 is derived from theirs by noting that Boltzmann's constant $k = R/$Avogadro's Number.

$$D_a = D_c/4.$$

One uncertainty in boundary conditions for this problem is that the concentration profile in olivine as it solidified is unknown, that is, some compositional zoning may have been present originally. Taylor and company treated the forsterite content of olivine as a step function, so that each grain originally had no compositional gradient other than a sharp boundary between magnesian core and iron-rich rim. Such a simplification means that the calculated cooling rate will be a *minimum* rate. In a later paper, Taylor et al. (1978) corrected this problem by estimating an "as-solidified" concentration profile. For simplicity, we will ignore this later refinement.

In the case of olivine, the limiting diffusion coefficient, D_a, is a function of both temperature (in turn, a function of time t) and composition. We can use the following one-dimensional diffusion equation to calculate the compositional profile:

$$\frac{\partial C}{\partial t} = \frac{\partial}{\partial x}\left\{D_a[C(x), T(t)]\frac{\partial C}{\partial x}\right\}, \tag{10-4}$$

where C is the iron concentration. Unlike the worked problem at the beginning of Chapter 5, which yielded easy, analytical solutions, this equation is nonlinear and must be solved by numerical iteration to approach the correct solution.

Taylor and his collaborators applied their solution to determination of the cooling rates for lunar basalts. Oxygen fugacities (which are temperature-dependent) appropriate for lunar rocks obey the relation

$$\log f_{O_2} = 0.015T(K) - 34.6.$$

This provides the information we need to determine D_a as a function of f_{O_2} for substitution into equation 10-4. The measured compositional profile in an olivine grain in lunar basalt 15555, obtained by electron microprobe traverse, is illustrated by circles in Figure 10.5. Calculated diffusion profiles for various cooling rates are shown for comparison. The minimum cooling rate for this basalt was approximately 5° C per day.

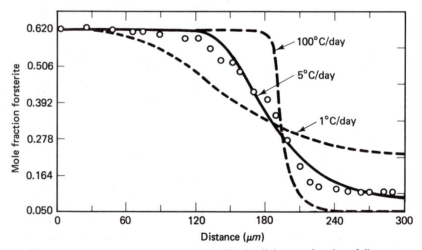

Figure 10.5 Calculated compositional profiles in olivine as a function of distance from the core-rim interface. These profiles are produced by interdiffusion of Fe and Mg during cooling. Circles are electron microprobe data for zoned olivines in lunar basalt 15555. (After Taylor et al. 1977.)

NUCLEATION

It is a wonder that crystals ever form. The most difficult step, from a thermodynamic perspective, is the initiation of a small volume of the product phase. This step is called *nucleation*. Crystal nuclei form because local thermal fluctuations in the host phase result in temporary, generally unstable clusters of atoms. This is an extremely difficult process to study, because no experimental technique has yet been devised that allows direct observation of the formation of crystal nuclei.

The driving force for nucleation, like other rate processes, is the deviation of the system from equilibrium conditions. A common expression for this driving force is *supersaturation,* the difference between the concentration of the component of interest and its equilibrium concentration. In studying nucleation in geochemical systems, it is often advantageous to define deviation from equilibrium in another way, that is, in terms of *undercooling* (ΔT). This parameter is the difference between the equilibrium temperature for the first appearance of a phase and the temperature at which it actually appears. For example, in magmatic systems crystallization commonly fails to occur at the liquidus temperature of a melt, where it should be expected. The magma is therefore said to be undercooled if it is at a temperature below its liquidus and crystallization is not yet initiated. The value of ΔT is equal to the liquidus temperature minus the actual temperature. In using this parameter, it is crucial that the equilibrium temperature of the actual system be used; the liquidus temperature of a crystallizing magma changes continuously as composition changes. The liquidus for the bulk parental magma is only useful at the start of crystallization.

Nucleation in Melts

The appearance of a new phase, as we learned in earlier chapters, is associated with a decrease in the free energy of the system. This change can be expressed as a total differential:

$$dG = \left(\frac{\partial G}{\partial T}\right)_{P,n_i} dt + \left(\frac{\partial G}{\partial P}\right)_{T,n_i} dP + \sum_i \left(\frac{\partial G}{\partial n_i}\right)_{P,T,n_{j \neq i}} dn_i,$$

in which the partial derivatives are written more familiarly as $-\bar{S}$, $\bar{V}$, and μ_i. In practice, we apply this equation by declaring that each phase in the system consists of a material whose thermodynamic properties are continuous. The boundary between one phase and another is marked by a discontinuity in thermodynamic properties, but we assume that it does not contribute any energy to the system itself. Calculations performed on this basis are generally consistent with observations of real systems.

We know, however, that the boundaries between phases are not simple discontinuities. The atoms on the surface of a droplet or a crystal are not surrounded by the same uniform network of bonds found in its interior. This irregular bonding environment leads to structural distortions that must increase the free energy of the system. The *surface free energy* contribution is generally quite small because the number of atomic sites near the boundary of a phase is much less than the number in its interior. We can predict, however, that it will increase in importance if the mean size of crystals or droplets in the system is vanishingly small, as it is during nucleation. Surface free energy, therefore, accounts for the kinetic inhibition we have described in the discussion of undercooling.

The extra energy associated with the formation of an interface is minimized if the surface area of the new particle is also minimized. Consequently, a growing particle must perform work to stretch the interface. This work, for a liquid droplet, is commonly defined in terms of a tensional force, σ, applied perpendicular to any line on the droplet surface. This concept of *surface tension* is only strictly applicable to liquid surfaces, but no adequate alternative has been formulated for solids. It is common practice to refer loosely to σ even when the nucleating particle is crystalline.

From the perspective of free energy, surface tension can be viewed as a measure of the change in free energy because of an increase in particle surface area A; that is,

$$\sigma = \left(\frac{\partial G}{\partial A}\right)_{T,P,n}.$$

The total differential of G thus can be modified to the form

$$dG = (dG)_{\text{vol}} + \left(\frac{\partial G}{\partial A}\right)_{T,P,n} dA$$

$$= (dG)_{\text{vol}} + \sigma \, dA,$$

in which $(dG)_{\text{vol}}$ is a shorthand notation for the familiar temperature, pressure, and composition terms that apply throughout the volume of the phase. For spherical nuclei with radius r, this becomes

$$\Delta G_{\text{total}} = \frac{4\pi r^3}{3} \Delta G_v + 4\pi r^2 \sigma, \tag{10-5}$$

where ΔG_v is the free energy change per unit volume. The surface energy term dominates at small values of r, and the volume term at large radii. Because σ is always positive, equation 10-5 implies that the total free energy of a nucleating particle increases with r until some critical radius (r_{crit}) is reached, after which the free energy decreases with further increase in r. Small nuclei with $r < r_{\text{crit}}$ are unstable and tend to redissolve, whereas nuclei with $r > r_{\text{crit}}$ will persist and grow. We can illustrate the importance of this observation with the following problem.

Worked Problem 10-3

How does the free energy of a nucleus vary with radius? To answer this, we will consider the formation of spherical nuclei of forsterite in Mg_2SiO_4 liquid. Assume an undercooling of 10°C and a surface energy of 10^3 ergs cm^{-2} ($= 2.39 \times 10^{-5}$ cal cm^{-2}). Calculate the critical radius for olivine nuclei under these conditions.

ΔG_v for olivine at a temperature of 2063 K (10° below the forsterite liquidus) is -1.738 cal cm^{-3}. Substitution of these values into equation 10-5 gives

$$\Delta G_{\text{total}} = \frac{4\pi}{3}(-1.738)r^3 + 4\pi(2.39 \times 10^{-5})r^2$$

$$= -7.276r^3 + 3.0 \times 10^{-4}r^2.$$

For $r = 2 \times 10^{-5}$ cm, ΔG_{total} is thus 6.18×10^{-14} cal. Substitution of other values for r gives a data set from which the solid line in Figure 10.6 is plotted. From this figure, we can see readily that increasing r initially causes an increase in total free energy of the system because the surface energy term dominates. At the point where the volume energy term begins to dominate, there is a downturn in total free energy and further crystal growth can take place spontaneously with a net decrease in free energy.

The peak of the free energy hump in this diagram corresponds to r_{crit}. This value can be calculated directly by recognizing that the reversal in slope corresponds to the point where the first derivative of equation 10-5 with respect to r equals zero. Solving for r under this condition yields

$$r_{\text{crit}} = -2\sigma/\Delta G_v. \tag{10-6}$$

Substitution of values given above for σ and ΔG_v gives r_{crit} for olivine $= 2.75 \times 10^{-5}$ cm under these conditions.

The total energy barrier to nucleation ΔG_{homo} (for *homogeneous* nucleation, defined below) can be found by combining the conditions necessary to form clusters of the critical radius (equation 10-6) with equation 10-5 to give

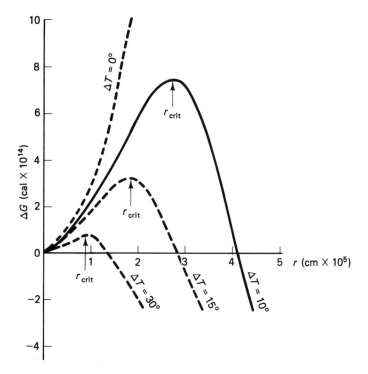

Figure 10.6 Calculated total free energy for nucleation of forsterite in a liquid of its own composition, as a function of nucleus radius and undercooling. The critical nucleus at each undercooling is r_{crit}.

$$\Delta G_{homo} = (16\pi\sigma^3)/(3\Delta G_v^2). \tag{10-7}$$

Also illustrated in Figure 10.6 by dashed lines are energy barriers to nucleation at $\Delta T = 0$, 15, and 30°. We can see that this barrier, as well as the critical radius for nuclei, decrease as undercooling increases.

What we have presented so far is an apparent paradox. A particle is not stable until its radius is greater than r_{crit}, but it can never reach r_{crit} because it must always begin as a nucleus too small to be stable. Worked Problem 10-3, however, suggests one way to avoid this paradox—by undercooling.

If nuclei persist once they reach critical radius, then the nucleation rate can be expressed by an Arrhenius-type equation as the product of the concentration of these nuclei and the rate at which atoms attach themselves to critical nuclei. The concentration of nuclei having the critical radius (N_r) at equilibrium is given by an expression of the Boltzmann distribution:

$$N_r = N_v \exp\left(-\Delta G_{total}/RT\right), \tag{10-8}$$

where N_v is the number of atoms per unit volume of the reactant phase, and the exponential term is the probability that they will have sufficient energy to react. The attachment frequency of atoms is the product of the number of atoms, n, located next to the critical cluster, the frequency, ν, with which atoms try to overcome the barrier, and the probability of atoms having sufficient energy to succeed. This is expressed by

$$f_{attach} = n\nu \exp\left(-\Delta G^*/RT\right), \tag{10-9}$$

where ΔG^* is the activation energy needed for attachment. The forms of equations 10-8 and 10-9 are similar to that of equation 10-1, the expression for the rate constant in terms of frequency factor and activation energy. We will see equations of this form over and over again in this chapter, because kinetic processes are governed by statistical probability. Multiplying equations 10-8 and 10-9 gives the nucleation rate, I:

$$I = n\nu N_v \exp\left(-\Delta G_{\text{total}}/RT\right) \exp\left(-\Delta G^*/RT\right). \tag{10-10}$$

For nuclei which form free in a liquid, ΔG_{total} in this equation is identical to ΔG_{homo} in equation 10-7. We can make the further approximation that $\Delta G_v = -\Delta S\,\Delta T$, where ΔS is the entropy change associated with nucleation and ΔT is the undercooling. By combining equations 10-7 and 10-10, therefore, we find that

$$I = n\nu N_v \exp\left(\frac{16}{3}\pi\sigma^3\,\Delta S^2\,\Delta T^2\,RT\right) \exp\left(\frac{-\Delta G^*}{RT}\right). \tag{10-11}$$

We can use equation 10-11 to make some generalizations about nucleation rate. Not surprisingly, undercooling exerts a major influence, and there is a rapid rise in nucleation rate as undercooling increases, that is, as the system moves farther away from equilibrium. As ΔT increases and T decreases, the two exponential terms compete for dominance, with the result that nucleation rate reaches a maximum at some critical undercooling and then decreases thereafter. In physical terms, the free energy advantage gained by undercooling is finally overwhelmed by the increasing viscosity of the liquid. If we were to plot I against ΔT, the result would be an asymmetrical bell-shaped curve.

Undercooling, then, reduces the kinetic inhibition to nucleation and offers a solution to the paradox. Nature provides another, more common solution as well. Equation 10-5 through 10-7 and equation 10-11 are written in terms of a free-standing spherical particle. This geometrical condition is known as *homogeneous nucleation*. Observation tells us, however, that nucleation more commonly takes place on a substrate. Frost forms on a windowpane, dew on spiderwebs, and crystals on other crystals. There must be an energetic advantage to this style of *heterogeneous nucleation* that makes it so widespread.

To see how nucleation on a substrate helps avoid the problem of a critical radius, consider Figure 10.7. Here, a particle of a phase C with an exceedingly small volume has been allowed to form from liquid phase L against a foreign surface S. It is impor-

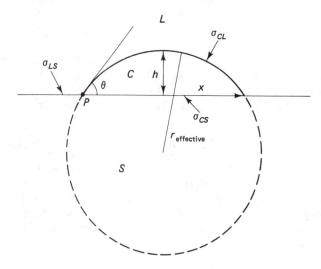

Figure 10.7 Schematic cross section through a cluster of atoms of phase C nucleating heterogeneously from phase L on a substrate S. The contact angle between the surface of the nucleus and S is θ. The effective radius of this nucleus is much greater than the critical radius of a spherical nucleus of the same volume formed by homogeneous nucleation. Distances x and h are used in problem 10-6 at the end of this chapter.

tant to note that the radius of this particle, if it had condensed homogeneously, would have been less than r_{crit}. Here, however, the same volume of phase C forms a spherical cap with an effective radius that is considerably larger than r_{crit}. Because the surface tension on the particle is a function of its degree of curvature, this nucleus now has a low surface free energy and is stable.

If the interfaces between phases C, L, and S are all mechanically stable, then the tensional forces perpendicular to any line along them must be perfectly balanced. This must also be true at the three-phase contact (point P), where there are three force vectors to resolve. Each is proportional to a surface tension in the system. σ_{LS} and σ_{CS} oppose each other along the flat boundaries L-S and C-S, respectively. The remaining vector, represented by σ_{CL}, is tangential to the curved C-L interface at point P. Because the sum of these three must be zero, we see that

$$\sigma_{LS} = \sigma_{CS} + \sigma_{CL} \cos \theta, \tag{10-12}$$

where θ is the angle between the tangent and interface C-S.

From equation 10-12, it is apparent that nucleation of a new particle is particularly favorable if σ_{CL} is nearly equal to σ_{LS}, and σ_{CS} is very small. This condition might easily be met if the substrate were a seed crystal of the nucleating phase itself. Under these conditions, θ is minimized, and the effective radius of the nucleus is at its largest. Even when the radius of a particle under homogeneous nucleation would be severely subcritical, the particle may still be stable.

By combining equation 10-12 with appropriate expressions for the surface area and volume of the spherical cap in Figure 10.7, we can express the energy barrier to heterogeneous nucleation ΔG_{hetero} as

$$\Delta G_{hetero} = \left(\frac{4\pi}{3} \sigma_{CL}^3 \, \Delta G_v^2 \right) (2 - 3 \cos \theta + \cos^3 \theta). \tag{10-13}$$

Compare this to equation 10-7, which describes the total energy barrier to homogeneous nucleation, and you will see that by nucleating against a substrate, the barrier has been reduced by a factor of $(2 - 3 \cos \theta + \cos^3 \theta)/4$. Clearly, then, nucleation against foreign objects is favored over nucleation of unsupported particles.

The rate of heterogeneous nucleation can be expressed by an equation similar to equation 10-11, with appropriate energy values substituted. We have not spent much time on these rate equations, for the simple reason that they do not appear to work very well for geologic systems. Theoretical models for nucleation qualitatively reproduce the general shapes of nucleation curves, but the temperatures for maximum nucleation rate and the peak widths are often different from experimentally determined curves. This difference between theoretical and experimental curves might be due to heterogeneous nucleation, but some researchers feel that classic nucleation theory is flawed. Therefore, these equations may require significant revision before they can be applied successfully to understand nucleation in silicate melts.

INTERNAL PRESSURE, BEER BUBBLES, AND PELE'S TEARS

Most people have opened a shaken can of beer or soda pop and experienced the fountain effect that follows rapid pressure release. Why does this happen? If not shaken, the same can emits only a gentle hiss when it is opened. To answer, we need to explore the relationship between surface tension and pressure.

The work required to transfer a small volume, dV, of material from liquid to the inside of a droplet, $(P_{int} - P_{ext})dV$, must equal the work required to extend its surface, $\sigma \, dA$, as follows:

Nucleation

$$(P_{\text{int}} - P_{\text{ext}})\, dV = \sigma\, dA.$$

Differentiating the expressions for the volume and area of a sphere gives

$$dV = 4\pi r^2\, dr$$

and

$$dA = 8\pi r\, dr.$$

Substituting these values into the above equation results in

$$(P_{\text{int}} - P_{\text{ext}})\, 4\pi r^2\, dr = \sigma\, 8\pi r\, dr,$$

or

$$P_{\text{int}} - P_{\text{ext}} = \frac{2\sigma}{r}. \tag{10-14}$$

Thus, the internal pressure of a vapor bubble must be greater than external pressure by an amount $2\sigma/r$.

In an unshaken can of beer, all of the vapor phase makes up a large "bubble" at the top. Its interface with the liquid is flat, so r in equation 10-14 is infinitely large. Under these circumstances, $P_{\text{int}} = P_{\text{ext}}$. If the can is shaken, however, the same volume of vapor is distributed in a very large number of small bubbles. The internal pressure on each is significantly higher than the external pressure, because r is a fraction of a millimeter. It is this vapor pressure that causes the explosive fountain.

The actual pressure, of course, depends on the magnitude of σ as well as the radii of vapor bubbles. It can be particularly high in some systems of geochemical interest. For example, the surface tension of basalt is about 350 dynes cm^{-1} at 1200° C and 1 bar. Highly fluidized basaltic ejecta sometimes form tiny glassy droplets and filaments called Pele's tears. The internal pressure of such droplets, which have radii on the order of 10^{-3} cm, can be calculated as follows, using equation 10-13:

$$P_{\text{int}} = 1 + (2 \times 350)/(10^{-3} \times 10^6) = 1.7 \text{ bars}.$$

The internal pressure is nearly double the external pressure!

Nucleation in Solids

The principles we have just discussed are also applicable to solid phases, but the process of nucleation is more complicated. A phase transformation in the solid state may involve a significant volume change that cannot be accomodated by flow of the reactant phase. This produces a "room problem" that induces strain energy into the system. This extra energy must be added to the energy barrier to nucleation. The free energy for nucleation thus becomes

$$\Delta G_{\text{total}} = \frac{4\pi}{3} r^3 (\Delta G_v + \Delta G_s) + 4\pi r^2 \sigma,$$

where ΔG_s is the strain energy.

Nucleation in metamorphic rocks is probably always heterogeneous. Grain boundaries, dislocations, and strained areas provide favorable nucleation sites. The formation of a nucleus at a boundary or dislocation involves the destruction of part of an existing surface, thereby lowering the free energy of the system. Nucleation at grain corners and triple points is also energetically favored.

An interesting example of solid-state nucleation is provided by exsolution, which commonly occurs in pyroxenes and feldspars. The upper part of Figure 10.8 shows the

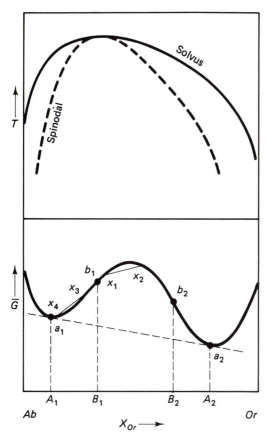

Figure 10.8 (a) T-X diagram of the subsolidus portion of the alkali feldspar system showing locations of the solvus and spinodal, after Waldbaum and Thompson (1969). (b) Schematic $\overline{G}$-X diagram for the alkali feldspar system at some temperature below the crest of the solvus. Tangent points a_1 and a_2 define the locations of solvus limbs, and inflection points b_1 and b_2 define spinodal limbs at this temperature.

experimentally determined solvus in the binary system $NaAlSi_3O_8$-$KAlSi_3O_8$ (Wald-baum and Thompson, 1969). In Chapter 9, we showed that subsolidus cooling of a homogeneous feldspar below the solvus will result in its separation into two phases with compositions defined by the limbs of the solvus. The exsolved phases nucleate, homogeneously or heterogeneously, within the solid state. The lower part of Figure 10.8 shows a schematic $\overline{G}$-X diagram for some temperature below the crest of the alkali feldspar solvus. Points a_1 and a_2 define the locations of the limbs of the solvus at this particular temperature, corresponding to phase compositions A_1 and A_2.

Another style of nucleation, which may be kinetically favored, results in the formation of exsolution zones with compositions that are metastable relative to A_1 and A_2. To show how this occurs, we have illustrated another curve, called the *spinodal*, in the upper part of Figure 10.8. Within the limits of the solvus, the free energy curve at the fixed temperature in the lower diagram has two inflections ($\partial^2 G/\partial X^2 = 0$), labelled b_1 and b_2, which define the positions of the spinodal arms. If any feldspar having an overall composition between B_1 and B_2 contains small local fluctuations in composition (as, for example, x_1 and x_2), its free energy will be less than that for a homogeneous feldspar. This situation also results in unmixing, but in this case the process is called *spinodal decomposition*. Within the limits of the spinodal (b_1 to b_2), coexisting compositions like x_1 and x_2 are stable relative to a homogeneous phase. Outside the spinodal a homogeneous phase is stable with respect to exsolved phases like x_3 and x_4. The energy difference between the homogeneous and exsolved phases is very small in either case, but only inside the spinodal is unmixing favored. Because the compositional difference between x_1 and x_2 is small, the structural mismatch between them is small and the strain energy is minimal. Therefore, spinodal decomposition involves no major nucleation hurdle and the interface between the unmixed phases is diffuse. In this way, exsolution lamellae with compositions between A_1 and A_2 can form metastably. Feldspars with bulk compositions that lie between the solvus and the spinodal (that is, between A_1 and B_1 or A_2 and B_2) can unmix only by solvus-controlled exsolution. In this case, the unmixed compositions are relatively far apart, requiring the accomodation of significant structural mismatch, so there is a high kinetic barrier to nucleation. For this reason, nucleation may be sluggish and exsolution phase boundaries are likely to be sharp.

The kinetics of exsolution processes can be rather complex, and it is convenient to summarize such behavior with the aid of a time-temperature-transformation (sometimes called *TTT*) plot, as illustrated in Figure 10.9. These plots provide a synthesis of kinetic data obtained from experiments carried out at different temperatures and cooling rates. The two sets of curves in Figure 10.9 correspond approximately to the beginning (1% product) and completion (99% product) of exsolution. Completion of exsolution occurs when equilibrium is reached.

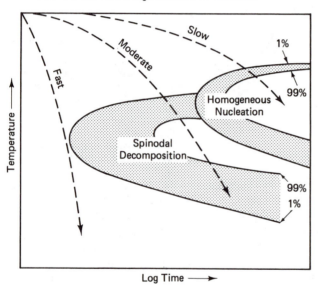

Figure 10.9 Time-temperature-transformation (TTT) plot illustrating how various cooling rates can control the degree of homogeneous nucleation and spinodal decomposition. Spinodal decomposition occurs more rapidly because nucleation is not required.

The behavior of a mineral under various thermal conditions can be determined using cooling curves such as those illustrated. For example, at slow cooling rates, homogeneous nucleation is favored over spinodal decomposition, whereas the reverse is true for moderate cooling rates. During fast cooling, even spinodal decomposition will be incomplete, and quenched samples show no exsolution at all. We may guess that heterogeneous nucleation will occur at higher temperature and at an earlier time than homogeneous nucleation, but it is not portrayed on a *TTT* plot because heterogeneous mechanisms are not generally experimentally reproducible.

GROWTH

Crystal growth is a complex activity involving a number of processes: (1) chemical reaction at the interface between the growing crystal and its surroundings; (2) diffusion of components to and from the interface; (3) removal of any latent heat of crystallization generated at the interface; and (4) flow of the surroundings to make room for the growing crystal. For most silicate minerals, only processes (1) and (2) are thought to be significant factors. These lead, then, to two end-member situations. When diffusion of components in the surroundings is fast relative to their attachment to the crystal nucleus, the rate of growth is controlled by the interface reaction. Conversely, when attachment of components to the crystal nucleus is faster than transport of components to it, the rate of growth is controlled by diffusion. For the special case in which crystals and surroundings have the same composition, no diffusion is necessary and growth is interface-controlled, though removal of latent heat of crystallization may be important in this situation. Buildup of latent heat may also be important where two growing crystals converge. Interface-controlled growth may dominate in igneous systems and diffusion-controlled growth is likely to be more important in metamorphic systems, but intermediate situations in which both processes play a role are also common in both environments.

In diffusion-controlled growth, migration of components to and from the area immediately surrounding the growing crystal cannot keep pace with their uptake or rejection at the interface, so that concentration gradients are formed. This situation is illustrated in Figure 10.10. The formation of these gradients obviously slows down the rate of growth. Fluid advection, if it occurs, tends to destroy the gradients and causes growth rate to increase. In interface-controlled growth, concentrations of components in the zone around the growing crystal are similar to those of the bulk surroundings, because the limitation on growth is not movement through the surroundings. As a consequence, growth rate under these conditions is unaffected by fluid movement.

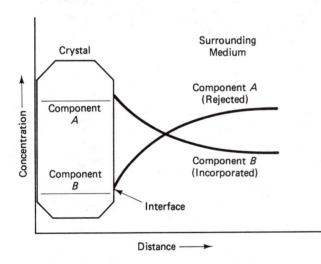

Figure 10.10 During diffusion-controlled growth, compositional gradients are established in the medium surrounding the growing crystal. Components rejected by the crystal cannot diffuse away fast enough, and incorporated components cannot diffuse to the interface fast enough, so gradients result.

Interface-Controlled Growth

The rate at which interface-controlled growth takes place is the difference between the rates at which atoms are attached to and detached from the crystal. The rate of attachment, r_a, can be expressed by a probability relationship similar in form to other equations we have already seen, namely:

$$r_a = \nu \exp\left(-\Delta G^*/RT\right),$$

where ν is the frequency at which the atoms vibrate and ΔG^* is the activation energy for attachment. Similarly, the rate of detachment, r_d, is given by

Growth **311**

$$r_d = \nu \exp\left[-(\Delta G' + \Delta G^*)/RT\right],$$

where $\Delta G'$ is the thermodynamic driving force. The growth rate, Y, is given by $r_a - r_d$ times the thickness per layer, a_o, and the fraction of surface sites to which atoms may successfully attach, f:

$$Y = a_o f \nu \exp(-\Delta G^*/RT)[1 - \exp(-\Delta G'/RT)]. \qquad (10\text{-}15)$$

Equation 10-15 exhibits the same qualitative variation with undercooling as that for nucleation rate, equation 10-11. At zero undercooling, the rate is zero because $\Delta G'$ is zero. As the system cools below the equilibrium temperature, growth rate increases because $\Delta G'$ increases. However, at large undercoolings the growth rate begins to decrease because ΔG^* becomes significant, reflecting the fact that atoms are less mobile as fluid viscosity increases. Therefore, with increasing undercooling, the growth rate first increases, passes through a maximum, and finally decreases, just as nucleation rate does.

Several mechanisms for interface-controlled growth, differing in the surface sites available for growth (f in equation 10-15), have been recognized. In the *continuous* model, the interface is rough and atoms can attach virtually anywhere. Consequently, f equals 1 for all undercoolings and growth is relatively fast. In *layer-spreading* models, the interface is flat except at steps, which provide the only locations where atoms can attach. Thus, f has some value less than 1. Two commonly observed layer-spreading mechanisms are illustrated in Figure 10.11. *Surface nucleation* forms one-atom-thick layers that spread laterally over the interface by adding atoms at the edges. If *screw dislocations* occur on the interface, atoms are attached to form spiral-shaped steps and renucleation is not required for each layer. One way to discriminate between these growth mechanisms is illustrated in Worked Problem 10-4.

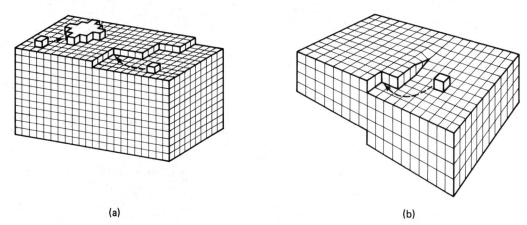

(a)   (b)

Figure 10.11 Diagramatic representations of layer-spreading mechanisms for crystal growth. (a) In surface nucleation, atoms are added only at the edges of a spreading layer of atoms. (b) In screw dislocations, atomsattached to the crystal surface form spiral stairs. (After A. C. Lasaga and R. J. Kirkpatrick, *Kinetics of geochemical processes.* Copyright 1981 by the Mineralogical Society of America. Reprinted with permission.)

Worked Problem 10-4

Kirkpatrick et al. (1976) made motion pictures to measure the growth rate of diopside from diopside melt, performing experiments in which a seed crystal was introduced into melt at various undercoolings. Using their data, presented in Figure 10.12a, can we deduce the growth mechanism for diopside?

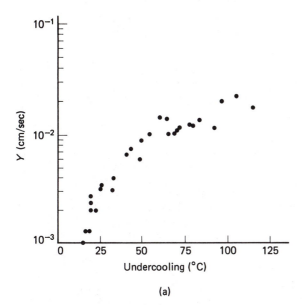

(a)

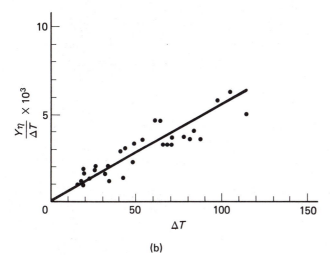

(b)

Figure 10.12 Experimental data on the growth of diopside from melt of the same composition. (a) Relationship between growth rate Y and undercooling T. (b) Calculated $Y\eta/\Delta T$ versus undercooling for data in (a), using viscosity measurements of Kirkpatrick (1974). The linear trend with positive slope indicates that a screw dislocation growth mechanism is dominant. (After R. J. Kirkpatrick, G. R. Robinson, and J. F. Hays, Kinetics of crystal growth from silicate melts: anorthite and diopside, *J. Geophys. Res.* 81:5715–20. Copyright 1976 by American Geophysical Union. Reprinted with permission.)

Because diopside is growing from a melt of its own composition, we know that growth rate will not be diffusion-controlled. The interface-controlled growth rate expression is difficult to solve in the form of equation 10-15 because we do not have appropriate values for ΔG^* and $\Delta G'$. There is a theoretical justification, known as the Stokes-Einstein relation, however, which allows us to replace the exponential terms in equation 10-15 with a viscosity (η) term. We will not produce the mathematical derivation for this step, because it is rather complex, but it should be intuitively reasonable to you. For any fixed degree of undercooling, the growth of crystals from a low-viscosity melt (low ΔG^* and $\Delta G'$) should be more rapid than from a high viscosity melt. This allows us to define a term $Y\eta/T$ that is, like f, a measure of available attachment sites on the crystal surface. A plot of $Y\eta/T$ against undercooling can be used to determine the growth mechanism (Kirkpatrick, 1975). This yields a horizontal line for continuous growth, a straight line with positive slope for growth by a screw dislocation mechanism, and a curve with positive curvature for growth by a surface nucleation mechanism.

The first step is to determine the parameter $Y\eta/\Delta T$ for each data point. Kirkpatrick (1974) tabulated viscosity data for liquid diopside at various degrees of undercooling. For example, at an undercooling of 50°, η equals 19 Poise (g cm^{-1} sec^{-1}). Substituting this value and the measured growth rate of 9×10^{-3} cm sec^{-1} at this undercooling, $Y\eta/\Delta T$ equals 3.4×10^{-3} g sec^{-2} deg^{-1}. A plot of similarly calculated values for all the data points in Figure 10.12a against ΔT is presented in Figure 10.12b. These data define a nearly straight line with positive slope, so diopside must grow by a screw dislocation mechanism.

Worked Problem 10-5

Crystallization under plutonic conditions may occur at a very small degree of undercooling relative to volcanic rocks. How long will it take to grow a one-centimeter diopside crystal at an undercooling of only 1° C?

The only growth rate data on diopside are those already presented in Worked Problem 10-3. It is not possible to measure the growth rate at small degrees of undercooling accurately, but this can be extrapolated from experimental data at larger undercoolings. The straight line passing through the point 0,0 in Figure 10-12b allows us to estimate Y at $\Delta T = 1°$. The equation for this line is

$$Y\eta/\Delta T = 6.6 \times 10^{-5}T.$$

Therefore, at $\Delta T = 1°$, $Y\eta = 6.6 \times 10^{-5}$ g sec^{-2}. Because $\eta = 15$ Poise (g cm^{-1} sec^{-1}) at 1° undercooling (Kirkpatrick, 1974), Y is on the order of 10^{-6} cm sec^{-1}. At this rate, it would take 10^6 sec, about 10 days, to grow a one-centimeter crystal.

For a number of reasons, this calculated rate must be higher than the growth of diopside in a real plutonic system. The experimental temperature (1390° C) is considerably higher than the liquidus temperatures of most magmas, and the viscosity of the experimental system is lower. Also, diffusion might play a role in a real, multicomponent magma.

Diffusion-Controlled Growth

Formulation of an equation for the rate of diffusion-controlled growth must start with either Fick's first or second law, depending on whether a steady state is reached or not. The growth rate is controlled by how rapidly atoms can diffuse to the interface. Consequently, growth rate decreases as time passes, because compositional gradients like those in Figure 10.10 extend further into the surroundings with time. The expressions for growth rate under these conditions involve combining a diffusion equation with a mass balance equation because the flux of atoms reaching the interface by diffusion must equal the atoms attached to the interface. Derivation of such expressions is rather complex and will not be attempted here. The equations for growth rate (Y) under diffusion-controlled conditions take the general form

$$Y = k\left(\frac{D}{t}\right)^{1/2}, \tag{10-16}$$

where k is a constant involving concentration terms, D is the diffusion coefficient of the rate-limiting species, and t is time.

The compositional or thermal gradients produced by diffusion can affect growth by causing crystal faces to break down. Cahn (1967) found that faceted crystals become unstable when their radii become smaller than D/Y. This ratio corresponds to the onset of diffusion-controlled growth. When a faceted crystal becomes unstable, protuberances begin to develop. Heat and components rejected during crystal growth flow away from these protuberances, hindering the development of additional structures in their vicinity. This can produce delicate skeletal, and dendritic morphologies such as those in Figure 10.13.

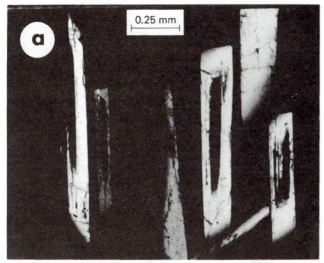

Skeletal

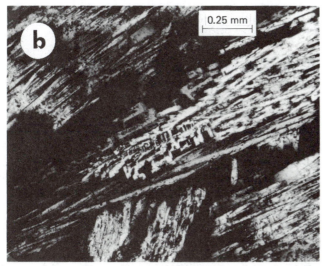

Dendritic

Figure 10.13 Photomicrographs of (a) skeletal and (b) dendritic plagioclase crystals formed in rapidly cooled experimental melts. Sluggish diffusion at large undercoolings causes breakdown of planar crystal surfaces. The fields of view for these pictures are approximately 1 mm in width. (After G. Lofgren, An experimental study of plagioclase morphology, *Amer. J. Sci.* 273:243–75. Copyright 1974 by *Amer. J. Sci.* Reprinted with permission.)

With increasing undercooling, D decreases and Y increases, so the ratio D/Y decreases. Lofgren (1974) illustrated by experiment how increasing undercooling, that is, decreasing D/Y, controls crystal morphologies. He studied the shapes of plagioclase crystals growing at 5 kbar in plagioclase melts at various degrees of undercooling, as shown in Figure 10.14. At nearly constant temperature intervals below the liquidus, various plagioclase compositions crystallized to form tabular, skeletal, dendritic, and spherulitic morphologies. The onset of diffusion-controlled growth in this system, that is, the point at which faceted crystals become unstable, is defined approximately by the boundary separating tabular and skeletal crystals in Figure 10.14. The periodicity of protuberances increases as D/Y decreases. The skeletal morphologies are instabilities with long wavelengths (in the sense of spacings between protuberances). With increas-

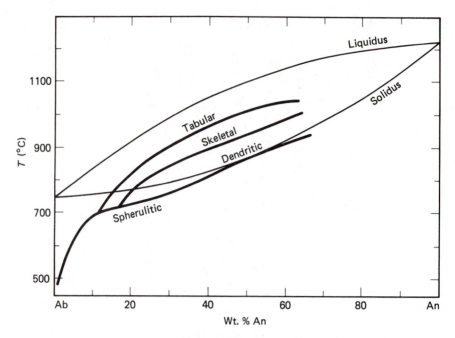

Figure 10.14 *T-X* diagram for the system albite-anorthite, illustrating how plagio-clase crystal morphologies change with undercooling. Skeletal and dendritic mor-phologies are illustrated in Figure 10.13.

ing undercooling, these give way to finer dendritic structures with shorter wavelengths. Similar morphological sequences are observed in olivine crystals in the interior and ex-terior parts of lava pillows.

KINETIC PROCESSES IN ROCKS: MESSY SYSTEMS

The characteristics of many real geologic systems reflect the interplay of several kinetic processes acting concurrently. We will now examine some examples of this complex behavior.

Textures of Igneous and Metamorphic Rocks

The texture of an igneous or a metamorphic rock depends on the relative sizes and shapes of crystals and their relationships to adjacent crystals, and is generally a product of both nucleation and growth processes. We have just seen how kinetic experiments on plagioclase crystallization permit us to estimate undercooling. Similar observations of more complex petrologic systems can also help us evaluate the role of kinetics in producing textures in rocks.

Consider a porphyritic volcanic rock containing euhedral olivine phenocrysts in a fine-grained groundmass. According to one traditional interpretation, we might infer that the phenocrysts crystallized relatively slowly at depth and were then carried up-ward by the ascending magma, and the remaining liquid was rapidly quenched. This two-stage kinetic model might be the correct interpretation in some cases. However, experimental studies of lavas have demonstrated that olivine nuclei grow faster than those of other phases. On the other hand, olivine nucleates less readily, so its growth proceeds from a small number of sites. Thus, a single-stage cooling history can produce porphyritic textures. A similar argument involving relative nucleation and growth rates may account for the development of large porphyroblasts in metamorphic rocks.

The maximum rates of nucleation and growth for a particular phase are likely to occur at different degrees of undercooling, as illustrated schematically in Figure 10.15. In the example shown, the maximum rate of nucleation occurs at a higher degree of undercooling than the maximum rate of growth. This is normally the case, but the positions of these curves could be reversed for some phases. Ophitic textures, in which small grains of plagioclase are enclosed by large, single crystals of pyroxene, may result because the nucleation of plagioclase was faster than its growth, whereas growth of pyroxene was faster than its nucleation at the same temperature.

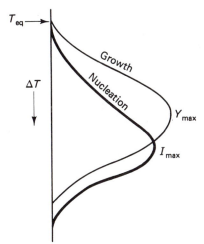

Figure 10.15 Schematic curves of nucleation and growth rates for a certain phase against undercooling ΔT.

Many metamorphic rocks have equigranular textures. The tendency for crystals of the same phase but of varying sizes to approach a common grain size during annealing arises from the necessity that the chemical potential of each component must be the same in all crystals for equilibrium. Because the contribution of surface energy to the net chemical potential varies with crystal size, the size difference must be compensated by a very slight variation in temperature between crystals. When a single temperature is imposed on the system, diffusion promotes the growth of larger crystals at the expense of smaller ones, a process known as *Ostwald ripening*.

Metamorphic rocks with preferred orientations of mineral grains form by crystallization during the application of stress. Nuclei of anisotropic crystals in a stress field presumably differ in stability according to their orientations. If this is true, nuclei in favorable orientations have lower free energy. These oriented nuclei would have smaller critical radii and would thus form and grow perferentially.

Deformation and Extent of Reaction

We have already discussed the fact that nucleation is inhibited in solid-solid reactions that involve volume changes. As a particular solid-state transformation progresses, we might expect this "room problem" to get worse and the reaction rate to diminish with time. As one example, Kunzler and Goodall (1970) calculated that volume changes induced by the reaction aragonite $\rightarrow$ calcite should inhibit the transformation after 19% reaction at 400° C, in qualitative agreement with kinetic experiments in this system (Davis and Adams, 1965). Their calculations assume that all strain is accomodated by elastic behavior in the host phase. In real geologic systems, though, strain may also be taken up by nonelastic deformation mechanisms, such as twinning, that could alleviate the room problem.

Deformation can enhance the kinetics of reactions in a variety of ways:

1. Volume diffusion is enhanced by increase in the density of point defects. Nucleation on crystal defects is also energetically more favorable.

2. Reduction in grain size by cataclasis or recrystallization increases the surface area available for reaction, and therefore increases reaction rate.

3. Stress gradients may increase chemical potential gradients, and thereby increase diffusion rates of certain components.

4. Deformation generally results in increased permeability to fluids. These fluids may act to transport dissolved material or may serve as catalysts for reactions.

The common observation that metamorphic reactions have taken place primarily along zones of deformation must indicate a kinetic effect. Experimental data confirm the effect of deformation on reaction rates. Figure 10.16 shows the extent of reaction for the aragonite → calcite transformation at 450° C and 1 bar as a function of time. The circles defining curve A represent runs of various duration, all collected after rapidly raising the temperature of the system to 450° C at 1 bar. The squares defining curve B are data obtained using a pressure vessel in which temperature was slowly increased to 450° C. Pressure in the vessel was initially increased to 10 kbar and then decreased rapidly to 1 bar to initiate the reaction. This pressure cycling resulted in high shear stresses in experiment B, inducing point and extended defects which, in turn, increased reaction rates. This experiment illustrates clearly how deformation enhances rates of reaction.

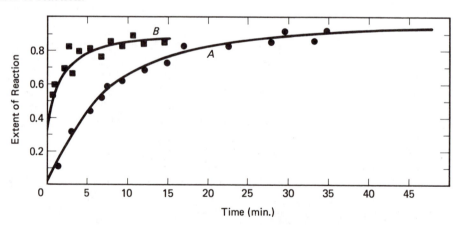

Figure 10.16 Two sets of kinetic data for the aragonite → calcite transformation at 450° C and 1 bar (Davis and Adams 1965). Data used to construct curve B were collected using strained materials, which accounts for the higher reaction rate.

Igneous Layering and Metasomatic Banding

Many petrologic processes involve coupled compositional diffusion and chemical reactions, some of which result in layering, or banding. For example, McBirney and Noyes (1979) attribute repetitious, inch-scale layering in the Stillwater igneous complex, Montana, shown in Figure 10.17, to this phenomenon. They envision that diffusion

Figure 10.17 Photograph of inch-scale layering in the Stillwater complex, Montana. Layering consists of a series of doublets of thin pyroxene layers in a host of anorthosite.

profiles are induced in the magma by thermal gradients. As diffusion profiles for various components intermingle, supersaturation occurs in certain critical locations. The resulting crystallization impoverishes the adjacent regions in the extracted components, and a repetitive pattern of mineral layers is produced. Orbicular structures of concentric bands that occur in some granitic rocks are probably a similar phenomenon.

Reaction zones between incompatible minerals or assemblages in metamorphic rocks also form by the interaction of diffusion and reaction. Thompson (1975) described the following zoning sequence of calc-silicate minerals produced by metasomatic reactions between limestone and shale:

marble | garnet | diopside + clinozoisite | amphibole | pelitic schist.

Metasomatic zones are also common in the blackwall at the borders of some ultramafic bodies. One common progression (Phillips and Hess, 1936) is

serpentinite | talc + magnesite | talc | actinolite | biotite | pelitic schist.

Zones like these may result largely from diffusion of one component. In Thompson's example, the calc-silicate bands are due primarily to CaO mobility. In other cases, several components may migrate at the same time. The sequence in the second example above requires a wider assortment of diffusing components, including SiO_2, MgO, and FeO. On what basis can we identify the diffusing components necessary to explain a particular reaction zone? Graphical analysis of system components and phases can provide such information, as illustrated in Worked Problem 10-6.

Worked Problem 10-6

The compositions of stable phases A through H in Figure 10.18a are shown in terms of components 1, 2, and 3; the tie-lines connect phases that are compatible at this fixed temperature. If two incompatible rocks with bulk compositions of X and Y are in contact and only component 3 is permitted to diffuse between them, what will be the sequence of zones produced?

We will follow the solution that Brady (1977) provided for this problem. From Figure 10.18a we see that rock X consists of phases $C + F$, and rock Y consists of phases $B + D$. Diffusion of component 3 must occur in response to a gradient in its chemical potential, and movement of 3 over time will tend to reduce these gradients until μ_3 varies monatonically along the diffusion path. Brady constructed Figure 10.18b (using principles similar to those we used in constructing the chemical potential diagram in Figure 9.16) to illustrate how stable phase assemblages vary with μ_3. This diagram consists predominantly of two-phase fields separated by three-phase horizontal lines. Vertical lines and points in

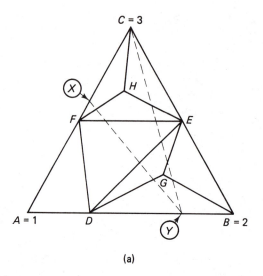

(a)

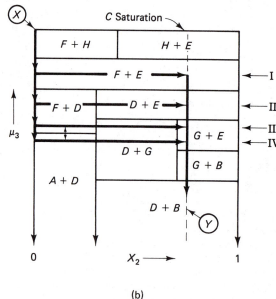

(b)

Figure 10.18 (a) A hypothetical system consisting of three components (1 through 3) and eight possible stable phases (*A* through *H*). Tie-lines connect compatible phases at one particular temperature and pressure. (b) A μ-*X* diagram for the system shown in (a) that illustrates the sequences of phases produced at various values of X_2 along a chemical potential μ_3 gradient. If component 3 is the only diffusing component, diffusion does not alter X_2. Arrows labelled I through IV on the righthand side represent the locations of original boundaries between rock *X* and *Y* in terms of μ_3 that correspond to sequences I through IV in the text. (After J. B. Brady, Metasomatic zones in metamorphic rocks, *Geochim. Cosmochim. Acta* 41: 113–25. Copyright 1977 by Pergamon Journals, Inc. Reprinted with permission.)

this diagram represent only one phase. The locations of rock compositions *X* and *Y* are also shown.

Possible sequences formed by diffusion of component 3 between *X* and *Y* are determined by moving down vertically from *X* and up vertically from *Y* in this diagram. The initial boundary between rocks *X* and *Y* will be marked by a discontinuity in the proportions of components 1 and 2, requiring an instantaneous shift (marked by a horizontal arrow) from one vertical progression to the other. The metasomatic zone grows in both directions from the original bedding plane between rock *X* and *Y*.

If component 3 diffuses rapidly, we expect most of the growth to take place in rock *Y*. We begin by examining the assemblage $F + C$ (rock *X*). Diffusion of component 3 will decrease μ_3 and cause phase *C* to disappear, leaving *F* only. If the value of μ_3 at the initial boundary lies within the $F + E$ field, at this point there is a discontinuous shift to the vertical line corresponding to rock *Y*, producing a zone of $F + E$ at the position of the initial interface. Moving downward into rock *Y*, we then produce in succession $D + E$, $G + E$, and $G + B$ before reaching $D + B$ (the pristine rock Y). We can represent the sequence in this way:

$$F + C \,|\, F \; \vdots \; F + E \,|\, D + E \,|\, G + E \,|\, G + B \,|\, D + B. \qquad \text{(sequence I)}$$

The position of the initial boundary between rock X and Y is given by the dashed vertical line. Other possible sequences, based on the initial boundary being located at lower values of μ_3, include

$$F + C \,|\, F \; \vdots \; D + E \,|\, G + E \,|\, G + B \,|\, D + B \qquad \text{(sequence II)}$$

$$F + C \,|\, F \; \vdots \; G + E \,|\, G + B \,|\, D + B \qquad \text{(sequence III)}$$

$$F + C \,|\, F \,|\, A \; \vdots \; G + E \,|\, G + B \,|\, D + B \qquad \text{(sequence IV)}$$

If component 3 diffuses slowly (relative to components 1 and 2), most of the growth occurs in rock X and we obtain sequence IV rather than I. Sequences II and III are intermediate cases. The differences between final assemblages will be the result of differences in the mobility of components 1, 2, and 3. Thermodynamics only defines how Figure 10.18b looks; kinetics determines the reaction sequence.

None of these zones have more than two phases. This is because component 3 cannot vary independently of the other components. Without further information, it is not possible to predict which of these four sequences would actually occur. Such a unique solution requires data on the relative rates at which each possible zone widens.

Quantifying Complex Kinetic Processes

Kinetic control of geologic processes may depend on the rate of diffusion, nucleation, or growth. A number of real geologic systems can be modelled using *combinations* of the rate equations for these and other kinetic processes. Rubie and Thompson (1985) provide a comprehensive summary of overall rate equations for different combinations of nucleation with various growth mechanisms. Worked Problem 10-7 illustrates the principle of handling one very complex system.

Worked Problem 10-7

What kinetic equations are appropriate for modelling the crystallization of lava? This problem was addressed by Kirkpatrick (1976), who devised the solution we will now describe.

Heat loss is obviously an important kinetic factor in a solidifying lava. If thermal conduction is the dominant heat transport mechanism, the following heat flow equation analogous to Fick's second law describes heat loss from this body of magma:

$$\partial T / \partial t = k \, \partial^2 T / \partial x^2.$$

In this equation, k is the thermal diffusivity and x is position. Because crystallization of the lava releases latent heat, the heat flow equation also requires an extra term describing the production rate for latent heat of crystallization. Therefore,

$$\partial T / \partial t = k \, \partial^2 T / \partial x^2 + (1/\rho C_p) \, dQ/dt, \qquad (10\text{-}17)$$

where ρ is the density of the magma (crystals plus melt), C_p is the heat capacity, and dQ/dt is the rate at which latent heat is generated per unit volume. The term dQ/dt can also be expressed in terms of the rate of change of the volume of crystals per unit volume dV/dt by

$$dQ/dt = dV/dt \, L_v,$$

where L_v is the latent heat of crystallization per unit volume.

The rate of crystallization is also obviously related in some way to the rates of nucleation and growth. Expressions for the rates of nucleation I and growth Y have already been given in equations 10-11 and 10-15, respectively. If we make the simplifying assumption that grains are spherical, dV/dt can be expressed as

$$dV/dt \mid_{t} = \frac{4\pi}{3} U_{t} I Y^{3} t^{3},\qquad(10\text{-}18)$$

where U_{t} is the volume fraction of melt remaining uncrystallized at time t. By combining rate equations 10-17 and 10-18, we can calculate temperature, rate of crystallization, fraction crystallized, nucleation rate, and growth rate as functions of time and position within a lava body.

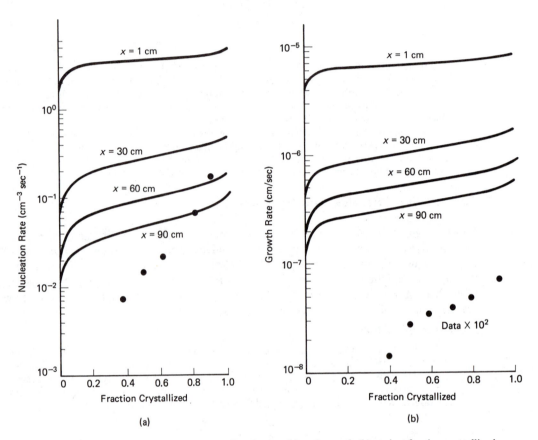

Figure 10.19 Calculated rates of nucleation (a) and growth (b) against fraction crystallized for a hypothetical dike. x corresponds to various distances from the cooling surface. Data points are nucleation and growth rates for plagioclase in the Makaopuhi lava lake, Hawaii. (Modified from Kirkpatrick 1976.)

Kirkpatrick substituted plausible values for some of these variables to calculate the rates of nucleation and growth as functions of the degree of crystallization, shown in Figures 10.19a and 10.19b. These data apply to a cooling dike 2 meters in thickness, and x in these figures indicates distance from the dike margin. They are not calculated specifically for a lava flow or lake, but they may illustrate conditions in extrusive magma bodies. Also shown as dots in Figures 10.19a and 10.19b are actual nucleation and growth rate data for plagioclase in a lava lake at Kilauea volcano, Hawaii, one of the few such data sets available. These results are qualitatively similar to the cooling rate calculations, in that nucleation and growth rates increase monatonically with fraction crystallized and there is an inflection point at about 50% crystallization.

Controlled Cooling-Rate Experiments

The challenge of unraveling kinetic controls on complex geologic systems may be more amenable to experiment than to calculation. Early in this chapter, we summarized some experimental results on the rate of dehydration of talc (Greenwood, 1963). There are very few kinetic data for metamorphic systems such as this. Experiments in which magma compositions are cooled at predetermined rates are more common.

Many of these experiments have been performed using lunar basalts, because oxygen fugacities for these anhydrous compositions are easier to control. One example is a programmed cooling-rate study of lunar rock 12002 by Walker et al. (1976), the results of which are summarized in Figure 10.20. Experimental charges were cooled at different rates and quenched from different temperatures within the crystallization range to construct the liquidus lines shown in the diagram. Liquidus temperatures from equilibrium experiments are illustrated by arrows on the left side of the figure.

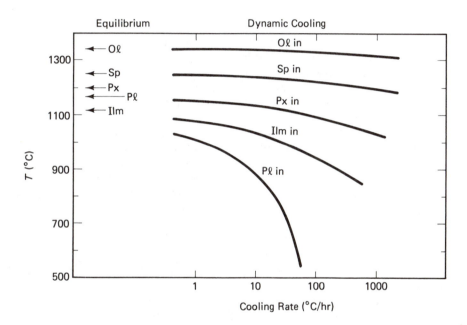

Figure 10.20 Results of controlled cooling rate experiments on lunar basalt 12002 by Walker et al. (1976). Equilibrium liquidi for various phases are shown by arrows on the left-hand side. The curves are contours of the first appearance of various phases at different cooling rates.

Several interesting features emerge from these data. First, the crystallization sequence is different under equilibrium and dynamic cooling conditions; specifically, the order of appearance of plagioclase and ilmenite reverses. We infer from the texture of the rock itself that ilmenite began to form before plagioclase, suggesting that the controlled cooling-rate experiments may be more applicable than equilibrium crystallization experiments to understanding the cooling history of this basalt. A second feature seen in Figure 10.20 is that all phases begin crystallizing below their liquidus temperatures in the programmed cooling experiments, and the amount of temperature suppression increases with cooling rate. Another conclusion, not apparent from this figure, is that the compositions of pyroxenes, in terms of both major and minor elements, are distinct in the experiments involving equilibrium and controlled cooling rate. Based on order of phase appearance, texture, and mineral compositions, Walker and his collaborators estimated that the actual cooling rate for lunar rock 12002 decreased from about 1° C per hour when olivine first crystallized to about 0.2° C per hour during crystallization of pyroxene. Decreasing cooling rate is consistent with its location near the margin of a lava flow.

Programmed cooling experiments can be quite sophisticated. Lofgren (1980) gives a thorough review of the extensive literature of studies that employ this technique. These experiments provide empirical methods by which to assess the mineralogical and textural features that characterize rapidly cooling igneous rocks and to quantify the physical conditions that produced them.

SUMMARY

In this chapter, we have seen that kinetic factors can be very important in understanding some geologic systems, even at high temperatures. The temperature dependence of rate processes is commonly expressed in terms of an activation energy barrier that must be overcome in order to initiate the process. This is reflected in supersaturation and undercooling, both ways of overstepping equilibrium crystallization conditions.

Three rate processes are particularly important in high-temperature systems. The first of these, diffusion, occurs in response to a gradient in chemical potential. Migration of components may occur along grain boundaries or through phases. Volume diffusion uses naturally occurring defects in crystal structures as pathways for movement. Diffusion rates can be calculated from Fick's first and second laws.

Nucleation, the second major rate-controlling process, is inhibited by surface free energy; this extra energy requirement is eliminated after the nucleus reaches a certain critical size. Nucleation may occur by homogenous or heterogeneous mechanisms. This process is very difficult to study experimentally, and theoretical models provide only qualitative agreement with experimental data. Exsolution requires nucleation and thus is rate-sensitive, but unmixing by spinodal decomposition occurs without nucleation of a new phase.

Growth of crystals, the third process, may be controlled by the rate of diffusion of components to the interface or their rate of attachment to the new phase. Interface-controlled growth may occur by continuous, surface nucleation, or screw dislocation mechanisms, each of which is characterized by a different relationship between growth rate and undercooling. The rate of diffusion-controlled growth slows with time because of chemical gradients that form in the medium surrounding the growing crystal. Compositional or thermal gradients near the growing interface also cause planar crystal surfaces to break down into skeletal or dendritic morphologies.

Many of the properties of real rocks, such as textures, some kinds of igneous layering, and metasomatic banding, can be explained by understanding the interplay between various rate processes. The relationships between deformation and kinetic processes are also important in real rocks. Complex geologic process may be modelled numerically by combining rate laws for diffusion, nucleation, growth, and other kinetic processes such as thermal conduction. Experiments under controlled cooling conditions are also useful in understanding kinetic behavior in complex systems.

SUGGESTED READINGS

*The literature on geochemical kinetics is not as volumi-
nous as that for equilibrium thermodynamics, and most
references are portions of books on other subjects.*

CARMICHAEL, I. S. E.; TURNER, F. J; and VERHOOGEN, J.
1974. *Igneous petrology.* New York: McGraw-Hill.
(Chapter 4 provides a discussion of kinetic factors applicable to magmatic systems.)

HENDERSON, P. 1982. *Inorganic geochemistry.* Oxford:
Pergamon. (Chapter 8 provides an easily understand-
able summary of kinetically controlled geochemical
processes.)

HOFFMANN, A. W.; GILLETTI, B. J.; YODER, H. S. JR.; and
YUND, R.A., eds. 1974. *Geochemical transport and
kinetics.* Washington, D.C.: Carnegie Inst. (A collection of rather technical papers relating to diffusion.)

KIRKPATRICK, J. R. 1975. Crystal growth from the melt:
A review. *Amer. Mineral.* 60:798–814. (An excellent
review paper on the kinetics of crystal growth in silicate melts.)

LASAGA, A. C., and KIRKPATRICK, R. J., eds. 1981. *Kinetics of geochemical processes.* Reviews in mineralogy 8. Washington, D.C.: Mineral. Soc. of Amer.,
(The best available reference on geochemical kinetics;
contains excellent chapters on rate laws, dynamic treatment of geochemical cycles, as well as applications to
metamorphic and igneous rocks.)

LOFGREN, G. E. 1980. Experimental studies of the dynamic crystallization of silicate melts. In *Physics of
magmatic processes,* ed. Hargraves, R. B., 487–551.
Princeton: Princeton Univ. Press. (A comprehensive
review of controlled cooling rate experimentation.)

SPRY, A. 1969. *Metamorphic textures.* Oxford: Pergamon. (This book discusses the development of many
metamorphic textures in terms of kinetic controls.)

THOMPSON, A. B., and RUBIE, D. C., eds. 1985. *Metamorphic reactions, kinetics, textures, and deformation.*
Advances in Physical Chemistry 4. New York:
Springer-Verlag. (This excellent reference includes
chapters on the kinetics of metamorphic reactions and
other chapters about the importance of crystal defects
and deformation on rate processes.)

VERNON, R. H. 1975. *Metamorphic processes: Reactions
and microstructure development.* New York: John Wiley & Sons. (Chapter 3 is a highly readable overview
of the importance of kinetics in metamorphic reactions.)

*The following papers were referenced in this chapter.
The interested student may wish to explore them more
thoroughly.*

BRADY, J. B. 1977. Metasomatic zones in metamorphic
rocks. *Geochim. Cosmochim. Acta* 41:113–25.

BUENING, D. K., and BUSECK, P. R. 1973. Fe-Mg lattice
diffusion in olivine. *J. Geophys. Res.* 78:6852–62.

CAHN, J. W. 1967. On the morphological stability of
growing crystals. In *Crystal growth,* ed. Peiser,
H. S., Oxford: Pergamon.

DAVIS, B. L., and ADAMS, L. H. 1965. Kinetics of the
calcite-aragonite transformation. *J. Geophys. Res.*
70:433–41.

FREER, R. 1981. Diffusion in silicate minerals and
glasses: A data digest and guide to the literature. *Contrib. Mineral. Petrol.* 76:440–454.

GREENWOOD, H. J. 1963. The synthesis and stability of
anthophyllite. *J. Petrol.* 4:317–51.

KIRKPATRICK, R. J. 1974. Kinetics of crystal growth in
the system $CaMgSi_2O_6$-$CaAl_2SiO_6$. *Amer. J. Sci.* 274:
215–42.

———. 1976. Towards a kinetic model for the crystallization of magma bodies. *J. Geophys. Res.* 81:
2565–71.

———, ROBINSON, G. R., and HAYS, J. F. 1976. Kinetics of crystal growth from silicate melts: Anorthite and
diopside. *J. Geophys. Res.* 81: 5715–20.

KUNZLER, R. H., and GOODALL, H. G. 1970. The arago-
nite-calcite transformation: A problem in the kinetics
of a solid-solid reaction. *Amer. J. Sci.* 269:360–91.

LASAGA, A. C. 1981. Transition state theory. In *Kinetics
of geochemical processes,* ed. Lasaga, A. C., and
Kirkpatrick, R. J., 135–69. Reviews in Mineralogy 8.
Washington, D.C.: Mineral. Soc. of Amer.

LOFGREN, G. 1974. An experimental study of plagioclase
morphology. *Amer. J. Sci.* 273: 243–73.

MATHEWS, A. 1980. Influences of kinetics and mechanism in metamorphism: A study of albite crystallization. *Geochim. Cosmochim. Acta* 44: 387–402.

McBIRNEY, A. R., and NOYES, R. M. 1979. Crystallization and layering of the Skaergaard intrusion. *J.
Petrol.* 20:487–554.

MUELLER, R. F., and SAXENA, S. K. 1977. *Chemical
petrology.* New York: Springer-Verlag.

PHILLIPS, A. H., and HESS, H. H. 1936. Metamorphic
differentiation at contacts between serpentinite and
siliceous country rocks. *Amer. Mineral.* 21:333–62.

RUBIE, D. C., and THOMPSON, A. B. 1985. Kinetics of
metamorphic reactions at elevated temperatures and
pressures: An appraisal of available experimental data.
In *Metamorphic reactions. kinetics, textures, and deformation.* Advances in Physical Chemistry 4. New
York: Springer-Verlag.

TAYLOR, L. A.; ONORATO, P. I. K.; and UHLMANN, D. R.
1977. Cooling rate estimations based on kinetic modelling of Fe-Mg diffusion in olivine. *Proc. Lunar Sci.
Conf. 8th,* 1581–92.

———; ONORATO, P. I. K.; UHLMANN, D. R.; and COISH,
R. A. 1978. Subophitic basalts from Mare Crisium:
Cooling rates. In *Mare Crisium: The view from Luna
24,* ed. Merrill, R. B., and Papike, J. J., 473–81. New
York: Pergamon.

THOMPSON, A. B. 1975. Calc-silicate diffusion zones between marble and pelitic schist. *J. Petrol.* 16:314–46.

WALDBAUM, D. R., and THOMPSON, J. B., JR. 1969. Mixing properties of sanidine crystalline solutions, Part IV:

Phase diagrams from equations of state. *Amer. Mineral.* 54:1274–1298.

WALKER, D.; KIRKPATRICK, R. J.; LONGHI, J.; and HAYS,

J. F. 1976. Crystallization history of lunar picritic basalt sample 12002: Phase equilibria and cooling rate studies. *Bull. Geol. Soc. Amer.* 87:646–56.

PROBLEMS

1. Given the following data, determine the activation energy for the reaction analcite + quartz → albite + water in sodium disilicate solution.

k	T (°C)
2.41×10^{-4}	345
2.68×10^{-3}	379
1.84×10^{-2}	404

2. Sketch schematic plots of concentration of element x in the medium surrounding a growing crystal as functions of distance from the interface for the following conditions:
   (a) diffusion-controlled growth,
   (b) interface-controlled growth,
   (c) growth controlled by both mechanisms.

3. Using equations 10-11 and 10-15, construct a plot that illustrates how the rates of nucleation and growth vary with undercooling.

4. (a) Given that $\Delta H = \Delta G + T\Delta S$, show that the surface enthalpy of a liquid droplet is correctly described by $\Delta H = \sigma - T(\partial\sigma/\partial T)_P$.
   (b) The surface tension of water against air at 1 atm has the following values at various temperatures:

T(°C)	20	22	25	28	30
σ (dynes cm^{-1})	72.75	72.44	71.97	71.50	71.18

   What is the surface enthalpy (in cal cm^{-2}) of water at 25° C?

5. Refer to Figure 10.18b. Suppose that the following sequence of mineral zones is observed at the contact between rock X and another rock Z:

$$F + C \mid F \mid A \mathbin{\vdots} D + G \mid D + B.$$

   Indicate by arrows a path in μ_3-X_2 coordinates that is compatible with this sequence. Speculate on the mobility of component 3 in this system relative to components 1 and 2.

6. (a) The volume of the spherical cap shown in Figure 10.7 can be calculated from

$$V = \frac{1}{6}\pi h(3x^2 + h^2).$$

   Using this formula and equations 10-6 and 10-1, derive expressions to calculate the volume (V) and height (h) of the smallest stable nucleus that can be formed on a substrate S, assuming that the quantities σ_{CL}, ΔG_v, and $(\sigma_{LS} - \sigma_{CS})$ can be measured experimentally.
   (b) What is the volume of the smallest stable nucleus of forsterite for which $(\sigma_{LS} - \sigma_{CS}) = 2.378 \times 10^{-5}$ cal cm^{-2} at 2063° C? Use the values for $\sigma (= \sigma_{CL})$ and ΔG_v for forsterite at 2063° C from Worked Problem 10-3. How does this minimal volume compare to the volume of the smallest homogeneously nucleated forsterite particle?

Chapter 11

The Crust and Mantle as a Geochemical System

OVERVIEW

In this chapter we explore how the crust and mantle of the earth can be considered as a geochemical system. First we must estimate the compositions of the crust and mantle, in terms of both chemistry and phase assemblages. We then investigate how geothermal gradients in the earth's interior relate to the partitioning of heat-producing elements. The metamorphism and melting caused by these gradients provide means by which crust and mantle interact. Because these geothermal gradients are also controlled in part by plate tectonics, we review the geochemical aspects of tectonic motions and discuss the generation of basaltic magma by partial melting of mantle rocks under various conditions of confining pressure and fluid pressure.

Ascent of magmas provides one mechanism by which mantle and crust interact. Most magmas experience processes in route to the surface that cause significant chemical changes, and it is necessary to read through these processes if we are to understand the crust-mantle system. Differentiation of magmas by fractional crystallization and liquid immiscibility are considered, and we attempt to quantify their geochemical consequences. The behavior of compatible and incompatible trace elements during melting and differentiation are also examined. We show that trace elements can be used to test geochemical models of magma evolution. We then consider the cycling of volatile elements within the crust-mantle system, paying particular attention to the generation and transport of fluids during metamorphism.

The *crust* is generally defined as the outer shell of the earth, in which P-wave seismic velocity is less than 7.7 km/sec. The lower boundary of the crust is the Mohorovicic discontinuity (the *Moho*), a layer within which seismic velocities increase rapidly or discontinuously from crustal values to mantle values of greater than 7.7 km/sec. The average thickness of the crust defined in this way is 6 km in ocean basins, about 35 km in stable continental regions, and ranges up to 70 km under mountain chains. Even so, the crust is only a miniscule portion of the mass of the earth, constituting only about 0.5 percent of it by weight.

Oceanic crust shows no major lateral and vertical changes in P-wave velocity. Thus, as a first approximation, it can be considered to have basaltic composition throughout. In contrast, the lateral heterogeneity of continental crust, so evident on geologic maps, apparently persists to great depths. In continental regions there are significant areal variations in subsurface P-wave velocities. This heterogeneity, of course, complicates any attempt to estimate the bulk composition of the crust as a whole. Even more worrisome are the vertical changes in seismic velocity below continents, presumably reflecting a different average composition for the lower crust, about which we know very little.

Continental crust is conventionally thought to be subdivided into a "granitic" upper layer and a "gabbroic" lower layer, separated by a seismic discontinuity (the Conrad discontinuity) that is not everywhere continuous. This view is certainly too simplistic and may be wrong altogether. The common mineral assemblage of gabbro, for example, is apparently not stable at the temperatures and pressures of the lower crust. Rocks of this composition should be transformed to eclogite, a mixture of garnet and clinopyroxene. The density of eclogite, however, is too high to fit the measured seismic velocities of the lower crust. If there is enough water in the lower crust, "gabbroic" rocks could instead consist of amphibolites (amphibole + plagioclase rocks), which do have the appropriate seismic properties. Another idea consistent with seismic constraints is that much of the lower crust under continents is composed of granulites of intermediate composition.

There are various ways to approach the thorny problem of determining an average composition for the crust, and all depend on assumptions that we make in describing crustal heterogeneity. We might choose to take the weighted average of the compositions of various crustal rocks in proportion to their occurrence using, for example, geologic maps as a basis. A less direct approach would be to analyze clays derived from glaciers which have sampled large continental areas, or to estimate the proportions of granitic and basaltic rocks by mixing their compositions in the proper ratio to reproduce the compositions of sediments derived by crustal weathering. All of these methods have been tried and generally give similar results. None of them incorporates the contribution from oceanic crust, although taking this into account does not seem to make a drastic difference. The most widely accepted crustal compositions involve averaging the compositions of rocks assigned to hypothetical crustal models, probably because they do take oceanic crust into account.

This last procedure was pioneered by Poldervaart (1955), and is essentially the same approach followed more recently by Ronov and Yaroshevsky (1969), using more refined data. The crustal models employed in both sets of computations are illustrated in Figure 11.1. The differences in the areal distributions and thicknesses of the various petrologic units represent improvements in available geophysical information during

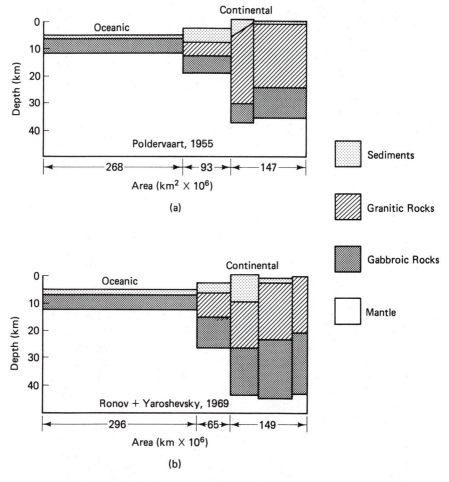

Figure 11.1 Models for the earth's crust used in calculations of its average composition. (a) Scheme of Poldervaart (1955), and (b) model of Ronov and Yaroshevsky (1969). (After P. J. Wyllie, *The dynamic earth*. Copyright 1971 by John Wiley & Sons, Inc. Reprinted with permission.)

the 14-year span between these studies. The compositions of the crust were calculated from these models and from average compositions for the various crustal units. The results are compared in Table 11-1.

These calculated crustal compositions (especially the more refined results of Ronov and Yaroshevsky) are remarkably similar to the composition of average andesite, also shown in Table 11-1. Ronov and Yaroshevsky also determined the average mineral abundances of the crust from the mineral assemblages of the various rock types in their crustal model. These (in volume %) are as follows: quartz (12), alkali feldspar (12), plagioclase (39), pyroxenes (11), micas (5), amphiboles (5), olivine (3), clay minerals (5), carbonates (2), and accessory minerals. These values are also grossly similar to the modal mineral abundances of andesites. Notice that both of the models presented here were based on a "gabbroic" lower crust. If the lower crust under continents has an intermediate composition, as suggested above, then the average crust may be slightly more felsic than either of these models.

TABLE 11-1 ESTIMATES FOR THE AVERAGE COMPOSITION
OF THE EARTH'S CRUST

Wt. % oxides	Poldervaart (1955)	Ronov and Yaroshevsky (1969)	Average andesites[*]
SiO_2	55.2	59.3	58.7
TiO_2	1.6	0.9	0.8
Al_2O_3	15.3	15.9	17.3
Fe_2O_3	2.8	2.5	3.0
FeO	5.8	4.5	4.0
MnO	0.2	0.1	0.1
MgO	5.2	4.0	3.1
CaO	8.8	7.2	7.1
Na_2O	2.9	3.0	3.2
K_2O	1.9	2.4	1.3
P_2O_5	0.3	0.2	0.2

[*]Average of 89 andesites from island arcs compiled by McBirney (1969).

COMPOSITION OF THE MANTLE

The *mantle,* that region of the earth's interior between the Moho and the core, is, of course, not directly accessible for study. Thus, what little we know about it comes only from indirect evidence, mostly geophysical. Seismic wave velocities as a function of depth can be used to calculate elastic parameters such as bulk modulus, shear modulus, and Poisson's ratio. All three parameters depend on the chemical and mineralogic composition of the mantle, as well as temperature and pressure. Comparison of measured seismic data with the results of laboratory measurements of the ultrasonic properties of various phases provides the information necessary for inferring the composition of the upper mantle. For the lower mantle, we must use theoretical relationships between velocity and density, and deduce compositions from shock wave experiments. Despite the inherent uncertainties in interpreting mantle seismic data, the mineralogic composition of the mantle seems reasonably well determined.

P-wave seismic velocities of 7.8–8.2 km/sec just below the Moho are consistent with an upper mantle having the bulk density of peridotite (olivine + orthopyroxene + clinopyroxene) or eclogite (defined above as clinopyroxene + garnet). Although both of these rock types occur in the mantle, peridotite is generally thought to predominate. As we will see, one of the constraints on petrologic modeling is that mantle rocks must be capable of generating basaltic liquids on partial melting. Peridotite can do this, but eclogite cannot. The simple reason is that eclogite already has a basaltic composition, and thus must melt completely in order to produce basaltic magma. Clearly, then, the mantle, consisting mostly of ultramafic rocks, must have a profoundly different bulk composition from the crust.

From a straightforward reading of the paragraph above, we would infer that the Moho is a chemical discontinuity, rather than a phase transition. However, the possibility that the Moho represents a physical discontinuity, that is, a difference in isochemical phase assemblages, has been a favored explanation of some geologists. Although current opinion is weighted on the side of a chemical discontinuity and arguments on the nature of the Moho may now appear to be somewhat dated, the importance of this question requires that we examine these ideas in more detail.

Worked Problem 11-1

Kennedy (1959) eloquently summarized the geological arguments against the Moho being a chemical discontinuity. He hypothesized that it is a phase transition boundary separating gabbroic rocks in the crust from eclogite in the mantle. Kennedy adopted a *P-T* phase diagram for the basalt-eclogite transformation and assumed geothermal gradients for continental and oceanic regions. (A *geothermal gradient* is the rate of increase of temperature with pressure, discussed in detail in the next section.) The intersections of these geothermal gradients with the transformation curve give the pressure (or depth) of the Moho in that region. Can we test Kennedy's idea using more up-to-date experimental data?

The phase diagram for gabbro-eclogite is illustrated in Figure 11.2a and 11.2b. In this example, we have recast the vertical (pressure) axes in terms of depth; the conversion is roughly 0.3 kbar per km of depth, assuming an average crustal density of 3 g cm^{-3}. The two experimentally determined bands correspond to the transition of (a) quartz-bearing gabbro to eclogite and (b) olivine-bearing gabbro to eclogite. Geothermal gradients for continental shields and ocean basins are also shown.

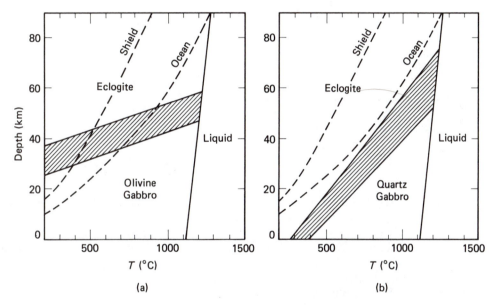

Figure 11.2 Phase transition for gabbro to eclogite. Hatched bands define the transition for (a) olivine-bearing gabbro and (b) quartz-bearing gabbro compositions, based on the experiments of Yoder and Tilley (1962) and Ringwood and Green (1966), respectively. Dotted lines are geothermal gradients for continental shields and ocean basins.

The first important point is that for quartz gabbro compositions, eclogite is stable throughout the lower crust in both continental and oceanic regions. Thus, the Moho cannot be a phase transformation for this composition. If the lower crust and mantle have the composition of olivine gabbro, then the intersection of the geothermal gradients with that transition curve give the depths to the Moho for those regions. But where do we mark this transition? Gabbro transforms to eclogite over a *range* of pressure, and it seems unlikely that such a smeared interval could produce a sharp seismic discontinuity. Even if we were to ignore this problem and arbitrarily adopt some sharp boundary within the band as the transition curve, we see that the transformation occurs nearer to the surface in continental regions than in ocean basins. This is the opposite of what we have already learned about the depth to the Moho in these two environments. This simple analysis, therefore, leads to the conclusion that the Moho must not be a phase transition.

Some crustal occurrences of tectonically emplaced ultramafic rocks may have been derived directly from the mantle. White (1967) argued that the most abundantly exposed ultramafic rocks (by volume) should be representative of the mantle. His preferred mantle composition, calculated from frequency histograms of 168 ultramafic rocks, is presented in column 1 of Table 11-2.

TABLE 11-2 ESTIMATES FOR THE AVERAGE COMPOSITION OF THE EARTH'S MANTLE

Wt. %	(1)	(2)	(3)	(4)	(5)
SiO_2	44.5	44.2	45.7	45.1	47.3
TiO_2	0.15	0.09	0.09	0.5	0.2
Al_2O_3	2.6	2.8	3.4	4.1	4.1
Fe_2O_3	1.5	1.2	-	2.0	-
Cr_2O_3	-	0.3	0.4	0.3	0.2
FeO	7.3	7.3	8.0	7.9	6.8
MnO	0.14	0.14	0.14	0.2	0.2
Ni	0.2	0.27	0.27	0.2	0.2
MgO	41.7	41.0	38.4	36.7	37.9
CaO	2.3	2.5	3.1	2.3	2.8
Na_2O	0.25	0.22	0.4	0.6	0.5
K_2O	0.02	0.04	-	0.02	0.2

(1) From frequency histograms of 168 ultramafic rocks (White, 1967).

(2) Mean of 5 mantle peridotite nodules (Harris et al., 1967).

(3) Pyrolite (Ringwood, 1975).

(4) Estimate of volatile-free upper mantle (Nicholls, 1967).

(5) 80% garnet peridotite + 20% eclogite (Anderson, 1980).

Another constraint on the composition of the mantle comes from *xenoliths* of mantle material carried upward by erupting basalt or kimberlite magmas. These consist either of peridotite or eclogite. We will focus on the peridotites, which we have already concluded are the best candidates for the bulk of the mantle. There are essentially two types: *spinel peridotites,* containing 60–70% magnesian olivine, 10–30% enstatite, 8–12% diopside, and 3–13% chromium-aluminum spinel; and *garnet peridotites,* containing 48–67% olivine, 28–47% enstatite, trace-3% diopside, and 2–5% pyrope garnet. In both cases, olivine and pyroxenes clearly predominate, but the presence of spinel or garnet is very important. Even though the pyroxenes in peridotite xenoliths are commonly aluminous, the assemblage olivine + orthopyroxene + clinopyroxene does not contain enough aluminum to provide the plagioclase component of basalts on partial melting. In rocks of ultramafic composition, plagioclase itself is the Al-bearing phase at low pressures, but this gives way to spinel and garnet at higher pressures. The spinel- and garnet-forming reactions can be represented by

$$2\ CaAl_2Si_2O_8 + 2\ Mg_2SiO_4 \longrightarrow CaMgSi_2O_6 \bullet x\ CaAl_2SiO_6$$

Anorthite   Forsterite   Aluminous Clinopyroxene

$$+\ 2\ MgSiO_3 \bullet x\ MgAl_2SiO_6 + (1 - x)\ CaAl_2Si_2O_6$$

Aluminous Orthopyroxene   Unreacted Anorthite

$$+\ (1 - x)\ MgAl_2O_4 \longrightarrow 2\ CaMg_2Al_2Si_3O_{12}.$$

Spinel   Garnet

One published example of the use of peridotite xenoliths as a measure of the composition of the mantle is illustrated in column 2 of Table 11-2.

One of the problems we encounter in estimating a composition for the mantle has to do with partial melting. A peridotite assemblage without spinel or garnet (and often without clinopyroxene as well) is *depleted,* because basaltic magma has already been extracted from it. Such a peridotite may also be called *barren,* because its calcium and aluminum contents are too low for the production of additional basaltic melt. Peridotite that can still produce basaltic melt is called *undepleted,* or *fertile.* But how do we determine what proportion of the mantle is barren and what proportion is fertile? This dilemma forms the basis for a class of hypothetical mantle models known as *pyrolite,* originally devised by Ringwood (1962). He postulated a primitive mantle material which was defined by the property "that on fractional melting it would yield basaltic magma and leave behind a residual refractory dunite-peridotite of alpine type." The composition of pyrolite (pyroxene-olivine rock) can be derived, therefore, by adding depleted peridotite and basalt in the proper proportions. This is a very flexible model for the mantle composition. Ringwood and his collaborators at various times have presented pyrolite compositions with peridotite : basalt ratios varying from 1 : 1 to 4 : 1, though the 3 : 1 ratio is most often quoted as a reasonable value. In another pyrolite-type modelling attempt, Nicholls (1967) added the outgassed volatile inventory, which we discussed in Chapter 4, as another component. These two model compositions for the mantle are compared with others already discussed in columns 3 and 4 of Table 11-2.

Not only is basalt removed from the mantle in the form of magma, but it also is recycled back into the mantle (in the form of eclogite) by subduction. Consequently, we might also envision a model composition for the mantle that mixes average eclogite and undepleted peridotite in the correct proportions. Anderson's (1980) model, presented in column 5 of Table 11-2, does just that.

All of these methods give comparable results, and the composition of the mantle seems reasonably well-constrained, at least for major elements. We should keep in mind, however, that these estimates are for the upper mantle, though they are commonly used to describe the composition of the entire mantle. Is there any basis for this extrapolation?

Ringwood (1975) has provided a wealth of information about the stability of phases at various depths in a mantle of pyrolite composition. Down to a depth of 600 km, the assumed mineral assemblages are based on the results of high-pressure experiments, but below this depth, Ringwood had to infer what phases would be stable from indirect evidence such as experimental data on germanate (GeO_4^{-4}) analog systems. If germanium is substituted for silicon, phase transitions that are experimentally inaccessible for silicates occur within pressure ranges that can be studied in the laboratory. The phase changes which Ringwood suggests should occur (see the box on page 334) correspond very well with observed seismic discontinuities in the mantle, so it seems plausible that, to a first approximation, the mantle has a pyrolite composition throughout. Thus, we can infer that most discontinuities within the mantle result from phase changes, in contrast to the discontinuity at the Moho, which is due to a change in chemical composition.

GEOTHERMAL GRADIENTS

The temperature distribution within the earth is one of the greatest uncertainties in geophysics. Surface heat flow measurements can be used to place some limits on the rate of temperature change with depth (the *geothermal gradient,* also called the

PHASE CHANGES IN THE MANTLE

Ringwood's suggested mantle mineralogy as a function of depth is summarized in Figure 11.3. The wavy diagonal line from upper left to lower right represents the density distribution with depth, inferred from P-wave velocities and corrected for the effect of compression due to the overlying rock. Starting near the top just below the Moho, we first encounter the low-velocity zone, a region of inferred partial melting in which seismic velocity decreases. The stable assemblage at this point is olivine + orthopyroxene + clinopyroxene + garnet. Just below 300 km, pyroxenes transform to a garnet crystal structure. One of the two major discontinuities in the mantle occurs at 400 km; this corresponds to the conversion of olivine, volumetrically the most important phase to this point, to a more tightly packed structure called the beta phase. At this point the mantle consists of β-olivine and a complex garnet solid solution. Another less pronounced seismic wiggle at about 500 km may indicate the transformation of the beta phase into a spinel (gamma) structure, as well as a second reaction in which the calcium component of garnet, possibly with some iron, separates as $(Ca, Fe)SiO_3$ in the dense perovskite structure.

The second major seismic discontinuity in the mantle is seen at 650–700 km. At this point γ-olivine is thought to disproportionate into its constituent oxides, MgO (periclase) + SiO_2 (stishovite). The $(Mg, Fe)SiO_3$-Al_2O_3 components of the garnet solid solution presumably also transform into the ilmenite structure. Below this transition, the mantle consists of periclase + stishovite + $MgSiO_3 \cdot Al_2O_3$ (ilmenite) + $(Ca, Fe)SiO_3$ (perovskite).

The density changes in mantle materials below about 700 km could be caused by further phase transformations to more tightly packed structures with Mg and Fe in higher coordination, but it is very difficult to judge what transformations may take place. Some additional phases that have been suggested to be stable in the lower mantle include $(Ca, Fe, Mg)SiO_3$ in the perovskite structure, $(Mg, Fe)O$ in the halite structure, and $(Mg, Fe)(Al, Cr, Fe)_2O_4$ in

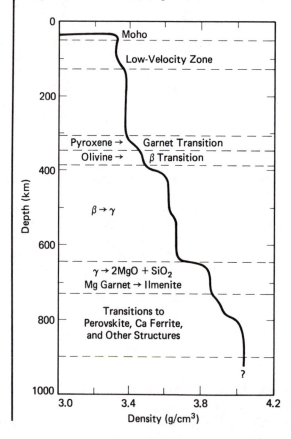

Figure 11.3 Summary of possible phase transformations in the mantle, modified from Ringwood (1975). The densities shown by the solid line are zero-pressure densities (corrected for depth) calculated for pyrolite; density discontinuities correspond to phase transformations inferred from seismic studies.

the calcium ferrite structure. From the generality of these formulas, we can see how speculative these phases are.

Another possible explanation for the higher densities in the deep mantle is an increased Fe/Mg ratio. Such a change in mantle composition would have important geochemical implications, but it is difficult to confirm because the behavior of iron in the mantle is so uncertain. It has been suggested that Fe^{2+} ions in silicates may undergo a contraction in radius due to spin-pairing of electrons below 1200 km depth. This is potentially important because low-spin Fe^{2+} probably would not substitute for Mg in solid solutions. Another hypothesis, with some experimental justification, is that Fe-bearing silicates might disproportionate to form some metallic iron, even at modest depths of about 350 km. This metal might then sink, stripping Fe from some parts of the mantle and enriching others.

geotherm), but only within the upper crust. How then can we determine this important variable for the remainder of the crust and mantle?

Temperatures and pressures estimated from metamorphic rocks can provide additional information on physical conditions in the earth's interior. In Chapter 8 we have already examined how mineral exchange geothermometers and geobarometers can be used to determine conditions of equilibration. Mantle conditions are harder to reconstruct than crustal conditions, but xenoliths of mantle peridotite may provide this information. For example, the ratio Ca/(Ca + Mg) in clinopyroxene in equilibrium with orthopyroxene is a function of temperature but is relatively insensitive to pressure, and the aluminum content of orthopyroxene in equilibrium with garnet varies with both temperature and pressure. Thus, mineral compositions for a set of garnet peridotite nodules in basalts can provide direct P-T information on the mantle source region. Representative data from mantle xenoliths are illustrated in Figure 11.4. Data like those for Lesotho appear to define an inflection in the mantle geotherm which, according to one interpretation, marks the boundary between the lithosphere and asthenosphere. However, this has been called into question by another calibration of the two-pyroxene geothermometer that eliminates this inflection. A summary of crustal geotherms from a number of metamorphic terranes is also given in Figure 11.4.

What is the meaning of such geotherms? Even if we neglect the considerable analytical uncertainties in geothermometers and geobarometers, understanding what these numbers represent remains a challenge. Most mineral assemblages record conditions close to the maximum temperature and pressure that they experienced. A collection of rocks that record these extreme conditions, however, does not necessarily define the geothermal gradient at any point in time. It is important to make the distinction between geothermometry-geobarometry data (so called P-T arrays) and actual pressure-temperature-time (P-T-t) paths. This distinction is illustrated graphically in Figure 11.5. As a rock is buried, its pressure increases "instantaneously," which is to say that it experiences the mass of overlying material as soon as it is applied, with virtually no time lag. Because the relatively low thermal conductivity of a rock places a limit on just how fast it can be heated, though, temperature rises less rapidly than pressure. If we follow a rock along path A during burial, then, it eventually reaches a P_{max} corresponding to its maximum depth of burial. Temperature is still rising at this point, and continues to increase even as the rock moves closer to the surface as a result of erosion. At some point, the rock reaches T_{max}, at which time reaction rates are most rapid and prograde metamorphism is at its peak. It is the temperature that is recorded in the rock and is later measured by geothermometry. Clearly, this is not the same environment in which the rock reached P_{max}. Also, a rock following path B or C to greater depths in the same region will not experience its T_{max} at the same time as a rock following path A. Therefore, the apparent geothermal gradient is not an isochronous line in P-T-t space. The

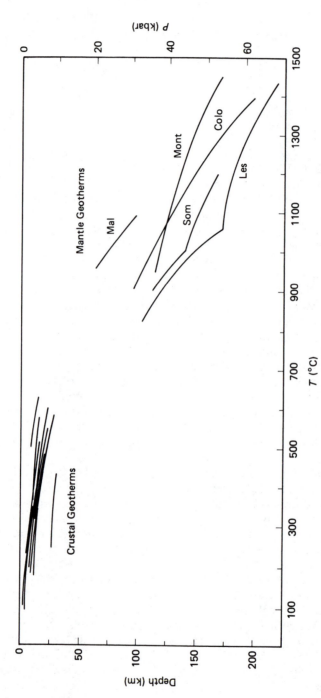

Figure 11.4 Mantle geotherms estimated from geothermometry and barometry of garnet periodotite xenoliths from various localities, after Barker (1983). Data from Les (Lesotho) and Som (Somerset island, Canadian Arctic) show inflections; Colo (Colorado), Mont (Montana), and Mal (Malaita) do not. Crustal geotherms from different metamorphic terrains in the upper left are from Turner (1981).

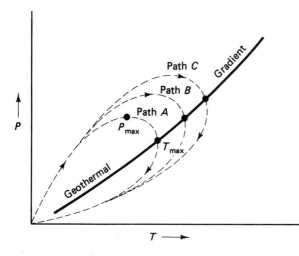

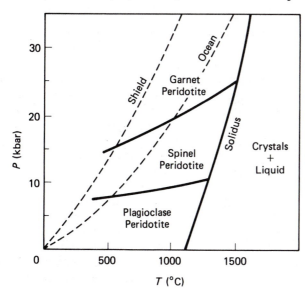

Figure 11.5 Diagram illustrating the difference between *P-T* arrays (solid line connecting the maximum temperature experienced) and *P-T-time* paths, showing *P-T* evolution of samples over time (dashed loops). Ideally, geotherms should be *P-T* arrays at a given time.

lesson to be learned is that gradients inferred from metamorphic assemblages may bear little resemblance to those which actually existed during the metamorphic event.

Geotherms, ideally, should be *P-T* arrays through isochronous points, rather than through points of maximum temperature. Average geotherms thought to be typical of modern continental and oceanic areas are shown in Figure 11.6. The phase boundaries for various types of peridotite are also illustrated for reference. Why should the continental and oceanic geotherms in this figure be so different? To answer this question, we must consider what produces heat in the earth's interior.

Much of the earth's heat production at the present time is the byproduct of the slow decay of long-lived radioactive isotopes of K, U, and Th. Because rocks are notoriously poor conductors of heat, the temperature at any point in the subsurface is largely controlled by the abundances of these isotopes in the vicinity. Even though the volume of the crust is almost insignificant compared to that of the mantle, the high crustal concentrations of radioactive isotopes dominate these geotherms. When we analyze mantle xenoliths for K, U, and Th, we find that these peridotites are significantly depleted in

Figure 11.6 Generalized phase diagram for mantle peridotite showing subsolidus transformations of aluminum-bearing minerals. Plagioclase, spinel, and garnet are stable with increasing pressure. Dotted lines are geotherms for continental shields and ocean basins.

Geothermal Gradients

heat-generating elements compared to continental crustal rocks. The lower heat production of mantle materials is reflected in the observation that continental and oceanic geotherms become shallower and finally converge at great depths.

In a later section of this chapter we will learn how these heat-producing elements have become concentrated in continental crust. The important point we wish to stress now is that these elements, so important in driving geologic processes, are strongly partitioned between crust and mantle. This chemical heterogeneity has caused geotherms to differ considerably between continental and oceanic regions. Specifying these geothermal gradients is an important task, because they allow us to reconstruct the stable phases at various depths within the earth.

INTERACTIONS BETWEEN CRUST AND MANTLE

Because the compositions of crust and mantle differ so greatly, it may seem puzzling to suggest that the two regions behave as one geochemical system. On a small scale, the crust and mantle are isolated, but through tectonics, they directly exchange materials. Let's review how this geochemical cycle, pictured in Figure 11.7, works.

In the mantle below divergent plate boundaries, partial melting of fertile peridotite produces basaltic magmas. The zone of fusion is the asthenosphere, which is surmounted by the lithosphere, composed of a subsolidus segment of mantle and the crust. Separation of melts leaves behind a solid residue of depleted peridotite. Upwelling of these mantle-derived magmas and extrusion on the ocean floor create new oceanic crust at spreading centers. This crust is carried laterally by plate tectonic motions, picking up a veneer of pelagic sediment as it goes.

At convergent plate margins, the lithosphere (mostly basaltic oceanic crust) is subducted and converted to eclogite by metamorphism at mantle temperatures and pressures. By this process, crustal rocks are reincorporated into the mantle. In fact, seismic studies suggest that portions of the mantle may be junkyards of stranded eclogite plates from now-abandoned subduction zones. At appropriate depths, the subducted plates or the overlying mantle peridotite partially melt, producing basaltic and andesitic magmas. If the overlying plate is oceanic lithosphere, these magmas may ultimately erupt to form volcanic island arcs. If the overlying plate contains continental crust, heat from the ascending magmas may cause crustal melting to produce granitic (calc-alkaline) magmas. These secondary melts may intermingle with basaltic magmas or erupt separately to create a volcanic mountain chain on the continent.

Not all igneous activity occurs at plate boundaries. Localized centers of melting in the mantle, reflected in intraplate volcanic islands and certain continental rift zones, also provide some communication between the mantle and crust. These melting centers, called *hot spots* or *mantle plumes,* generate basaltic magmas, the compositions of which suggest that they tap deeper parts of the mantle than the basalts at plate boundaries. Such hot spots may spawn volcanic chains, each volcano dying out while another is created nearby as overlying plates shift positions.

In Figure 11.7, it appears that the only interactions between crust and mantle are in the form of ascending magmas generated in the mantle or subducting lithospheric (predominantly crustal) slabs. As we first concluded in Chapter 4, however, the recycling of volatile elements can play a significant role in crust-mantle evolution. Much, but not all, of the movement of volatile elements occurs in response to the processes we have just examined. Volumetrically important gases in the system H-C-O-S, as well as more exotic volatile elements like Ne, Ar, Kr, and Xe, are brought from the mantle

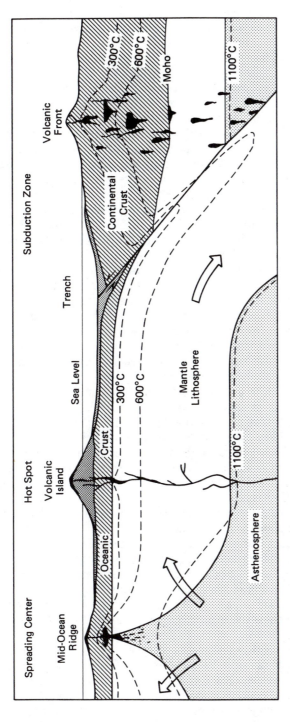

Figure 11.7 Schematic diagram showing divergent and convergent plate boundaries. Tholeiitic basalts erupting at the spreading center create new oceanic crust which is rafted away from the midocean ridge. The lithospheric slab is ultimately reincorporated into the mantle in the subduction zone. Calc-alkaline magmas are produced in this environment. Isotherms (dashed lines) are deflected upward by ascending magmas in both locations, and downward in the subduction zone itself; these control the *P-T* regime of metamorphism. Hot-spot igneous activity is illustrated by intraplate volcanism located spatially between these two tectonic environments.

and expelled into the oceans and atmosphere by volcanic exhalations. Some fraction of gases is ultimately trapped in the lithosphere in the form of carbonates, amphiboles and other hydrous phases, and sulfides. This may happen as newly erupted basalts are hydrated by interaction with sea water, as sulfides are precipitated on the ocean floor or in subsurface plutons, or as carbonate reefs and the tests of calcareous organisms are added to the basaltic substrate. As part of the oceanic lithosphere, these may be subducted in some cases and thus returned to the mantle.

Even in the absence of igneous activity and subduction, fluids composed of compounds of these volatile elements may also move vertically through the crust and mantle during metamorphism. As a result, the characteristic continental and oceanic geotherms that we described in the previous section are perturbed at plate boundaries to produce anomalously high or low temperature gradients. For example, the heat carried by ascending magmas at spreading centers and under subduction zone volcanic fronts increases the local surface heat flow and subsurface geothermal gradients and promotes near-surface metamorphism. Conversely, subducted slabs typically have low geothermal gradients because the slabs are subducted faster than they can achieve thermal equilibration with their surroundings. This situation allows these plates to descend deeper into the mantle before they dehydrate to form eclogite. Deflections of isotherms at plate boundaries are illustrated in Figure 11.7.

It is interesting to speculate about the extent to which the interaction between crust and mantle is either cyclic or directional. Put another way, are these exchanges at a steady-state condition, or is the earth slowly evolving? Most geochemists would probably favor the latter interpretation, because at least some portion of the mantle appears to have passed through an irreversible geochemical cycle. Once primary mantle peridotite has been transformed into the depleted state by partial melting, not much else can happen to it, because further melting would require temperatures higher than can be achieved with present geothermal gradients.

If we accept the premise that mantle differentiation is directional, then what portion of the mantle is barren and what portion fertile? Dickinson and Luth (1971) divided the mantle into two regions—the *asthenosphere,* extending to a depth of 650 km, and the *mesosphere,* the mantle region between the bottom of the asthenosphere and the core. They estimated the rate of generation of oceanic lithosphere through time, and from this value calculated the necessary mass of complementary depleted mantle. This mass, 2.8×10^{27} g, is almost identical to the mass of the mesosphere. Thus, they hypothesized that the mesosphere consists of depleted material, which was subducted to depths previously thought to be inaccessible and which displaced unfractionated mantle material upward.

Dickinson and Luth's model suggests that approximately 70% of the entire mantle has been irreversibly differentiated in this way. Gast (1972) reached a similar conclusion from another line of evidence. The extreme concentration of certain trace elements (including the heat-producing elements uranium and thorium) in the crust implies a pervasive differentiation process. Assuming that these elements were once uniformly distributed throughout the mantle, he estimated that between 30 to 70% of the mantle has been differentiated.

The production of continental crust is probably also one-directional. The aggregate density of "granitic" continental crust is too low for this material to be subducted. Consequently, once continental crust forms, it is likely to remain at or near the earth's surface. This is in marked contrast to oceanic crust, which is consumed at convergent plate boundaries at almost the same rate as it is generated at divergent boundaries. In fact, the only way that oceanic crust can escape reincorporation in the mantle is by its accidental accretion onto continents by tectonic processes (producing "ophiolites"). Although the mass of continental crust must therefore increase over time, its total bulk composition probably remains relatively constant. This is because its composition is determined by the same partial melting and differentiation processes that have operated over all of geologic history.

We can therefore conclude that the crust and mantle constitute a grand geochemical system. Much of the interaction between crust and mantle occurs through plate tectonic processes, although hot spot igneous activity and migration of fluids during metamorphism also play a role. The earth is continuing to differentiate irreversibly in such a way that the mantle is depleted in fusible components which are ultimately added to continental crust. In the following sections, we will examine in more detail how these interactions between mantle and crust take place.

MELTING IN THE MANTLE

The transformation from the crystalline to the liquid state is not very well understood at the atomic or molecular level. One (largely chemical) description of the melting process is that it results from increasing the vibrational stretching of bonds between atoms to the point at which the weakest bonds are severed. Another (primarily structural) view of melting is that it parallels a change from long-range crystallographic order to the short-range order characteristic of quasi-crystalline melts. X-ray diffraction studies of minerals at low and high temperatures have shown no significant distortions of crystal structures as melting temperatures are approached, but unit cell volumes increase. However, more work needs to be done before the physics of the melting process can be specified.

Thermodynamic Effects of Melting

Melting, like other phase transformations in the mantle and crust, results in discontinuous changes in the thermodynamic properties of the system. A sudden increase in entropy, reflecting the higher state of atomic disorder in liquids relative to crystals, is probably the most obvious thermodynamic change. More important from the geochemist's point of view, however, is the increase in enthalpy that accompanies melting. The thermal energy necessary to convert rock at a particular temperature to liquid at that same temperature is the *enthalpy of melting,* also called the *latent heat of fusion.* This is thermal energy required beyond that necessary to raise the temperature of the rock to its melting point.

The enthalpy of melting can be measured by dropping crystals or glass into a calorimeter at temperatures just above and just below the melting point. (The principles of calorimetry were discussed earlier in Chapter 2.) From the rise in temperature of the calorimeter bath, or change in the proportion of crystals and liquid, the energy released by each state of the material can be evaluated. It is possible to calculate the enthalpy of melting for more complex systems that consist of several crystalline phases, provided that the phase diagram relating these phases has been determined and $\Delta \overline{H}_m$ for the individual phases are known, as illustrated in Worked Problem 11-2.

Worked Problem 11-2

How much thermal energy is required to melt a eutectic mixture of diopside and anorthite, initially at 0°C?

This calculation is performed in two steps. First, we must determine the energy required to raise the temperature of this material from 0°C to the melting point, which is

1274°C for a eutectic composition in this system. Recall that the eutectic point represents the composition of the last liquid to crystallize or the first liquid to form during partial melting. You should turn back to Chapter 9 and review the diopside-anorthite phase diagram, illustrated at the bottom of Figure 9.7. From it, we find that the eutectic composition (in weight %) is $Di_{58}An_{42}$. The heat capacity $(\Delta\bar{C}_P)$ for diopside is 100.40 J mol^{-1} K^{-1} = 23.99 cal mol^{-1} K^{-1}, and that for anorthite is 146.30 J mol^{-1} K^{-1} = 34.97 cal mol^{-1} K^{-1}.[1] We will convert these heat capacities into units of cal g^{-1} K^{-1} because the eutectic composition is given in units of weight. Division by the gram formula weights for diopside (216.55 g mol^{-1}) and anorthite (278.21 g mol^{-1}) gives $\Delta C_P = 0.1109$ cal g^{-1} K^{-1} for diopside and 0.1257 cal g^{-1} K^{-1} for anorthite. The energy required to raise the temperature of this mixture by 1274 degrees is therefore

$$(0.1109 \text{ cal g}^{-1} \text{ K}^{-1})(1274 \text{ K})(0.58)$$

$$+ (0.1257 \text{ cal g}^{-1} \text{ K}^{-1})(1274 \text{ K})(0.42) = 149.21 \text{ cal g}^{-1}.$$

The second step is to calculate the enthalpy of melting of this eutectic mixture. $\Delta\bar{H}_m$ for diopside is 77.40 kJ mol^{-1} = $18,345$ cal mol^{-1} and for anorthite is 81.00 kJ mol^{-1} = $19,359$ cal mol^{-1}. Division by the appropriate gram formula weights gives ΔH_m for diopside of 84.71 cal g^{-1} and for anorthite of 69.58 cal g^{-1}. Multiplying these values by the appropriate fractions for the eutectic composition yields

$$(84.71 \text{ cal g}^{-1})(0.58) + (69.58 \text{ cal g}^{-1})(0.42) = 78.36 \text{ cal g}^{-1}.$$

The heats of mixing of these liquids are so small that they can be neglected. Thus, the total thermal energy necessary for this melting process is the sum of the energy required to raise the temperature of a eutectic mixture of diopside and anorthite to the melting point, plus the energy required to transform this mixture into a liquid at that same temperature:

$$149.21 \text{ cal g}^{-1} + 78.36 \text{ cal g}^{-1} = 227.51 \text{ cal g}^{-1}.$$

A realistic value for the enthalpy of melting of mantle peridotite is difficult to obtain, because of the unknown effects of high pressure and of other solid solution components in pyroxenes, olivine, and garnet or spinel. One commonly quoted value for garnet peridotite at 40 kbar pressure is 135 cal g^{-1}. To have melting in the mantle, this amount of thermal energy must be supplied in excess of the heat necessary to bring mantle rock to the melting temperature.

Types of Melting Behavior

If the temperature of mantle peridotite is increased to the melting point and some extra thermal energy is added for the enthalpy of melting, magma will be produced. As in the case of eutectic melting in the diopside-anorthite system discussed in Worked Problem 11-2, fusion will begin at some invariant point so that several phases will melt simultaneously. The melting process can be described in several ways. *Equilibrium melting* (also called *batch melting*) is a relatively simple process in which the liquid remains at the site of melting in chemical equilibrium with the solid residue until mechanical conditions allow it to escape as a single "batch" of magma. *Fractional melting* involves continuous extraction of melt from the system as it forms, thereby preventing reaction with the solid residue. Fractional melting can be visualized as a large number of infinitely small equilibrium melting events. *Incremental batch melting* lies between these two extremes, with melts extracted from the system at discrete intervals. *Partial melting* is fusion of some portion less than the whole and can therefore refer to either batch or fractional melting or something between.

[1] Data from Robie et al. (1978).

Worked Problem 11-3

To understand the distinction between equilibrium and fractional melting, let's examine fusion processes in the two-component system forsterite-silica, shown in Figure 11.8. Describe melting of various compositions in this system under equilibrium and fractional conditions.

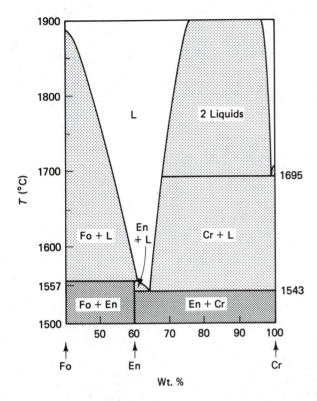

Figure 11.8 Phase diagram for the system forsterite (Mg$_2$SiO$_4$)-silica (SiO$_2$). The silica-rich side contains a region of silicate liquid immiscibility.

This system contains a peritectic point between the compositions of forsterite (Fo) and enstatite (En) and a eutectic between the compositions of enstatite and cristobalite (Cr). If enough heat is available to raise the temperature of the system to 1542°C and offset the enthalpy of melting, any composition between En and Cr will begin melting at the eutectic, with both solid phases entering the melt. For small degrees of partial melting under either equilibrium or fractional conditions, the melt will have the eutectic composition. At higher degrees of partial melting, one phase will usually be exhausted from the solid residue before the other. Begin, for example, with a composition with 63% silica in Figure 11.8. Cr will be exhausted from the residue first, and as melting continues, the composition of the liquid will leave the eutectic and follow the liquidus upward until all of the En is melted. Notice that temperature must increase for melting to continue after one phase has been exhausted. Also, after one phase is exhausted from the residue, fractional melting will produce a series of melts with different compositions, whereas equilibrium melting will result in one homogeneous batch of melt.

For compositions between Fo and En, equilibrium and fractional melting will produce different liquids from the onset of melting. For example, under equilibrium melting conditions, a solid composition with 55% silica will yield a liquid with the peritectic composition at 1557°C as En melts incongruently to form Fo + liquid (L). The liquid will remain at the peritectic until En is exhausted, and then with more advanced melting will follow the liquidus upward as Fo melts. Under fractional melting conditions, however, the liquid formed at the peritectic temperature and separated from the solid residue has a composition richer in silica than is the case under equilibrium melting conditions; this occurs because the melt is not constrained to remain in contact with Fo. Such a liquid will ultimately crystallize to form En + Cr.

Some details of the melting relationships of mantle peridotite are illustrated schematically in Figure 11.9. Phase changes for the stable aluminum-bearing mineral in fertile peridotites are illustrated at temperatures below the solidus. Melting of peridotite begins as its temperature is raised just above the solidus. As temperature increases further, clinopyroxene, spinel, and garnet are completely melted and thereby exhausted from the solid residue (as indicated by lines labelled Cpx-out, Sp-out, and Gar-out, respectively). This all occurs within about 50° of the solidus, producing a wide *P-T* field for the coexistence of olivine + orthopyroxene with liquid. The olivine + orthopyroxene residue is depleted peridotite that has experienced the irreversible differentiation discussed earlier.

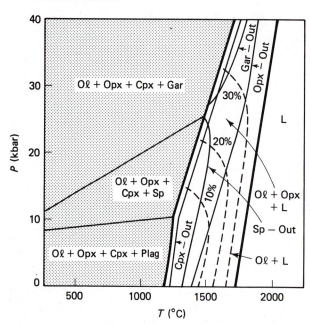

Figure 11.9 Schematic phase diagram for mantle peridotite, illustrating the effects of partial melting. The solidus and liquidus are indicated by heavy lines. With increasing temperature, first clinopyroxene (Cpx) and then spinel (Sp) or garnet (Gar) are exhausted from the residue, leaving olivine (Ol) + orthopyroxene (Opx) or olivine alone. The dashed lines are contours of the percentage of dissolved (normative) olivine in the melt; they show that with increasing pressure, magmas become more olivine-rich. (Modified from Wyllie 1971.)

The composition of the basaltic liquid produced by partial melting of peridotite changes with pressure, becoming less silica-rich at higher pressure. Lower silica contents of liquids increase the proportion of olivine that can ultimately crystallize from the melt on cooling; the percentage of potential ("normative") olivine in the melt is contoured as dashed lines in Figure 11.9. Variation in pressure (or depth) of magma generation thus produces melts ranging from tholeiitic basalt (the common lava composition at divergent plate boundaries) at modest pressures to less silica-rich basalts (alkali olivine basalts and basanites, common at hot spots) at higher pressures.

Causes of Melting

In an earlier section of this chapter, we noted that the elastic properties of the mantle low-velocity zone are consistent with partial melting of this region. It is difficult to specify the degree of melting, but published estimates range from less than 1% to about 8%. This zone is commonly thought to be the source region for many basaltic magmas. However, a look at the relative positions of geotherms and the peridotite solidus in Figure 11.6 indicates that there is no apparent reason for incipient melting in this zone or, for that matter, anywhere else in the mantle. How, then, can we account for magma generation? There is no simple answer to this question, but there are at least three intriguing possibilities: localized temperature increase, decompression, and addition of volatiles. Let us see how each of these could cause melting.

Simply raising the temperature of a rock (the $+T$ path in Figure 11.10) is the most obvious mechanism by which melting can occur; this has been called the "hot plate" model. However, in considering this model, we should keep in mind the difference in magnitude between the enthalpy of melting and the heat capacity for silicate minerals. C_P values are generally several hundred times lower than corresponding ΔH_m values (compare the values for diopside and for anorthite in Worked Problem 11-2), so it is much easier to raise the temperature of a rock than to produce a significant amount of partial melt from it. Melting consumes large quantities of thermal energy, moderating temperature variations within the system in its melting range. Even so, it may be possible to have localized mantle heating that produces magma. Frictional heating of subducted lithospheric slabs or dissipation of tidal energy due to the gravitation attraction of the sun and moon have been suggested as possible causes of localized heating in the mantle. Local heating of lower crustal materials may also occur in response to underplating by mantle-derived magmas.

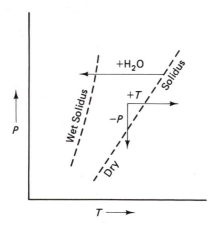

Figure 11.10 *P-T* diagram illustrating three possible ways that mantle rocks may melt. Increase in temperature ($+T$ path) could result from localized concentration of heat-producing radionuclides or underplating by other magmas. Decrease in confining pressure ($-P$ path) may occur because of diapiric upwelling of ductile mantle rock. Addition of a fluid phase ($+H_2O$ path) depresses the melting curve to lower temperature for the same pressure.

A more plausible mechanism for melting mantle rocks may be pressure release. The effect of decreasing confining pressure is generally to lower solidus temperatures. One example, in the system diopside-anorthite, is illustrated in Figure 11.11. The eutectic temperature, which is the point at which partial melting begins, is depressed by more than 100°C in going from 10 kbar to 1 atm pressure.[2] The effect of decreasing pressure on the solidus of mantle peridotite can be seen in Figure 11.9. Because of the positive slope of this melting curve in *P-T* space, decompression of solid rock can induce melting, as illustrated by the $-P$ path in Figure 11.10. Pressure release has been suggested to occur because of diapiric uprise of hot, plastic mantle material in a convection cell. Ascent is assumed to be adiabatic; otherwise, the diapir might not intersect its solidus. Based on the slope dP/dT of the mantle peridotite solidus, it has been estimated that 30% melting, postulated to be necessary for the generation of basaltic

---

[2] The *position* of the eutectic point also changes, because the melting point of diopside increases much further for a given pressure change than that of anorthite. Thus the Di field is enlarged with increasing pressure at the expense of the An field.

Melting in the Mantle

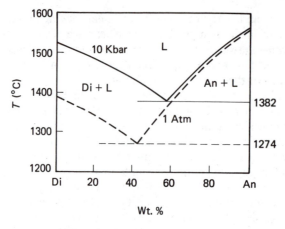

Figure 11.11 Phase diagram for the system diopside (CaMgSi₂O₆)-anorthite (CaAl₂Si₂O₈), demonstrating the effect of increasing confining pressure on melting relationships. Lowering pressure from 10 kbar to 1 atm results in a decrease in the eutectic melting temperature of over 100 degrees. (Based on experiments by Presnall et al. 1978.)

magmas at divergent plate boudaries, would require that mantle diapirs rise roughly 60 km after intersecting the solidus.

Changes in the composition of a rock system due to gain or loss of volatiles, principally H_2O and CO_2, are also potentially very important in facilitating melting. Water lowers the solidus of peridotite, as illustrated in Figure 11.12 by the wet solidus curve. The addition of water to a dry peridotite that is already above the wet solidus will cause spontaneous melting, as illustrated by the $+H_2O$ path in Figure 11.10. An influx of water into the mantle system could be caused by dehydration reactions in a subducted slab of hydrated oceanic lithosphere. The effect of the addition of CO_2 to mantle peridotite is less drastic than for H_2O, but still significant. The solidus of peridotite in the presence of CO_2 is illustrated in Figure 11.12 by the line labelled CO_2. When the fluid phase contains both H_2O and CO_2, the effect on the peridotite melting curve is complicated. A solidus for a mixed fluid composition of $X_{CO_2} = 0.6$ is also illustrated in Figure 11.12. At modest pressures of 10 to 20 kbar, water predominates because it is most soluble in the melt, but at higher pressures the effect of CO_2 becomes marked as its solubility increases.

Both of these volatile components also strongly influence the compositions of the

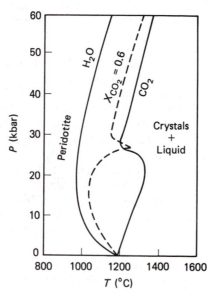

Figure 11.12 Phase diagram showing the solidus in the system peridotite-CO_2-H_2O. Melting in the presence of pure water occurs at lower temperatures than for pure CO_2. The dashed line labelled $X_{CO_2} = 0.6$ corresponds to peridotite melting with a mixed fluid of that composition. Water is the most important volatile species at pressures of less than about 20 kbar, but CO_2 becomes increasingly important at higher pressures because of its increased solubility in the melt. (Modified from Wyllie 1971.)

magmas that are generated. Water selectively dissolves silica out of mantle phases before melting occurs; as soon as the melt forms, it effectively swallows this fluid and thereby becomes rich in silica. CO_2, on the other hand, favors the production of alkalic magmas poor in silica (nephelinites and kimberlites).

Before we conclude this section on melting, it may be useful to review some of these concepts by discussing current ideas about how mantle-derived magmas are formed in various tectonic settings. Partial melting of relatively dry peridotite probably accounts for the vast outpourings of tholeiitic basalt at divergent plate margins. This style of melting may be initiated through pressure release due to diapiric rise of undepleted mantle under spreading centers. Hydrated slabs of oceanic lithosphere subducted at convergent plate boundaries transform into eclogite at appropriate depths and thus release water into the overlying wedge of mantle peridotite. This addition of H_2O may cause melting to produce calc-alkaline magmas such as andesites. At substantially deeper levels in the mantle, localized heating at hot spots generates alkalic magmas. These appear to have formed in contact with fluids rich in CO_2.

DIFFERENTIATION IN MELT-CRYSTAL SYSTEMS

In the preceding section, we have painted a broad picture of melting processes in the mantle. In reality, the picture is much fuzzier than we have led you to believe, because very few mantle-generated magmas arrive at the earth's surface in pristine condition. Most have been altered in composition (differentiated) to various degrees. We will now see how that happens.

Fractional Crystallization

The most common mechanism by which magmas are differentiated is *fractional crystallization,* the physical separation of crystals and liquid so that these phases can no longer maintain equilibrium. Crystals may be isolated from the major part of their parent magma through gravity settling (or rarely, floating), or they may be carried to the bottom of a magma chamber and deposited by convection currents. A pile of accumulated crystals may be purified further by squeezing out of some of the interstitial liquid as the overburden of additional crystals accumulates. Rapid crystallization, in which the reaction of melt with solid phases cannot keep up with falling temperature and advancing solidification, can also produce compositionally zoned crystals, isolating their interiors kinetically to promote fractional crystallization. Several igneous rocks affected by these fractional crystallization processes are shown in Figure 11.13. All of these magmatic processes act to obscure the information about the mantle that may be carried in igneous rocks.

Because equilibrium crystallization and melting are simply the reverse of each other, it may seem logical to assume that fractional crystallization and melting also share this characteristic. This is not the case, however. To see how fractional crystallization differs from equilibrium crystallization, let's consider two binary phase diagrams that involve olivine.

Worked Problem 11-4

Fractional melting in the system forsterite-silica has already been discussed in Worked Problem 11-3. Using the phase diagram for this system in Figure 11.8, trace the fractional crystallization paths of melts of various compositions.

Cooling of any melt composition to the liquidus will cause the appropriate solid

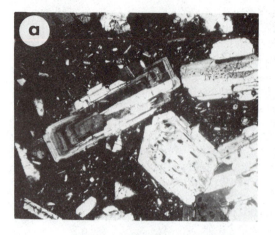
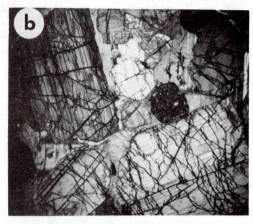

Figure 11.13 Photomicrographs of rock affected by fractional crystallization. Width of each photograph corresponds to approximately 5 mm. (a) Zoned plagioclase crystals were partly isolated from the melt by kinetic factors during growth. These phenocrysts occur in andesite from Hakone volcano, Japan. (b) Cumulate orthopyroxenes were concentrated on the bottom of a magma chamber, and most of the interstitial liquid was squeezed out of the accumulated pile. This sample is from the Ultramafic Zone of the Stillwater layered complex, Montana.

phase to appear. If we begin with a composition of 63% silica, removal of En crystals as they form will drive the residual liquid composition toward the eutectic. When this point is reached, both En and Cr will separate from the melt. The liquid trajectory is, in this case, no different from that for equilibrium crystallization, but in fractional crystallization, the bulk composition of the magma changes because crystals are removed. Under equilibrium conditions, these crystals remain suspended in the magma and become part of the resulting erupted lava and igneous rock.

Now consider a liquid having 55% silica. The first solid phase to form is now forsterite. Under equilibrium conditions, the liquid should follow the Fo liquidus to intersect the peritectic, at which point it reacts with already crystallized Fo to produce En. Under fractional crystallization conditions, Fo is physically removed from the system as it forms, so there is none of this phase to react with the melt at the peritectic temperature. Thus, there is no Fo + L → En reaction to cause the liquid to delay at the peritectic; it is as if the peritectic were not there. The cooling liquid slides through the peritectic and immediately begins to crystallize En rather than Fo. The melt composition continues toward the eutectic. Under conditions of perfect fractionation, the liquid will reach the eutectic, but such ideal conditions do not exist in nature, so that the liquid will be exhausted at some point before the eutectic is reached.

Olivine that crystallizes from real magmas is, of course, not pure forsterite, but a solid solution of Mg- and Fe-rich end-members. The phase diagram for the system forsterite-fayalite (Fo-Fa), shown in Figure 11.14, is very similar to the one for the plagioclase system that was used in Chapter 9 to illustrate solid solution behavior. Crystallization of a liquid of composition *a* in Figure 11.14 initially produces olivine of composition *b*. If equilibrium were achieved, this olivine would then react continuously with the melt to form more Fe-rich compositions, ultimately having composition *d* as the last drop of liquid is exhausted. Under fractional crystallization conditions, however, the early-formed Mg-rich olivine is removed from the melt, so this reaction cannot take place. Hence, the liquid crystallizes a range of olivines that are progressively more Fe-rich, from composition *b* to *d* and beyond. Again, under conditions of perfect fractionation, the liquid should ultimately reach the composition of pure fayalite, but in real systems we cannot specify the exact point at which the last liquid will solidify, only that the liquid composition will continue past point *c*.

Fractional crystallization affects ternary systems in an analogous manner. Reac-

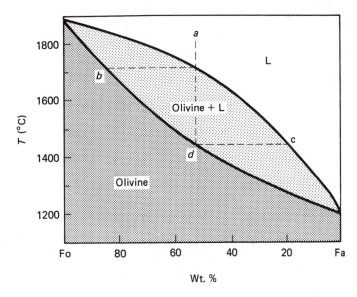

Figure 11.14 Phase diagram for the system forsterite (Mg$_2$SiO$_4$)-fayalite (Fe$_2$SiO$_4$), illustrating how fractional crystallization affects a solid solution series. Olivine crystals forming from a melt with initial composition *a* will range in composition from *b* to some point past *d*.

tion curves, the ternary equivalents of binary peritectic points, no longer define the compositions of fractionating liquids, because there are no solid phases to react. Fractional crystallization in a more complex system is described in Worked Problem 11-5.

Worked Problem 11-5

Phase relationships in the system forsterite-anorthite-silica are shown in Figure 11.15. This is actually a pseudoternary rather than a ternary system, because the primary phase field of spinel appears in the diagram but the composition of spinel lies outside of this plane. The only complication this situation presents is that we cannot specify the crystallization path of liquids *within* the spinel field, because we do not know where on this diagram the projection of spinel plots. We have already applied Alkemade's Theorem to identify the various subtraction and reaction curves in this diagram. Describe fractional crystallization paths for liquids of composition *x* and *y* in Figure 11.15.

Liquid *x* cools to the Fo liquidus and then changes in composition away from the forsterite apex as this phase is removed. When the liquid composition intersects the subtraction curve, Fo + An crystallize and are separated together. The liquid moves to and through the tributary reaction point, because there is no olivine in contact with melt to react to form enstatite. The liquid continues toward the ternary eutectic point between En, An, and Tr. We cannot specify the exact point where the liquid phase will be exhausted.

A liquid with composition *y* will fractionate Fo until it reaches the reaction curve for Fo to En. Because there is no Fo to react, the melt will step over this curve into the enstatite primary phase field. At this point, En will begin to crystallize, and the liquid composition will move directly away from En. Continued En fractionation will drive the melt to the subtraction curve, and An will join the fractionating assemblage. The liquid will then follow the subtraction curve toward the ternary eutectic.

Consideration of these experimentally derived phase diagrams provides some indication of the ways in which magmas *might* differentiate, but what is the evidence that fractional crystallization actually occurs as a widespread phenomenon in magmatic systems? To answer this question, we must examine real rocks for the effects of this process. Experimental studies of most basaltic rocks indicate that they are "multiply-

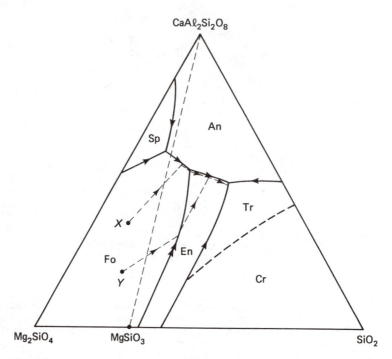

Figure 11.15 Phase diagram for the pseudoternary system forsterite (Mg₂SiO₄)-anorthite (CaAl₂Si₂O₈),-silica (SiO₂). The compositions of liquids derived by fractional crystallization of melt compositions *x* and *y* are illustrated by dashed lines.

saturated" at low pressures, that is, several minerals will begin to crystallize at virtually the same temperature. We would expect that kind of behavior for partial melts formed at low pressure, because melting begins at invariant points (such as eutectics, the lowest temperature points at which liquids are stable). However, basaltic magmas are mantle-derived and could not have formed at low pressures, so multiple-saturation must have another explanation. The fact that most basalts are already multiply-saturated at low pressure suggests that they have been differentiated by fractional crystallization near the surface or, in other words, have evolved into eutectic compositions. Many petrologists argue that fractional crystallization also occurs under high pressure as magmas ascend, and that such polybaric differentiation can account for much of the observed diversity in mantle-derived magmas.

It has also been recognized that many basalts tend to cluster at certain preferred compositions. This observation is difficult to interpret if these are primitive magmas, unless the preferred compositions represent those of liquids in equilibrium with mantle peridotite at invariant points. An alternative view is that they represent points along a differentiation path that have special properties that allow them to erupt. In fact, Stolper and Walker (1980) noted that the density of basaltic liquids first decreases and then increases during fractional crystallization. One preferred composition for basalts corresponds to that which has the minimum density, and thus may be the very composition that would be most likely to ascend to the surface. This implies that the crust may actually act as a filter that stops unfractionated magmas.

Widespread fractionation has obvious ramifications for what we can hope to learn about the mantle from igneous rocks. Crystallization experiments on lavas that have experienced fractional crystallization serve no purpose, except to understand the crystallization path itself. Luckily, there are a few apparently unfractionated magmas that somehow reach the surface, so judicious selection of samples for experiments may yet provide useful information on mantle melting processes.

Chemical Variation Diagrams

The chemical changes wrought by fractional crystallization can be modelled quantitatively using mixing calculations, which are most easily visualized from element *variation diagrams*. Figure 11.16 illustrates the predicted path of a liquid, called the *liquid line-of-descent*, as it undergoes fractional crystallization of different minerals. In (a), crystals of composition A separate from liquid L_1, with the result that the liquid composition is driven toward L_2. In (b) and (c), mixtures of two or three minerals crystallize simultaneously. The bulk composition of the extracted assemblage is given in each case by A, and the liquid line-of-descent follows the path $L_1 \rightarrow L_2$. In (d), magma L_1 first crystallizes mineral D and then D plus E, with the bulk composition of the fractionating assemblage represented by A. In this case, the liquid line-of-descent contains a kink where the additional phase joins the crystallization sequence. A more complex case in which the fractionating phase is a solid solution is illustrated in (e). Mineral AB is a solution of end-members A and B, and the proportions of these end-members change as temperature drops during the crystallization sequence. The trajectory of the liquid at any point is defined by a tangent to the liquid composition drawn from the composition of crystallizing AB. For example, the liquid path when $A_{80}B_{20}$ and $A_{50}B_{50}$ are extracted is shown by the dashed tangents from these compositions. In this case, the liquid line-of-descent traces a curve. In all of these diagrams, the relative proportions of fractionating phases can be determined by application of the lever rule.

In actual practice, these problems must be worked in reverse. A suite of volcanic rocks representing various points along a liquid line-of-descent is analyzed, and from these we must determine the fractionated phases. To test possible solutions to this problem, we require rock bulk compositions and the identities and compositions of extracted minerals. The latter may be obtained from petrographic observations and microprobe analyses of phenocrysts in the rocks. An example of such an exercise is given in Worked Problem 11-6.

Worked Problem 11-6

How can we test for fractional crystallization in real rocks? Assume that we have collected a suite of volcanic rock samples that we believe are part of a fractionation sequence. Furthermore, we have obtained bulk chemical analyses of the rocks. Petrographic observations indicate that three phases—olivine (ol), plagioclase (plg), and clinopyroxene (cpx)— exist as phenocrysts in this suite of these rocks, and we have also performed microprobe analyses of these minerals. Because these three phases were obviously on the liquidus at some point, we will determine whether fractional crystallization of some combination of them could have produced the observed residual liquids.

Figure 11.17 shows schematically the mineral compositions and the liquid line-of-descent $(L_1 \rightarrow L_2)$ defined by the analyzed bulk rock compositions using three different variation diagrams. The information gained from any of them individually is ambiguous, but we can compare them to obtain a unique answer. In (a), we see that the liquid line-of-descent is consistent with fractionation of ol + plg, ol + cpx, or ol + plg + cpx. This diagram thus rules out cpx + plg as a possibility. In (b), ol + cpx or ol + plg + cpx are permissible assemblages that could be extracted to produce this sequence but ol + plg, allowed in (a), is not allowed. The final distinction between ol + cpx and ol + plg + cpx is made using (c). This diagram indicates that fractionation of ol + cpx is the only internally consistent model. The testing of all interrelationships between elements is an important part of this kind of study; otherwise, we could have incorrectly assumed that all of the observed phenocrysts could have fractionated to produce this liquid line-of-descent. Notice also that the relative proportions of olivine and plagioclase (approximately 1 : 1) in all the

Differentiation in Melt-Crystal Systems

diagrams are the same, as deduced from application of the lever rule to the point at which the extrapolated liquid line-of-descent crosses the olivine-plagioclase tie-line. Not only the identity of the fractionating phases but also their relative proportions should be consistent in this set of variation diagrams.

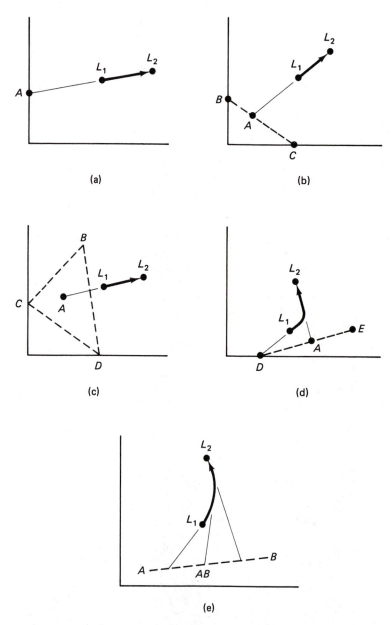

Figure 11.16 Schematic chemical variation diagrams (axes can be any two elements or oxides) showing the liquid line-of-descent ($L_1 \rightarrow L_2$) during fractional crystallization of one (a), two (b), or three phases (c). In (d), the line-of descent has an inflection due to late entry of a second phase. In (e), the line-of-descent is curved because of continuous change in the composition of an extracted solid solution phase.

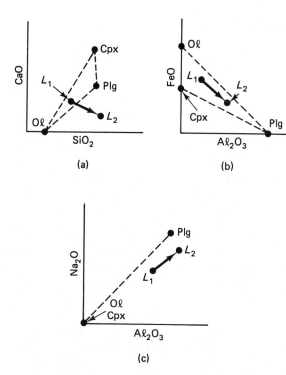

Figure 11.17 Schematic chemical variation diagrams for a suite of basalts containing phenocrysts of olivine (ol), clinopyroxene (cpx), and plagioclase (plg). The liquid line-of-descent defined by bulk chemical compositions of these basalts can be derived by fractional crystallization of ol + plg.

As the number of possible fractionating phases or the number of analyzed elements increases, it becomes more difficult to apply these graphical methods. Numerical methods offer a real advantage in such situations. Several computer programs, commonly called *petrologic mixing programs,* are available for use on small computers to perform calculations of this type. These compute the composition of the parent magma by mixing the residual liquid with fractionating phases in various proportions. The answer is a best fit to the least-squares regression for the analyzed liquid line-of-descent. Two widely used programs are MODII (Wright and Doherty, 1970) and IGPET (available from M. J. Carr, Rutgers University).

Liquid Immiscibility

Fractional crystallization is not the only mechanism by which magmas can differentiate. Experimental petrologists have recognized for half a century that the binary systems $CaO-SiO_2$, $FeO-SiO_2$, and $MgO-SiO_2$ contain fields in which silicate melts will separate spontaneously into two liquids. The phase diagram for the forsterite-silica system was shown earlier in Figure 11.8; a stability field for two liquids is shown near the SiO_2 compositional end member. This region of unmixing behaves exactly like a solvus, except that it occurs above the solidus. Liquid phase separation, or *immiscibility,* can cause magmatic differentiation, provided that the two melts have different densities. In such a case, globules of the more dense liquid tend to coalesce and sink to the bottom of the magma chamber, in a similar manner to the segregation of crystals by gravity fractionation.

Liquid immiscibility obviously indicates highly nonideal mixing of melt components. However, it may be helpful to recognize that the separation of an immiscible liquid fraction is really no different than the formation of a crystal, at least in the thermodynamic sense. In both cases, the parental melt becomes saturated with respect to a new phase as temperature decreases. However, there is a difference in the ease with which equilibrium can be achieved in these two situations. Crystal growth and separation of an additional liquid phase both proceed by diffusion of ions through the melt to the sites of growth and across the interface. Diffusion in liquids is generally many orders of magnitude faster than in crystals, so immiscible melts are more likely to maintain equilibrium with falling temperature. Consequently, a cooling magma may form zoned plagioclase crystals under nonequilibrium conditions while at the same time segregating an immiscible liquid fraction under equilibrium conditions.

Although many examples of silicate liquid immiscibility are known from melting experiments on simple systems, most of these occur at very high temperatures and have geologically unreasonable compositions (see, for example, the very high temperature and SiO_2 content of immiscible liquids in the system forsterite-silica in Figure 11.8). The only experimentally known example of silicate liquid immiscibility that may have a natural analog is in the system fayalite (Fe_2SiO_4)-leucite ($KAlSi_2O_6$)-silica. As shown in Figure 11.18, this system contains two immiscibility fields, one along the Fa-silica sideline and the other, of more geochemical interest, near the center of the diagram. The upper surface of this immiscibility field slopes downward from a maximum temperature at point A to a minimum at B. You will realize with a moment's reflection that the size of this solvus must increase with decreasing temperature, so we should contour this surface at different temperatures to portray it accurately. For clarity, we have not done that here; the dotted line represents only one of these contours, the 1180°C isotherm. This two-liquid field is a reasonable compositional model for immiscibility in some uncommon dike rocks ("lamprophyres") and in the residual liquids formed during the last stages of basalt crystallization (*mesostasis*). It is unlikely, however, that silicate liquid immiscibility produces large volumes of differentiated magmas, because these iron-rich compositions are produced at the end of fractional crystallization sequences when little liquid is left.

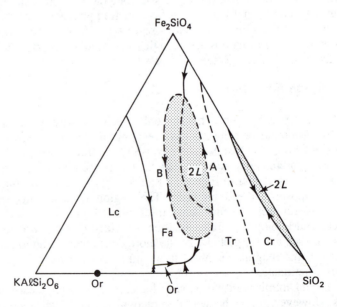

Figure 11.18 Phase diagram for the ternary system fayalite (Fe_2SiO_4)-leucite ($KAlSiO_4$)-silica (SiO_2), illustrating two shaded fields of silicate liquid immiscibility. (Modified from Roedder 1979.)

Liquid immiscibility seems to be most effective when the structures of the two melts are radically different. This can be achieved most readily if melt components with nonsilicate compositions are considered. One important example involves the separation of carbonate and silicate melts in CO_2-bearing systems. A photomicrograph of globules of carbonate-rich magma suspended in alkalic basalt is illustrated in Figure 11.19. Immiscibility in such systems has been well documented experimentally and almost certainly accounts for the production of the carbonate magmas ("carbonatites") that are sometimes associated with alkalic silicate magmas. Another example is the separation of immiscible sulfide melts from basaltic magmas on cooling. The sulfide melts are more dense than silicate magma and tend to sink, in some cases forming sulfide ore deposits.

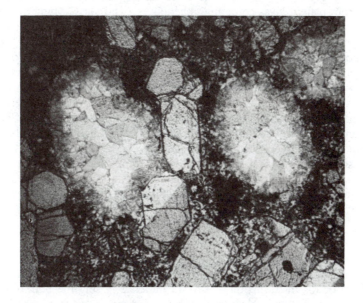

Figure 11.19 Photomicrograph of immiscible globules of carbonatite magma in alkali basalt. This sample is from the Kaiserstuhl volcanic complex, Germany. Width of the photograph corresponds to approximately 5 mm.

THE BEHAVIOR OF TRACE ELEMENTS

Much of the gemstone industry depends on the chance incorporation of minute quantities of certain elements into otherwise "ho-hum" minerals. Small proportions of foreign impurities can turn ordinary corundum into valuable rubies or sapphires. For the geochemist, elements present in trace concentrations can serve another purpose—to test petrogenetic models.

Trace elements are commonly defined as those occurring in rocks in concentrations of a few tenths of a percent or less by weight. The mixing behavior of trace components in crystals or melts is often highly nonideal; consequently, different minerals may concentrate or exclude trace elements much more selectively than they do major elements. Trace element distributions can therefore provide quantitative constraints on the processes of partial melting and differentiation that are not possible from consideration of major elements.

The Behavior of Trace Elements

Fractionation of Trace Elements during Melting and Crystallization

In melt-crystal systems under equilibrium conditions, elements are partitioned among phases according to their activities in those phases. For trace elements, we normally assume Henry's law behavior, $a_i = h_i X_i$. The *distribution coefficient*, K_D, derived earlier in Chapter 8, is given by K_D = concentration in mineral/concentration in liquid. Trace element distribution coefficients are based on the assumption that the trace component obeys Henry's law in the phase of interest. Experimental studies suggest that the linear Henry's law region for many trace elements extends at least to several weight percent, so it must be obeyed for the concentration ranges measured in natural rocks.

In principle, K_D can be measured from partially crystalline experimental charges or from natural lavas containing phenocrysts and glass. Many of the available distribution coefficients were determined from natural rocks, but there is unfortunately no assurance that these reached equilibrium. There are also analytical problems in obtaining concentrates of pure phases (that is, crystals with no adhering or included glass, and vice versa), because trace elements are usually below the detection limits for *in situ* measurements. Another problem with K_D's is that they are dependent on the compositions of the phases. In other words, their distributions are nonideal; usually enough information is not at hand to determine the appropriate activity coefficients. Despite these potential pitfalls, distribution coefficients are widely used to solve geochemical problems, and their compositional dependence can be minimized by selecting K_D values for phases similar in composition to those in the problem of interest.

The *bulk distribution coefficient*, D is the K_D value adjusted for systems containing more than one solid phase. It is calculated from the weight proportions w_ϕ of each mineral by

$$D = \sum_\phi w_\phi K_{D\phi}. \tag{11-1}$$

D must be calculated if we intend to handle models for partial melting of mantle peridotite or for fractionation of several minerals. Using the bulk distribution coefficient, we can predict trace element distributions during various magmatic processes.

We will now derive expressions that describe changes in the concentrations of trace elements during solidification of a magma.[3] We will consider two ideal cases, equilibrium crystallization and perfect fractional crystallization (the latter is sometimes called *Rayleigh fractionation*). We will assume that more than one phase is crystallizing.

In the case where equilibrium between crystals and liquid is maintained,

$$X_{i\,\text{solid}}/X_{i\,\text{melt}} = D. \tag{11-2}$$

If a fraction of magma crystallizes, the fraction of melt remaining F is given by

$$F = n_{\text{melt}}/n_{\text{melt}}^o, \tag{11-3}$$

where n_{melt}^o is the number of moles of all components in the original magma and n_{melt} is the total number of moles remaining after crystallization commences. The ratio of solid to melt remaining is therefore

$$(n_{\text{melt}}^o - n_{\text{melt}})/n_{\text{melt}} = 1/F - 1. \tag{11-4}$$

---

[3] These derivations follow closely those of Wood and Fraser (1977).

If we define y_i as the number of moles of trace component i in the system, equations 11-2 and 11-4 can be combined to give

$$X_{i\,\text{solid}}/X_{i\,\text{melt}} = [y_{i\,\text{solid}}/(n^o_{\text{melt}} - n_{\text{melt}})]/(y_{i\,\text{melt}}/n_{\text{melt}}) = D.$$

Because $y_{i\,\text{solid}} = y^o_{i\,\text{melt}} - y_{i\,\text{melt}}$,

$$D = [(y^o_{i\,\text{melt}}/y_{i\,\text{melt}}) - 1][(n_{\text{melt}}/(n^o_{\text{melt}} - n_{\text{melt}}))].$$

Rearranging and dividing by $n^o_{\text{melt}}/n_{\text{melt}}$ gives

$$(y^o_{i\,\text{melt}}/n^o_{\text{melt}})/(y_{i\,\text{melt}}/n_{\text{melt}}) = \{D[(n^o_{\text{melt}} - n_{\text{melt}})/n_{\text{melt}}] + 1\}n/n^o.$$

Substituting equation 11-3 into the expression above yields

$$X_{i\,\text{melt}}/X^o_{i\,\text{melt}} = 1/(D - FD + F). \tag{11-5}$$

Expression 11-5 gives the concentration of trace element i in the liquid as a function of the original concentration, the bulk distribution coefficient, and the fraction of liquid remaining for a system undergoing equilibrium crystallization. The same expression also holds for equilibrium melting.

Now let's see how this differs from fractional crystallization. Using the same symbols as above, the mole fraction of trace component i in the system is $X_i = y_i/n$. If fractionation of a mineral containing i occurs, then n becomes $n - dn$, and y becomes $y - dy$. The compositions of solid and melt are therefore

$$X_{i\,\text{solid}} = dy/dn \tag{11-6a}$$

and

$$X_{i\,\text{melt}} = (y - dy)/(n - dn). \tag{11-6b}$$

If trace component i obeys Henry's law, then

$$X_{i\,\text{solid}} = dy/dn = DX_{i\,\text{melt}}. \tag{11-7}$$

However, dy/dn can also be expressed in terms of $X_{i\,\text{melt}}$. If we neglect dy in comparison with much larger y and dn in comparison with n in equation 11-6b, we see that $X_{i\,\text{melt}}$ can be approximated as y/n and y as $n(X_{i\,\text{melt}})$. Differentiating the latter expression with respect to n gives

$$dy/dn = n(dX_{i\,\text{melt}}/dn) + X_{i\,\text{melt}}.$$

Substitution of this expression for dy/dn into equation 11-7 gives

$$DX_{i\,\text{melt}} = n(dX_{i\,\text{melt}}/dn) + X_{i\,\text{melt}},$$

which can be recast into the form

$$[1/X_{i\,\text{melt}}(D - 1)]dX_{i\,\text{melt}} = (1/n)\,dn. \tag{11-8}$$

The change in concentration of i during crystallization can now be obtained by integration of expression 11-8 between X^o_i and X_i and between the initial and final moles of melt n^o and n, as follows:

$$\ln(n/n^o) = [1/(D - 1)] \ln(X_{i\,\text{melt}}/X^o_{i\,\text{melt}}),$$

or

$$X_{i\,melt}/X^o_{i\,melt} = (n/n^o)^{D-1}.$$

Because (n/n^o) equals F, the fraction of liquid remaining,

$$X_{i\,melt}/X^o_{i\,melt} = F^{D-1}. \tag{11-9}$$

Thus, expressions 11-5 and 11-9 describe the behavior of trace elements under the extreme conditions of equilibrium and fractional crystallization, respectively.

Using the same principles, analogous expressions can be derived for partial melting models. These are given below without derivation. For equilibrium melting

$$X_{i\,melt}/X^o_{i\,solid} = 1/(D - FD + F), \tag{11-10}$$

and for fractional melting

$$X_{i\,melt}/X^o_{i\,solid} = 1/D(1 - F)^{(1/D)-1}. \tag{11-11}$$

Because fractional melting involves continuous removal of melt as it is generated, this process probably does not occur in nature, but it can be considered as an end-member for intermediate melting processes. Geochemical models for magma generation have become very complex and are beyond the scope of this book; Shaw (1977) gives an excellent account of this topic.

The equations we have just derived can be expressed equally well in units of concentration other than mole fractions, as long as the distribution coefficients are defined in those same terms. In fact, most trace element analyses are expressed in units of weight—for example, parts per million (ppm). We can substitute these concentration units for mole fractions in equations 11-5 and 11-9 if D is expressed as a weight ratio.

The changes in trace element concentration for residual liquids during equilibrium and fractional crystallization are illustrated graphically in Figure 11.20. Curves labelled $K_D = 0$ are limiting cases, and trace element enrichments greater than these cannot be produced by crystallization processes alone.

Compatible and Incompatible Elements

It is useful to distinguish between trace elements that are *compatible* in the crystallographic sites of minerals and those that are *incompatible* with such minerals. Compatible elements have K_D values greater than 1, whereas incompatible elements have K_D's much less than 1. Consequently, compatible elements will be preferentially retained in the solid residual on partial melting or extracted from the liquid during fractional crystallization; incompatible elements will show opposite behavior. This can readily be seen in Figure 11.20. Of course, as new minerals begin to melt or crystallize during the evolution of a magma, a particular trace element may alter its behavior. For example, phosphorus may act as an incompatible element during magmatic crystallization until the point at which apatite starts to form, after which it behaves compatibly. For applications involving mantle-derived magmas, it is often convenient to define compatible and incompatible behavior in terms of the minerals that constitute mantle rocks—olivine, pyroxenes, garnet, and spinel. Trace elements that are incompatible with these phases are sometimes called *hygromagmatophile* or *large-ion lithophile*. Whether a particular trace element is compatible or incompatible is determined primarily by its ionic size and electrostatic charge. Incompatible elements are generally those with ionic radii too large to substitute for the more abundant elements, or those with charges of +3 or higher.

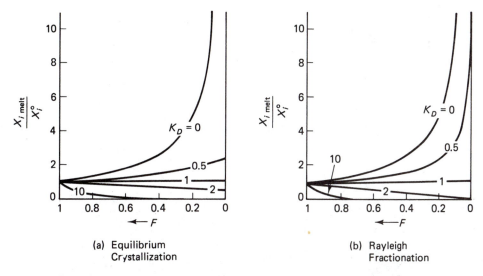

Figure 11.20 Changes in trace element concentrations in residual liquids for different values of K_D during (a) equilibrium crystallization and (b) Rayleigh fractionation. Element concentrations are given in terms of mole fractions of each element i relative to its concentration in the parent magma. F is the fraction of liquid remaining. (After Wood, B. J. and Fraser, D. G., *Elementary thermodynamics for geologists*. Copyright 1977 by Oxford University Press. Reprinted with permission.)

PREDICTING TRACE ELEMENT BEHAVIOR: TRANSITION AND RARE EARTH ELEMENTS

Chemical physics provides powerful ways to predict trace element behavior. We will illustrate this approach by considering several types of *transition elements,* which are characterized by inner d or f atomic orbitals that are incompletely filled by electrons. First we will see how chemical bonding models can predict relative distribution coefficients, and then we will utilize the physical properties of atoms.

Elements of the first transition series (Sc, Ti, V, Cr, Mn, Fe, Co, Ni and Cu) have incompletely filled 3d orbitals, and cations are formed when the 4s and some 3d electrons are removed. Divalent ions in this series are compatible in common ferromagnesian minerals because of similarities to Mg^{2+} in ionic size and charge. However, the degree of compatibility differs markedly and can be explained by the effect of crystal structure on d orbitals. *Crystal field theory* assumes that electrostatic forces originate from the anions surrounding the crystallographic site in which a transition metal ion may be located, and that these control its substitution in the structure.

A transition metal ion has five 3d orbitals which have different spatial geometries, as illustrated in Figure 11.21a. In an isolated ion, the d electrons have equal probability of being located in any of these orbitals, but they will attempt to minimize electron repulsion by occupying orbitals one electron at a time. For ions with more than five d electrons, this is obviously not possible, so a second electron can be added to each orbital by reversing its spin.

When a transition metal ion is placed in a crystal field (for example, in octahedral coordination with surrounding anions), the five 3d orbitals are no longer degenerate; that is, they no longer have the same energy. Electrons in all five orbitals are repelled by the negatively charged anions, but electrons in the d_{z^2} and $d_{x^2-y^2}$ orbitals, which are oriented parallel to bonds between the cation and the anions around it, are affected to a greater extent than those in the other three. In octahedral coordination, therefore, these two orbitals will have a higher energy level. The reverse is true for ions in tetrahedral coordination.

The energy separation between the two groups of orbitals is called *crystal-field splitting*. This is illustrated in Figure 11.21b. Ions with electrons in only the lower energy orbitals

The Behavior of Trace Elements

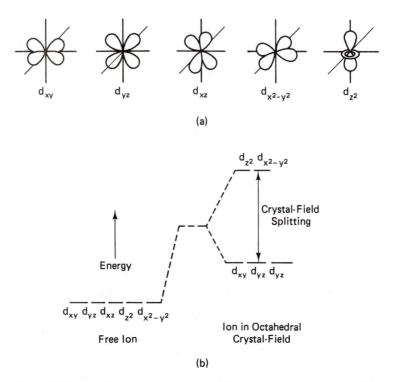

Figure 11.21 (a) Sketch of the geometric orientations of the five d orbitals in transition metal ions. As shown in (b), when the free ion is placed in an octahedral mineral site, its total energy increases. Some orbitals have higher energy because they experience greater electron repulsion; the magnitude of the energy difference between the two groups of orbitals is the crystal-field splitting.

for a particular site geometry will be stabilized in that site, whereas electrons in the higher energy orbitals will destabilize the ion. Of course, the number of electrons depends on which transition metal the cation is made from and on its oxidation state, so some transition elements will prefer octahedral sites more than others. For example, the predicted octahedral site preference energies for divalent cations decrease in the order Ni > Cu > Co > Fe > Mn, and for trivalent cations Cr > V > Fe. As olivine or spinel (which offer octahedral sites) crystallizes from a magma, we should expect to see rapid uptake of Ni and Cr in these minerals relative to other compatible transition metals. Tetrahedral sites in minerals produce a different but predictable order of cation site preferences. Crystal-field effects, thus, provide an explanation for the differing K_D's for transition metals. The geochemical and mineralogical applications of crystal-field theory are treated comprehensively by Burns (1970).

The *lanthanide* or *rare earth elements,* commonly abbreviated as REE, include La, Ce, Pr, Nd, Pm, Sm, Eu, Gd, Tb, Dy, Ho, Er, Tm, Yb, and Lu. In this part of the periodic table, increasing atomic number adds electrons to the inner 4f orbitals, so this is a somewhat similar case to the transition metals we just discussed. The f electrons in REE, however, do not participate in bonding. The valence electrons occur in higher level orbitals, and loss of three of these results in a +3 charge for most REE cations. These are large ions and this, coupled with their high charges, makes them incompatible.

Europium (Eu) is unique in that its third valence electron does not enter the 5d orbital as in other REEs, but enters a 4f orbital, which then becomes exactly half full. This is a particularly stable configuration, and more energy is required to remove the third electron. A consequence of this is that under reducing conditions, such as on the moon or within the earth's mantle, Eu may exist in the divalent and trivalent states (just how much will depend on the redox state of the system). Eu^{2+} has about the same size and charge as Ca^{2+}, for which it readily substitutes, so europium, at least in part, may behave as a compatible element.

Under oxidizing conditions, such as in the marine environment, cerium (Ce) is oxidized to Ce^{4+}. This results in a significant contraction in its ionic radius; thus Ce^{4+} has a very different behavior from trivalent ions of the other rare earth elements. Cerium in the oceans precipitates in manganese nodules, consequently leading to a dramatic Ce depletion in seawater. Marine carbonates formed in this environment mimic this geochemical characteristic. Metalliferous sediments deposited at midocean ridges also exhibit Ce depletions, pointing to a seawater source for the hydrothermal solutions that produced them.

The natural abundances of rare earths vary by a factor of 30. To simplify comparisons, REE concentrations in rocks are normally divided by their concentrations in another rock

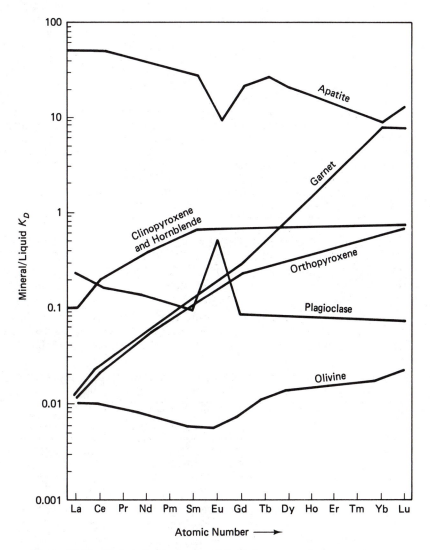

Figure 11.22 Typical distribution coefficients for rare earth elements between various minerals and basaltic melt, after Zielinski (1975). REE concentrations are normalized to values in chondritic meteorites. All of these phases are intolerant of REE (that is, have K_D values less than 1) except apatite (and garnet for the heavy rare earths). The positive Eu anomaly in plagioclase occurs because some of this element is in the divalent state and can substitute for Ca^{2+} ions.

The Behavior of Trace Elements

sample.[4] Normalization of REE data using chondritic meteorites or shales produces relatively smooth patterns that enable comparisons between samples, accentuating relative degrees of enrichment or depletion and contrasting behavior between neighboring elements in the lanthanide series. Normalized REE concentrations are conventionally shown on diagrams like Figure 11.22, plotted by increasing atomic number. The elements on the left side of the diagram thus have lower atomic weights and are called the light rare earth elements (LREE), to distinguish them from their heavier counterparts (HREE) on the right. What is important from the standpoint of geochemical behavior, however, is that the lanthanides show a regular contraction in ionic radii from La^{3+} to Lu^{3+}.

The rare earth elements, except for Eu^{2+} in reduced systems and Ce^{4+} in oxidized systems, are all incompatible but, like the first series transition metals, their behavior varies in degree. Some minerals can tolerate certain REE more easily than others, as shown in Figure 11.22. For example, apatite can readily accomodate all rare earth elements. Some minerals discriminate between different rare earths, resulting in their fractionation. Garnet has a phenomenal preference for heavy rare earths because of their smaller ionic radii, and pyroxenes show a similar but less pronounced affinity. The extra Eu in plagioclase, accounting for the spike, or *positive europium anomaly*, is Eu^{2+} substituting for calcium. Thus K_D's for the rare earth elements in various phases are controlled by size, rather than by bonding characteristics.

Studies of REE have found many uses in geochemistry. They constitute important constraints on petrogenetic models for partial melting in the mantle and magmatic fractionation. Because REE are relatively insoluble under conditions of metamorphism, they provide insights into the protoliths for metamorphic rocks. The lanthanides are also insoluble under conditions at the earth's surface, so their abundances in sediments reflect those of the average crust. The short residence times for REE in seawater also allow them to be used to track oceanic mixing processes. More exhaustive treatments of REE geochemistry can be found in Henderson (1983) and Taylor and McLennan (1987).

A knowledge of compatible and incompatible trace element distribution coefficients can provide a powerful test for geochemical models involving partial melting or crystallization. Some representative K_D values for trace elements in minerals in equilibrium with basaltic magma are presented in Table 11-3. An example of such a test is given in Worked Problem 11-7.

We are now in a position to evaluate why the crust-mantle system has become so stratified in terms of its heat-producing radioactive elements, as discussed in the earlier section in this chapter on geothermal gradients. Potassium, uranium, and thorium are

TABLE 11-3 REPRESENTATIVE TRACE ELEMENT K_D VALUES FOR MINERALS IN EQUILIBRIUM WITH BASALTIC MAGMA[*]

	Ni	Rb	Sr	Ba	Ce	Rare earth elements Sm	Eu	Yb
Olivine	10	0.001	0.001	0.001	0.001	0.002	0.002	0.002
Clinopyroxene	2	0.001	0.07	0.001	0.1	0.3	0.2	0.3
Orthopyroxene	4	0.001	0.01	0.001	0.003	0.01	0.01	0.05
Garnet	0.04	0.001	0.001	0.002	0.02	0.2	0.3	4.0
Amphibole	3	0.3	0.5	0.4	0.2	0.5	0.6	0.5
Plagioclase	0.01	0.07	2.2	0.2	0.1	0.07	0.3	0.03

* From Cox et al. (1979).

[4] Chondritic meteorites, which are thought to contain unfractionated element abundances, are the most commonly used standard for REE normalization. The geochemical significance of these meteorities will be considered in Chapter 13. Shales also generally have uniform lanthanide patterns, and several sets of shale REE abundances are used for normalization in some applications. As we have already discussed, shales have geochemical significance because their compositions may represent that of the average crust.

all very large ions and behave as incompatible elements. Consequently, any partial melts produced in the mantle will scavenge these elements, and over time they will become depleted from mantle residues and concentrated in the crust by ascending magmas. This differentiation of heat-producing elements has had a profound effect on the earth's thermal history.

Worked Problem 11-7

How can trace element abundances in basalts be used to constrain models for their petrogenesis? A geochemical study of volcanic rocks from Reunion Island in the Indian Ocean (Zielinski, 1975) provides an excellent example, Measured trace element abundances in eight samples are shown in Figure 11.23 on page 364. All of these samples contain phenocrysts of olivine, clinopyroxene, plagioclase, and magnetite in a groundmass of the same minerals plus apatite. Zielinski used the major element compositions of whole rocks and phenocrysts to construct a fractionation model. Using a petrologic mixing program, he found that lavas 3 through 8 could be formed by fractionation of these phenocrysts from assumed parental magma 2. Basalt 1 was not a suitable parental magma for this suite, even though it has the highest Mg/Fe ratio; this rock probably contains cumulus olivine and pyroxene. Zielinski then tested this model using trace elements.

The chondrite-normalized rare earth element patterns (Figure 11.23a) are parallel and increase in the order 1 to 8. This is just as expected; incompatible REE should have higher concentrations in residual liquids, and none of the solid phases fractionates light from heavy REE appreciably. Using the K_D values shown in Figure 11.22 and the proportions of fractionating phases from the mixing program, it is possible to calculate the bulk distribution coefficient (equation 11-1) for each sample. The mixing program also gives the fraction of liquid F. Assuming equilibrium between the entire separated solid and melt, substitution of these values into equation 11-5 gives the ratio $C_{i\,\text{melt}}/C_{i\,\text{melt}}^{o}$. Assume, as Zielinski did, that rock 2 is the parental magma; its trace element concentrations are then $C_{i\,\text{melt}}^{o}$. Thus we can calculate the expected REE pattern for liquids derived from 2. When Zielinski did this exercise, he was able to duplicate the measured REE patterns fairly accurately. The negative Eu anomaly in rock 8 is caused by the removal of large quantities of plagioclase at the end of the sequence.

The abundances of some other trace elements in these rocks are shown in Figure 11.23b. U and Th are even more incompatible than the rare earths and increase progressively in the sequence. Ba is also incompatible, but it and Sr can substitute in sodic plagioclase, so both drop towards the end of the sequence as feldspar fractionation becomes dominant. In contrast, Ni and Cr are compatible and show marked depletions in fractionated rocks. Their contents in rock 1 are much higher than expected for a primary magma, and this observation lends support to the idea that this rock contains cumulus phases. Using appropriate D and F values, we could also predict the abundances of these elements in the hypothetical fractionation sequence. Zielinski was also able to model these fairly well.

Even though Zielinski's modeling may appear convincing, we should note that this is not a unique solution to this problem. The same geochemical patterns might also be produced by various degrees of partial melting of mantle peridotite. Rock 8 would represent a magma formed during a small amount of melting. As melting advanced, the incompatible elements in the melt phase would be progressively diluted, ultimately to produce magma 2. Actually, Zielinski recognized this possibility, but argued against it on the basis of field evidence for shallow differentiation. Like many other tools of geochemistry, trace element modeling may give ambiguous answers, and is at its best when used in combination with other investigative methods.

CYCLING OF VOLATILE ELEMENTS

Another way in which the crust and mantle can interact geochemically is by exchange of volatile elements. Hydrogen, oxygen, carbon, sulfur, and other volatiles in the earth's interior may exist as a discrete fluid phase. We use the term *fluid* because in

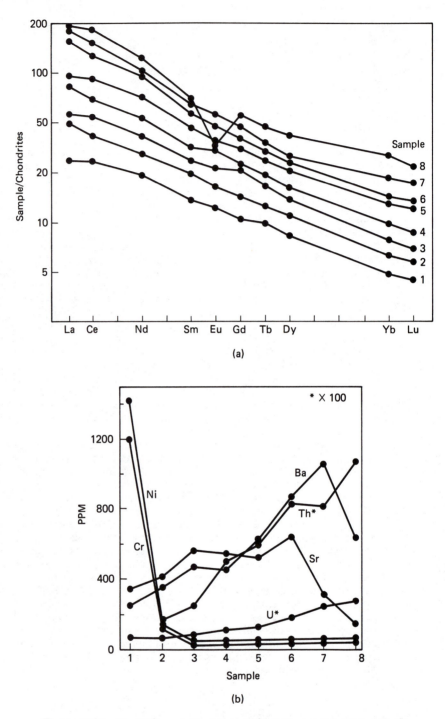

Figure 11.23 Chondrite-normalized rare earth element patterns (a) and other trace element variations (b) for a suite of volcanic rocks analyzed by Zielinski (1975). The patterns are consistent with a model whereby samples 3 through 8 were formed by progressive fractional crystallization of a liquid having the composition of sample 2. Sample 1 appears to be a cumulate rock.

general, *P-T* conditions are such that this phase is above its critical point, where the distinction between liquid and gas is meaningless. Before we can investigate how fluids are cycled in the crust-mantle system, we should examine their compositions.

Metamorphic Fluid Compositions

The compositions of fluids in the crust and mantle can be ascertained from three different lines of evidence. The first is direct chemical analysis of waters from geothermal fields. Active metamorphism is occurring in some of these environments, like the Salton Sea, California (McDowell and Elders, 1980), and deep wells in these locations can provide fluid samples. In Chapter 6, we showed how computer programs like EQ6 can be used to model the chemical evolution of natural waters. Sophisticated methods like this have been used to verify that the measured compositions of such fluids are in equilibrium with the metamorphic minerals in core samples, supporting the idea that these are contemporary metamorphic fluids. It is often difficult, however, to assess the extent to which such fluids may have been contaminated by meteoric water or drilling muds, and sampling of fluids from depths of more than a few kilometers is out of the question.

Fluid inclusions trapped in the minerals of crustal and mantle rocks provide another method of direct observation. Although these inclusions contain only minute quantities of fluids, they can sometimes be extracted for chemical analysis. More commonly, their compositions are inferred from heating and freezing experiments *in situ*. The experimental manipulation of fluid inclusions is considered further in an accompanying box.

EXTRACTING THE INFORMATION IN FLUID INCLUSIONS

Fluids commonly are stranded in interstitial spaces as mineralization proceeds. They may be trapped along old crystal boundaries, in fractures, or in pore spaces. With time, these gradually change shape to minimize their surface energies. The result is a population of more or less equally dimensioned spherical fluid inclusions. Because the minimal surface energy may be best attained by adapting inclusion shape to the structure of the host mineral, some inclusions have *negative crystal* shapes. The fluid at the time of trapping is usually a single phase liquid or supercritical fluid. The fluid will shrink on cooling, because the coefficient of thermal expansion for the host mineral is much less than for the fluid. As the pressure in inclusions drops below the vapor pressure of the fluid components, vapor bubbles will nucleate and grow. Most fluid inclusions, therefore, contain both a liquid and a vapor phase. Solid phases are also commonly encountered in inclusions which have become saturated with respect to one or more salts on cooling. These *daughter crystals* may be valuable clues to the compositions of the parent fluid. It is also common, particularly in sedimentary basin environments, to find inclusions with a second, immiscible fluid (usually organic). Some of these features are illustrated in the inclusion in Figure 11.24.

Experimental manipulation of fluid inclusions can provide information used for geothermometry. Phase transformations that occur with changing temperature can be observed by fitting a heating stage to a microscope. As an inclusion is heated, the vapor bubble shrinks and finally disappears. The temperature of homogenization is a measurement of the temperature at which the fluid was trapped, provided that it was very near the boiling point curve. If the trapped fluid was supercritical, its trapping temperature may be estimated from the homogenization temperature and an independent estimate of confining pressure. Fluid inclusion geothermometry is based on the assumption that the original sealed fluid was a single, homogeneous phase; an observational test of this assumption is that all inclusions trapped simultaneously should have the same bulk composition and therefore the same volumetric ratio of phases at room temperature.

Cycling of Volatile Elements

Figure 11.24 Photomicrograph of a fluid inclusion in fluorite from southern Illinois. The inclusion measures about 200 μm in its longest dimension, and consists of a NaCl-rich brine and a gas bubble. Note that the inclusion has a shape that mimics the host fluorite crystal. (Courtesy of C. K. Richardson.)

The fluids trapped in large inclusions can be sampled by drilling into them and extracting them with a micropipette. Their compositions can be determined directly by various standard analytical techniques. Fluids in small inclusions present a much greater challenge. If only one generation of inclusions is present, the fluid they contain may be released by crushing, but some information can also be gleaned from nondestructive observations. Salinity can be determined by measuring the freezing point of a fluid. This can be done on the same microscope stage used for the heating experiments, usually by passing liquid nitrogen through a cooling jacket on the stage. Freezing point depression is a function of total salinity and the identity of the salt in solution. Generally, it is assumed that inclusions are primarily NaCl solutions, and salinity is calculated in terms of weight percent *NaCl equivalent*. More sophisticated studies may involve spectroscopic measurements of inclusions to provide further information on fluid compositions. Daughter crystals may be identified by using optical properties such as color, pleochroism, and crystal habit, as well as melting points.

As with most geochemical techniques, there are many potential pitfalls in fluid inclusion work. Leakage of fluid into or out of inclusions after trapping is always of great concern. Care must be taken to identify multiple generations of inclusions by optical methods, because each generation will presumably have its own specific composition and trapping temperature. Recrystallization of the host mineral may cause several generations of inclusions to coalesce into large inclusions, or separation of large inclusions into smaller ones ("necking"). This latter process is critical if inclusions consist of more than one phase, because the phases may not be redistributed uniformly. These potential problems can be overcome in most cases, however, so that fluid inclusions studies offer a unique source of information on fluids from the earth's interior.

The third method of determining fluid compositions is by studying metamorphic assemblages which depend on fluid composition. Characterization of fluids in this case is based on the thermodynamic principles that we have already considered in Chapter 8. For example, for the equilibrium reaction

$$\text{Al}_2\text{Si}_4\text{O}_{10}(\text{OH})_2 \longleftrightarrow \text{Al}_2\text{SiO}_5 + 3\text{SiO}_2 + \text{H}_2\text{O},$$

pyrophyllite   andalusite   quartz   fluid

we know that

$$\mu_{\text{Al}_2\text{Si}_4\text{O}_{10}(\text{OH})_2,\text{pyroph}} = \mu_{\text{Al}_2\text{SiO}_5,\text{and}} + 3\mu_{\text{SiO}_2,\text{qtz}} + \mu_{\text{H}_2\text{O},\text{fluid}}.$$

Evaluation of the chemical potentials in the solid phases allows calculation of $\mu_{\text{H}_2\text{O},\text{fluid}}$. This can be converted into fluid composition $(X_{\text{H}_2\text{O},\text{fluid}})$ if the relationship between chemical potential and composition is known from an appropriate equation of state for the fluid. Worked Problem 11-8 illustrates how this is done.

Worked Problem 11-8

Ferry (1976) determined the composition of fluids during regional metamorphism in south-central Maine. We will examine his calculations for zoisite-bearing rocks, in which the following reaction occurs:

$$2 \text{ zoisite} + \text{CO}_2 = 3 \text{ anorthite} + \text{calcite} + \text{H}_2\text{O}.$$

At equilibrium,

$$3(\Delta\overline{G}^o_{\text{An}} + RT \ln a_{\text{An,plag}}) + \Delta\overline{G}^o_{\text{Cc}} + RT \ln a_{\text{Cc,calcite}} + \Delta\overline{G}^o_{\text{H}_2\text{O}} + RT \ln a_{\text{H}_2\text{O,fluid}}$$

$$= 2(\Delta\overline{G}^o_{\text{Zo}} + RT \ln a_{\text{Zo,zoisite}}) + \Delta\overline{G}^o_{\text{CO}_2} + RT \ln a_{\text{CO}_2,\text{fluid}}. \qquad (11\text{-}12)$$

In this equation, the $\Delta\overline{G}^o_i$ terms refer to standard free energy values at P and T. By recalling that

$$(\Delta\overline{G}^o_i)_{P,T} = (\Delta\overline{G}^o_i)_{1,T} + \int_1^P \Delta\overline{V}^o_i \, dP,$$

we can rewrite expression 11-12 as

$$(3\Delta\overline{G}^o_{\text{An}} + \Delta\overline{G}^o_{\text{Cc}} + \Delta\overline{G}^o_{\text{H}_2\text{O}} - 2\Delta\overline{G}^o_{\text{Zo}} - \Delta\overline{G}^o_{\text{CO}_2}) + \int_1^P (3\overline{V}^o_{\text{An}} + \overline{V}^o_{\text{Cc}} - 2\overline{V}^o_{\text{Zo}}) \, dP$$

$$+ RT \ln [(a^3_{\text{An,plag}} a_{\text{Cc,calcite}})/a^2_{\text{Zo,zoisite}}] - RT \ln f_{\text{CO}_2} + RT \ln f_{\text{H}_2\text{O}} = 0. \qquad (11\text{-}13)$$

This is an equation of the general form

$$\Delta\overline{G}^o + \Delta\overline{V}^o_{\text{solids}}(P - 1) + RT \ln K - RT \ln f_{\text{CO}_2} + RT \ln f_{\text{H}_2\text{O}} = 0. \qquad (11\text{-}14)$$

The first two terms in this equation can be evaluated from

$$\Delta\overline{G}^o + \Delta\overline{V}^o_{\text{solids}}(P - 1) = \Delta\overline{H}^o - T\Delta\overline{S}^o + P\Delta\overline{V}^o_{\text{solids}},$$

because the difference between P and $P - 1$ is negligible in geologic problems. P and T can be estimated from mineral assemblage stability fields or from geothermometers and barometers. The compositions of coexisting minerals can be used to determine K; however, in this case we require two reactions to be solved simultaneously for the two unknowns f_{CO_2} and $f_{\text{H}_2\text{O}}$.

For zoisite + calcite + plagioclase rocks in Maine, Ferry (1976) estimated metamorphic P and T values of 3,500 bar and 798 K to 710 K, respectively. The ratio $f_{\text{CO}_2}/f_{\text{H}_2\text{O}}$ can be determined from the reaction already discussed above. Using estimated thermodynamic data and K values obtained from coexisting mineral compositions at various temperatures, Ferry determined $f_{\text{CO}_2}/f_{\text{H}_2\text{O}}$ ratios ranging from 0.53 to 5.10 throughout his field area. These rocks coexist with other assemblages that can be used to estimate either f_{CO_2} or $f_{\text{H}_2\text{O}}$, and thus both variables can be determined by solving simultaneous equations.

During prograde metamorphism, devolatilization reactions commonly release enough fluid that we can safely assume the following condition:

$$P_{\text{fluid}} = P_{\text{H}_2\text{O}} + P_{\text{CO}_2} = P_{\text{total}}.$$

It follows that

$$P_{H_2O} + P_{CO_2} = f_{H_2O}/\gamma_{H_2O} + f_{CO_2}/\gamma_{CO_2} = 3{,}500 \text{ bars.} \qquad (11\text{-}15)$$

If we assume that the gas mixture was ideal,

$$f_i = f_i^\circ X_i = \gamma_i^\circ P X_i = \gamma_i^\circ P_i. \qquad (11\text{-}16)$$

From these relationships, Ferry determined values for $X_{CO_2, \text{fluid}}$ ranging from 0.06 to 0.32 for his assemblages. This range of values indicates that gradients in fluid composition existed on an outcrop scale, possibly due to mixing of several fluids.

The compositions of fluids determined from fluid inclusions and calculated from metamorphic equilibria can generally be represented by the system H-O-C-S, although chloride complexes may also be present locally. The major species in naturally occuring fluids are H_2O, CO_2, CH_4, and H_2S.[5] The relative proportions of these species shift constantly as intensive variables for the system change. Figure 11.25 illustrates how the species in a fluid of fixed composition alter in response to changing temperature.

Fluids in the upper crust are typically H_2O-rich except in carbonate rocks, where CO_2 may predominate. Granulites from the middle and lower crust have fluid inclusions dominated by CO_2 and to a lesser extent by CO and CH_4. The most abundant volatile species in fluid inclusions from mantle xenoliths is CO_2, although some inclusions from the same specimens may be dominated by H_2O. Magmatic gases in tholeiitic basaltic lavas have higher H_2O/CO_2 ratios than those from alkaline lavas that may have formed by melting at deeper levels.

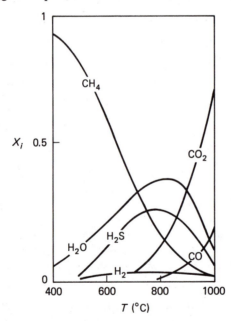

Figure 11.25 The effect of temperature on speciation of fluids in the system H-C-O-S coexisting with graphite at 2 kbar pressure. The fugacities of O_2 and S_2 are fixed by appropriate buffers. (Based on calculations by Holloway 1981.)

Fluid-Rock Interactions

Understanding how fluids and rocks interact depends on knowing the proportions of fluid and rock that constitute the reacting system. Making quantitative statements about the proportions of fluids that cycle through crustal and mantle rocks is very difficult,

---

[5] Although the species O_2 and S_2 actually occur in vanishingly small quantities, it is common practice to characterize fluid compositions by defining the fugacities of these species. These variables can serve as useful monitors for oxidation/reduction and sulfidation/desulfidation processes in such fluids.

but from the meager information available, fluid/rock ratios are surprisingly large, sometimes exceeding unity. This information comes from the assessment of the progress of continuous reactions in metamorphic rocks.

We must first recognize that while metamorphic fluids can be added from outside the system, they also can be generated internally. Most crustal rocks contain significant amounts of volatiles that can be liberated as fluids during metamorphism. For example, Walther and Orville (1982) calculated that the H_2O and CO_2 released from an average shale would occupy 12 volume percent of the rock at 500°C and 5 kbar. The pressure exerted by this fluid would readily fracture the rock, permitting most of the fluid to be transported to other areas. Fluid-rock interactions in which the fluid is all internally generated by dehydration/decarbonation reactions are termed *Rayleigh distillation* models. If minerals interact with an externally derived fluid flowing through the rock, this is called *infiltration*.

The computation scheme for determining fluid : rock ratios in both cases involves preserving mass balance for small increments of reaction progress. The procedure can employ either chemical or stable isotopic data. During each increment of reaction, chemical equilibrium is maintained between minerals and fluid; however, equilibrium is not maintained between minerals in successive increments of reaction. If enough of these infintesimal increments are modeled, the extent of the irreversible reaction (reaction progress) can be approximated. The number of steps necessary to reproduce the observed conditions is a measure of the amount of fluid, and in some cases it is also possible to determine whether Rayleigh distillation or infiltration was dominant. We have considered chemical and isotopic approaches to water-rock interaction problems in Chapters 6 and 7, making simplifying assumptions in each case. You may want to review Worked Problems 6-2 and 7-5 to see how these are handled. In general, however, tracing the progress of water-rock reactions in metamorphic rocks is much more difficult than these earlier problems. Readers interested in this subject should see papers by Ferry (1980) and Nabelek et al. (1984).

Although it is highly desirable to have this kind of quantitative information on fluids, very few studies of this kind have been done. It is more common to express fluid composition and proportion in terms of a fluid pressure relative to P_{total}. Many metamorphic reactions are critically dependent on P_{fluid}, either because they involve some component of the fluid directly or the fluid acts as a catalyst. For example, the transformation of aluminosilicate polymorphs is very sluggish, as indicated by the experimental uncertainty in their transition curves, but the presence of water increases reaction rates by eight orders of magnitude. This catalytic effect occurs because the fluid promotes nucleation and growth by aiding in transport of components to the sites of reaction, as noted earlier in Chapter 10.

Hydration/dehydration reactions are also significantly displaced in *P-T* space by dilution of an aqueous fluid with other species. One example is shown in Figure 11.26. The calculated *P-T* stability fields for assemblages involving paragonite (Pg), zoisite (Zo), kyanite (Ky), and quartz (Q) are illustrated as functions of P_{H_2O}/P_{total}; these fields are diminished in size and reduced to lower pressures and temperatures with decreasing P_{H_2O} (increasing dilution of the aqueous fluid). Displacement of reaction curves in Figure 11.26 can be readily understood by writing each of them as a stoichiometric relationship. For example,

$$NaAl_3Si_3O_{10}(OH)_2 + SiO_2 \longrightarrow NaAlSi_3O_8 + Al_2SiO_5 + H_2O$$

$$\text{paragonite} \qquad \text{quartz} \qquad \text{albite} \qquad \text{kyanite} \qquad \text{fluid}$$

is displaced in the direction of Ab + Ky + V as P_{H_2O}/P_{total} (X_{H_2O} in the fluid) is re-

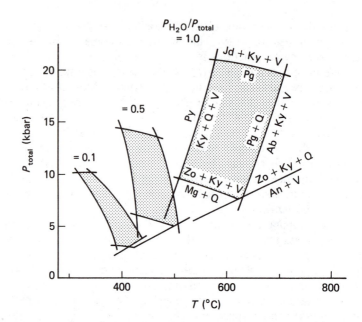

Figure 11.26 Calculated stability fields for assemblages containing zoisite (Zo), paragonite (Pg), kyanite (Ky), and quartz (Q) coexisting with fluid. The rectangular field changes size and shifts position with P_{H_2O}. Other phases are albite (Ab), jadeite (Jd), margarite (Mg), and pyrophyllite (Py). (Modified from Vernon 1975.)

duced from 1.0 to 0.5. If we treat all solid phases as ideal, pure phases, then $K_{eq} = P_{H_2O}$. Decreasing P_{H_2O} makes K_{eq} smaller and, because $\Delta \overline{G}_r = \Delta \overline{G}_r^\circ + RT \ln K$, makes $\Delta \overline{G}_r$ for this reaction more negative. Therefore, a system containing Pg + Q at $X_{H_2O} = 1.0$ (say, for example, at 600°C and 15 kbar), reacts to produce Ab + Ky + V as X_{H_2O} approaches 0.5.

Cycling Mechanisms

Volatile elements can be cycled (in both directions) between the crust and mantle most readily by plate tectonic processes. They are transported most easily as dissolved fluid species in magmas. Significant quantities of mantle-derived fluids are expelled from magmas erupting at spreading centers and subduction zones. The movement of fluids in this way is critically dependent on their solubilities in magmas. H_2O is very soluble for all magma compositions, but CO_2 solubility is more dependent on magma composition and pressure. Most magmas will become saturated with respect to CO_2 during ascent, causing the formation of a discrete fluid phase at some point. Once a fluid phase has formed, all of the dissolved volatiles will partition between it and the magma, and the fluid phase is likely to escape into the surrounding crust. Magmas provide transport for volatiles in only one direction: upward.

Metamorphism occurs at both divergent and convergent plate boundaries. Crustal fluids trapped in metamorphic rocks formed at both environments are ultimately carried downward into the mantle in subducted slabs. At depths appropriate for the eclogite transformation, dehydration reactions occur and the fluids are released. These may rise into the mantle wedge above the subducted plate and promote melting. The ensuing magmas may carry some of these volatiles again into the upper crust.

Near the beginning of this chapter, we noted that granulites may constitute a major part of the lower crust. The CO_2-rich fluid inclusions that characterize granulites, as well as the evidence for low a_{H_2O}, suggest that these rocks may form when the lower crust is invaded by CO_2. Newton et al. (1980) have suggested that such CO_2-dominated fluids are due to mantle degassing, possibly providing another mechanism (independent of plate tectonics) for the cycling of fluids between mantle and crust. They determined that mantle carbon, as found in diamonds, is isotopically light ($\delta^{13}C = -6\%$) relative to carbon in crustal limestone ($\delta^{13}C = +20\%$), and carbon from CO_2 inclusions in granulites is near the mantle isotopic value. During metamorphism, continuous fluxing with CO_2 must take place, because dehydration reactions produce water which would otherwise dilute the fluid. Significant quantities of CO_2 are required over the metamorphic episode, but the actual amount of CO_2 in the rocks at any time may be small because the fluids occur as films at grain boundaries.

Fluids circulating within the mantle itself may play an important role in petrogenesis. A local influx of fluids may lead to *mantle metasomatism*, because the fluids may carry quantities of dissolved components that may be added or subtracted from the system. An example of metasomatic exchange between a solid phase and fluid was studied by Orville (1963). Alkali feldspars exchange sodium and potassium readily if immersed in a NaCl or KCl solution, according to the reaction

$$KAlSi_3O_8 + Na^+ \longleftrightarrow NaAlSi_3O_8 + K^+.$$

This exchange is temperature-dependent, so that sodium in the fluid phase migrates toward the hotter portions in a thermal gradient. Consequently, a solid body of rock with nonuniform temperature distribution may develop alkali feldspars of different compositions as a fluid circulates through it. This metasomatic reaction is not one that occurs in the mantle, but it serves to illustrate the principle. Many geochemists believe that at least localized parts of the mantle have been affected by similar alteration processes. Fertile mantle rocks can yield magmas highly depleted in silica only when a small amount of partial melting has occurred. However, the observed high concentrations of incompatible trace elements in such magmas could only be produced by high degrees of partial melting because olivine, pyroxenes, and garnet contain so little of these elements. This dilemma can be resolved if the incompatible elements are concentrated in accessory phases that melt at modest temperatures. Many xenoliths of mantle peridotite contain apatite, phlogopite, and amphibole enriched in incompatible elements which were probably introduced by fluids during mantle metasomatism.

SUMMARY

The compositions of the crust and mantle are so different that it may be tempting to surmise that they do not interact, but we have seen that geochemical exchange does occur on a global scale. Ascending magmas formed by partial melting of mantle peridotite carry mantle components into the crust. Mantle melting is promoted by local temperature increases, depressurization, or fluxing by fluids. Incompatible and volatile elements are partitioned into these melts, so that they are ultimately concentrated in the crust. One important consequence of this irreversible differentiation is that heat-producing radionuclides are localized in the crust, where they affect geothermal gradients. The geochemical record in igneous rocks is commonly obscured by fractional crystallization or other processes that alter magma compositions during ascent to the surface.

Crustal materials are carried downward in subducted plates, where they may be stranded indefinitely. However, volatile elements in these slabs may be recycled because of dehydration/decarbonation reactions during metamorphism in the mantle. Only oceanic crust is subducted; the amount of buoyant continental crust appears to have increased with time.

With this background in the processes and pathways by which materials in the earth's interior interact, we are now in a position to explore another tool for formulating and testing geochemical interactions in these regions. In the next chapter on unstable isotopes, we will build on the information just presented.

SUGGESTED READINGS

The wealth of excellent books and papers on the subjects in this chapter makes selection of a few difficult. We recommend these enthusiastically.

BURNS, R. G. 1970. *Mineralogical applications of crystal field theory.* Cambridge: Cambridge Univ. Press. (A complete work on the behavior of transition metals in geologic systems. Chapter 2 discusses crystal field theory, and the remaining chapters describe applications to spectra, thermodynamic properties, and element distribution patterns.)

COX, K. G.; BELL, J. D.; and PANKHURST, R. J. 1979. *The interpretation of igneous rocks.* London: Allen and Unwin. (Chapter 6 gives a superb account of element variation diagrams, and chapter 14 discusses trace element fractionation.)

ERNST, W. G. 1976. *Petrologic phase equilibria.* San Francisco, Calif.: Freeman. (Chapter 5 gives a concise description of mantle mineralogy, melting processes, and crust-mantle differentiation in relation to plate tectonics.)

FERRY, J. M., ed. 1982. *Characterization of metamorphism through mineral equilibria.* Reviews in Mineralogy 10. Washington, D.C.: Mineral. Soc. of Amer. (An up-to-date manual of the quantitative geochemical aspects of metamorphism. Chapter 6, by Ferry and D. M. Burt, and chapter 8, by D. Rumble, III, provide very readable information on fluids in metamorphic systems.)

FYFE, W. S.; PRICE, N. J.; and THOMPSON, A. B. 1978. *Fluids in the earth's crust.* Amsterdam: Elsevier. (Chapter 2 summarizes the compositions of naturally occurring fluids, and other chapters describe their generation and transport.)

HARGRAVES, R. B., ed. 1980. *Physics of magmatic processes.* Princeton: Princeton Univ. Press. (A high-level but very informative book about the quantitative aspects of magmatism. Chapter 4, by S. R. Hart and C. J. Allegre, presents trace element constraints on magma genesis, and chapter 5, by E. R. Oxburgh, describes the relationship between heat flow and melting.)

HENDERSON, P. 1983. *Rare earth element geochemistry. Developments in geochemistry, 2.* Amsterdam: Elsevier. (A book devoted to research developments in the field of REE. Chapter 4, by L. A. Haskin, provides an in-depth look at petrogenetic modelling using these elements.)

RINGWOOD, A. E. 1975. *Composition and petrology of the earth's mantle.* New York: McGraw-Hill. (An exhaustive and authoritative monograph on the nature of the mantle. Chapter 5 describes the pyrolite model, and later chapters describe experimental constraints on mantle mineralogy.)

WALTHER, J. V., and WOOD, B. J., eds. 1986. *Fluid-rock interactions during metamorphism.* Advances in Geochemistry 5. New York: Springer-Verlag. (Chapter 1, by M. L. Crawford and L. S. Hollister, provides an up-to-date summary of fluid inclusion research.)

WOOD, B. J., and FRASER, D. G. 1977. *Elementary thermodynamics for geologists.* Oxford: Oxford Univ. Press. (Chapter 6 presents derivations of equations for trace element behavior.)

WYLLIE, P. J. 1971. *The dynamic earth.* New York: John Wiley and Sons. (An excellent, unintimidating book that should be required reading. Chapters 6 and 7 describe the composition of the mantle and crust, and chapter 8 discusses magma generation.)

YODER, H. S. JR. 1976. *Generation of basaltic magma,* Washington, D.C.: Nat. Acad. of Sciences. (Almost everything you might want to know about melting in the mantle.)

YODER, H. S. JR., ed. 1979. *The evolution of the igneous rocks, 50th anniversary perspectives.* Princeton: Princeton Univ. Press. (An updated version of the classic treatise by Bowen. Chapter 2, by E. Roedder, and Chapter 3, by D.C. Presnall, describe liquid immiscibility and fractional crystallization and melting, and Chapter 17, by P. J. Wyllie, considers magma generation.)

The following papers were also cited in this chapter, and provide much more detail for the interested student:

ANDERSON, O. L. 1980. The temperature profile of the upper mantle. *J. Geophys. Res.* 85: 7003–10.

BARKER, D.S. 1983. *Igneous rocks.* Englewood Cliffs, N.J.: Prentice Hall.

DICKINSON, W. R., and LUTH, W. C. 1971. A model for plate tectonic evolution of mantle layers. *Science* 174: 400–04.

FERRY, J. M. 1976. P, T, f_{CO_2}, and f_{H_2O} during metamorphism of calcareous sediments in the Waterville-Vassalboro area, south-central Maine. *Contrib. Mineral. Petrol.* 57: 119–43.

———. 1980. A case study of the amount and distribution of heat and fluid during metamorphism. *Contrib. Mineral. Petrol.* 71: 373–85.

GAST, P. W. 1972. The chemical composition of the earth, the moon and chondritic meteorites. In *The nature of the solid earth.* ed. E. C. Robertson, 19–40. New York: McGraw-Hill.

HARRIS, P. G.; REAY, A.; and WHITE, I. G. 1967. Chemical composition of the upper mantle. *J. Geophys. Res.* 72: 6359–69.

HOLLOWAY, J. R. 1981. Compositions and volumes of supercritical fluids in the earth's crust. In *Fluid inclusions: Applications to petrology,* ed. Hollister, L. S. and Crawford, M. L., 13–38. Calgary, Canada: Mineral. Assoc. of Canada.

KENNEDY, G. C. 1959. The origin of continents, mountain ranges, and ocean basins. *Amer. Sci.* 47: 491–504.

McBIRNEY, A. R. 1969. Compositional variations in Cenozoic calc-alkaline suites of Central America. *Oregon Dept. Geol. Mineral Ind. Bull.* 65: 185–89.

NABELEK. P. I.; LABOTKA, T. C.; O'NEIL, J. R.; and PAPIKE, J. J. 1984. Contrasting fluid/rock interaction between the Notch Peak granitic intrusion and argillites and limestones in western Utah: Evidence from stable isotopes and phase assemblages. *Contrib. Mineral. Petrol.* 86: 25–34.

NICHOLLS, G. D. 1967. Geochemical studies of the ocean as evidence for the composition of the mantle. In *Mantles of the earth and terrestrial planets.* ed. S. K. Runcorn, 285–304. New York: Wiley-Interscience.

NEWTON, R. C.; SMITH, J. V.; and WINDLEY, B. F. 1980. Carbonic metamorphism, granulites and crustal growth. *Nature* 288: 45–50.

ORVILLE, P. M. 1963. Alkali ion exchange between vapor and feldspar phases. *Amer. J. Sci.* 261: 201–37.

POLDERVAART, A. 1955. Chemistry of the earth's crust. In *Crust of the earth,* ed. A. Poldervaart, Geol. Soc. of Amer. Spec. Paper 62, 119–144. Baltimore: Waverly Press.

PRESNALL, D.C.; DIXON, S. A.; DIXON, J. R.; O'DONNELL, T. H.; BRENNER, N. L.; SCHROCK, R. L.; and DYCUS, D. W. 1978. Liquidus phase relations on the join diopside-forsterite-anorthite from 1 atm to 20 kbar: Their bearing on the generation and crystallization of basaltic magma. *Contrib. Mineral. Petrol.* 66: 203–20.

RINGWOOD, A. E. 1962. A model for the upper mantle. *J. Geophys. Res.* 67: 857–66.

———— and GREEN, D. H. 1966. An experimental investigation of the gabbro-eclogite transformation and some geophysical implications. *Tectonophys.* 3: 383–427.

ROEDDER, E. 1979. Silicate liquid immiscibility. In *The evolution of the igneous rocks, 50th anniversary perspectives,* ed. H. S. Yoder, ed., 15–57. Princeton: Princeton Univ. Press.

RONOV, A. B., and YAROSHEVSKY, A. A. 1969. Chemical composition of the earth's crust. In *The earth's crust and upper mantle.* ed. P. J. Hart, Amer. Geophys Union Monograph 13, 37–57. Washington: American Geophysical Union.

SHAW, D. M. 1977. Trace element behavior during anatexis. *Oregon Dept. Geol. Mineral Industries Bull.* 96: 189–213.

STOLPER, E., and WALKER, D. 1980. Melt density and the average composition of basalt. *Contrib. Mineral. Petrol.* 74: 7–12.

TAYLOR, S. R., and McLENNAN, S. M. 1987. The significance of the rare earths in geochemistry and cosmochemistry. *Handbook on the physics and chemistry of rare earths* 13,. Amsterdam: North Holland Pub. Co.

TURNER, F. J. 1981. *Metamorphic petrology: Mineralogical, field and tectonic aspects.* 2d ed. New York: McGraw-Hill.

VERNON, R. H. 1975. *Metamorphic processes, reactions and microstructure development.* New York: John Wiley and Sons.

WALTHER, J. V., and ORVILLE, P. M. 1982. Volatile production and transport in regional metamorphism. *Contrib. Mineral. Petrol.* 79: 252–57.

WHITE, I. G. 1967. Ultrabasic rocks and the composition of the upper mantle. *Earth Planet. Sci. Lett.* 3: 11–18.

WRIGHT, T. L., and DOHERTY, P. C. 1970. A linear programming and least squares computer method for solving petrologic mixing problems. *Geol. Soc. of Amer. Bull.* 81: 1995–2008.

YODER, H. S., and TILLEY, C. E. 1962. Origin of basalt magmas: An experimental study of natural and synthetic rock systems. *J. Petrol.* 3: 342–52.

ZIELINSKI, R. A. 1975. Trace element evaluation of a suite of rocks from Reunion Island, Indian Ocean. *Geochim. Cosmochim. Acta* 39: 713–34.

PROBLEMS

1. Describe the fractional crystallization path for a liquid with 45% SiO_2 in the system forsterite-silica (see Figure 11.8).

2. Describe the equilibrium crystallization path for a liquid with 72% SiO_2 in the system forsterite-silica (see Figure 11.8).

3. Describe the phase assemblages in the melting sequence under equilibrium and fractional melting conditions for a composition of Fo_{50} in the system forsterite-fayalite (Figure 11.14).

4. **(a)** Sketch qualitative chemical variation diagrams for CaO versus MgO and Al_2O_3 versus SiO_2 for liquid lines-of-descent during fractionation of compositions x and y in Figure 11.15. **(b)** How would these variation diagrams be affected if the system also contained FeO (hint: what effect would this have on olivine and orthopyroxene?) **(c)** Describe the sequence of rocks produced by perfect fractional crystallization of these two compositions.

5. A basaltic magma has the following trace element abundances: Ni—1100 ppm, Rb—100 ppm, Sr—300 ppm, Ba—220 ppm, Ce—82 ppm, Sm—9.4 ppm, Eu—3.3 ppm, Yb—3.3 ppm. Fractional crystallization of 20 wt. % olivine, 12 wt. % orthopyroxene, and 10 wt. % plagioclase from this magma results in a residual liquid that erupts on the surface. Using the distribution coefficients in Table 11-3, calculate the trace element contents of this erupted melt.

6. The following are rare earth element abundances in chondritic meteorites: Ce—0.616 ppm, Sm—0.149 ppm, Eu—0.056 ppm, Yb—0.159 ppm. Construct a chondrite-normalized REE pattern for the parent magma and residual liquid in problem 5, following the example in Figure 11.23. How would you explain the general shapes of these patterns?

7. A spinel periodotite has a solidus temperature of 1500°C at 20 kbar. **(a)** Using Figure 11.9, describe the phases involved in its melting as a mantle diapir of this material rises adiabatically to a depth of 40 km. **(b)** How does the liquid composition, in terms of normative olivine, change over this interval?

8. Calculate the free energy of the reaction paragonite + quartz $\rightarrow$ albite + kyanite + H_2O at 600°C and 10 kbar for $P_{H_2O}/P_{total} = 1.0$ and 0.5. Compare your results with Figure 11.26.

Chapter 12

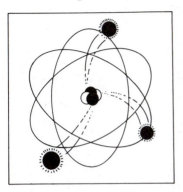

Using Radioactive Isotopes

OVERVIEW

In the geochemist's arsenal of techniques for unraveling geological problems, the study of unstable isotopes has become very prominent over the last several decades. This chapter begins with a discussion of nuclide stability and decay mechanisms. After equations that describe radioactive decay are derived, we examine the utility of certain naturally occurring radionuclide systems (K-Ar, Rb-Sr, Sm-Nd, U-Th-Pb) in geochronology. We discuss the concept of extinct radionuclides that were present in the early solar system; the principles behind both mass spectrometry and fission track techniques; and how induced radioactivity can be used to solve geochemical problems. Examples include neutron activation analysis, ^{40}Ar-^{39}Ar dating, and the use of cosmogenic nuclides (^{14}C and others) for dating geological and archeological materials and for monitoring cosmic ray exposure in meteorites.

We then explore how radionuclides can be used as geochemical tracers in such problems as assessing mantle heterogeneity, identifying metallogenic provinces, recognizing crustal contamination of mantle-derived magmas, determining cycling of material between crust and mantle at subduction zones, revealing global oceanic mixing patterns, and understanding outgassing of the earth's interior to produce the atmosphere.

PRINCIPLES OF RADIOACTIVITY

Nuclide Stability

In Chapter 7, we learned how stable isotopes can be used to solve geochemical problems. The study of unstable (that is, *radioactive*) nuclides, considered in this chapter, is an equally powerful technique. Because only about 260 of the 1700 known nuclides are stable, we can infer that nuclide stability is the exception rather than the rule. Figure

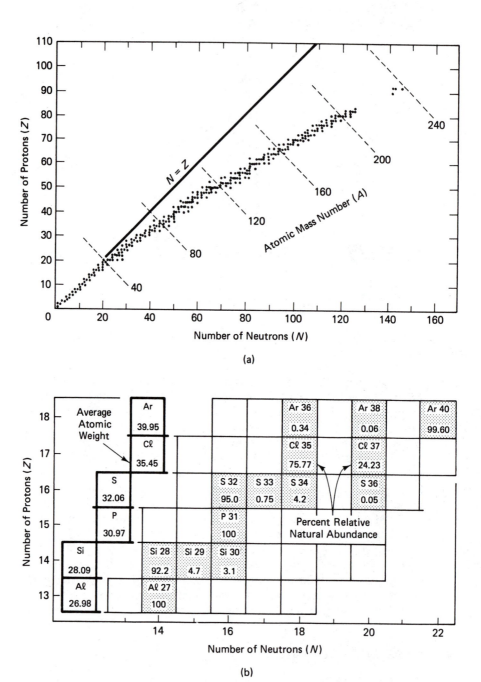

(a)

(b)

Figure 12.1 (a) Number of protons (Z) versus neutrons (N) in stable nuclides. The band of stability is illustrated by dots. Atomic mass number (A) is N + Z. (b) Expanded segment of the nuclide chart in (a) showing stable nuclides (patterned boxes) and their percentages of relative natural abundances. The average atomic weight for each element (shown in the boxes to the left of the figure) is the sum of the weights of the various isotopes times their respective relative abundances.

12.1a, a plot of the number of neutrons (N) versus protons (Z), illustrates this point. Stable nuclides occur only within a thin diagonal band in which N is slightly greater than Z, which is flanked on both sides by unstable nuclides. Careful inspection of a segment of this band (Figure 12.1b) will show that in most cases, the stable nuclides have an even number of neutrons and protons; stable nuclides with either N or Z as odd numbers are much less common, and those with both N and Z as odd numbers are generally unstable. This is clearly indicated by these counts of stable nuclei:

	N even	N odd	Total
Z even	156	48	204
Z odd	50	5	55

Most of the known radioactive isotopes do not occur in nature. Although some of these may have occurred naturally in the distant past, their decay rates are so rapid that they have long since been exhausted. In most cases, radioactive isotopes that are of interest to geochemists require very long times for decay or are produced continually by natural nuclear reactions.

In Chapter 7, we explored how variations in stable isotopes are caused by mass fractionation during the course of chemical reactions. With the exception of radioactive ^{14}C and a few other nuclides, the atomic masses of most of the unstable isotopes of geochemical interest are very large, so that mass differences with other nuclides of the same element are miniscule. Consequently, these isotopic systems can be considered to be immune to mass fractionation processes. Thus, all of the measured variations in these nuclides are normally ascribed to radioactive decay.

Decay Mechanisms

Radioactivity is the spontaneous transformation of an unstable nuclide (the *parent*) into another nuclide (the *daughter*). The transformaton process, called *radioactive decay,* results in changes in N and Z of the parent atom, so that another element is produced. Such processes occur by emission or capture of a variety of nuclear particles. Isotopes produced by the decay of other isotopes are said to be *radiogenic*. The radiogenic daughter may be stable or unstable; if it is unstable, the process continues until a stable nuclide is produced.

Beta decay involves the emission of negatively charged beta particles (electrons emitted by the nucleus), commonly accompanied by radiation in the form of gamma rays. This is equivalent to the transformation of a neutron into a proton and an electron. As illustrated in Figure 12.2, Z increases by one and N decreases by one for each beta particle emitted.

Another type of radioactivity occurs as a result of *positron decay*. A positron is a positively charged electron expelled from the nucleus. This can be regarded as the conversion of a proton into a neutron, a positron, and a neutrino (a particle with appreciable kinetic energy, but without mass). Positron decay produces a nuclide with N increased by one and Z decreased by one, as illustrated in Figure 12.2.

Inspection of Figure 12.1a allows us to predict which nuclides tend to transform by beta decay or positron decay. Unstable atoms that lie below the band of stability, and therefore have excess neutrons, are likely to decay by emission of beta particles, so that their N numbers are reduced. Similarly, nuclides that lie above the band of stability, having excess protons, may experience positron decay. In each case, these pro-

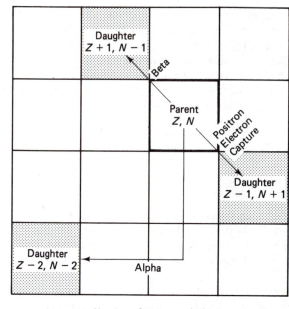

The y-axis is labeled "Number of Protons (Z)" and the x-axis is labeled "Number of Neutrons (N)".

Labels within the chart: "Daughter Z + 1, N − 1", "Beta", "Parent Z, N", "Positron Electron Capture", "Daughter Z − 1, N + 1", "Daughter Z − 2, N − 2", "Alpha".

Figure 12.2 A portion of the nuclide chart illustrating the N-Z relationships for daughter nuclides formed from a hypothetical parent by emission of beta particles, positrons, and alpha particles, or by electron capture.

cesses occur in such a way that the resulting daughter nuclides fall within the band of nuclear stablity.

An alternate type of decay mechanism is *electron capture*. In this case, the nuclide increases its N and decreases its Z by addition of an orbiting electron from outside the nucleus. A neutrino is also liberated during this process. The daughter nuclide produced during electron capture occupies exactly the same position relative to its parent in Figure 12.2 as that produced during positron decay.

Alpha decay proceeds by the emission of heavy alpha particles from the nucleus. An alpha particle consists of two neutrons and two protons. Therefore, the N and Z values for the daughter nuclide both decrease by two. This decay mechanism is summarized in Figure 12.2.

To be complete, we should also add to this list *spontaneous fission,* which is an alternate mode of decay for heavy atomic nuclei. These atoms may break apart, possibly as a result of electrostatic repulsion. Fission products generally have excess neutrons and tend to decay further by beta emission. For most geochemical applications, this can be considered a relatively minor side effect.

The decay of natural radionuclides may be much more complex than suggested by the simple decay mechanisms just introduced. A particular transformation may employ several of the these decay mechanisms simultaneously, so that parent atoms form more than one daughter. Such a process is called *branched decay*. As an example, let us consider the decay of ^{10}K. Some atoms of the parent nuclide decay by positron emission and electron capture into ^{40}Ar. ^{40}Ca is produced from the remainder by beta decay. Branched decay decreases the yield of daughter atoms of a particular nuclide, requiring more sensitivity in measurement.

Worked Problem 12-1

How many atoms of ^{40}Ar would be produced by the complete decay of ^{40}K in a 1 cm^3 crystal of orthoclase? To answer this question, we must first calculate the number of atoms of parent ^{40}K in this sample. We can convert volume of orthoclase to weight by multiplying by density, and weight to moles by dividing by the gram formula weight of orthoclase, as follows:

$$1 \text{ cm}^3 \, (2.59 \text{ g cm}^{-3})/278.34 \text{ g mol}^{-1} = 9.30 \times 10^{-3} \text{ mol.}$$

Each mole of orthoclase contains one mole of potassium, so there are 9.30×10^{-3} moles of K in this sample. Multiplication of this result by Avogadro's number gives the number of atoms of potassium:

$$9.30 \times 10^{-3} \text{ mol } (6.023 \times 10^{23} \text{ atoms mol}^{-1}) = 5.60 \times 10^{19} \text{ atoms.}$$

The natural abundance of ^{40}K in potassium (before decay) is 0.01167%. Therefore, this orthoclase sample contains

$$5.60 \times 10^{19} \text{ atoms } (1.167 \times 10^{-4}) = 6.535 \times 10^{15} \text{ atoms } ^{40}K.$$

We have already found that ^{40}K undergoes branched decay, and we are concerned here with only one of the daughter nuclides. A total of 11.16% of the ^{40}K atoms decay to ^{40}Ar by electron capture and positron decay, the remainder transforming to ^{40}Ca. Consequently, the number of atoms of ^{40}Ar produced by complete decay is

$$6.535 \times 10^{15}(0.1116) = 7.293 \times 10^{14} \text{ atoms } ^{40}Ar.$$

Rate of Radioactive Decay

In each of the decay mechanisms just discussed, the rate of disintegration of a parent nuclide is proportional to the number of atoms present. In more quantitative terms, the number of atoms (N, not to be confused with the symbol for the number of neutrons used in earlier paragraphs) remaining at any time t is

$$-dN/dt = \lambda N, \tag{12-1}$$

where λ is the constant of proportionality, normally called the *decay constant*. The value of the decay constant is characteristic for a particular radionuclide, and it is expressed in units of reciprocal time. Rearranging the terms of equation 12-1 and integrating gives the result

$$-\int dN/N = \lambda \int dt,$$

or

$$-\ln N = \lambda t + C, \tag{12-2}$$

where C is the constant of integration. If $N = N_o$ at time $t = 0$, then

$$C = -\ln N_o.$$

By substituting this value into equation 12.2, we obtain

$$-\ln N = \lambda t - \ln N_o,$$

or

$$\ln N/N_o = -\lambda t.$$

This is more commonly expressed as

$$N/N_o = e^{-\lambda t}. \tag{12-3}$$

Equation 12-3 is the basic relationship that describes all radioactive decay processes. With it, we can calculate the number of parent atoms (N) that remain at any time t from an original number of atoms (N_o) present at time $t = 0$.

Principles of Radioactivity

We can also express this relationship in terms of atoms of the daughter nuclide rather than the parent. If no daughter atoms (D_o) are present at time $t = 0$ and none are added to or lost from the system during decay, then the number of radiogenic daughter atoms produced by decay $(D*)$ at any time t is

$$D* = N_o - N. \qquad (12\text{-}4)$$

By rearranging equation 12-3 and substituting it into 12-4, we see that

$$D* = N e^{\lambda t} - N,$$

or

$$D* = N(e^{\lambda t} - 1). \qquad (12\text{-}5)$$

Equation 12-5 gives the number of daughter atoms produced by decay at any time t as a function of parent atoms remaining. If some atoms of the daughter nuclide were present initially (D_o), then

$$D = D_o + D*.$$

Combining this equation with 12-5 produces the useful result

$$D = D_o + N(e^{\lambda t} - 1). \qquad (12\text{-}6)$$

This important equation is the basis for geochronology. Both D and N are measureable quantities, and D_o is a constant whose value can be determined.

It is common practice to express radioactive decay rates in terms of *half-lives*. The half-life $(t_{1/2})$ is defined as the time required for one-half of a given number of atoms of a nuclide to decay. Therefore, when $t = t_{1/2}$, it follows that $N = 1/2N_o$. Substitution of these values into equation 12-3 gives

$$\tfrac{1}{2} N_o/N_o = e^{-\lambda t_{1/2}},$$

or

$$\ln \tfrac{1}{2} = -\lambda t_{1/2}.$$

This is equivalent to

$$\ln 2 = \lambda t_{1/2}.$$

Solving for $t_{1/2}$ gives

$$t_{1/2} = \ln 2/\lambda = 0.693/\lambda. \qquad (12\text{-}7)$$

This equation shows the relationship between the half-life of a nuclide and its decay constant.

The equations above show that the disintegration of a parent radionuclide and the generation of daughter nuclides are both exponential functions of time. This is illustrated in the following Worked Problem.

Worked Problem 12-2

Follow the decay of parent ^{87}Rb and growth of daughter ^{87}Sr in a granite sample over the course of six half-lives. Assume that the granite initially contains 1.2×10^{20} atoms of ^{87}Rb and 0.3×10^{20} atoms of ^{87}Sr.

The half-life of ^{87}Rb is 48.8×10^9 years. The decay constant for this radionuclide can be calculated by using equation 12-7:

$$\lambda = 0.693/48.8 \times 10^9 = 1.42 \times 10^{-11} \text{ yr}^{-1}.$$

Using equation 12-3, we can determine the number of atoms of parent ^{87}Rb (N) remaining after one half-life:

$$N/(1.2 \times 10^{20}) = e^{-(1.42\times 10^{-11})(48.8\times 10^{9})}$$

$$N = 6.00 \times 10^{19} \text{ atoms after 1 half-life.}$$

To calculate the number of atoms of the parent nuclide remaining after two half-lives, we substitute for N_o in the same equation the value of N just calculated:

$$N/(6.00 \times 10^{19}) = e^{-(1.42\times 10^{-11})(48.8\times 10^{9})}$$

$$N = 3.00 \times 10^{19} \text{ atoms after 2 half-lives,}$$

and so forth.

To calculate the number of atoms of daughter ^{87}Sr (D) after one half-life, we use equation 12-6:

$$D = 0.3 \times 10^{20} + 6.00 \times 10^{19}[e^{(1.42\times 10^{-11})(48.8\times 10^{9})} - 1]$$

$$D = 9.00 \times 10^{19} \text{ atoms after 1 half-life.}$$

We follow a similar procedure as above for calculating daughter atoms after successive half-lives.

The easiest way to summarize these calculations is by means of a graph, plotting the number of atoms of parent (N) or daughter (D) versus time in units of half-life. Such a diagram is show in Figure 12.3. The exponential character of this decay is obvious. After six half-lives, N approaches zero asymptotically and, for most practical purposes, the production of daughter atoms ceases.

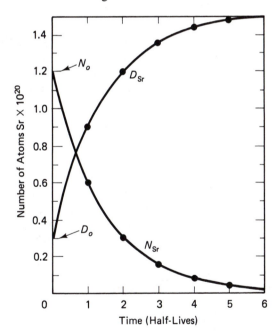

Figure 12.3 The decay of radioactive ^{87}Sr (N) into its stable radiogenic daughter, ^{87}Rb (D), as a function of time measured in half-lives. N_o and D_o represent the number of atoms of parent and daughter present initially.

Decay Series and Secular Equilibrium

The calculations above presume that a radioactive parent decays directly into a stable daughter. However, many radionuclides decay to a stable nuclide by way of transitory, unstable daughters; that is, a parent N_1 may decay to a radioactive daughter N_2, which in turn decays to a stable daughter N_3. Naturally occurring decay series arising from ra-

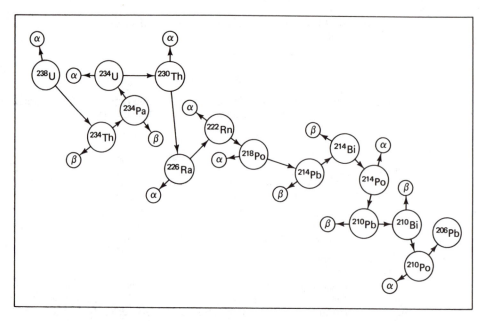

Figure 12.4 Representation of the radioactive decay of ^{204}U to ^{206}Pb through a series of daughter isotopes. Alpha and beta particles produced at each step are labelled.

dioactive isotopes of uranium and thorium behave in this manner. For example, ^{238}U produces 13 separate daughters before finally arriving at stable ^{206}Pb, as shown in Figure 12.4. Equations giving the number of atoms of any member of a decay series at any time t are presented by Faure (1986), and will not be considered here. It is interesting, however, to examine the special case in which the half-life of the parent nuclide is very much longer than those of the radioactive daughters. In this situation, it can be shown that after some initial time interval has passed,

$$\lambda_1 N_1 = \lambda_2 N_2 = \lambda_3 N_3 = \lambda_n N_n,$$

where $\lambda_{1,2,..n}$ are the decay constants for each nuclide in the series. This condition, in which the rate of decay of the daughters equals that of the parent, is known as *secular equilibrium*. It allows an important simplification of decay series calculations, because the system can be treated as if the original parent decayed directly to the stable daughter, without intermediate steps. Secular equilibrium is assumed in uranium prospecting methods that utilize radiation measurements to calculate uranium abundances.

PROSPECTING AND WELL-LOGGING TECHNIQUES THAT USE NATURAL RADIOACTIVITY

Uranium prospecting tools rely principally on detection of gamma radiation, because alpha and beta particles cannot penetrate an overburden cover of even a few centimeters. The Geiger counter is commonly used for field surveys, although it is a rather inefficient detector. The instrument consists of a sealed glass tube containing a cathode and anode with a voltage applied. The tube is filled with a gas that is normally nonconducting, but when gamma radiation passes through the gas, it is ionized and the ions accelerate toward the electrodes. The resulting current pulses are recorded on a meter or heard as "clicks."

The scintillation counter is a more efficient tool for gamma detection. Certain kinds of crystals *scintillate*, that is, they emit tiny flashes of visible light when they absorb gamma radiation. These scintillations are detected by photomultiplier tubes and converted to electrical pulses that can be counted. This instrument can be used in ground or airborne surveys.

A commonly used well-logging tool employs natural radioactivity to mark lithologic boundaries in drill holes. The gamma log depends on scintillation detection of gamma rays produced by decay of ^{40}K and daughter products in the uranium and thorium decay series. These large ions are incompatible in most crystal structures, but are accomodated in clay minerals. Therefore, increases in gamma radiation are ascribed to clay concentrations (that is, shale formations) and, conversely, decreased radiation to cleaner sandstone or limestone units. An example of a portion of a gamma well log is shown in Figure 12.5. Gamma rays from decay of any one nuclide have a particular energy, and the energy spectrum of this radiation is resolved in some gamma logs. This permits estimates to be made of the concentrations of potassium, uranium, and thorium (assuming secular equilibrium), and the K/U or K/Th ratios may provide unique geochemical signatures for certain shale horizons, which may be useful in stratigraphic correlations.

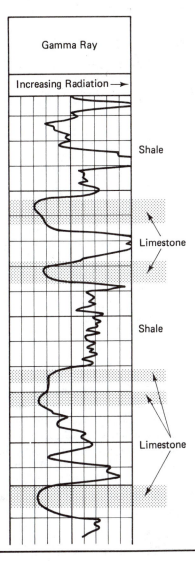

Figure 12.5 An example of a gamma-ray well log used in identifying lithologic variations. The horizontal scale is radioactivity (increasing to the right) and the vertical scale is depth in the bore hole. Limestone horizons have low radioactivity relative to shale units because of lower contents of K, U, and Th.

One of the prime uses of radiogenic isotopes is, of course, to determine the ages of rocks. All radiometric dating systems assume that certain condtions are satisfied: (1) The system, which may be defined as the rock or as an individual mineral, must have remained closed so that neither parent nor daughter atoms were lost or gained except as a result of radioactive decay. (2) If atoms of the daughter nuclide were present before system closure, it must be possible to assign a value to this initial amount of material. (3) The value of the decay constant must be known accurately. (This is particularly critical for nuclides with long half-lives, because a small error in the decay constant will translate into a large uncertainty in the age.) (4) The isotopic compositions of analyzed samples must be representative and must be measured accurately.

Several naturally occurring radionuclides are in current use for geochronology; some of the most common are listed in Table 12-1 and discussed below. Each of these is employed for its special properties, such as rate of decay, response to heating and cooling, or concentration range in certain rocks.

TABLE 12-1 RADIONUCLIDES USED IN GEOCHRONOLOGY

Parent	Stable daughter	Half-life (yr)	Decay constant (yr^{-1})
Long-lived radionuclides			
^{40}K	^{40}Ar	1.25×10^9	5.54×10^{-10}
^{87}Rb	^{87}Sr	4.88×10^{10}	1.42×10^{-11}
^{147}Sm	^{143}Nd	1.06×10^{11}	6.54×10^{-12}
^{232}Th	^{208}Pb	1.40×10^{10}	4.95×10^{-11}
^{235}U	^{207}Pb	7.04×10^8	9.85×10^{-10}
^{238}U	^{206}Pb	4.47×10^9	1.55×10^{-10}
Short-lived radionuclides			
^{14}C	^{14}N	5.73×10^3	1.21×10^{-4}
Extinct radionuclides			
^{26}Al	^{26}Mg	7.20×10^5	9.60×10^{-7}
^{107}Pd	^{107}Ag	6.50×10^6	1.07×10^{-7}
^{129}I	^{129}Xe	1.60×10^7	4.30×10^{-8}
^{244}Pu	$^{131-136}$Xe*	8.20×10^7	8.50×10^{-9}

*Many other fission products are also produced.

Potassium-Argon System

We have already seen that radioactive ^{40}K undergoes branching decay to produce two different daughter products. Its decay into ^{40}Ca is not a useful chronometer because ^{40}Ca is also the most abundant stable isotope of calcium. As a result, the addition of radiogenic ^{40}Ca will increase its abundance only slightly. Even though ^{40}Ar is the most abundant isotope of atmospheric argon, potassium-bearing minerals do not commonly contain argon. The radioactive decay of ^{40}K can therefore be monitored through time by measuring ^{40}Ar.

Each branch of the decay scheme has its own separate decay constant. The total decay constant λ is therefore the sum of these:

$$\lambda = \lambda_{Ca} + \lambda_{Ar}.$$

The commonly adopted value for the total decay constant and the corresponding half-life are given in Table 12-1. The proportion of ^{40}K atoms that decay to ^{40}Ar is given by the ratio of their decay constants, $\lambda_{Ar}/\lambda_{Ca}$, commonly called the *branching ratio*. In this case the branching ratio is 0.117.

We can use equation 12-6 to specify the increase in ^{40}Ar through time:

$$^{40}\text{Ar} = {}^{40}\text{Ar}_o + 0.117 \; {}^{40}\text{K}(e^{\lambda t} - 1).$$

Notice that N, the number of atoms of ^{40}K, is multiplied by the branching ratio to correct for the fact that most of them will decay to another daughter nuclide. Because most minerals contain virtually no argon initially, $^{40}\text{Ar}_o = 0$ and the equation above reduces to

$$^{40}\text{Ar} = 0.117 \; {}^{40}\text{K}(e^{\lambda t} - 1). \tag{12-8}$$

Application of equation 12-8 then requires only the measurement of ^{40}Ar and potassium in the sample; ^{40}K is calculated from total potassium using its relative isotopic abundance of 0.01167%.

The K-Ar system is useful for dating igneous and, in a few cases, sedimentary rocks. Sediments containing authigenic glauconite and certain clay minerals may be suitable for K-Ar chronology. However, because the daughter is a gas, it may diffuse out of many minerals, especially if they have been buried deeply or have protracted thermal histories. Metamorphism appears to reset the system in most cases, and K-Ar dates may record the time of metamorphism in cases where cooling was relatively rapid. During slow cooling, argon diffusion will continue until the system reaches some critical temperature (the *blocking temperature*) at which it becomes closed to further diffusion. Different minerals in the same rock may have different blocking temperatures and thus yield slightly different ages.

Rubidium-Strontium System

^{87}Rb produces ^{87}Sr by beta decay, at a rate given in Table 12-1. Unlike the K-Ar system, the radiogenic daughter is added to a system that already contains ^{87}Sr. Thus, the difficulty in applying this geochrometer lies in determining how much of the ^{87}Sr measured in the sample was produced by decay of the parent.

For the Rb-Sr system, equation 12-6 takes the form

$$^{87}\text{Sr} = {}^{87}\text{Sr}_o + {}^{87}\text{Rb} \, (e^{\lambda t} - 1).$$

Another isotope of strontium, ^{86}Sr, is stable and is not produced by decay of a naturally occurring isotope of another element. Therefore, if we divide both sides of the above equation by the number of atoms of ^{86}Sr in the sample (a constant), we will not affect the equality:

$$^{87}\text{Sr}/^{86}\text{Sr} = {}^{87}\text{Sr}_o/^{86}\text{Sr} + {}^{87}\text{Rb}/^{86}\text{Sr} \, (e^{\lambda t} - 1). \tag{12-9}$$

The initial ratio of $^{87}\text{Sr}_o/^{86}\text{Sr}$ in this equation can be derived from graphical analysis of a suite of samples. Equation 12-9 is the expression for a straight line (of the form $y = mx + b$). Let us examine a plot of $^{87}\text{Sr}/^{86}\text{Sr}$ versus $^{87}\text{Rb}/^{86}\text{Sr}$; that is, y versus x. Both of these ratios are readily measurable in rocks. We can assume that, at the time of crystallization ($t = 0$), all minerals in a given rock will have the same $^{87}\text{Sr}/^{86}\text{Sr}$ value, because they cannot discriminate between these relatively heavy isotopes. The various minerals will have different contents of rubidium and strontium, however, and thus different values of $^{87}\text{Rb}/^{86}\text{Sr}$.

This situation is illustrated schematically by the line labelled $t = 0$ in Figure 12.6. After some time has elapsed, a fraction of the ^{87}Rb atoms in each mineral will have decayed to ^{87}Sr. Obviously, the mineral which had the greatest initial concentration of ^{87}Rb will now have the greatest concentration of radiogenic ^{87}Sr. The position of each mineral in the rock will shift along a line with a slope of -1, as shown by arrows in Figure 12.6. The slope of the resulting straight line is $(e^{\lambda t} - 1)$, and its intercept on the y axis is the initial strontium isotopic ratio, ^{87}Sr$_o$/^{86}Sr. This line is called an *isochron*,[1] and its slope increases with time.

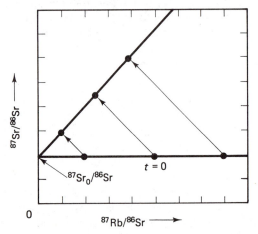

Figure 12.6 Schematic Rb-Sr isochron diagram illustrating the isotopic evolution of three samples with time. All samples have the same initial ^{87}Sr/^{86}Sr ratio but different ^{87}Rb/^{86}Sr ratios at time $t = 0$. The isotopic composition of each sample moves along a line of slope -1 as ^{87}Rb decays to ^{87}Sr. After some time has elapsed, the samples will define a new isochron, the slope of which is $(e^{\lambda t} - 1)$. Extrapolation of this isochron to the abcissa gives the initial Sr isotopic composition.

Isochron diagrams like Figure 12.6 can be constructed by using data from coexisting minerals in the same rock (a *mineral isochron*) or from fractionated comagmatic rocks which concentrate different minerals to varying degrees (a *whole-rock isochron*). Metamorphism may rehomogenize rubidium and strontium isotopes, so that the timing of peak metamorphism may be recorded by this geochronometer. A mineral isochron is more likely than a whole-rock isochron to be reset by metamorphism, because it is easier to reequilibrate adjacent minerals than portions of a rock body. Authigenic minerals in some sedimentary rocks may also have high enough rubidium contents that the age of diagenesis can be measured. The half-life of ^{87}Rb is long, so that this system complements the K-Ar system in terms of the accessible time range to which it can be applied.

Worked Problem 12-3

Nyquist et al. (1979) obtained the following isotopic data for a whole-rock (WR) sample and for mineral separates of plagioclase (Plag), pyroxene (Px), and ilmenite (Ilm) from Apollo 12 lunar basalt 12014.

	^{87}Rb/^{86}Sr	^{87}Sr/^{86}Sr
WR	0.0296	0.70096
Plag	0.00537	0.69989
Px	0.0492	0.70200
Ilm	0.1127	0.70490

What is the age of this rock?

[1] The fit of data points to a straight line is never perfect, and a linear regression is required. Analytical errors that lead to dispersion of data about the line give rise to corresponding uncertainties in ages. York (1966) gives the procedure for least-squares regression that is commonly used to determine isochrons. For a simplified discussion, see Appendix A.

We first determine the initial Sr isotopic ratio, $^{87}Sr_o/^{86}Sr$, by plotting these data on a isochron diagram (Figure 12.7). A least squares regression through the data can then be calculated, as illustrated by the line in the figure. The y intercept of this regression line corresponds to an initial isotopic ratio of 0.69964.

Equation 12-9 can be solved for t as follows:

$$t = 1/\lambda \ln \{[(^{87}Sr/^{86}Sr) - (^{87}Sr_o/^{86}Sr)]/(^{87}Rb/^{86}Sr) + 1\}. \qquad (12\text{-}10)$$

If we now insert the initial Sr isotopic ratio and the measured isotopic data for one sample (we will use WR) into this expression, we find $t = 1/(1.42 \times 10^{-11}) \ln \{[(0.70096 - 0.69964)/(0.0296)] + 1\} = 3.09$ b.y. This time corresponds to the crystallization age for the basalt. It differs slightly from the age reported by Nyquist and coworkers because they used an older decay constant for ^{87}Rb. The estimated uncertainty in age based on analytical error is ±0.11 b.y.

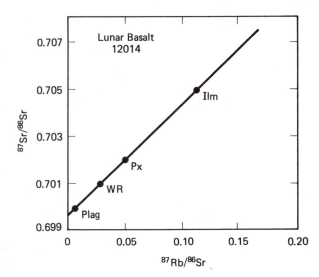

Figure 12.7 Rb-Sr isochron for lunar basalt 12014. Isotopic data for the whole-rock (WR), pyroxene (Px), plagioclase (Plag), and ilmenite (Ilm) were obtained by Nyquist et al. (1979). The slope of this isochron corresponds to an age of 3.09 b.y.

Samarium-Neodymium System

The rare earth elements samarium and neodymium form the basis for another geochronometer. ^{147}Sm decays to ^{143}Nd at a rate indicated in Table 12-1. The complication in this system is the same as that in the Rb-Sr system: some ^{143}Nd is already present in rocks before radiogenic ^{143}Nd is added, so we must find a way to specify its initial abundance. This can be done by graphical methods after normalizing parent and daughter to a stable isotope of neodymium (the conventional choice is ^{144}Nd), in the same way we normalized to Sr^{86} in the Rb-Sr system. An expression for the Sm-Nd system analogous to equation 12-9 can be written, therefore, as

$$^{143}Nd/^{144}Nd = {}^{143}Nd_o/^{144}Nd + {}^{147}Sm/^{144}Nd \; (e^{\lambda t} - 1) \qquad (12\text{-}11)$$

where the subscript o indicates the initial isotopic composition of neodymium at the time of system closure. On an isochron plot of $^{143}Nd/^{144}Nd$ versus $^{147}Sm/^{144}Nd$, the y intercept is the initial isotopic ratio $^{143}Nd_o/^{144}Nd$ and the slope of the line becomes steeper with time.

The Sm-Nd radiometric clock is very similar in its mathematical form to the Rb-Sr geochronometer, but the two systems can be applied to different kinds of rocks. Basalts and gabbros usually cannot be dated precisely using the Rb-Sr system, because their contents of rubidium and strontium are so low. The Sm-Nd system is ideal for crystallization age determinations on mafic rocks. This system may also be useful in some cases to "see through" a metamorphic event to the earlier igneous crystallization. Because both parent and daughter are relatively immobile elements, they are less likely to be disturbed by thermal overprints.

Uranium-Thorium-Lead System

The U-Th-Pb system is actually a set of independent geochronometers that depend on the establishment of secular equilibrium. ^{238}U decays to ^{206}Pb, ^{235}U to ^{207}Pb, and ^{232}Th to ^{208}Pb, all through independent series of transient daughter isotopes. The half-lives of all three parent isotopes are much longer than those of their respective daughters, however, so the prerequisite condition for secular equilibrium is present and the production rates of the stable daughters can be considered to be equal to the rates of decay of the parents at the beginning of the series. The half-lives and decay constants for the three parent isotopes are given in Table 12-1. We will refer to the decay constant for ^{238}U as λ_1, that for ^{235}U as λ_2, and that for ^{232}Th as λ_3.

In addition to these three radiogenic isotopes, lead also has a non-radiogenic isotope (^{204}Pb) that can be used for reference.[2] We can express the isotopic composition of lead in minerals containing uranium and thorium by the following equations, analogous to equations 12-9 and 12-11:

$$^{206}Pb/^{204}Pb = ^{206}Pb_o/^{204}Pb + ^{238}U/^{204}Pb \, (e^{\lambda_1 t} - 1), \qquad (12\text{-}12a)$$

$$^{207}Pb/^{204}Pb = ^{207}Pb_o/^{204}Pb + ^{235}U/^{204}Pb \, (e^{\lambda_2 t} - 1), \qquad (12\text{-}12b)$$

$$^{208}Pb/^{204}Pb = ^{208}Pb_o/^{204}Pb + ^{232}Th/^{204}Pb \, (e^{\lambda_3 t} - 1). \qquad (12.12c)$$

The subscript o in each case denotes an initial lead isotopic composition at the time the clock was set.

Each of these isotopic systems gives an independent age. In an ideal situation, these dates should be the same. In many instances, though, these dates are not concordant, because the minerals do not remain completely closed to the diffusion of uranium, thorium, lead, or some of the intermediate daughter nuclides. To simplify this complicated situation, let us consider a system without thorium, so that we only have to deal with two radiogenic lead isotopes. From equation 12-5, we know that the amount of radiogenic ^{206}Pb at any time must be

$$^{206}Pb^* = ^{238}U \, (e^{\lambda_1 t} - 1),$$

or

$$^{206}Pb^*/^{238}U = e^{\lambda_1 t} - 1. \qquad (12\text{-}13a)$$

In the expression above, the superscript * denotes radiogenic lead, that is, $^{206}Pb - ^{206}Pb_o$. Similarly, the amount of radiogenic ^{207}Pb at any time can be expressed as

$$^{207}Pb^*/^{235}U = e^{\lambda_2 t} - 1. \qquad (12\text{-}13b)$$

If a uranium bearing mineral behaves as a closed system, we know that equations 12-13a and 12-13b should yield concordant ages (that is, the same value of t). Logically, then, we can reverse the procedure and calculate compatible values for $^{206}Pb^*/^{238}U$ and $^{207}Pb^*/^{235}U$ at any given time, t. Figure 12.8 shows the results of such a calculation.

---

[2] ^{204}Pb is actually weakly radioactive, but it decays so slowly that it can be treated as a stable reference isotope.

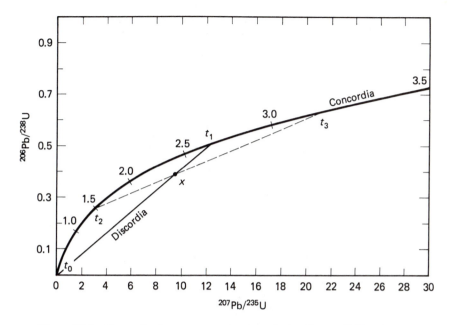

Figure 12.8 Concordia diagram comparing isotopic data for two U-Pb systems. An undisturbed system formed at time t_0 will move along the concordia curve; the age at any time after formation is indicated by numbers (in b.y.) along the curve. If episodic lead loss occurs at time t_1, the position of the system will be displaced toward the origin by some amount, as illustrated by point x. Other samples that lose different amounts of lead will be displaced along the same chord and will give discordant dates. With further elapsed time, the discordia line will subsequently be rotated (dashed line) as t_0 moves to t_2 and t_1 moves to t_3.

The curve in this figure, called the *concordia* curve by Wetherill (1956), represents the locus of all concordant U-Pb systems.

At the time of system closure (t_o), no decay has occurred yet, so there is no radiogenic lead. The system's isotopic composition, therefore, plots at the origin of Figure 12.8. At any time thereafter, the system is represented by some point along the concordia curve; numbers along the curve indicate the elapsed time in billions of years since formation of the system. If, instead, the system experiences loss of lead at time t_1, the system composition shifts along a chord connecting t_1 to the origin. A complete loss of radiogenic lead would displace the isotopic composition all the way back to the origin, but this rarely happens. If some fraction of lead is lost, as is the more common case, the system composition is represented by a point, which we will represent by x, along the chord. A series of samples with discordant ages that have experienced varying degrees of lead loss will define the chord, called a *discordia* curve, more precisely. When the disturbing event is over, each point x on the discordia moves along an evolutionary path similar in form but not coincident with the concordia curve.

The result is that the discordia curve retains its linear form but changes slope, intersecting the concordia curve at a new set of points. The time of lead loss is translated along the concordia curve to t_3, and t_o is similarly translated to t_2, the overall effect being a rotation of discordia about point x. This is illustrated by the dashed discordia curve in Figure 12.8.

If the discordia chord can be adequately defined, then this diagram may yield two geologically useful dates: t_2, the time of formation of the rock unit, and t_3, the time of a subsequent disturbance (possibly metamorphism or weathering) that caused lead loss.

This interpretation assumes that loss of lead was instantaneous; if lead was lost by continuous diffusion over a long period of time, then t_3 is a fictitious date without geologic significance. Unfortunately, it is not possible to distinguish between these alternatives without other geological information.

It is possible to compensate for the effect of lead loss on U-Pb ages by calculating ages based on the $^{207}Pb/^{206}Pb$ ratio. This ratio is, of course, insensitive to lead mobilization, because it is practically impossible to fractionate isotopes of the same heavy element. By combining equations 12-12a and 12-12b, we obtain

$$[^{207}Pb/^{204}Pb - {}^{207}Pb_o/^{204}Pb]/[^{206}Pb/^{204}Pb - {}^{206}Pb_o/^{204}Pb]$$

$$= [^{235}U (e^{\lambda_2 t} - 1)]/[^{238}U (e^{\lambda_1 t} - 1)]. \qquad (12\text{-}14)$$

The left side of this equation is equivalent to the ratio of radiogenic ^{207}Pb to radiogenic ^{206}Pb; that is, to $(^{207}Pb/^{206}Pb)^*$. This is determined by subtracting initial $^{207}Pb/^{204}Pb$ and $^{206}Pb/^{204}Pb$ values from the measured values for these ratios. Consquently, ages can be determined solely on the basis of the isotopic ratios in the sample, without having to determine lead concentrations. Moreover, because the ratio $^{235}U/^{238}U$ is constant $(1/137.8)$ for all natural materials at the present time, ages can be calculated without measuring uranium concentrations. Equation 12-14 can therefore be expressed as

$$(^{207}Pb/^{206}Pb)^* = 1/137.8 [(e^{\lambda_2 t} - 1)/(e^{\lambda_1 t} - 1)]. \qquad (12\text{-}15)$$

This is the expression for the $^{207}Pb/^{206}Pb$ age, sometimes called the *lead-lead* age. The only difficulty in applying this method is that equation 12-15 cannot be solved for t by algebraic methods. One possible solution is to make use of tabulated values of $(^{207}Pb/^{206}Pb)^*$ and t (Stacey and Stern, 1973), interpolating for any intermediate value. The equation can also be solved by a method of successive approximations, like the Newton-Raphson method, until a solution is obtained within an acceptable level of precision. A program called PBPB, based on this technique, is provided on the diskette available with this book.

The U-Th-Pb dating system is applicable to rocks containing minerals like zircon, apatite, monazite, and sphene. These phases provide large structural sites for uranium, thorium, lead, and other incompatible elements. Igneous or metamorphic rocks of granitic composition are most suitable for dating by this method. Several generations of zircons, recognizable by distinct morphologies, have been found to give distinct ages in some rock units. The possibility of phases inherited from earlier events further complicates the interpretation of U-Th-Pb ages.

Worked Problem 12-4

Aleinikoff et al. (1985) obtained the following Pb isotopic data for a zircon from the Sebago granitic batholith in Maine (units are atom %): $^{204}Pb = 0.010$, $^{206}Pb = 90.6$, $^{207}Pb = 4.91$, $^{208}Pb = 4.43$. The measured U concentration was 1,396 ppm, and that for Pb was 62.2 ppm. What are the $^{206}Pb/^{238}U$, $^{207}Pb/^{235}U$, and $^{207}Pb/^{206}Pb$ ages for this specimen?

We begin by determining the proportions of ^{235}U and ^{238}U in this sample. All natural uranium has $^{235}U/^{238}U = 1/137.8$, corresponding to 99.27% ^{238}U. Aleinikoff and coworkers assumed that the nonradiogenic lead in this sample had an isotopic composition of $204 : 206 : 207 : 208 = 1 : 18.2 : 15.6 : 38$, corresponding to an average atomic weight of 207.23. From these proportions, we can calculate the following ratios:

$$^{238}U/^{204}Pb = (1,396/62.2)(207.23/238.03)(99.27/0.010) = 193,970$$

$$^{235}U/^{204}Pb = 193,970/137.8 = 1,407.6$$

$$^{206}Pb/^{204}Pb = (90.6/0.010) - 18.2 = 9,041.8$$

$$^{207}Pb/^{204}Pb = (4.91/0.010) - 15.6 = 475.4$$

$$^{206}Pb_o/^{204}Pb = 18.2$$

$$^{207}Pb_o/^{204}Pb = 15.6$$

We can now solve equation 12-12 for t. The expression for the $^{206}Pb/^{238}U$ age, corresponding to equation 12.12a, is

$$t_{206} = 1/\lambda_1 \ln \{[(^{206}Pb/^{204}Pb) - (^{206}Pb_o/^{204}Pb)]/[^{238}U/^{204}Pb] + 1\}.$$

Substitution of the ratio above into this equation gives

$$t_{206} = 1/(1.55 \times 10^{-10}) \ln [(9,041.8 - 18.2)/(193, 970) + 1] = 293 \text{ m.y.}$$

The $^{207}Pb/^{235}U$ age, calculated from the analogous expression derived from equation 12-12b, is

$$t_{207} = 1/(9.85 \times 10^{-10}) \ln [(475.4 - 15.6)/(1,407.6) + 1] = 287 \text{ m.y.}$$

Subtracting the initial ratios for the isotopic composition of nonradiogenic lead gives

$$^{207}Pb/^{204}Pb = 491 - 15.6 = 475.4,$$

and

$$^{206}Pb/^{204}Pb = 9060 - 18.2 = 9041.8.$$

Dividing these results gives

$$(^{207}Pb/^{206}Pb)^* = 475.4/9041.8 = 0.0526.$$

We substitute this ratio into equation 12-15 and solve for t using Program PBPB to find that the $^{207}Pb/^{206}Pb$ age is 308 m.y.

These ages are similar and date the approximate time of crystallization of the Sebago batholith. This sample would plot on the concordia curve, because ages determined by both U/Pb systems are nearly the same.

Extinct Radionuclides

All of the naturally occurring radionuclides that we have considered so far have very long half-lives, so that an appreciable portion of the parent isotope remains. However, some meteorites contain evidence of the former presence of now extinct radionuclides that had short half-lives. The stable daughter products and rates of decay for ^{26}Al, ^{107}Pd, ^{129}I, and ^{244}Pu are given in Table 12-1. That these isotopes occurred at all as "live" radionuclides in some meteorites demands that such samples formed very soon after the event that created the isotopes. Measurement of the daughter products incorporated as parent nuclides in meteorites provides a means of determining the elapsed time since nucleosynthesis, called the *formation interval*. To prove that these daughter isotopes were produced after the meteorite formed, it is necessary to demonstrate that the daughter is now an anomaly in a mineral which originally concentrated the parent element. Because of their rapid rates of decay, extinct radionuclides may have been important sources of heating during the early stages of planet formation. The implications of the former presence of ^{26}Al in meteorites will be considered further in Chapter 13.

Fission Tracks

In observations of micas and cordierite under the microscope, you may have noticed zones of discoloration around tiny inclusions of zircon and other uranium-bearing minerals. These *pleochroic haloes* are manifestations of radiation damage produced by alpha particles. The crystal lattices of larger grains of minerals that contain uranium also

MASS SPECTROMETRY

In discussing the K-Ar, Rb-Sr, Sm-Nd, and U-Th-Pb geochronometers, we have so far avoided explaining how the isotopic data are actually obtained. The modern *mass spectrometer* is based on a classic design by Nier (1940). The instrument consists of three parts: (1) a source from which a beam of ions is emitted; (2) an electromagnet or other means of separating ions by mass; and (3) an ion collector. A simplified sketch of a single-focusing mass spectrometer is illustrated in Figure 12.9. If the sample is a gas like argon, it is allowed to leak into the source and is ionized by electron bombardment. If the sample is a solid like strontium or lead, a salt of the element is deposited on a filament in the source. The filament is then heated to a temperature at which the salt is volatilized. In both cases, the resulting ions are accelerated by an adjustable voltage and collimated into a beam by plates with slits.

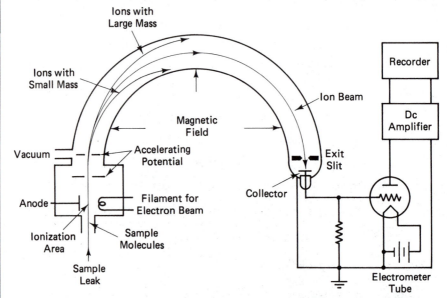

Figure 12.9 Schematic diagram of a single-focusing mass spectrometer. An ion beam generated in the source is deflected by the electromagnet to varying degrees, depending on mass and magnetic field strength. A particular ion beam is then aimed into the collector, where its voltage is recorded.

The ion beam then enters the magnetic field generated by an electromagnet, where the ions are caused to travel in a curved path, the radius of which depends on velocity and the square root of mass, as well as field strength. Ion beams that separate according to mass in this way then travel to the collector. The path of a particular ion beam can be controlled by adjusting the accelerating voltage and magnetic field. Each ion beam, in turn, is aimed through the collecting slit and its voltage is recorded in the collector cup. The voltage is proportional to the abundance of that isotope, so the ratio of any two isotopes can be determined very precisely.

Other instrument designs are also possible. The ability of the single-focusing mass spectrometer to discriminate between small mass differences is limited by the small variations in kinetic energy of ions as they leave the source. A double-focusing instrument focuses the ion beam twice. An electrostatic field first focuses only ions of the same kinetic energy on a slit, which in turn serves as the source for the magnetic separator. Ions passing through the slit are then focused on the collector by the electromagnet. The double-focusing instrument has much greater resolving power than its single-focusing cousin.

Ion separation can also be achieved without an electromagnet by accelerating them with an electric field pulse and measuring the time it takes for them to travel to the collector. All accelerated ions enter a field-free tube with the same kinetic energies; thus, their velocities in the tube vary inversely with their masses, so that lighter ions arrive at the collector earlier than heavier ones. Time-of-flight instruments have poor resolution, but they are rugged and suitable for mobile applications.

The mass spectrometer can also be used to measure absolute concentrations of elements like rubidium and strontium. This is done by an analytical technique called *isotope dilution*. In this procedure, we use a *spike,* consisting of a solution containing an element whose isotopic composition has been altered by enrichment of one of its naturally occurring isotopes. A known quantity of the spike is mixed with the sample to be analyzed. The sample, of course, contains an unknown quantity of the element of interest, but its isotopic composition has been measured. Therefore, when the spike is added, the isotopic composition of the mixture of spike plus sample can be used to calculate the concentration of element in the sample. The numerical manipulations for the isotope dilution technique are explained by Faure (1986).

record the disruptive passage of alpha particles in the form of *fission tracks*. Although the tracks are only about 10 microns long, they can be enlarged and made optically visible by etching with suitable solutions, because the damaged regions are more soluble. A microscopic view of fission tracks can be seen in Figure 12.10. The density of fission tracks in a given area of sample can be determined by counting under a petrographic microscope.

These tracks, except in rare instances, are produced solely by spontaneous fission of ^{238}U. The observed track density is proportional to the concentration of ^{238}U in the sample and to the amount of time over which tracks have been accumulating: the age of

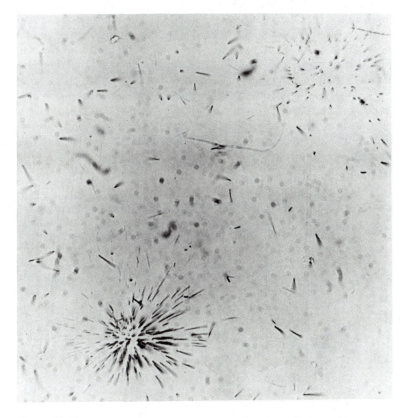

Figure 12.10 Photomicrograph of fission tracks in mica. Radiating tracks emanate from point sources such as small zircon inclusions; isolated tracks probably formed from single U or Th atoms. (Courtesy of G. Crozaz.)

the host mineral. Thus, if a means can be found to measure the uranium concentration in the sample, fission tracks can be used for geochronology. Uranium can be determined by measuring the density of tracks induced by irradiating the sample in a reactor, where bombardment with neutrons causes more rapid decay (induced radioactivity is explained more fully in the next section). Price and Walker (1963) formulated a solution for the fission track age equation (presented here without derivation):

$$t = \ln \left[1 + (\rho_s/\rho_i)(\lambda_2 \phi \sigma I/\lambda_F)\right]1/\lambda_2. \tag{12-15a}$$

Explaining the terms in this equation will illustrate how a fission track age is determined. λ_2 and λ_F are the decay constants for ^{238}U and spontaneous fission (8.42 $\times$ 10^{-17} yr^{-1}), respectively. The area density of fission tracks in the sample, ρ_s, is a function of age and uranium content. Thus, a specific age can be determined if the uranium content can be determined. Uranium concentration is measured by placing the sample (after ρ_s has been visually counted) into a nuclear reactor, which induces rapid ^{238}U decay and increases the number of fission tracks in the sample. The remaining terms in equation 12-15a are related to this irradiation process: ρ_i is the area density of induced tracks, ϕ is the flux of neutrons passed through the sample, σ is the target cross-section for the induced fission reaction, and I is the (constant) atomic ratio $^{235}U/^{238}U$.

Annealing of the sample at elevated temperatures causes fission tracks to fade, as the damage done to the crystal structure is healed. The annealing temperatures for most minerals used in fission track work are quite modest (several hundred degrees C or less), so the ages derived by this method must be interpreted as the points in time when rocks cooled through temperatures at which annealing ceased, commonly called *cooling ages*. Phases like apatite, sphene, and epidote are useful for the interpretation of cooling ages, because they begin to retain fission tracks at temperatures that are different from the blocking temperatures for K-Ar or Rb-Sr geochronometers. Fission track annealing temperatures, like geochronometer blocking temperatures, can be extrapolated from experimental data. Fleischer et al. (1975) have summarized annealing data for a number of minerals. The determination of cooling history using a combination of fission tracks and other isotopic methods is illustrated in the following Worked Problem.

Worked Problem 12-5

How can we determine cooling history from geochronological data? The elucidation of cooling history usually requires integration of several different kinds of isotopic data, tempered with thoughtful geologic reasoning. Nielson et al. (1986) characterized the temperature history of a Cenozoic igneous intrusive complex in Utah in this way. Critical observations for their interpretation of the cooling history are:

1. The K-Ar age of a hornblende sample from the complex is 11.8 m.y. The argon blocking temperature for hornblende is approximately 525°C.
2. The K-Ar age of a biotite sample from the complex is 10.8 m.y. The approximate argon blocking temperature for biotite is 275°C.
3. The fission track ages for zircons in these rocks range from 8.3 to 8.9 m.y. The temperature at which this phase begins to retain tracks depends on the cooling rate, but is generally about 175°C.
4. The fission track ages for apatites range from 8.1 to 9.1 m.y. Apatite begins to retain tracks at approximately 125°C.

Using these data, we can construct a temperature versus time plot that will describe the thermal history of this complex. This is the diagram shown in Figure 12.11. We can also speculate about the geologic controls on the cooling path. The temperature-time curve in

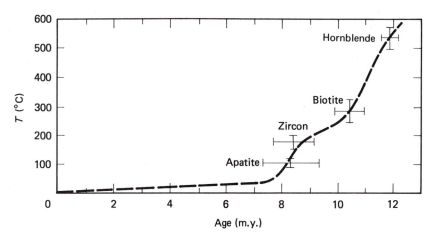

Figure 12.11 The cooling history of an igneous complex, inferred from argon blocking temperatures for hornblende and biotite and from fission track ages for zircon and apatite. These data were reported by Nielson et al. (1986).

Figure 12.11 implies that this region cooled from above 500°C through about 100°C within a span of only 4 m.y. This rapid cooling could not have taken place very deep in the crust, and must indicate uplift and erosion. Nielson and coworkers calculated that uplift proceeded at a rate between 0.6 and 1.1 mm yr^{-1} for this interval, based on an assumed geothermal gradient of 30° km^{-1}.

GEOCHEMICAL APPLICATIONS OF INDUCED RADIOACTIVITY

In medieval times, alchemists toiled endlessly to turn various metals into gold. They were unsuccessful, of course, but only because they did not yet know about nuclear irradiation. It is now possible to make gold, but it is still not economical because the starting material must be platinum.

Nuclei that are bombarded with neutrons, protons, or other charged particles are transformed into different nuclides. In many cases, the nuclides produced by such irradiation are radioactive. Monitoring their decay provides useful analytical information. Nuclear irradiation can occur naturally or can be induced artificially. Below we will consider some geochemical applications of both situations.

Neutron Activation Analysis

Nuclear reactions caused by neutron bombardment are commonly used to perform quantitative analyses of trace elements in geological samples. Fission of ^{235}U in a nuclear reactor produces large quantities of neutrons. These are emitted at high velocities and must be slowed down by a *moderator* in order to sustain the chain reaction in the reactor. We can use these neutrons to induce nuclear changes in a mineral or rock sample. Slow neutrons are readily absorbed by the nuclei of most stable isotopes, transforming them into heavier elements. Because many of the irradiation products are radioactive, this procedure is called *neutron activation*.

In earlier sections, we described radioactivity in terms of N, the number of atoms remaining. In neutron activation analysis, radioactivity is monitored using *activity A*, defined as

$$A = c\lambda N, \tag{12-16}$$

where c is the detection coefficient (determined by calibrating the detector) and λN is the rate of decay. The easiest way to determine the concentration of an element by neutron activation is by irradiating a standard containing a known amount of that element at the same time as the unknown. After irradiation, the activities of both standard and unknown are measured at intervals using a scintillation counter to detect emitted gamma rays.[3] The activities are then plotted as shown in Figure 12.12; the exponential decay curves are transformed into straight lines by plotting $\ln A$. Extrapolation of these lines back to zero time gives values of A_o, the activities when the samples were first removed from the reactor. The amount of element X in the unknown can then be determined from the following relationship:

$$A_{o\,unkn}/A_{o\,std} = \text{amount of } X_{unkn}/\text{amount of } X_{std}.$$

The concentration of element X in the unknown can be calculated by dividing its amount by the measured weight of sample.

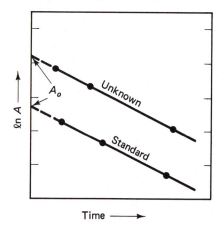

Figure 12.12 Plot of the measured activity A of an unknown and a standard after neutron irradiation, as a function of time. Extrapolation of these straight lines gives A_o, the activity at the end of irradiation. These data are then used to determine the concentration of the irradiated element in the unknown.

It is also possible to calculate A_o theoretically, using the following relationship:

$$A_o = cN_o\phi\sigma(1 - e^{-\lambda t_i}), \tag{12-17}$$

where c is the detection coefficient, N_o is the number of target nuclei, ϕ is the neutron flux in the reactor, σ is the neutron capture cross section of target nuclei, and t_i is the time of irradiation. This calculation, normally done by computer, makes it easier to handle many elements simultaneously in complex samples.

^{40}Argon-^{39}Argon System

One of the disadvantages of the K-Ar geochronometer is that gaseous argon tends to diffuse out of many minerals, even at modest temperatures. Loss of the radiogenic daughter would thus lead to an erroneously young radiometric age. ^{40}Ar-^{39}Ar measurements provide a way to get around the problem of argon loss. Let's examine how this technique works.

---

[3] This step may be complicated for geological samples because they contain so many elements. It is necessary to screen out radiation at energies other than that corresponding to the decay of interest. This can normally be done by adjusting the detector to filter unwanted gamma radiation, but in some cases, chemical separations may be required to eliminate interfering radiation.

We might expect that the outer portions of crystals would lose argon first because of shorter diffusion distances, but the centers of crystals could retain argon longer. In this case, the ratio of radiogenic ^{40}Ar to ^{39}K, a stable isotope of potassium, will be highest in crystal centers and lower in crystal rims. Irradiating the sample with neutrons in a reactor causes some ^{39}K to be converted to ^{39}Ar. Consequently, the ratio $^{40}Ar/^{39}K$ becomes $^{40}Ar/^{39}Ar$, which is readily analyzed using mass spectrometry.

The number of ^{39}Ar atoms produced during the irradiation is given by

$$^{39}Ar = {}^{39}K\, t_i \int \phi\sigma\, d\epsilon, \tag{12-18}$$

where t_i is the irradiation time, ϕ is the neutron flux in the reactor, σ is the neutron capture cross-section, and the integration is carried out over the neutron energy spectrum $d\epsilon$. The number of radiogenic ^{40}Ar atoms produced by decay of ^{40}K is given by equation 12-8. Dividing equation 12-8 by 12-18 gives the $^{40}Ar/^{39}Ar$ ratio in the sample after irradiation:

$$^{40}Ar/^{39}Ar = (0.124/t_i)(^{40}K/^{39}K)(e^{\lambda t} - 1)/(\int \phi\sigma\, d\epsilon). \tag{12-19}$$

Because equation 12-19 can be solved for t, measurement of the $^{40}Ar/^{39}Ar$ ratio of a sample defines its age.

In practice, the sample is heated incrementally, and the isotopic composition of argon released at each temperature step is measured. Figure 12.13 illustrates a series of $^{40}Ar/^{39}Ar$ ages, calculated as a function of total ^{39}Ar released, using equation 12-19, for samples of mica and amphibole in a blueschist sample from Alaska. Each step represents a temperature increment of $50°$, starting at $550°C$. If this sample had remained closed to argon loss since its formation, all of the ages would have been the same. The spectrum of ages that we see in Figure 12.13 results from ^{40}Ar leakage over time. Argon from the centers of crystals is released at the higher temperature steps, and the

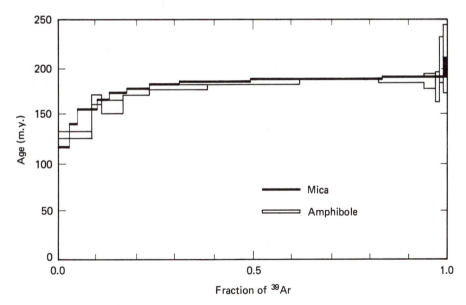

Figure 12.13 ^{40}Ar-^{39}Ar age versus cumulative ^{39}Ar released for mica and amphibole samples from blueschist, after Sisson and Onstott (1986). The argon was released at successive temperature steps; each step in this figure corresponds to a $50°$ increment, starting at $550°C$. The plateau of ages at 185 m.y. corresponds to radiogenic argon from the interiors of crystals.

Geochemical Applications of Induced Radioactivity

plateau of ages at approximately 185 m.y. seen at high temperatures represents the time of metamorphism for this sample, because the mica blocking temperature is approximately equal to the peak metamorphic temperature.

Cosmic-Ray Exposure

In both of the examples we have just considered, the samples were irradiated intentionally by man. Natural irradiation can also have useful geochemical consequences. Cosmic rays, consisting mostly of high-energy protons but also of neutrons and other charged particles generated outside the solar system, offer a natural irradiation source. Radiogenic isotopes produced by interaction of matter with cosmic rays are called *cosmogenic* nuclides.

Radioactive ^{14}C is produced continuously in the atmosphere by the interaction of ^{14}N and cosmic-ray neutrons. This isotope is then incorporated into carbon dioxide molecules, which in turn enter plant tissue through photosynthesis. The concentration of cosmogenic ^{14}C in living plants (and the animals that eat them) is maintained at a constant level by atmospheric interaction and isotope decay. When the plant or animal dies, addition to its tissues of ^{14}C from the atmosphere stops, so the concentration of ^{14}C decreases due to decay. Measurement of the activity of ^{14}C in plant or animal remains thus provides a way to determine the time elapsed since death.

The activity A of carbon from a plant or animal that died t years ago is given by

$$A = A_o e^{-\lambda t}, \tag{12-20}$$

where A_o is the ^{14}C activity during life (13.56 dpm g^{-1}), and λ is given in Table 12-1.[4] Radiocarbon ages have only limited geologic utility because of the short half-life of ^{14}C, but they are very useful for archaeological dating.

Meteoroids orbiting in space are continuously irradiated by exposure to cosmic-ray protons, but cosmogenic nuclides are produced in only the outer meter or so because rock is an effective shield for cosmic radiation. It is therefore likely that cosmogenic isotopes preserved in most meteorite specimens do not begin to accumulate until the host meteoroid has been broken into meter-sized chunks. Production of these isotopes then continues until the meteorite falls to earth, where the atmosphere blocks all but the most energetic cosmic rays. Although many cosmogenic nuclides are produced, only a few can be detected because their abundances in meteorites prior to irradiation were small or nonexistent. ^3He and ^{21}Ne are the most easily resolvable, using mass spectrometry. By measuring the amounts of these cosmogenic nuclides, it is possible to determine how long meteorites orbited as meter-size bodies in space. These time intervals are called *cosmic-ray exposure ages*. Measured concentrations of cosmogenic nuclides are converted to exposure ages by knowing nuclide production rates, which can be determined experimentally.

RADIONUCLIDES AS TRACERS
OF GEOCHEMICAL PROCESSES

In biological research, certain organic molecules may be "tagged" with a radioactive isotope so that their participation in reactions or progress through an organism can be followed. It is difficult to design analogous experiments to trace geochemical processes

---

[4]Compare equation 12-20 with equation 12-16. Both express activity in terms of λN times a constant, c or A_o. Activity is commonly expressed in units of disintegrations per minute per gram of sample (dpm g^{-1}).

that have already taken place within the earth, but radionuclides can sometimes be used for this purpose. If the geochemical behavior of elements that have radioactive isotopes is known, measured variations in isotopic abundances may provide constraints on such processes. In this section, we will consider a variety of examples.

Heterogeneity of the Earth's Mantle

Two of the rare earth elements, samarium and neodymium, partition in different ways between continental crust and mantle rocks. Neodymium shows a marked preference for the crust, and samarium concentrates in the mantle. This occurs because of different distribution coefficients for these elements between minerals and melts. ^{147}Sm decays to ^{143}Nd over time, however, so the ratio of ^{143}Nd to stable ^{144}Nd changes in both crust and mantle as one loses radioactive samarium and the other gains radiogenic neodymium. It is thought that the mantle originally contained samarium and neodymium in the same proportions as in chondritic meteorites.[5] The deviation between ^{143}Nd/^{144}Nd measured in a mantle-derived rock and the same ratio in a chondrite of identical age is called ϵ_{Nd}, and can be used to estimate the degree to which the mantle may have evolved from its pristine state.

If basalts are derived from melting of mantle material, their initial Nd isotopic ratios should be the same as that of their mantle sources at the time of melting. ϵ_{Nd} values for young basalts from various tectonic settings are summarized in Figure 12.14. Almost all of the basalts that occur in oceanic regions have positive ϵ_{Nd} values; that is, they have high ^{143}Nd/^{144}Nd ratios which are complementary to those of continental crust. From this, we could inferred that the mantle source regions for oceanic basalts have been differentiated and have lost material to form the continents. In contrast, con-

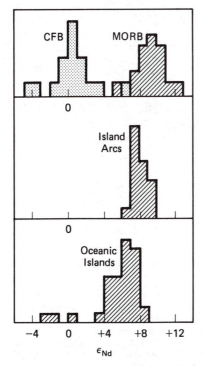

Figure 12.14 The isotopic composition of neodymium in young basalts from various tectonic settings, in terms of ϵ_{Nd}. Abbreviations are: CFB = continental flood basalts, MORB = midocean ridge basalts. (After D.J. DePaolo, Neodymium isotopic studies: Some new perspectives on earth structure and evolution. *EOS* 62: 137–40. Copyright 1981 by American Geophysical Union. Reprinted with permission.)

[5] The evidence that chondrites were possible planetary building blocks is considered in Chapter 13.

tinental basalts and a few oceanic island basalts have ϵ_{Nd} values close to zero, indicating that their mantle source regions are evolving like chondrites and are thus undifferentiated. What do these data tell us about the nature of the mantle?

Wasserburg and DePaolo (1979) inferred that these differences result from melting processes within a layered mantle. Their model is depicted schematically in Figure 12.15. The mantle is visualized as consisting of two layers, the lower portion undifferentiated with respect to Sm and Nd (and other incompatible elements as well), and the upper mantle highly fractionated with a positive ϵ_{Nd} value. The lower mantle has presumably remained in its primitive condition since its formation, sampled only infrequently at hot spots to produce continental basalts and a few oceanic islands. The upper mantle has been continuously cycled by the formation and subduction of oceanic crust, as well as the production of continental crust made in magmatic arcs at subduction zones. The various intermediate values of ϵ_{Nd} in oceanic basalts from different tectonic settings arise from mixing materials from the two mantle layers.

The $+12\epsilon_{Nd}$ value of the upper mantle is counterbalanced by the -15 value for continental crust. Knowing these values and the mass of the continental crust, it is possible to perform a mass balance calculation to determine the proportion of the differentiated upper mantle. The mass of the continental crust is about 2×10^{25} g and its Nd abundance is approximately 50 times higher than the upper mantle. Thus the mass of

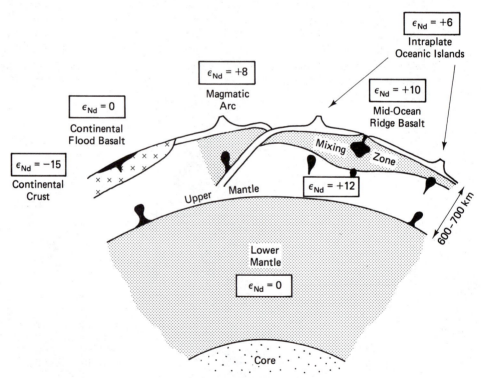

Figure 12.15 A model of the earth's internal structure based on Nd isotopic data. Continental crust and the upper mantle have complementary compositions. The isotopic composition of the lower mantle lies on a chondritic evolution curve, and thus it does not participate to any great extent in the crust-forming process. Plumes of magma from the lower mantle produce intraplate volcanism, erupting as continental flood basalts and forming some oceanic islands. Magmas derived from the upper mantle occur at plate boundaries, and their isotopic variability can be explained by mixing of materials in the upper mantle.

the upper mantle is 100×10^{25} g. This corresponds to a thickness on the order of 600-700 km. Comparison with Figure 11.3 shows that this depth range coincides with a prominent seismic discontinuity, from which we might infer that at least some of the seismic structure of the mantle might result from chemical differences rather than phase transformations.

If this view of the mantle is correct, other isotopic systems should be correlated with Sm-Nd data. In fact, ϵ_{Sr} and ϵ_{Pb}, the deviations in $^{87}Sr/^{86}Sr$ and $^{206}Pb/^{204}Pb$ from chondritic evolution, do vary systematically with ϵ_{Nd}. This has led to the concept of a characteristic Nd-Sr-Pb *mantle array*. Inclusion of additional isotopic systems appears to require that two-component compositional models for the mantle be expanded to as many as four or five components (White, 1985; Hart et al., 1986).

Metallogenic Provinces

Heterogeneity in the geographic distribution of metals has spawned the concept of *metallogenic provinces* with distinct kinds of economic deposits. Isotopic data provide information from which the sources for ores can be discerned. Using lead isotopes, Zartman (1974) recognized three such provinces in the western United States.

The lead isotopic compositions of Mesozoic to Tertiary igneous rocks (representing a time span of about 200 m.y.) in this region are summarized in Figure 12.16. By comparing only young rocks, Zartman avoided the uncertainties introduced by making isotopic corrections for *in situ* decay of uranium and thorium in rocks since their crystallization. As illustrated by the lead evolution curves in Figure 12.16, this 200 m.y. interval is much less than that necessary to account for the observed isotopic variations. Leads from hydrothermal ore deposits in these regions were found to be similar in isotopic composition to the igneous rocks, although some small deposits in area I plot outside this field.

In order to interpret these results, we must understand how U, Th, and Pb are distributed within the lithosphere. In the lower crust, uranium is depleted relative to lead, but in the upper crust uranium is enriched. The distribution of thorium within the crust is less well understood, but all three elements are depleted in the mantle relative to the crust.

Leads in area I vary widely in isotopic composition, but are consistently less radiogenic than in areas II and III. Area I is underlain by a thick Precambrian basement. Leads derived from the lower crustal portions of this basement, which presumably are characterized by low U/Pb ratios, should be depleted in those isotopes derived from uranium (^{207}Pb and ^{206}Pb).

Area II contains a thick sequence of miogeosynclinal sediments of late Precambrian to Phanerozoic age, presumably eroded from upper crustal material in the basement. Because upper crustal rocks are enriched in uranium, they should develop more radiogenic lead with time, as observed. The small range in isotopic composition for rocks in area II may be accounted for by mixing of upper crustal materials to form a relatively homogeneous sediment.

Area III contains a thick pile of eugeosynclinal rocks of Mesozoic to Tertiary age. The lead isotopic compositions of the igneous rocks in this terrain resemble those of many island arcs. Derivation of lead from subducted oceanic crust, mantle, and associated trench deposits may explain, at least in part, the isotopic character of this province. Using strontium (Kistler and Peterman, 1978) and neodymium (Bennett and DePaolo, 1987) isotopes, further studies have confirmed, at least in part, that these metallogenic provinces represent major tectonic blocks.

Identification of lead isotopes with their source materials offers the possibility of

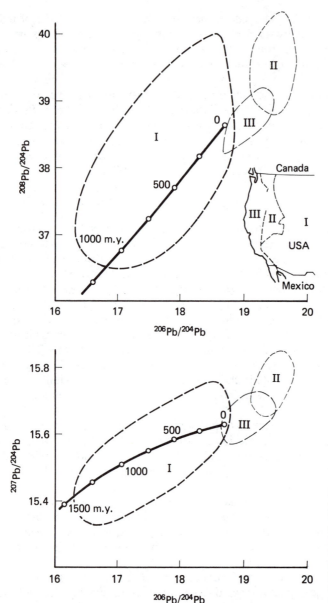

Figure 12.16 Lead isotopic ratios of young igneous rocks of the western United States (see inset map) define three metallogenic provinces I, II, and III. Calculated lead evolution curves (labelled with time in m.y.) demonstrate that these variations cannot be explained by *in situ* decay of U and Th after crystallization of these rocks. (Modified from Zartman 1974.)

understanding the depths involved in extracting this metal and the migration routes traversed by hydrothermal fluids. More studies like Zartman's may provide sufficient data with which to quantify ore-forming processes. This information, plus the mapping of metallogenic provinces, should guide future exploration activities for lead and other resources.

Magmatic Assimilation

In some cases, ascending magmas are contaminated by assimilation of crustal materials, with the result that their compositions are altered. Radiogenic isotopes provide the most definitive way of recognizing this process. In uncontaminated basalts, the initial

^{87}Sr/^{86}Sr ratio is the same as that of the mantle source region. This is always a relatively low number (ranging from about 0.699 to 0.704, depending on the age of the rock—the mantle accumulates radiogenic ^{87}Sr over time), because the amount of parent rubidium relative to strontium in mantle rocks is low. Rb/Sr ratios in continental crustal rocks are much higher, and therefore crustal rocks contain more radiogenic ^{87}Sr. We can infer, then, that magmas derived from mantle sources should have low initial ^{87}Sr/^{86}Sr ratios. Those derived from, or contaminated with, crustal sources should have higher initial ^{87}Sr/^{86}Sr ratios. A corollary is that rocks related through fractional crystallization of some parental magma should all have identical ^{87}Sr$_o$/^{86}Sr ratios.

Other isotopic systems can also be used to test for assimilation of crustal materials. Contaminated basalts should have lower initial ^{143}Nd/^{144}Nd ratios than uncontaminated rocks. It is not possible to predict the exact isotopic compositon of crustal lead, but if Pb isotopes can be analyzed in plausible contaminating materials, its mixing effects can be assessed.

If successive pulses of magma utilize the same passageway repeatedly, the metamorphosed wall rocks may eventually insulate magmas from chemical contamination. This scenario predicts that the amount of assimilation would diminish with time for one eruptive center. This appears to be borne out by isotopic measurements in arc volcanics like the Aleutians (Myers et al., 1985), where sequential magmas use the same conduits. Other studies of radioactive isotopes in volcanic rocks indicate correlations between the degree of crustal contamination and local crustal thickness (Barreiro and Clark, 1984; Nohda and Wasserburg, 1986). This observation suggests that the amount of assimilation may be controlled by the ease or difficulty with which magma ascends through the overlying rocks.

Assimilation is actually a rather complicated process, because a source for the necessary heat (that required to raise the temperature of the wall rocks to their melting points, plus the enthalpy of fusion of the wall rocks) must be identified. Refer to Chapter 11 for a review of this problem. Most magmas are already at or below their liquidus temperatures, so the only plausible source of heat is exothermic crystallization of the magma itself. Consequently, we might expect fractional crystallization to occur simultaneously whenever assimilation takes place. Taylor (1980) and DePaolo (1981b) derived equations to model the change in isotopic composition of a magma experiencing both processes. A magma body of mass M^o assimilates a mass of rock at rate M_a; the rate at which crystals separate from the magma is M_c. At any time during this combined process, the mass of magma remaining is M. The parameter $F = M/M^o$ is the mass of magma relative to the mass of the original magma. Any isotopic ratio in the magma I_m resulting from the combined effects of assimilation and fractional crystallization is given by

$$I_m = \{(r/[r-1])(C_a/z)(1 - F^{-z})I_a + C_m^o F^{-z}I_m^o\}/$$
$$\{(r/[r-1])(C_a/z)(1 - F^{-z}) + C_m^o F^{-z}\}. \qquad (12\text{-}21)$$

In this equation, r is defined as M_a/M_c, C_a is the concentration of the element in the assimilated rock, C_m^o is its initial concentration in the original magma, and I_m^o and I_a are the initial isotopic ratios of the magma and assimilated rocks, respectively. The term z is defined by

$$z = (r + D - 1)/(r - 1),$$

where D is the bulk distribution coefficient between the fractionating crystal phases and the magma. Equation 12-21 can be used to predict the isotopic composition of a hybrid magma resulting from assimilation and fractionation. I_m could be any isotope ratio, like ^{87}Sr/^{86}Sr, or any normalized parameter describing such a ratio, like ϵ_{Sr}.

Strontium isotopic compositions of modern andesites in the Peruvian Andes were reported by James et al. (1976). These authors concluded that the elevated $^{87}Sr/^{86}Sr$ ratios in these volcanic rocks could not be explained by assimilation of crustal materials. However, by considering the combined effects of assimilation and fractional crystallization, DePaolo (1981b) showed that their data are, in fact, consistent with a contamination model for Andean volcanism.

The inferred isotopic compositions of the initial magma for the Arequipa volcanics and possible contaminants are illustrated in Figure 12.17. The volcanic rocks clearly do not lie on a straight mixing line connecting these two compositions. Models for the combined effects of assimilation and fractional crystallization can be constructed using equation 12-21. To illustrate, let us calculate the isotopic composition from the following parameters: the initial magma had 400 ppm Sr, 10 ppm Rb, and $^{87}Sr/^{86}Sr = 0.7030$. The contaminant had 500 ppm Sr, 40 ppm Rb, and $^{87}Sr/^{86}Sr = 0.71025$. We will assume that $r = 0.8$, $D_{Sr} = 0.5$, and $F = 2$. The corresponding value for z would then be -1.5. Substitution of these parameters into equation 12-21 gives

$$I_m = \{(0.8/-0.2)(500/-1.5)(1 - 2^{1.5})(0.71025) + (400)(2^{1.5})(0.7030)\}/$$

$$\{(0.8/-0.2)(500/-1.5)(1 - 2^{1.5}) + (400)(2^{1.5})\}$$

$$= 0.7078.$$

This calculated value is only one point on a curve like the curves in Figure 12.17, which illustrate how these variables change during the combined processes. DePaolo (1981b) calculated such a diagram for Arequipa volcanics, which are the curves actually shown in Figure 12.17. For simplicity, he assumed that $r = 1$ and $D_{Sr} = 0, 0.25, 0.5$, and 1.25. The measured Rb/Sr ratios and isotopic composition of Sr fall within the range of values calculated using these parameters. It is expected that D_{Sr} would vary during fractional crystallization/assimilation as different phases crystallized and the liquid composition changed. This reasonably good correspondence suggests that assimilation of crustal materials is a valid petrogenetic model for these volcanic rocks.

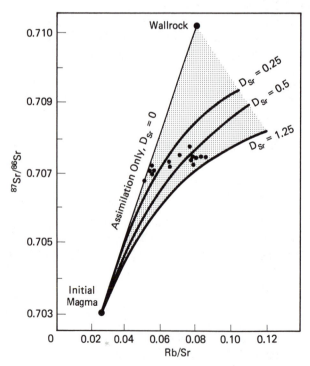

Figure 12.17 Strontium isotopic data for volcanic rocks of the Peruvian Andes, after DePaolo (1981b). The data points do not lie on a simple mixing line connecting the isotopic compositions of initial magma and wallrock, but do lie near calculated mixing lines for an assimilation-fractional crystallization model with various values of D_{Sr}. It is expected that D_{Sr} would increase in the latter stages of magma evolution. This model appears to explain the observed isotopic data fairly well.

Geochemical Cycling of Sediments
in Subduction Zones

There are simpler mixing events than that just considered which do not require that the effects of several processes be monitored concurrently. One example is mixing of components that subsequently experience melting. This may occur in subduction zones.

The possible subduction of sediments in oceanic trenches and subsequent incorporation into arc volcanics is a controversial subject. Radiogenic isotopes that are preferentially incorporated into such sediments may serve as tracers to follow their fate in this tectonic setting. These sediments characteristically have high $^{208}Pb/^{204}Pb$ and $^{207}Pb/^{204}Pb$ ratios. We know that U and Th are incompatible elements that are concentrated in crustal rocks, and the high radiogenic lead contents of these sediments are the result of their derivation from old U- and Th- enriched rocks on continents. Armstrong (1971) noted that the isotopic compositions of lead in some arc volcanics and in abyssal sediments are very similar, but are so distinct from those of oceanic basalts that recycling of crustal material must occur. Mass balance calculations suggest that a ratio of melted oceanic basalt : sediments of 90 : 10 could account for the observed data.

Another isotope that may shed light on this problem is ^{10}Be. This is a cosmogenic nuclide, formed by interaction of cosmic rays with oxygen and nitrogen in the atmosphere. ^{10}Be is transferred to the earth's surface by precipitation, where it is incorporated into sediments. The half-life of this isotope is only 1.5 m.y., a time scale similar to that required for subduction and magma generation. Tera et al. (1986) measured the concentrations of this isotope in arc volcanics and in ocean floor basalts. Lavas in some island arcs clearly had higher abundances of ^{10}Be, suggesting that sediments were subducted and participated in the melting process to produce arc volcanics.

Both lines of geochemical evidence point to a role for admixed sediments in the petrogenesis of arc volcanics, in contrast to some geological and geophysical studies that stress the mechanical difficulty of subducting low-density sediments into the mantle. Volcanic rocks in some arcs show no isotopic effects that would be attributable to sediment mixing, however, so this process does not occur in all oceanic trenches. An important test for this hypothesis is that high radiogenic Pb isotope and ^{10}Be abundances should occur in the same samples. This correlation has not yet been demonstrated, however.

Isotopic Composition of the Oceans

The isotopic composition of ocean water depends on input from various sources. These are primarily (1) submarine weathering of and magmatic interaction with young basaltic rocks of the ocean floor, (2) recycling of marine carbonate rocks that now reside on continents, and (3) weathering of crystalline continental crust, mostly granitic, igneous, and metamorphic rocks. We will briefly consider the behavior of strontium and neodymium in the oceanic system.

All oceans at the present time have a $^{87}Sr/^{86}Sr$ ratio of 0.7090. In contrast, the measured $^{143}Nd/^{144}Nd$ ratios of ocean water indicate that each ocean has a distinct and characteristic compositional range, as summarized in Figure 12.18 in terms of ϵ_{Nd}. What causes this difference in isotopic behavior?

It is thought that much of the strontium in seawater is derived from weathering of marine carbonates on continents. Peterman et al. (1970) determined that the $^{87}Sr/^{86}Sr$ ratio in these carbonates has varied during Phanerozoic time, decreasing to a minimum of 0.70675 during the Jurassic and increasing again to the current high value of 0.7090.

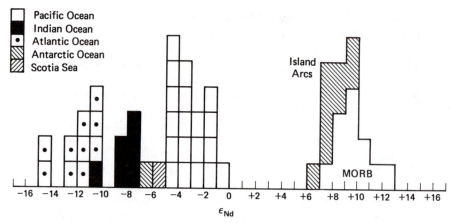

Figure 12.18 The neodymium isotopic compositions of seawater and ferromagnesian nodules from different oceans, expressed as ϵ_{Nd}. These compositions are distinct from each other and from oceanic volcanic rocks. (After D. J. Piepgras, G. J. Wasserburg, and E. J. Dasch, The isotopic composition of Nd in different ocean masses. *Earth Planet. Sci. Lett.* 45: 223–36. Copyright 1979 by Springer-Verlag. Reprinted with permission.)

The isotopic composition of marine carbonates mimics that of the ocean in which they formed, so the oceans have varied in Sr isotopic composition over time. Different oceans drain continents with different carbonate sources, so the fact that oceans have the same Sr isotopic composition everywhere must relate to a long residence time for this element in seawater. Interoceanic circulation patterns are rapid compared to Sr residence, so that a worldwide isotopic composition results at any one time.

Continental crystalline rocks probably supply most of the neodymium to the oceans. Differences in the ages of the continental rocks that provide the weathering products that are drained into the oceans produce distinct radiogenic isotopic signatures in river waters. We might expect that, as with Sr isotopes, these differences would be erased by global circulation, but they are not. The Atlantic is supplied by very old continental regions, with high $^{143}Nd/^{144}Nd$, whereas the Pacific is ringed by younger regions with correspondingly lower values of Nd isotopic ratios. Waters in the two oceans retain these isotopic signatures. This can be explained only by assuming that the residence time for neodymium in seawater is short, that is, that Nd entering the oceans is lost to the seafloor before interoceanic circulation patterns can homogenize the isotopic composition between oceans.

Radiogenic isotopic studies of seawater thus offer the possibility of shedding light on interoceanic mixing rates, provided that residence times can be fixed. Neodymium isotope measurements may even be able to identify the location of currents responsible for such mixing. Strontium isotopes also provide information on the kinds of rocks exposed to weathering on continents over geologic time.

Worked Problem 12-7

What are reasonable values for input into the modern oceans from oceanic volcanics, marine carbonates, and crystalline continental crustal rocks? Faure (1986) presented a mass-balance model based on the premise that the strontium isotopic composition of the oceans could be explained by input from these three sources. The $^{87}Sr/^{86}Sr$ ratio of seawater (*sw*) is then given by

$$({}^{87}Sr/{}^{86}Sr)_{sw} = ({}^{87}Sr/{}^{86}Sr)_v V + ({}^{87}Sr/{}^{86}Sr)_m M + ({}^{87}Sr/{}^{86}Sr)_c C,$$

where $({}^{87}Sr/{}^{86}Sr)_{v,m,c}$ are strontium isotopic ratios contributed by volcanic, marine carbonate, and continental crustal rocks, respectively, and the coefficients V, M, C are the fractions of the total input contributed by these sources. By substituting reasonable current values for the strontium isotopic compositions of each of these sources, we obtain

$$({}^{87}\text{Sr}/{}^{86}\text{Sr})_{sw} = 0.704V + 0.708M + 0.720C. \qquad (12\text{-}22)$$

If we then plot $({}^{87}\text{Sr}/{}^{86}\text{Sr})_{sw}$ versus V, as illustrated in Figure 12.19, we can contour values of M and C from equation 12-22. Plausible solutions are limited by the restriction that no source can supply a negative amount of material, that is , V, M, and C must be equal to or greater than zero. The isotopic composition of modern seawater is illustrated in Figure 12.19 by a horizontal dashed line at ${}^{87}\text{Sr}/{}^{86}\text{Sr} = 0.7090$. From this figure we can see that V cannot be greater than 0.68. However, V cannot be very close to this upper bound, because marine carbonates must make a substantial contribution. The high Sr contents and susceptibility to chemical weathering demand an important role for these rocks in this system. This model does not provide a unique solution to the problem, unless the input from any one of the sources can be fixed unambiguously. However, it does allow us to assess the relative contributions from these sources if a reasonable value for one of them can be guessed. For example, if we assume that M has the value 0.7, then the contributions from other sources are $V = 0.16$ and $C = 0.14$.

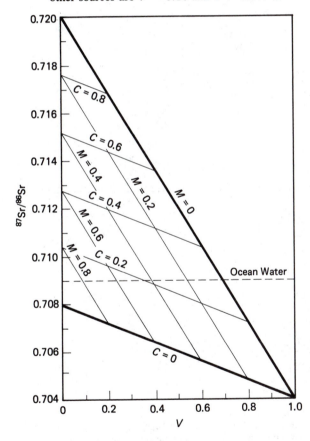

Figure 12.19 A model to explain the strontium isotopic composition of seawater. Mixture of various proportions of volcanic rocks (V), marine carbonates (M), and continental crustal rocks (C) can account for the ${}^{87}\text{Sr}/{}^{86}\text{Sr}$ ratio of modern ocean water (dashed horizontal line at 0.709). (After G. Faure, *Principles of isotope geology*, 2d ed. Copyright 1986 by John Wiley and Sons. Reprinted with permission.)

Degassing of the Earth's Interior to Form the Atmosphere

The atmosphere is thought to have formed by degassing of the interior, and isotopic variations in gases dissolved in basalts may provide information on this process. Noble gases are particularly useful in understanding atmospheric evolution, because all except helium are too heavy to escape from the atmosphere to space. Radioactive decay results in the production of many noble gas isotopes; this is easily monitored by considering ratios of radiogenic to stable nuclides. For example, variations observed at the present time in $^4\text{He}/^3\text{He}$ ratios reflect the balance between radiogenic ^4He formed by decay of ^{232}Th, ^{235}U, and ^{238}U and primordial (stable) ^3He. Similarly, the $^{40}\text{Ar}/^{36}\text{Ar}$ ratio is controlled by the progressive decay of ^{40}K to ^{40}Ar. The isotopic composition of xenon reflects several processes. Radiogenic ^{129}Xe was produced by decay of now-extinct ^{129}I.

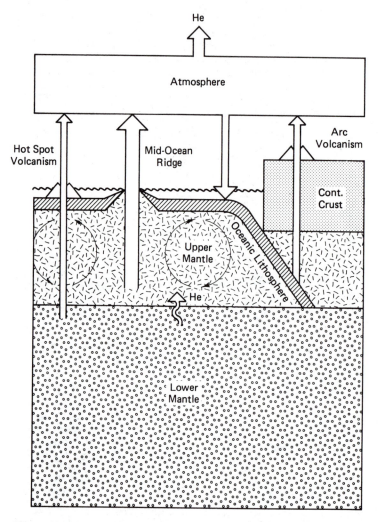

Figure 12.20 Schematic representation of noble gas reservoirs and fluxes in the mantle-crust-atmosphere system. Outgassing of the upper mantle occurs through volcanism at spreading centers, and outgassing of a portion of the lower mantle results from hot spot volcanic activity. In this model, helium diffuses between mantle reservoirs, but they behave as closed systems to other noble gases. Subducted oceanic crust contains atmospheric gases, but these are only temporarily recycled because arc volcanism returns them to the atmosphere. Continental crust is a significant reservoir for radiogenic gases because their parent isotopes are fractionated into this site and remain for long periods of time. Helium is the only noble gas that escapes from the atmosphere to space. (Modified from Allegre et al. 1986–1987.)

Fission of ^{238}U also generates ^{134}Xe, ^{136}Xe, and several other xenon isotopes, creating variations in the ratios of these isotopes to stable ^{130}Xe.

Two different geological processes also affect these isotopic ratios. Outgassing occurs by volcanic and hydrothermal activity. Decreases in gas concentrations in the mantle reservoir affect both radiogenic and stable nuclides, but not nongaseous parent nuclides. For example, degassing will lower ^{36}Ar (the isotope to which radiogenic ^{40}Ar is normalized) but will not affect ^{40}K (which generates ^{40}Ar). Fractionation of incompatible elements will also affect these ratios. We have already learned that K, U, and Th are fractionated into the crust by magmatic processes. Because some isotopes of these elements are parent nuclides for noble gases, this process leads to depletion of radiogenic gas isotopes in mantle source regions.

Outgassing and crustal formation have occurred on completely different time scales, and the parent nuclides which produce the isotopic variations in noble gases have very different half-lives. Consequently, these isotopic systems permit us to trace these geologic processes and to determine their time dependence.

Allegre et al. (1986–1987) modelled outgassing of the mantle using mass balance calculations that employed argon and xenon isotopic ratios. Their model employs four noble gas reservoirs: atmosphere, continental crust, upper mantle, and lower mantle. Their calculations indicate that approximately half of the earth's mantle is 99% outgassed. Based on other isotopic evidence such as Nd isotopes discussed earlier, this is a larger mantle proportion than that corresponding to the differentiated upper mantle. This disagreement can be rationalized by assuming that some of the primitive lower mantle has also lost a portion of its complement of gases.

Helium was not used in the mass balance calculations because it has limited residence time in the atmosphere. However, this becomes an advantage if we want to calculate the gas flux from the mantle. Assuming a steady-state condition in which the mantle releases as much He as is lost from the atmosphere to space, Allegre and co-workers calculated He fluxes in the mantle-atmosphere system. Fluxes of Ar and Xe can also be estimated based on correlations of their isotopic ratios with ^{4}He/^{3}He. The model of Allegre and his collaborators is summarized in Figure 12.20. From this diagram, we can see that the only recycling occurs at subduction zones, and that this recycling is temporary because gases in subducted lithosphere are returned to the atmosphere by arc volcanism. This study thus allows a reasonable description of both the extent and kinetics of mantle outgassing.

SUMMARY

In this chapter, we have seen that isotope stability is the exception rather than the rule. Radionuclides may decay by emission of beta particles, alpha particles, positrons, and capture of electrons, or by spontaneous fission. Branched decay involving several of these decay mechanisms occurs in some cases. The rate of radioactive decay is exponential and is most easily expressed in terms of half-life. We have derived equations that allow determination of the isotopic evolution of a system as a function of time. These equations permit the age-dating of rocks, based on the decay of certain naturally occurring radionuclides such as ^{40}K, ^{147}Sm, ^{87}Rb, ^{232}Th, ^{235}U, and ^{238}U. In such systems, it is possible to determine the age of rocks even though some daughter isotope may have been present in the system initially. Uranium and thorium isotopes decay through a series of intermediate radioactive daughters, but the half-lives of transitory daughters are short enough that each system can be treated as the decay of parent directly to the stable daughter. Independent U-Th-Pb geochronometers can give the same or different results, but useful information can be gained even in cases of discordancy. The interpretation of geochronological data can be complicated by the failure of rock systems to remain closed, as well as by other factors. The earlier presence of now extinct radionuclides in meteorites indicates that they must have formed very soon after nucleosynthesis.

The abundances of radiogenic isotopes can be determined from mass spectrometry or from fission track techniques. The production of radioactive isotopes can be induced artificially in a nuclear reactor. This capability forms the basis for quantitative elemental analysis by neutron activation. The ^{40}Ar-^{39}Ar geochronometer also depends on induced radioactivity. The same effect is produced naturally by cosmic ray exposure, which forms ^{14}C, making it possible to date recent events, and other radionuclides in meteorites, making it possible to date orbital time in space.

If the geochemical behavior of radionuclides is known, these provide powerful tracers for geologic processes. Uses include understanding such phenomena as mantle heterogeneity, metallogenic provinces, magmatic contamination, cycling of materials between crust and mantle reservoirs, global oceanic mixing, and outgassing of the earth's interior to form the atmosphere.

The importance of radiogenic isotopes will be considered further in the following chapter on the early solar system and processes at planetary scales.

SUGGESTED READINGS

There are unfortunately few textbooks that treat isotope geochemistry in any detail. Faure's book is required reading for anyone seriously interested in this subject, and the other references provide further elaboration on certain aspects of radiogenic isotopes.

DePaolo, D. J. 1981a. Nd isotopic studies: Some new perspectives on earth structure and evolution. *EOS* 62, 14: 137–140. (A lucid account of advances in isotope geochemistry using Nd isotopes as tracers.)

Faure, G. 1986. *Principles of isotope geology.* 2d ed. New York: John Wiley and Sons. (The best available text on isotope geochemistry. This is an excellent source for detailed information of radiogenic isotopes; geochronology is covered more exhaustively than other aspects of this subject.)

Jager, E., and Hunziker, J. C., eds. 1979. *Lectures in isotope geology.* Berlin: Springer-Verlag. (A series of lectures given in Switzerland by respected isotope geochemists. The first half of the book deals with geochronology.)

Schaeffer, G. A., and Schaeffer, O. A. 1977. ^{39}Ar-^{40}Ar ages of lunar rocks. *Proc. 8th Lunar Sci. Conf.,* 2253-3300. (This paper provides a nice description of the ^{40}Ar-^{39}Ar technique.)

Walker, F. W.; Miller, D. G; and Feiner, F. 1983. *Chart of the nuclides.* San Jose, Calif.: General Electric Company. (A real bargain that contains physical constants, conversion factors, and periodic table. Order from General Electric Nuclear Energy Operations, 175 Curtner Ave., M/C 684, San Jose, CA 95125.)

The following references were cited in this chapter. These provide more thorough treatment of the applications of radionuclides in solving geochemical problems.

Aleinikoff, J. N.; Moench, R. H.; and Lyons, J. B. 1985. Carboniferous U-Pb age of the Sebago batholith, southwestern Maine: Metamorphic and tectonic implications. *Geol. Soc. Amer. Bull.* 96: 990–96.

Allegre, C. J.; Staudacher, T.; and Sarda, P. 1986–1987. Rare gas systematics: Formation of the atmosphere, evolution and structure of the earth's mantle. *Earth Planet. Sci. Lett.* 81: 127–50.

Armstrong, R. L. 1971. Isotopic and chemical constraints on models of magma genesis in volcanic arcs. *Earth Planet. Sci. Lett.* 12: 137–42.

Barreiro, B., and Clark, A. H. 1984. Lead isotopic evidence for evolutionary changes in magma-crust interaction, Central Andes, southern Peru. *Earth Planet. Sci. Lett.* 69: 30–42.

Bennett, V. C., and DePalo, D. J. 1987. Proterozoic crustal history of the western United States as determined by neodymium isotopic mapping. *Geol. Soc. Am. Bull.* 99, 674–85.

Dallmeyer, R. D., and VanBreeman, O. 1981, Rb-Sr whole-rock and ^{40}Ar-^{39}Ar mineral ages of the Togus and Hallowell quartz monzonite and Three Mile Pond granodiorite plutons, south-central Maine: Their bearing on post-Acadian cooling history. *Contrib. Mineral. Petrol.* 78: 61–73.

Dallmeyer, R. D.; Wright, J. E.; Secor, D. T. Jr.; and Snoke, A. W. 1986. Character of the Alleghanian orogeny in the southern Appalachians : Part II. Geochronological constraints on the tectonothermal evolution of the eastern Piedmont in South Carolina. *Geol. Soc. of Amer. Bull.* 97: 1329–44.

DePaolo, D. J. 1981b. Trace element and isotopic effects of combined wallrock assimilation and fractional crystallization. *Earth Planet. Sci. Lett.* 53: 189–202.

FLEISCHER, R. L.,; PRICE, R. B.; and WALKER, R. M. 1975. *Nuclear tracks in solids. Principles and applications.* Berkeley: Univ. of California Press.

HART, S. R.; GERLACH, D. C.; and WHITE, W. M. 1986. A possible Sr-Nd-Pb mantle array and consequences for mantle mixing. *Geochim. Cosmochim. Acta* 50: 1551–57.

KISTLER, R. W.; and PETERMAN, Z. E. 1978. Reconstruction of crustal blocks of California on the basis of initial strontium isotopic compositions of Mesozoic granitic rocks. *U. S. Geol. Soc. Prof. Paper* 1071, 1–17.

MYERS, J. D.; MARSH, B. D.; and SINHA, A. K. 1985. Strontium isotopic and selected trace element variations between two Aleutian volcanic centers (Adak and Atka): Implications for the development of arc volcanic plumbing systems. *Contrib. Mineral. Petrol.* 91: 221–34.

NIELSON, D. L.; EVANS, S. H.; and SIBBETT, B. S. 1986. Magmatic, structural, and hydrothermal evolution of the Mineral Mountains intrusive complex, Utah. *Geol. Soc. Amer. Bull.* 97: 765–77.

NOHDA, S., and WASSERBURG, G. J. 1986. Trends of Sr and Nd isotopes through time near the Japan Sea in northeastern Japan. *Earth Planet. Sci. Lett.* 78: 157–67.

NYQUIST, L. E.; SHIH, C.-Y.; WOODEN , J. L.; BANSAL B. M.; and WISEMAN, H. 1979. The Sr and Nd isotopic record of Apollo 12 basalts: Implications for lunar geochemical evolution. *Proc. 10th Lunar Planet. Sci. Conf.:* 77–114.

PETERMAN, Z. E.; HEDGE, C. E.; and TOURTELOT, H. A. 1970. Isotopic composition of strontium in sea water throughout Phanerozoic time. *Geochim. Cosmochim. Acta* 34: 105–120.

PIEPGRAS, D. J.; WASSERBURG, G. J.; and DASCH, E. J. 1979. The isotopic composition of Nd in different ocean masses. *Earth Planet. Sci. Lett.* 45: 223–36.

PRICE, P. B., and WALKER, R. M. 1963. Fossil tracks of charged particles in mica and the age of minerals. *J. Geophys. Res.* 68: 4847–62.

SISSON, V. B., and ONSTOTT, T. C. 1986. Dating blueschist metamorphism: A combined ^{40}Ar/^{39}Ar and electron microprobe approach. *Geochim. Cosmochim. Acta* 50: 2111–17.

STACEY, J. S., and STERN, T. W. 1973. *Revised tables for the calculation of lead isotope ages.* Springfield, Va.- U.S. Dept. Commerce, Nat. Tech. Info. Serv. PB-20919.

TAYLOR, H. P. JR., and SHEPPARD, S. M. F. 1986. Igneous rocks: I. Processes of isotopic fractionation and isotope systematics. In *Stable isotopes in high temperature geological processes*, ed. J. W. Valley; H. P. Taylor Jr; J. R. O'Neil, 227–72. Reviews in minerology 16. Washington, D.C.: Mineral. Soc. of America.

TERA, F.; BROWN, L.; MORRIS, J.; SACKS, I. S.; KLEIN, J.; and MIDDLETON, R. 1986. Sediment incorporation in island-arc magmas: Inferences from ^{10}Be. *Geochim. Cosmochim. Acta* 50: 535–50.

WASSERBURG, G. J., and DEPAOLO, D. J. 1979. Models of earth structure inferred from neodymium and strontium isotopic abundances. *Proc. Nat. Acad. Sci. USA* 76: 3594–98.

WHITE, W. M. 1985. Sources of oceanic basalts: Radiogenic isotopic evidence. *Geology* 13: 115–18.

YORK, D. 1966. Least-squares fitting of a straight line. *Can. J. Phys.* 44: 1079–86.

ZARTMAN, R. E. 1974. Lead isotopic provinces in the Cordillera of the western United States and their geologic significance. *Econ. Geol.* 69: 792–805.

PROBLEMS

1. Given the following Rb-Sr isotopic data for whole-rock samples of the Togus monzonite intrusion in Maine (Dallmeyer and VanBreeman 1981), determine the initial strontium isotopic ratio and age of this body.

Sample No.	^{87}Rb/^{86}Sr	^{87}Sr/^{86}Sr
1	11.86	0.7718
2	7.66	0.7481
3	6.95	0.7436
4	9.68	0.7587
5	6.54	0.7413
6	9.69	0.7599
7	3.74	0.7259

2. (a) Solve equation 12-8 for t. (b) Use the equation you just derived to determine the age of a biotite sample, for which the following data have been obtained: K = 7.10 wt. %, ^{40}Ar = 1.5×10^{12} atoms/g.

3. A zircon grain initially contained 1000 ppm ^{235}U and 100 ppm ^{207}Pb. How many atoms of ^{207}Pb will this zircon contain after 1 b.y.?

4. Using the data below and Figure 12.8, construct a concordia diagram. Are these samples concordant? Interpret your results, explaining any assumptions that you make.

Sample No.	$^{206}Pb/^{238}U$	$^{207}Pb/^{235}U$
1	0.500	15.93
2	0.543	18.77
3	0.565	20.00

5. Dallmeyer et al. (1986) reported the following geochronological data for a portion of the southern Appalachians: A gneiss unit gives whole-rock Rb-Sr and zircon U-Pb ages of 315 m.y. Mineral geothermometers in this unit give a peak metamorphic temperature of 530°C. ^{40}Ar-^{39}Ar gas retention ages are 295 m.y. for hornblende and 283 m.y. for biotite. Fission track ages for apatites range from 130–150 m.y. Using these data, construct a time-temperature path illustrating the late Paleozoic thermal history of this part of the Appalachian orogen.

6. The measured activity of a sample of charcoal found at an archaeological site is 5.30 dpm g^{-1}. What is the age of the sample?

7. How much heat energy would be generated in 100,000 years by a sample of rock containing live ^{26}Al? The half-life of ^{26}Al is given in Table 12-1. Assume that the sample initially contains 100 g of aluminum with a $^{26}Al/^{27}Al$ atomic ratio of 0.1. Also assume that the energy of decay for ^{26}Al is 10^{-14} cal/atom.

8. A magma containing 200 ppm Sr is contaminated by crustal rocks containing 1000 ppm Sr. The magma simultaneously undergoes fractional crystallization of phases with a bulk distribution coefficient $D_{Sr} = 0.75$. The initial Sr isotopic ratios of the magma and crustal rocks are 0.704 and 0.713, respectively. Assume that $r = 0.5$. Use equation 12-21 to calculate the isotopic composition of the resulting hybrid magma as a function of mass of magma relative to original mass (F), and illustrate on a graph.

9. Construct a plot of ($^{207}Pb/^{206}Pb$)* versus time, from 4.5 billion years ago to the present, using Program PBPB on the diskette available with this text.

Chapter 13

Stretching Our Horizons: Cosmochemistry

OVERVIEW

In this chapter we explore the rapidly emerging field of cosmochemistry. This subject involves geochemical aspects of systems of planetary or solar system scale. We first consider nucleosynthesis processes in stars and use this foundation to understand cosmic elemental abundances. Chondritic meteorites are discussed next, as samples of average solar system material stripped only of the lightest elements. Analyses of these meteorites provide information on the behavior of various elements in the early solar nebula; from these we learn that elements were fractionated according to both geochemical affinity and volatility. Chondritic fractionation patterns are also important for understanding planetary compositions.

We then consider the evidence for proposed cosmochemical processes in the early solar nebula, such as condensation, evaporation, and infusion of nucleosynthetic products from a nearby supernova. The characteristics and origin of extraterrestrial organic compounds and ices will also be explored. We then introduce various geochemical models for planets, based on condensation, heterogenous accretion, and mixing of chondrite-like materials. Finally, we consider ways to model the compositions of planets by taking into account core formation.

WHY STUDY COSMOCHEMISTRY?

In earlier chapters, we have occasionally used observations from other planetary bodies to illustrate processes or pathways in geochemistry. Perhaps these seemed exotic or even esoteric at the time you read them. From the geologist's uniformitarian perspective, however, the planets and other extraterrestrial bodies provide a larger laboratory in which to study the chemical behavior of our own planet. Historically, the study of cosmochemistry has paralleled (and, in some cases, spurred) the development of geochemistry as a discipline, and has involved many of the same scientists. It is significant that the leading journal in geochemistry is also the premier journal in cosmochemistry (*Geochimica et Cosmochimica Acta*). In this chapter, we examine several of the large-scale problems that attract geochemists to study the universe beyond the earth, and observe how some of the answers may improve our understanding of the world under our feet.

Our solar system consists of a single star surrounded by nine planets and a large retinue of smaller bodies, including their satellites, as well as asteroids, comets, and numerous small debris. As geochemists, much of our attention is drawn by chemical differences and similarities among these bodies. To what extent have they evolved separately, and what characteristics have they inherited from a common origin? Which processes are unique in a given body, and which processes are likely to affect all of them? How did the solar system form, and how has it reached its present state?

More than 99% of the mass of our solar system is concentrated in the sun, so many of the questions regarding the bulk composition of the solar system and its early history must first address the behavior and composition of stars. Here, we rely on astrophysicists to construct models that infer the structure and evolutionary development of stars in general. These models are based on observations of the sun and distant stars and on the principles of thermodynamics and nuclear chemistry we have discussed in earlier chapters, although the language of astrophysics is different.

From the sun, we turn to the planets, which are usually divided into two groups. Those closest to the sun (Mercury, Venus, Earth, and Mars) are called *terrestrial* planets. The larger planets beyond (Jupiter, Saturn, Uranus, and Neptune) are known as *Jovian* planets. Pluto, a small icy body at the outer reaches of the solar system, may be an escaped moon from one of the Jovian planets. Geochemists who study planetary bodies are concerned primarily with the terrestrial planets, because they are most like the earth. In this chapter, in fact, those are the only bodies we discuss in any detail. The presence of Jovian planets, however, places important restrictions on any models for the history of the solar system as a whole. We consider some of these restrictions in this chapter.

The terrestrial bodies include not only the four inner planets, but also the moons of earth and Mars, and several thousand asteroids. Observations of these, particularly of our own moon, have helped geochemists to determine how planetary size affects differentiation, outgassing, and other global-scale processes. Some of the most valuable information about terrestrial bodies, however, comes from meteorites. Most of these are fragments of asteroids, some of which experienced differentiation and core formation, and others which remained almost unchanged after they had accreted. From meteorites, we obtain observations that help model the development of planetary cores and mantles. We also use the nearly pristine nature of some classes of meteorites to deduce the chemical and petrologic nature of the materials from which larger terrestrial planets were formed.

We begin our survey of cosmochemistry by considering the sun as a star, to find chemical clues to its origin and the origin of other solar system materials. As we proceed, our focus is drawn more to the development of planets, and finally back to the earth itself. In this tour, you will recognize many familiar geochemical concerns and, we hope, see the place of cosmochemistry in understanding how the earth works.

ORIGIN AND ABUNDANCE OF THE ELEMENTS

Nucleosynthesis in Stars

The universe is believed to have begun in a cataclysmic explosion—the *Big Bang*—and has been expanding ever since. At that beginning stage, matter existed presumably in the form of discrete protons and electrons or as simple combinations of these (hydrogen atoms). The Big Bang itself may have produced some other nuclides, but only ^{4}He was formed in any abundance. In chemical terms, this was a pretty dull universe. How then did all of the other elements originate?

Local concentrations of matter periodically coalesce to form stars. Burbidge et al. (1957) first argued that other elements are formed in stellar interiors by nuclear reactions, with hydrogen as the sole starting material. When hydrogen atoms are heated to sufficiently high temperatures and held together by enormous pressures such as occur in the deep interior of the sun, *fusion* reactions occur. The proton-proton chain, the dominant energy-producing reaction in the sun, is illustrated in Figure 13.1. This fusion process, commonly called hydrogen burning, produces helium. When the hydrogen fuel in the interior begins to run low and the stellar core becomes dense enough to promote reactions at higher pressures, another nuclear reaction, in which helium atoms are fused to make carbon and oxygen, may take place. While helium burning proceeds in the core of a star, a hydrogen-burning shell works its way toward the surface; a star at this evolutionary stage expands into a *red giant*. This fate is all our sun can aspire to.

A star more massive than the sun, however, can employ other fusion reactions, successively burning carbon, neon, oxygen, and silicon. The ashes from one burning stage provide the fuel for the next. The internal structure of such a massive, highly evolved star would then consist of many concentric shells, each of which produces fusion products that are burned in the adjacent inward shell. The ultimate products of such fusion reactions are elements near iron in the periodic table (V, Cr, Mn, Fe, Co, and Ni). Fusion between nuclei cannot produce nuclides heavier than the iron-group

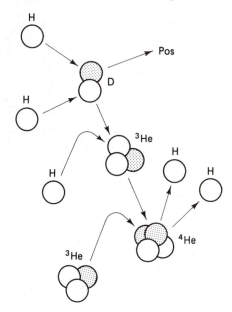

Figure 13.1 The proton-proton chain, in which hydrogen atoms are fused into helium, is the dominant energy-producing reaction in the sun. Open circles are protons, filled circles are neutrons, and *pos* is a positron. For each gram of ^{4}He produced, 175,000 KWh of energy are released.

elements. For elements lighter than iron, the energy yield is higher for fusion reactions than for fission, but for elements heavier than iron, the energy yield for fission is greater. When a star reaches this evolutionary dead-end, a series of both disintegrative and constructive nuclear reactions occur among iron-group nuclei. After some time a steady state, called *nuclear statistical equilibrium*, is reached; the abundances of iron-group elements reflect this process.

Elements heavier than the iron group can be formed by addition of neutrons to iron-group seed nuclei. We have already seen in Chapter 12 that nuclei with neutron/proton ratios greater than the band of stability will undergo beta decay to form more stable nuclei. Helium burning produces neutrons that are captured by iron-group

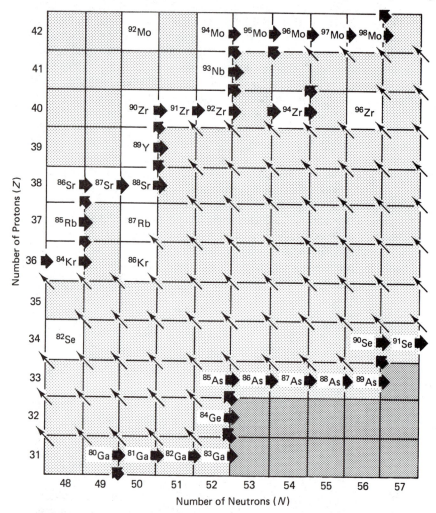

Figure 13.2 A portion of the nuclide chart illustrating how the *s*-process (upper set of heavy arrows) and the *r*-process (lower set of heavy arrows) produce new elements by neutron capture. White boxes represent stable nuclides, whereas gray boxes represent unstable nuclides that undergo beta decay, which shifts them up and to the left (small arrows). Dark gray boxes represent extremely unstable nuclides that transform rapidly by neutron capture. (After J. A. Wood, *The solar system.* Copyright 1979 by Prentice-Hall. Reprinted with permission.)

nuclei in a slow and rather orderly way, called the *s-process*. This process is slow enough that nuclei, if they are unstable, experience beta decay to stable nuclei before additional neutrons are added. This is illustrated in Figure 13.2, in which stable and unstable nuclei are shown as white and gray boxes, respectively. A portion of the *s*-process track is illustrated by the upper set of heavy arrows. Neutrons are added until an unstable nuclide is produced; then beta decay occurs, resulting in a decrease of N and an increase of Z by 1, as indicated by small arrows. The resulting nuclide can then capture neutrons until it beta decays again, and the process continues. Notice that a helium-burning star producing heavy elements by this process would have to be at least a second-generation star, having somehow inherited iron-group nuclei from an earlier star whose material had been recycled.

Fusion, nuclear statistical equilibrium, and *s*-process neutron-capture reactions in massive stars thus provide mechanisms by which elements heavier than hydrogen can be produced. But how are these nuclides placed into interstellar space for later availability in making planets, oceans, and organic life? Processed stellar matter is lost continuously as fluxes of energic ions, such as the solar wind. Other elements are liberated by *supernovae*—stellar explosions that scatter matter over vast distances. At the same time, supernovae provide additional nucleosynthetic pathways by which elements are generated. Very rapid addition of neutrons to seed nuclei (the *r-process*) results in a chain of unstable nuclides.

One example is illustrated by the set of heavy arrows at the bottom of Figure 13.2. The nuclides in dark gray boxes decay even more rapidly than the seconds or minutes required for nuclides in light gray boxes, so a nucleus shifted into a dark gray box by neutron capture will immediately transform by beta decay as shown. When the supernova event ends, the *r*-process nuclides will transform more slowly by successive beta decays into stable nuclides. Many of these stable isotopes are the same as produced by the *s*-process but some, such as ^{86}Kr, ^{87}Rb, and ^{96}Zr, can be reached only by the *r*-process. In an analogous manner, nuclides like ^{92}Mo (Figure 13.2) on the proton-rich side of the *s*-process band can be produced during supernovae by addition of protons to seed nuclei (the *p-process*), followed by positron emission or electron capture.

Cosmic Abundance Patterns

The relative abundances of the elements in a star, then, are controlled by a combination of nucleosynthetic processes. Elemental abundances in the sun, relative to 10^6 atoms of silicon, are given in Table 13-1 and illustrated graphically in Figure 13.3. This pattern is commonly called the *cosmic abundance of the elements*. This is a misnomer, of course; the composition of our sun is not necessarily representative of that of the universe. Each star may have its own peculiar composition because of different contributions from fusion, *s*- and *r*-processes, and other nucleosynthesis mechanisms.

Let us inspect the cosmic abundance pattern in Figure 13.3 to see why the sun has this particular composition. Clearly, hydrogen and helium are the dominant elements. This is because only these two emerged from the Big Bang. Elements between helium and the iron-group were formed, in most cases, by fusion reactions. These show a rapid exponential decrease with increasing atomic number, reflecting decreasing production in the more advanced burning cycles. The sun is presently burning only hydrogen, so these elements must be inherited from an earlier generation of stars. Exceptions are the elements lithium, beryllium, and boron, which have abnormally low abundances. The abundances of these three elements may have been reduced by bombardment with neu-

TABLE 13-1 COSMIC ABUNDANCES OF THE ELEMENTS, BASED ON
10^6 SILICON ATOMS[*]

Atomic number	Element	Symbol	Abundance
1	Hydrogen	H	2.72×10^{10}
2	Helium	He	2.18×10^9
3	Lithium	Li	59.7
4	Beryllium	Be	0.78
5	Boron	B	24
6	Carbon	C	1.21×10^7
7	Nitrogen	N	2.48×10^6
8	Oxygen	O	2.01×10^7
9	Fluorine	F	843
10	Neon	Ne	3.76×10^6
11	Sodium	Na	5.70×10^4
12	Magnesium	Mg	1.075×10^6
13	Aluminum	Al	8.49×10^4
14	Silicon	Si	1.00×10^6
15	Phosphorus	P	1.04×10^4
16	Sulfur	S	5.15×10^5
17	Chlorine	Cl	5240
18	Argon	Ar	1.04×10^5
19	Potassium	K	3770
20	Calcium	Ca	6.11×10^4
21	Scandium	Sc	33.8
22	Titanium	Ti	2400
23	Vanadium	V	295
24	Chromium	Cr	1.34×10^4
25	Manganese	Mn	9510
26	Iron	Fe	9.00×10^5
27	Cobalt	Co	2250
28	Nickel	Ni	4.93×10^4
29	Copper	Cu	514
30	Zinc	Zn	1260
31	Gallium	Ga	37.8
32	Germanium	Ge	118
33	Arsenic	As	6.79
34	Selenium	Se	62.1
35	Bromine	Br	11.8
36	Krypton	Kr	45.3
37	Rubidium	Rb	7.09
38	Strontium	Sr	23.8
39	Yttrium	Y	4.64
40	Zirconium	Zr	10.7
41	Niobium	Nb	0.71
42	Molybdenum	Mo	2.52
43	Technetium[**]	Tc	—
44	Ruthenium	Ru	1.86
45	Rhodium	Rh	0.344
46	Palladium	Pd	1.39
47	Silver	Ag	0.529
48	Cadmium	Cd	1.69
49	Indium	In	0.184
50	Tin	Sn	3.82
51	Antimony	Sb	0.352
52	Tellurium	Te	4.91
53	Iodine	I	0.90
54	Xenon	Xe	4.35
55	Cesium	Cs	0.372
56	Barium	Ba	4.36
57	Lanthanum	La	0.448
58	Cerium	Ce	1.16
59	Praseodymium	Pr	0.174

TABLE 13-1 (cont.)

Atomic number	Element	Symbol	Abundance
60	Neodymium	Nd	0.836
61	Promethium**	Pm	—
62	Samarium	Sm	0.261
63	Europium	Eu	0.0972
64	Gadolinium	Gd	0.331
65	Terbium	Tb	0.0589
66	Dysprosium	Dy	0.398
67	Holmium	Ho	0.0875
68	Erbium	Er	0.253
69	Thulium	Tm	0.0386
70	Ytterbium	Yb	0.243
71	Lutetium	Lu	0.0369
72	Hafnium	Hf	0.176
73	Tantalum	Ta	0.0226
74	Tungsten	W	0.137
75	Rhenium	Re	0.0507
76	Osmium	Os	0.717
77	Iridium	Ir	0.660
78	Platinum	Pt	1.37
79	Gold	Au	0.186
80	Mercury	Hg	0.52
81	Thallium	Tl	0.184
82	Lead	Pb	3.15
83	Bismuth	Bi	0.144
84–89	Unstable elements		—
90	Thorium	Th	0.0335
91	Protactinium**	Pa	—
92	Uranium	U	0.0090

*From Anders and Ebihara (1982).
**Unstable element.

trons and protons over the sun's lifetime, although the production processes just discussed tend to bypass these elements. The peak corresponding to the iron-group elements represents nuclides formed by nuclear statistical equilibrium, also an addition from earlier stars. The abundances in the sun of elements heavier than iron reflect additions of s-, r-, and p-process material.

Superimposed on the peaks and valleys in Figure 13.3 is a peculiar sawtooth pattern, caused by higher abundances of elements with even rather than odd atomic numbers (Z). A reason for this is that nuclides with even atomic numbers are more likely to be stable, as already noted in Chapter 12. This answer, however, begs the question: Why are even-numbered atomic numbers more stable? Whenever nuclear particles, either protons or neutrons, can pair with spins in opposite directions, they can come closer together and the forces holding them together will be stronger. This stronger nuclear binding force results in higher stability.

The chemical composition of the sun, then, can be understood primarily in terms of nucleosythesis reactions from an earlier generation of stars whose materials have been recycled. Some helium has been produced at the expense of hydrogen during the sun's lifetime. Because the sun contains more than 99% of the mass of the solar system, its composition is approximately equivalent to that of the whole system.

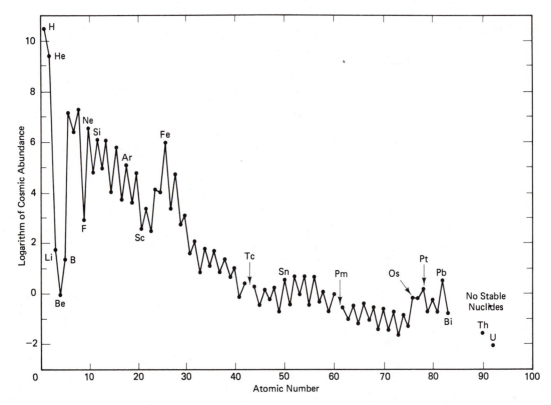

Figure 13.3 Cosmic abundances of the elements, relative to 10^6 silicon atoms.

MEASURING THE COMPOSITION OF A STAR

The spectrum of light emitted by a star's photosphere (the region of the stellar atmosphere from which radiation escapes) provides a means for determining its composition. Superimposed on the continuous spectrum of a star are absorption bands, or missing wavelengths. These bands, produced by electron transitions in various atoms, appear as dark lines when photographed through a spectrograph. These are sometimes called Fraunhofer lines, after the German physicist who discovered them in 1817.

Each Fraunhofer line corresponds to a particular element, and its abundance can be determined from the intensity of the absorption. What is actually measured is the width of the absorption band, because it is broadened with increasing concentration of its correponding element. In practice, allowance must be made for the prevailing conditions of temperature and pressure in the star's photosphere before converting line width to element abundance. This is because the width of absorption depends not only on the elemental abundance, but also on the fraction of atoms that are in the right state of ionization and excitation to produce the band, which is in turn a function of temperature and pressure. From observations of the continuous spectrum, temperature and pressure can be modelled theoretically for various depths within the photosphere. It is then possible to combine the contributions from atoms at all depths throughout the photosphere to predict how the line width will vary with abundance.

Absorption bands corresponding to any one element are not necessarily visible in all stars. This is probably more a function of the varying temperatures and pressures of stars, which control the ability of a particular element to absorb radiation, than to the absence of that element. Measurements at certain wavelengths are also blocked by the earth's atmosphere, although this problem has now been circumvented by spacecraft observations. Finally, we should also note that the composition of the photosphere may be quite different from that of the bulk star, unless convective mixing takes place.

Chondrites are the most common type of meteorite. These objects are essentially ultramafic rocks, and petrographic studies suggest that they have never been geologically processed by melting and differentiation. Radiometric age determinations indicate that chondrites are ancient, approximately 4.5 billion years old. Because these are the oldest samples that have ever been dated, it is thought that they represent materials surviving from the early solar system.

Chondrites occupy a singularly important niche in geochemistry: they are the baseline from which the effects of most geochemical processes can be gauged. The reason for this is quite simple. Of all known materials that can be studied directly, chondrites most closely match the composition of the sun. This is illustrated in Figure 13.4, a plot of the abundance of elements in the solar atmosphere against that in one type of chondrite (C1), all relative to atoms of silicon. A perfect correspondence is given by the diagonal line in this figure. The correspondence may even be better than indicated, because of some uncertainties in measuring solar composition. To be fair, we should note that the scales in this figure are logarithmic but, with the exception of a few elements, the abundances are certainly within a factor of 2 of each other (Anders and Ebihara, 1982). The gaseous elements—carbon, nitrogen, oxygen—are more abundant in

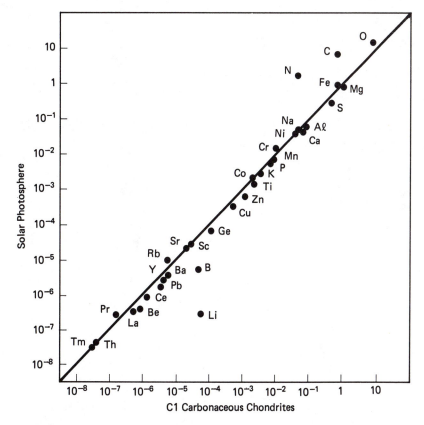

Figure 13.4 Comparison of the abundances of condensable elements in C1 chondritic meteorites with those in the solar photosphere. All elements are relative to 10^6 silicon atoms. This excellent correspondence indicates that chondrites have approximately cosmic compositions.

the sun, so it may be reasonable to visualize chondrites as a solar sludge from which gases have been distilled. In contrast, lithium, boron, and possibly beryllium have higher abundances in chondrites than in the sun. Recall that these three elements are systematically destroyed in the sun by interactions with nuclear particles. In this regard, then, chondrites may record the composition of the ancient sun even better than the present-day sun does.

It should now be obvious why chondrites are used as a basis for normalization of geochemical data: their nearly cosmic elemental abundances provide the justification for this common practice. As an example, let us consider chondrite normalization of rare earth element patterns, which we introduced in Chapter 11. Rare earths in both chondrites and basalts exhibit the sawtooth pattern of even-odd atomic numbers in Figure 13.5. However, division of the abundance of each rare earth element in basalt by the corresponding value in chondrites gives the smoothed, normalized values indicated by triangles. Without this step, the interpretation of rare earth fractionations would be very cumbersome.

The ancient ages and nearly cosmic compositions of chondrites suggest that these are nearly pristine samples of early solar system matter (this is actually true only to a degree; see the accompanying box). There is ample evidence that most chondrites have

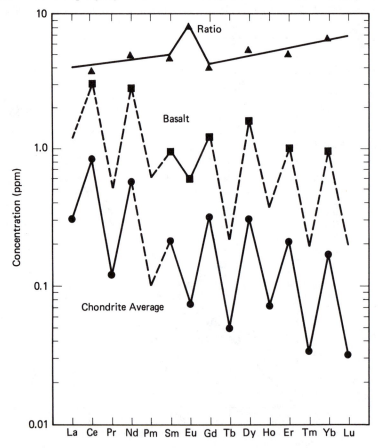

Figure 13.5 Measured abundances of rare earth elements in a basalt and in chondrites. Normalization of the basalt data to chondrites removes the zigzag pattern of odd-even abundances, revealing the pattern imposed during the igneous history of this rock.

been stored in asteroids, bodies generally too small to sustain geologic processes.[1] It is not unreasonable, therefore, to infer that chondrites, or something much like them, were the building blocks of planets. The central role played by chondrites in cosmochemistry thus becomes clear.

CHONDRITE PETROLOGY AND CLASSIFICATION

Chondrites take their name from *chondrules,* round, millimeter-sized, quenched droplets of silicate melt that are particularly abundant in these meteorites. Chondrules are typically composed of olivine, pyroxenes, and glass or feldspar. These objects are cemented together by a fine-grained *matrix,* consisting of mixtures of olivine, pyroxene, feldspathoids, graphite, magnetite, and other minor phases. These components are illustrated in Figure 13.6. Irregularly shaped, white bodies called *Ca, Al-rich inclusions* also occur in some chondrites. These are composed mostly of melilite, spinel, clinopyroxene, and anorthite. Larger grains of iron-nickel *metal* and *sulfides* are also scattered throughout these meteorites. These diverse kinds of materials have accreted together in space to form a kind of cosmic sediment. Despite extensive research, the origins of all of these components are obscure.

There are actually many classes of chondrites, differing slightly in petrography and chemical composition. The ordinary chondrites are by far the most common extraterrestrial material that falls to earth. The H, L, and LL groups of ordinary chondrites contain varying proportions of iron, and differ in the oxidation state of the iron they contain. (H chondrites contain the highest concentration of iron, mostly as metal; LL chondrites contain the least iron, mostly in oxidized form in silicates.) The enstatite chondrites are fully reduced, containing iron only as metal, and can also be subdivided by iron content into EH and EL groups. Carbonaceous chondrites consist of at least four distinct chemical groups, called CI, CM, CO, and CV. Most of these have high oxidation states relative to other chondrite types. The CI chondrites differ from other groups in that they contain no chondrules, but their chemical compositions are clearly chondritic.

Most ordinary and enstatite chondrites and a few carbonaceous chondrites have experienced thermal metamorphism. This has resulted in recrystallization, which blurs the distinctive chondrule textures. The major phases of chondrites (olivines and pyroxenes) were preserved, although their original scattered compositions were homogenized, and glasses were devitrified to form feldspars. The response of chondrites to heating was very different from that of terrestrial ultramafic rocks, because of the absence of fluids in most chondrite parent bodies. Judging from the relative abundances of metamorphosed and unmetamorphosed chondrites in meteorite collections, asteroids must consist mostly of thermally sintered material. Van Schmus and Wood (1967) formulated a classification scheme for chondrites that takes into account their compositions and thermal histories. Increasing metamorphic grade is indicated by numbers from 1 to 6; these are appended to chemical group symbols, so that a particular chondrite may be classified, for example, as H3 or EL6.

Many carbonaceous chondrites have been affected by a different process, aqueous alteration by fluids. The matrices of CM and CI chondrites were transformed to complex mixtures of phyllosilicates, and veins of carbonates and sulfates permeate some samples. Van Schmus and Wood classified these meteorites as metamorphic grades 1 and 2, inferring that they had experienced the least thermal effects. McSween (1979) reinterpreted grades 1 and 2 to reflect varying degrees of aqueous alteration. Thus, we are left with a somewhat confusing situation in which type 3 chondrites are the most primitive.

Despite the distorting effects of these secondary processes in chondrites, their bulk chemical compositions seem little affected. In fact, CI1 (often abbreviated to C1, a convention we will follow here) chondrites provide the best match for the solar composition. Some elements, however, were apparently mobilized by heating and were redistributed within ordinary chondrite parent bodies, and the stable isotopic compositions of CM2 and C1 chondrites were altered by exchange with fluids.

---

[1] Some asteroids have experienced melting and differentiation. Meteoritic samples of such asteroids are *achondrites* (igneous rocks of basaltic or ultramafic composition), *iron meteorites* (mostly differentiated cores), and *pallasites* (stony-iron meteorites that formed at core-mantle interfaces).

Figure 13.6 Photomicrograph of the Tieschitz (Czechoslovakia) ordinary chondrite. The rounded objects are chondrules. The sample is approximately 2 cm long.

COSMOCHEMICAL BEHAVIOR OF ELEMENTS

Controls on Cosmochemical Behavior

The cosmic abundance of an element is controlled by its nuclear structure, but its geochemical character is a function of its electron configuration. In 1923 V. M. Goldschmidt first coined the terms *lithophile, chalcophile,* and *siderophile* to identify elements with affinity for silicate, sulfide, and metal, respectively. Geochemical affinity is actually a qualitative assessment of the relative magnitudes of free energies of formation for various compounds of the element under consideration. For example, suppose that metallic calcium were exposed to an atmosphere containing oxygen and sulfur. Two potential reactions for calcium are

$$Ca + 1/2\ O_2 \longrightarrow CaO$$

and

$$Ca + 1/2\ S \longrightarrow CaS.$$

Of the two, the first is dominant because $\Delta \overline{G}_f^o$ for the oxide is more negative than $\Delta \overline{G}_f^o$

of the sulfide. Therefore, calcium is lithophile. Of course, we must specify the conditions on the system in order for this conclusion to have meaning.

There was little direct evidence on which to speculate about geochemical behavior when this idea was put forward, but Goldschmidt recognized that chondrites represent a natural experiment from which element behavior could be deduced. He and his coworkers measured the abundances of elements in coexisting silicate, sulfide, and metal from chondrites, and this pioneering work has been supplemented by studies of metal, sulfide, and silicate slag from smelters. Geochemical affinity is important in understanding element partitioning in any systems containing more than one of these phases—for example, the formation of magmatic sulfide ores or the differentiation of core and mantle.

This is not, however, the only control of element behavior in cosmochemical systems. Another important factor is *volatility,* by which we mean the temperature range in which an element will condense from a gas of solar composition. Volatility depends on several factors, such as elemental vapor pressure, cosmic abundance (which controls partial pressure), total pressure of the system, and element reactivity (which determines whether it occurs in pure form or in a compound). *Refractory* elements are those that condense at temperatures in excess of 1400 K; *volatile* elements generally condense at temperatures below 1200 K.

Chemical Fractionations Observed in Chondrites

In chondrites, groups of elements appear to have been fractionated together. If one element of the group is enriched or depleted, all of the others are as well, and by nearly the same factor. At least four types of chemical fractionations, relative to cosmic (and C1 chondrite) composition, have been determined in chondrites. These are:

1. fractionation of refractory elements (for example, Ca, Al, Ti), resulting in high refractory element abundances in carbonaceous chondrites and lower abundances in ordinary and enstatite chondrites;
2. fractionation of siderophile elements (for example, Ir, Ni, Au) to produce depletions in some ordinary and enstatite chondrites;
3. fractionation of moderately volatile elements (for example, Sb, Ga, Zn), leading to depletions in carbonaceous, ordinary, and enstatite chondrites; and
4. fractionation of highly volatile elements (for example, Pb, In, Tl) to produce severe depletions in ordinary and enstatite chondrites.

For three of the four of these fractionated groups, the only common property is volatility, which must have been a very important factor in controlling element behavior in the early solar system. This is illustrated graphically in Figure 13.7. Elements are plotted in order of increasing volatility from left to right; degree of depletion in these chondrite classes clearly increases with increasing volatility. The fractionation of siderophile elements is often referred to as *metal-silicate fractionation,* a reasonable term in view of the fact that chondrites contain both metal and silicate phases.

How could such fractionations have been accomplished? The separation of siderophile elements may have resulted from the distinct densities of metal and silicates, possibly leading to differences in accretion rates. Fractionation of elements by volatility requires thermal processing in space. In the next section, we will consider volatile behavior in a more quantitative fashion.

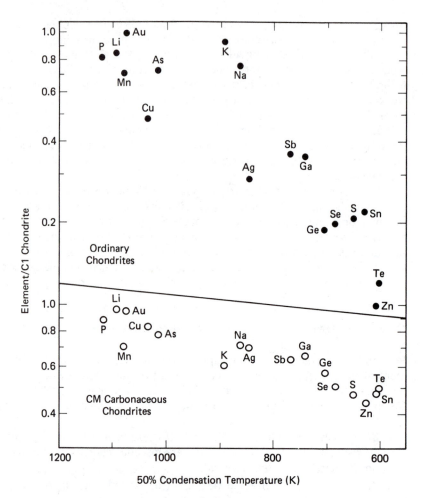

Figure 13.7 The C1 chondrite-normalized elemental abundances of various chondrite groups show depletion patterns related to element volatility. However, the degrees of depletion, reflected in the slopes of these trends, are distinct for different groups. In this diagram, volatility increases from left to right. Modified from Wasson (1985).

CONDENSATION OR EVAPORATION?

Stars are so hot that all of the matter they contain is in the gaseous state, and any molecules are broken down into their constituent elements. However, as matter is expelled from stars, it cools and many elements may condense as solid metals or compounds. No liquids are produced, at least under equilibrium conditions, because pressures in space are so low.

When interstellar dust and gas subsequently aggregate to form new stars, heating again occurs. Physical and dynamic conditions in the collapsing disk-shaped cloud that ultimately formed the sun, called the *solar nebula,* have been modelled. Early astrophysical models suggested a hot nebula, with temperatures and pressures reaching 2000 K and 10^{-2} atm at the center but decreasing outward. These models led cosmochemists to conclude that all of the presolar solid matter was vaporized, which was followed by recondensation as the nebula cooled after the sun formed. More recent theoretical treatments indicate that the solar nebula never got this hot, achieving perhaps only half of this temperature at comparable pressures. Under these conditions, partial rather than complete evaporation of solids probably occurred, though recondensation of the more volatile elements would still have been possible. In either case, condensation could have been an important process in establishing the chemical characteristics of the early solar system.

How Equilibrium Condensation Works

Condensation from a gas of cosmic composition has been modelled quantitatively by a number of researchers; here we will consider the calculations of Grossman (1972). The condensation temperature of an element is the temperature at which the most refractory solid phase containing that element first becomes stable relative to a solar gas. The calculation of this temperature involves two steps. First, the distribution of a particular element among possible gaseous molecules as a function of temperature must be determined. This step is reminiscent of the calculation performed in Worked Problem 4-7. Next, the condensation temperatures of all potential solid phases that contain the element must be assessed. Comparison of these two sets of calculations yields the most stable phase—gas, solid, or some combination of the two. As with other types of thermodynamic calculations, it is necessary to consider all possible phases if the results are to be meaningful. Because such phases normally consist of several elements, the condensation temperature of any element may depend on the concentrations of other elements in the system. As a consequence, previously condensed phases may exert some control on the condensation temperatures of elements not contained in such phases.

Worked Problem 13-1

How does one calculate the condensation temperature of corundum from a gas of solar composition at a total pressure of 10^{-4} atm? We have already seen that hydrogen has by far the highest cosmic abundance, and H_2 is much more abundant that any other H-containing molecule below 2000 K. Therefore,

$$P_{H_2} = P_{total},$$

where P_{total} is the pressure of some region of the nebula, in this case 10^{-4} atm. The ideal gas law, which is applicable at these low pressures, gives

$$N_{H_2} = \frac{P_{H_2}}{RT},$$

where N_{H_2} refers to the number of moles per liter of H_2. Because $2N_{H_2} = N_H$, we can express the concentration of any element X (in this case Al or O) in the gas as

$$N_x = \left(\frac{A_x}{A_H}\right) N_H. \tag{13-1}$$

In this equation, A_X and A_H are the molar cosmic abundances of element X and hydrogen, respectively.

A mass balance equation can now be written for each element X that describes its partitioning into various gaseous molecules. For example, the total number of oxygen atoms can be expressed as

$$N_O + N_{H_2O} + N_{CO} + 2N_{CO_2} + N_{MgO} + \cdots = N_O^{total}. \tag{13-2}$$

All of these gaseous compounds were assumed to have formed by reaction of monatomic elements. In the example above, water formed by the reaction

$$2H(g) + O(g) \longleftrightarrow H_2O(g).$$

From the free energies of the species in this reaction at the temperature of interest, an equilibrium constant K can be calculated, such that

$$K = \left(\frac{P_{H_2O}}{P_H^2 P_O}\right). \tag{13-3}$$

Expression 13-3 assumes, of course, that activities of these components are equal to partial pressures, probably a valid supposition at these low pressures. Substituting equation 13-3 into the ideal gas law and rearranging gives

$$N_{H_2O} = KN_H^2 N_O(RT)^2. \tag{13-4}$$

Equations of this form can be derived for each gaseous species and substituted into the mass balance equation 13-2. Other analogous mass balance equations for all elements of interest are then generated. The result is a series of n simultaneous equations in n unknowns, representing the monatomic gaseous elements such as N_O^{total}. These equations can then be solved, using a method of successive approximations, for the species composition of the gas phase at any temperature and a nebular pressure of 10^{-4} atm.

For the second step, we write an equation representing equilibrium between solid corundum and gas:

$$Al_2O_3(s) \longleftrightarrow 2Al(g) + 3O(g),$$

for which

$$\log K_{eq} = 2 \log P_{Al} + 3 \log P_O - \log a_{Al_2O_3}.$$

Because the activity of pure crystalline corundum is unity, this reduces to

$$\log K_{eq} = 2 \log P_{Al} + 3 \log P_O. \tag{13-5}$$

Values for $\log K_{eq}$ at various temperatures can be determined from appropriate free energy data.[2] Log K_{eq} values for corundum calculated in this way are shown by circles in Figure 13.8. The line connecting these points divides the figure into regions in which solid corundum and gas are stable. We then calculate $\log K_{eq}$ for the same reaction using the partial pressures P_{Al} and P_O from the mass balance equations. These values are shown by squares in Figure 13.8. The temperature at which the two values of $\log K_{eq}$ are the same (that is, the point at which the two lines in Figure 13.8 cross) is the temperature at which corundum condensation occurs, in this case 1758 K. The problem could also be solved numerically by finding the temperature at which the two values of $\log K_{eq}$ are equal.

Determining condensation temperatures for subsequently condensed phases is a little tricky. Once the condensation temperature for any element X is reached, the equation

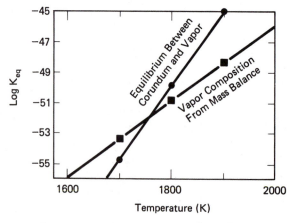

Figure 13.8 Plot of K_{eq} versus temperature, illustrating how the condensation temperature of corundum is obtained. The circles represent calculated conditions of equilibrium between corundum and vapor, and the squares represent the nebula composition, determined by solving the system of mass balance equations. The intersection of these lines marks the condensation point.

[2] It is easy to make an error at this step, because tabulated values of free energy of formation from the elements for corundum do not use Al(g) and O(g) as the standard states. Free energy data for the formation of gaseous monatomic species can be found in JANAF (1971).

analogous to 13-5 must be solved again, this time adding terms for the concentration of the crystalline phase in the appropriate mass balance equations. If we require that the condensed phase be in equilibrium with the gas as it cools, the gas composition will have a lower P_X than would have been the case had not the solid phase appeared. Therefore, the affected mass balance equations must be corrected for the appearance of each new condensed phase.

It seems likely that nebular condensation could have occurred under nonequilibrium conditions, because kinetic factors may have been important. Complexities such as barriers to nucleation, the formation and persistence of metastable condensates, and supersaturation by condensable species in the gas phase are not presently understood and cannot be modelled. Treating the solar nebula as if it were in thermodynamic equilibrium provides much insight, but this may serve only as a boundary condition for the condensation process.

The Condensation Sequence

The equilibrium condensation sequence for a gas of solar composition at 10^{-4} atm, calculated by Grossman (1972), is illustrated schematically in Figure 13.9. Some solid phases condense directly from the vapor. Others form by reaction of vapor with previously condensed phases. This latter situation is the cosmic equivalent of a peritectic reaction in solid-liquid systems.

Corundum is the first solid phase to form that contains a major element, although trace elements like osmium and zirconium may condense at higher temperatures. Corundum is followed by perovskite, which may contain uranium, thorium, tantalum, niobium and rare earth elements in solid solution. Corundum then reacts with vapor to form spinel and melilite, which in turn react to produce diopside at a lower temperature. Iron metal, containing nickel and cobalt, condenses by 1375 K. Pure magnesian forsterite appears shortly thereafter; it later reacts with vapor to form enstatite. Anorthite then forms by reaction of previously condensed phases with vapor. All of these refractory phases appear above 1250 K.

Below this temperature, iron metal reacts with vapor and becomes enriched in germanium, gallium, and copper. Similarly, anorthite reacts to form solid solutions with the alkalis—potassium, sodium, and rubidium. Condensation of these moderately volatile elements occurs between 1200 K and 600 K. These elements are somewhat depleted in chondrites relative to cosmic values (see Figure 13.7) and are sometimes called the *normally depleted* elements. Below 750 K iron begins to oxidize, and the iron contents of olivine and pyroxene increase rapidly as temperature drops further. This leads to the interesting conclusion that oxidation state changes during condensation. Iron occurs only in metallic form at high temperatures, and only in oxidized form at the end of the condensation sequence. Metallic iron also reacts with sulfur in the vapor to form troilite (FeS) below 700 K.

The highly volatile elements, which are *strongly depleted* in most chondrites, condense in the interval 600 K to 400 K. Examples are lead, bismuth, and thalium. Magnetite becomes a stable phase at 405 K, and olivine and pyroxene react with vapor to form hydrated phyllosilicates at lower temperatures.

Evidence for Condensation in Chondrites

The mineralogy of chondrites is very similar to that predicted from condensation theory. The most refractory material occurs in the form of Ca, Al-rich inclusions, one of which is shown in Figure 13.10. These occur most frequently in carbonaceous chon-

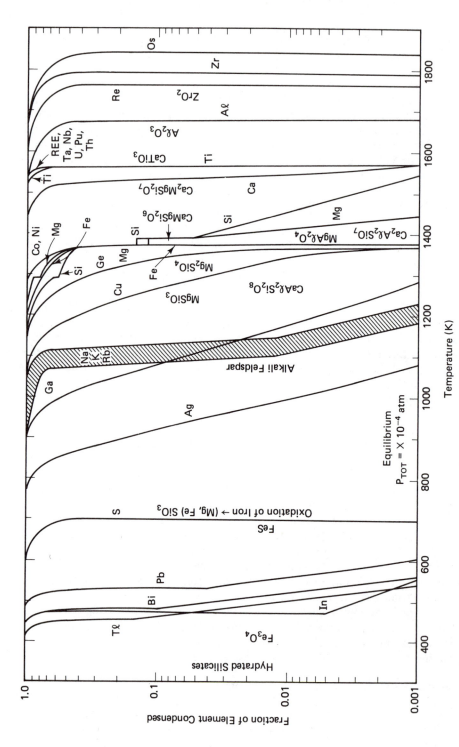

Figure 13.9 Summary of the calculated condensation sequence for a gas of solar composition at 10^{-4} atm, as a function of temperature and fraction of element condensed. (After L. Grossman and J. W. Larimer, Early chemical history of the solar system. In *Rev. Geophys. Space Phys.* 12:71–101. Copyright 1974 by American Geophysical Union, reprinted with permission.)

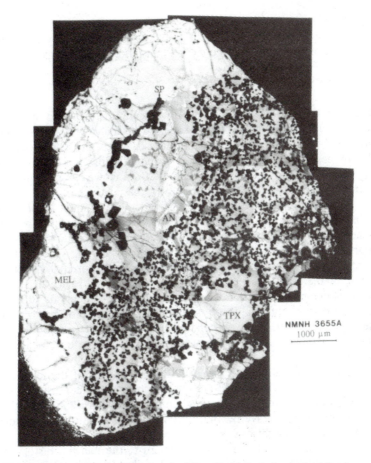

Figure 13.10 SEM photomicrograph of Ca, Al-rich inclusion in the Allende carbonaceous chondrite. Mineral abbreviations are MEL (melilite), SP (spinel), AN (anorthite), and TPX (Ti-bearing pyroxene). (Courtesy of J. M. Paque.)

drites and consist largely of melilite, spinel, diopside (actually fassaite, a titanium-rich variety), anorthite, and perovskite, all phases predicted to condense at high temperatures. Hibonite ($CaO \cdot 9Al_2O_3$), not included in Grossman's original calculations because of the absence of appropriate thermodynamic data, appears to take the place of corundum as an early condensate. These inclusions sometimes contain tiny nuggets of refractory platinum-group metals as well. Geochemical studies of Ca, Al-rich inclusions demonstrate that they are markedly enriched in other refractory trace elements.

Olivine and pyroxene are the major constituents of chondrules and unaltered fine-grained matrix, and metallic iron occurs in abundance in most chondrites. Most silicate grains in chondrules are clearly not condensates, because they crystallized from melted droplets. However, the chondrules themselves may have formed by remelting of solid condensate material. A few relict olivine grains in chondrules appear to have survived the melting process; these have higher refractory trace element contents than other olivines and could be condensates. The phyllosilicate matrix phases of many carbonaceous chondrites, commonly mixed with magnetite, could possibly represent the low-temperature end of the condensation sequence, although there is considerable evidence that these phases formed by aqueous alteration in asteroids.

Thus, chondrites might be viewed as physical mixtures of material formed at various stages in the condensation sequence. These were aggregated in various proportions, accounting for the differences in abundances of refractory and volatile elements observed in ordinary, carbonaceous, and enstatite chondrite groups (Figure 13.7).

Another View: Evaporation

The most compelling evidence for condensation is provided by the mineralogy and refractory element geochemistry of Ca, Al-rich inclusions, which match those properties predicted from equilibrium condensation calculations. The formation of inclusions by this process would require complete evaporation of all solids in the nebula, necessitating very high temperatures of at least 1800 K. However, reasonable maximum temperatures in current astrophysical models for the nebula are much lower.[3]

It is clear that other tests are needed. For example, if all the solid matter in the nebula had been vaporized, it is likely that the gas composition would have been homogeneous. Any elemental and isotopic differences that may have existed in presolar dust grains and gas would have been obliterated by mixing of vaporized elements at high temperatures. Thus, the condensation model would predict uniform isotopic compositions for materials that condensed from this vapor. However, the individual components of chondrites, such as Ca, Al-rich inclusions and chondrules, commonly have distinct isotopic signatures. These isotopic variations and their possible cause are discussed in the next section. Their existence argues that the elegant condensation theory we presented earlier provides too simplistic a model for the cosmochemical behavior of the nebula.

Detailed observations and experimental studies of the textures of Ca, Al-rich inclusions suggest to many workers that such objects may instead be refractory residues formed during heating to modest temperatures. Distillation of the volatile elements would presumably leave behind partially melted clumps of refractory compounds that, when solidifed, might look suspiciously like condensates.

Cosmochemistry is intimately linked to the thermal history of the solar nebula, but this connection is unfortunately not well-constrained. Whether partial or complete evaporation took place depends on the maximum temperatures achieved, and chondrites should be useful in determining what these were. Unfortunately, various investigators differ in their interpretations of Ca, Al-rich inclusions as either high-temperature condensates or refractory residues from evaporation. Much of the evidence is ambiguous, and roles for both processes can be envisioned. Even if temperatures were lower than that required to vaporize the refractory elements, the volatile elements that were distilled off had to recondense at some point. Condensing solids might nucleate on surviving refractory grains, so that some chondrite components might properly be viewed as products of both partial evaporation and condensation.

COSMOCHEMICAL EVIDENCE FOR THE ORIGIN OF THE SOLAR SYSTEM

Among the billions of stars in our galaxy, only two or three undergo supernova explosions each century. Yet supernova nucleosynthesis accounts for the abundances of many of the elements in the solar system. Are supernovae debris flung so far and wide that

---

[3] The maximum temperature depends on location within the nebula. For example, Cameron (1978) estimated temperatures exceeding 1000 K at the formation location of Mercury, but only about 325 K at the asteroid belt. If chondrites formed in the inner solar system, nebular conditions might be bracketed by these temperatures.

they eventually get incorporated into all stars, or are there other factors at work? Before we can answer this question, we must consider some isotopic evidence for the infusion of supernova products into the solar nebula.

Isotopic Diversity in Meteorites

Variability in isotopic composition has been demonstrated for many elements in most geological samples. In Chapters 7 and 12, we learned that such diversity can arise from (1) isotope fractionation processes, (2) mixing, (3) radioactive decay, and (4) interaction with cosmic rays. Given this variability, how could we use isotopes to search for and recognize materials that were made outside the solar system? We know that Ca, Al-rich inclusions in chondrites, whether they formed by condensation or evaporation, are probably some of the oldest surviving relicts of the early solar nebula. This conclusion is corroborated by radiometric dating of these objects. For example, U-Th-Pb dating methods indicate ages of about 4.55×10^9 yr, and initial $^{87}Sr/^{86}Sr$ isotopic compositions are the lowest known of any kinds of materials. If we wish to search for supernova products that have not been completely blended with other solar system matter, these inclusions provide a good place to start.

The oxygen isotopic compositions of Ca, Al-rich inclusions were first analyzed by Clayton et al. (1973). Figure 13.11 illustrates the relationship between $^{17}O/^{16}O$ and $^{18}O/^{16}O$ in inclusions and terrestrial rocks. The terrestrial mass fractionation line has a slope of $+1/2$, reflecting the fact that any separation of ^{17}O from ^{16}O (a difference of 1 atomic mass unit) will be half as effective as the separation of ^{18}O from ^{16}O (a difference of 2 atomic mass units). The trend of the inclusion data, clearly distinct from the mass fractionation line, must signify some other process. Clayton and coworkers interpreted this as a mixing line, joining "normal" solar system oxygen on the terrestrial mass fractionation line to pure ^{16}O (to which the mixing line in this figure extrapolates). Pure ^{16}O is produced by explosive carbon burning in supernovae. Thus, it was argued that Ca, Al-rich inclusions must contain an admixture of interstellar grains containing supernova-generated oxygen. This exotic component may have served as a substrate for nucleation of other material during condensation, or it may have survived evaporation in refractory mineral sites.

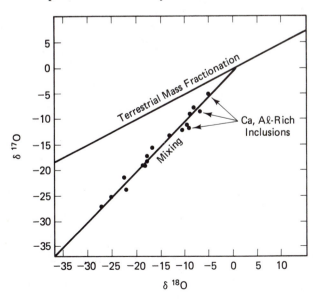

Figure 13.11 The relationship between ^{16}O, ^{17}O, and ^{18}O in Ca, Al-rich inclusions and terrestrial samples. Mass fractionation will produce a line of slope $+1/2$ in this diagram, as observed for terrestrial materials. The Ca, Al-rich inclusions define an apparent mixing line between "normal" solar-system material, such as that along the terrestrial mass fractionation line, and pure ^{16}O. The δ values are per mil, relative to a carbonaceous chondrite reference standard.

The exciting discovery of oxygen isotopic diversity initiated a flurry of activity to find anomalies in the isotopic compositions of other elements. Nucleosynthesis theory leads us to expect anomalies in magnesium and silicon in samples with large concentrations of ^{16}O, but the results so far are puzzling. Although nuclear anomalies in magnesium and silicon, as well as titanium, calcium, barium, and strontium, have been found, these occur in only a very few Ca, Al-rich inclusions and are not generally correlated.

One of the most interesting isotopic anomalies in these objects provides evidence for the former existence of now extinct radionuclides. Lee et al. (1977) discovered excess amounts of ^{26}Mg in these inclusions, which they attributed to radioactive decay of ^{26}Al. The half-life of ^{26}Al is only 7.2×10^5 yrs, so its incorporation as a "live" radionuclide indicates that formation of the solar nebula occurred very soon after nucleosynthesis. In fact, after just a few million years had elapsed, the abundance of this nuclide should have decreased to the point at which it would be undetectable. Like ^{16}O, ^{26}Al is produced by explosive carbon burning. The occurrence of this isotope provides a time scale for addition of supernova material to the solar nebula.

Worked Problem 13-2

How would one demonstrate that ^{26}Al was incorporated as a "live" radionuclide? First let us review the evidence that ^{26}Al, dead or alive, was added to Ca, Al-rich inclusions. The decay product of this radionuclide is ^{26}Mg, so it is only necessary to look for an excess of this isotope. This turns out to be a rather difficult measurement, however, because ^{26}Mg is already an abundant isotope, constituting about 11% of normal magnesium. Only in samples with a high original proportion of ^{26}Al would the addition of a relatively small amount of extra radiogenic daughter ^{26}Mg be detectable.

Lee et al. (1977) found enhancements of ^{26}Mg that changed the relative abundance of this isotope in inclusions up to about 11.5%. Isotopic fractionation can be ruled out as a cause for this anomaly because there should have been a similar but smaller effect in ^{25}Mg relative to ^{24}Mg, which is not observed. Thus, it is clear that the ^{26}Mg anomaly must be nuclear in origin, but there are nuclear reactions other than ^{26}Al decay that could give rise to this product.

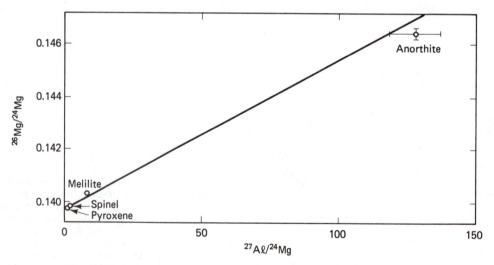

Figure 13.12 Correlation between ^{26}Mg and ^{27}Al, both normalized to ^{24}Mg, in a Ca, Al-rich inclusion. This relationship suggests that ^{26}Mg excess was produced by decay of the extinct radionuclide ^{26}Al.

The origin of the excess ^{26}Mg can be established by demonstrating that ^{26}Mg is correlated with aluminum in individual phases of the inclusions. Lee et al. separated anorthite, melilite, spinel, and diopside, each of which contains different amounts of aluminum and magnesium. The results of analyzing these phases are illustrated in Figure 13.12.[4] This diagram shows that phases which initially had high total aluminum and low total magnesium contents incorporated more ^{26}Al and therefore now have the highest ^{26}Mg. Spinel and diopside, with very low Al/Mg ratios, have virtually no excess ^{26}Mg. Melilite, which contains both elements, has a 4 % ^{26}Mg excess, and anorthite, with a high Al/Mg ratio, has a 9 % excess. Because the radiogenic ^{26}Mg now occurs in crystallographic sites occupied by aluminum, it must have been incorporated as ^{26}Al into these phases.

From the slope of the correlation line in Figure 13.12, it is possible to calculate the initial isotopic ratio ^{26}Al/^{27}Al at the time the inclusion formed. This value is approximately 5×10^{-5}.

A Supernova Trigger?

The presence of now extinct radionuclides in chondrites demands that these meteoritic materials formed soon after nucleosynthesis. Because of the long travel times across vast interstellar distances, these data further suggest that a supernova occurred in our cosmic backyard at about the time that the solar system formed. Does that not seem highly fortuitous? That would indeed be the case if there were no connection between these events. Cameron and Truran (1977) turned the argument around, proposing that the supernova precipitated the formation of the solar nebula. They suggested that the advancing shock wave generated by a supernova explosion triggered the collapse of gas clouds surrounding it to form nebulae and ultimately new stars. The isotopic anomalies found in chondrites would be a natural consequence of the blending of small amounts of supernova products with matter indigenous to the gas-dust cloud. Any evidence of this addition in materials other than chondrites would have been subsequently obliterated by geological processing.

Astronomers have subsequently observed the formation of new stars at the leading edges of expanding supernova remnants, providing independent evidence that such a process is possible. The idea of a supernova trigger neatly explains some of the most vexing questions in cosmochemistry. It is also an interesting demonstration of the utility of geochemistry in solving interdisciplinary problems.

THE MOST VOLATILE MATERIALS: ORGANIC COMPOUNDS AND ICES

It is a sobering thought to realize that the carbon atoms in our bodies were produced by nucleosynthesis in stars. But how did they arrive at their present molecular complexity? Did all chemical evolution of organic compounds occur in the earth's atmosphere and oceans, or was the process initiated at some other location? And for that matter, where did the gases that make up the atmosphere and oceans come from? We now turn to the cosmochemistry of the most volatile elements—H, C, N, O—which are depleted in chondrites relative to cosmic abundances.

---

[4] A word of caution is in order here. This diagram is *not* analogous to the plots utilized for Rb-Sr and Nd-Sm chronology in Chapter 12. ^{27}Al is not the radioactive parent, but the stable isotope of aluminum. If the ordinate were ^{26}Al/^{24}Mg, this would be an isochron diagram and the slope of the regression line would be a function of time. Such a diagram could only be constructed during the first few million years after nucleosynthesis, while ^{26}Al was still measureable.

Extraterrestrial Organic Compounds

Much of the interstellar medium is an inhospitable place for organic compounds, because any molecules that form are promptly dissociated by ultraviolet radiation. Dense interstellar clouds, however, are relatively opaque to such radiation and organic compounds apparently form in them. When such a cloud collapses during star formation, the nebula inherits biogenic elements already in some state of molecular complexity, although simpler compounds like CO and CH_4 may dominate. These molecules probably formed mantles on interstellar silicate dust grains.

When the solar nebula formed, organic compounds in what is now the inner solar system were heated and partially decomposed into gases plus more refractory residues, a process called *pyrolysis*. Complex molecules could have survived in the outer portions of the nebula, but gas and dust may have been cycled between warmer and cooler regions. During subsequent cooling of the nebula, condensation of the vaporized fraction of biogenic elements ensued. These are extremely volatile elements, so condensation occurred at low temperatures. As temperature dropped, it became possible to stabilize and preserve organic compounds in regions where high temperatures had previously prevented their condensation, so we can envision these components as infiltrating the inner solar system late in the evolutionary history of the nebula. These compounds may have been relict interstellar materials transported from greater radial distance, or local condensates, or both.

Let us first consider how organic compounds condense. Under equilibrium conditions, the dominant carbon and nitrogen species in the nebula should have been CO and N_2 at high temperatures, converting to CH_4 and NH_3 at low temperatures. Condensation of messy organic compounds cannot be predicted because of inadequate thermodynamic data, and kinetics must have played a major role anyway.

One condensation model involves the Fischer-Tropsch type process, based on a commercial synthesis of hydrocarbon fuels from coal. As the nebula cooled below 600 K, adsorption of CO, NH_3, and H_2 onto previously condensed mineral grains led to synthesis of a variety of compounds. The Fischer-Tropsch synthesis depends on catalytic activity on the mineral substrate. It has been proposed that complex organic compounds, as well as H_2O and CO_2, formed in this way. This is a very difficult model to test. Laboratory experiments indicate that the expected compounds can be produced, but only under very different conditions from those in the nebula. Because the dominant nitrogen species at nebular temperatures is N_2 rather than NH_3, the Fischer-Tropsch synthesis would need to have been very efficient. The formation of complex organic molecules may also have been accomplished by other processes, such as photochemical reactions driven by ultraviolet radiation.

Chondritic meteorites, especially carbonaceous chondrites, provide the only direct sampling of organic compounds from the nebula. Two categories of carbonaceous materials have been isolated from these samples. One consists of amorphous carbon mixed with macromolecular organic material similar to terrestrial kerogen. At least some fraction of this acid-insoluble component contains noble gases whose elemental and isotopic compositions suggest that it may be relict interplanetary material. The smaller category consists of readily soluble organic compounds. Among these are many of the compounds present in living systems, such as carboxylic and amino acids. At one time, it was argued that such compounds were biologic in origin. However, the molecular structures of compounds in chondrites do not show the preferred forms that characterize those produced biologically. Amino acids, for example, are racemic mixtures rather than L-isomers. Table 13-2 gives the distribution of carbonaceous materials in one well-characterized chondrite.

TABLE 13-2 DISTRIBUTION OF CARBON IN THE MURCHISON CARBONACEOUS CHONDRITE*

Species	Abundance
Acid-insoluble carbonaceous phase	1.3–1.8%
CO_2 and Carbonate	0.1–0.5%
Hydrocarbons	
Aliphatic	12–35 ppm
Aromatic	15–28 ppm
Acids	
Monocarboxylic (C_2–C_8)	170 ppm
Hydroxy (C_2–C_5)	6 ppm
Amino Acids	10–20 ppm
Alcohols (C_1–C_4)	6 ppm
Aldehydes (C_2–C_4)	6 ppm
Ketones (C_3–C_5)	10 ppm
Ureas	2 ppm
Amines (C_1–C_4)	2 ppm
N-heterocycles	1.1–1.5 ppm
	Sum: 1.43–2.35%

*From Wood and Chang (1985).

These two carbonaceous components have very different stable isotopic compositions. For the kerogen component, measured mean values of δD, $\delta^{13}C$, and $\delta^{15}N$ are +800, −12, and +20, respectively; corresponding values for the soluble organic component are +250, +25, and +75 (Wood and Chang, 1985). These major differences appear to require either more than one source or more than one production mechanism for meteoritic carbonaceous materials. Such data are at least consistent with the idea that some organic components may be relict interstellar phases.

Ices: The Only Thing Left

What happened to the simple gaseous molecules, such as CH_4, NH_3, and H_2O, that were not used in the production of more complex organic compounds? Under equilibrium nebular conditions, some of the water should have been incorporated into hydrous silicates, but there was no obvious sink for methane and ammonia. At temperatures below about 200 K, these finally condensed into ices. This probably did not occur in the inner solar system, but evidence of abundant ices is obvious at greater radial distances. The Jovian planets and their satellites consist of rocky cores surrounded by great thicknesses of these ices. Comets are probably dirty snowballs containing considerable ice. It is generally thought that comets are relatively pristine samples of condensed matter, containing even the most volatile elements in their cosmic proportions.

It should come as no surprise that ices are associated physically with organic compounds, because both condensed at low temperatures. We know this from spectroscopic observations of cometary comas. It is not possible to reconstruct the primitive organic molecules that may have occurred in comet nuclei, however, because solar irradiation breaks down all but the simplest molecules before they are blown away into the tail.

Worked Problem 13-3

What are the relative weight proportions of ices to rock in a fully condensed solar nebula? We will restrict our calculations to those elements with cosmic abundances greater than 10^4 atoms per 10^6 atoms of silicon (Table 13-1). Of these, we will assume that the following elements condense as oxides to form rock: Na, Mg, Al, Si, Ca, Fe, and Ni. The following elements condense as ices: C, N, O, Ne, S, Ar, and Cl. We will model all of these ices except the noble gases as hydrides. In the case of oxygen, correction must be made to adjust for the amount already partitioned into oxides. The element He will not condense, but will remain in the gas phase; some H will be consumed in hydride ices, and the remainder will be gaseous.

Rock components are calculated in the following way: For sodium, convert atomic abundance to moles by dividing by Avogadro's number:

$$5.7 \times 10^4 \text{ atoms}/6.023 \times 10^{23} \text{ atoms mol}^{-1} = 9.46 \times 10^{-20} \text{ mol.}$$

Now multiply half this number of moles by the gram formula weight of Na_2O:

$$(9.46 \times 10^{-20}/2 \text{ mol})(61.98 \text{ g mol}^{-1}) = 2.93 \times 10^{-18} \text{ g } Na_2O.$$

If we follow the same procedure for each of the other elements condensable as rock, these sum to 3.08×10^{-16} g of oxides.

Now repeat the procedure for elements condensable as ices, using the gram formula weight of the hydride for all but the noble gases. For oxygen, we obtain

$$2.01 \times 10^7 \text{ atoms}/6.023 \times 10^{23} \text{ atoms mol}^{-1} = 3.34 \times 10^{-17} \text{ mol.}$$

In the case of oxygen, an adjustment must be made to correct for the amount already partitioned into oxides. Summing the amount of oxygen needed for Na_2O, MgO, Al_2O_3, SiO_2, CaO, FeO, and NiO, we find that 7.03×10^{-18} mol of oxygen are required to combine with the condensable elements to make rock. Subtracting this from the total moles of oxygen available leaves 2.64×10^{-17} mol of oxygen that can be condensed as ice. Multiplying this result by the gram formula weight of the hydride gives

$$(2.64 \times 10^{17} \text{ mol})(18.016 \text{ g mol}^{-1}) = 4.76 \times 10^{-16} \text{ g } H_2O.$$

Repeating for other elements condensable as ice, we obtain a total of 1.03×10^{-15} g. The ratio of these two numbers is ice/rock = $1.03 \times 10^{-15}/3.08 \times 10^{-16} = 3.34$. Thus, potential ices are greater than three times more abundant by weight than potential rock. From this comparison, it is easy to see why the outer planets consist mostly of gases derived from ices.

ESTIMATING BULK COMPOSITIONS OF PLANETS

Mean Densities As Geochemical Data

The mean densities of the planets range from 3.9–5.5 g cm^{-3} for the inner (terrestrial) planets to 0.7–1.7 g cm^{-3} for the outer (Jovian) planets. Most of this difference can be explained by the fact that varying amounts of rocky and icy material were incorporated to form these bodies, due to radial temperature differences within the solar nebula. Within the terrestrial planet group there are density variations, less dramatic but still of sufficient magnitude to suggest important chemical differences. In order to compare these, we must correct mean densities for the effects of self-compression due to gravity. Such *uncompressed mean densities* (in g cm^{-3}) for the terrestrial planets, calculated for a pressure of 10 kbar, are as follows: Mercury, 5.30; Venus, 3.96; Earth, 4.07; Mars 3.73; Moon, 3.40. Even after the effects of self-compression have been compensated, significant density variations remain.

Several explanations have been put forward to explain these data. Cosmic abundance considerations indicate that planetary density variations must involve iron, the only abundant, heavy element that occurs in minerals with different densities. Urey (1952) proposed that varying proportions of silicate and metal, with densities of 3.3 and 7.9 g cm^{-3} respectively, were mixed to produce the inner planets. Density would then be a straightforward function of metal/silicate ratio. Urey envisioned that metal-silicate fractionation occurred in the nebula, due to differences in the physical properties of these materials. Ringwood (1959) appealed to variations in oxidation state to explain density differences. The effect of redox state on chondrites illustrates the principle. If all the iron and nickel in these meteorites were converted to oxides, the density would be 3.78 g cm^{-3}. If FeO were fully reduced to metallic iron, the density would increase to 3.99 g cm^{-3}. This density range is not sufficient to encompass the mean densities of all the planets, so it is necessary to invoke some metal-silicate fractionation as well.

Both of these proposals find some support in chondrites. The various chondrite groups can be distinguished on the basis of both Fe/Si and Fe^0/FeO ratios. Whatever the origin of these differences in planetary mean density, they provide a fundamental test that acceptable geochemical models for planetary bulk compositions must pass.

The Equilibrium Condensation Model

The condensation sequence already described defines a succession of equilibrium assemblages that form from a cooling gas of solar composition. An important feature of this sequence is that most phases form by reaction of previously condensed phases with vapor. Therefore, even in thermal regimes where evaporation rather than condensation occurred, the mineralogy of condensed phases can be the same as predicted by condensation calculations, assuming of course that these phases reach equilibrium with the vapor.

The basis for the equilibrium condensation model for planet formation (Lewis, 1972; Goettel and Barshay, 1978) is that solids (of whatever origin) were thermally equilibrated with the enclosing nebula gas, and thus had compositions dictated by condensation theory. The temperature in the nebula is assumed to have been highest near the protosun and to have declined outward. Whatever solids were stable at particular radial distances were accreted to form the various planets and thereby isolated from further reaction with the gas. Any uncondensed materials were somehow flushed from the system.

At the location of Mercury, the temperature (1400 K) was such that refractory elements like calcium and aluminum fully condensed and all iron was metallic. Magnesium and silicon had only partly condensed at this point. The high mean density of Mercury is thus explained by the occurrence of iron only in its most dense, metallic state, as well as the high iron abundance relative to magnesium and silicon. The temperature (900 K) at the Venus site allowed complete condensation of magnesium and silicon, but no oxidation of iron. Earth formed in a somewhat cooler (600 K) region, in which some iron reacted to form sulfide and ferrous silicate. The addition of sulfur, with an atomic weight greater than the mean atomic weight of the other condensed elements, resulted in a mean density for the earth that is higher than that of Venus. The low nebular temperature (450 K) for Mars permitted oxidation of all of the remaining iron and its incorporation into silicates. Its mean density was further reduced by hydration of some olivine and pyroxene to form phyllosilicates. Conditions in the asteroid belt were appropriate for the formation of carbonaceous chondrites and, further outward, ices were stable with silicates to form the Jovian planets.

Worked Problem 13-4

How can we estimate a planetary bulk composition using the equilibrium condensation model? This is a very long and complex problem, so we will simply illustrate how it is done by examining how to "set up" the problem, using a procedure described by BVSP

Estimating Bulk Compositions of Planets

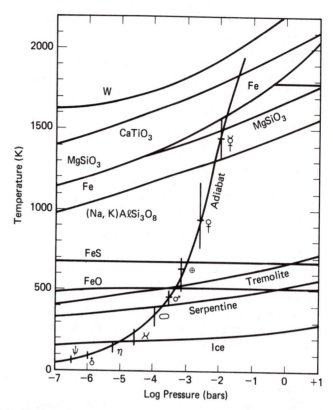

Figure 13.13 Equilibrium condensation in the solar nebula can account for the compositions of the terrestrial and Jovian planets, if temperature and pressure decreased radially away from the sun, as shown by the curved line. Planetary symbols begin with Mercury at high temperature and progress radially outward to Neptune. (After S. S. Barshay and J. S. Lewis, Chemistry of primitive solar material, *Ann. Rev. Astron. Astrophys.* 14: 81–94. Copyright by Annual Reviews. Reprinted with permission.)

(1981). To begin, we must define the dimensions of a "feeding zone" from which each planet will draw in condensed material during accretion. These annular zones can be the exclusive property of each planet (Lewis, 1972) or can be overlapping (Weidenschilling, 1976).

The condensation sequence of Grossman (1972), with which we are already familiar, was calculated at one pressure. However, pressure and temperature decreased outward in the solar nebula. Lewis (1972) generalized the condensation sequence by constructing a plot showing how some of these reactions varied with temperature *and* pressure, illustrated in Figure 13.13. Lewis adopted a *P-T* gradient, also shown in this figure, that produced mineral assemblages at the locations of planets whose densities corresponded with those of the planets. The temperature at which a mineral first becomes stable was taken from the intersection of its reaction line with the *P-T* gradient in Figure 13.13. The calculations of Grossman (1972) were then used to estimate the degree of condensation below that temperature. The composition of each planet was defined by integrating, over equal increments of distance, the compositions of solid material within its feeding zone.

This model satisfactorily accounts for planetary mean densities, but a number of other problems are evident. First, it is not possible to explain the large difference in mean density of the earth and its moon by using equilibrium condensation. Also, it is unrealistic to assume that a large planet would accrete only condensates formed at a single temperature. Mixing would have been unavoidable. At some point the kinetic barriers to gas-solid equilibrium must have also played a role, which would have resulted in further mixing of materials with different thermal histories. Finally, at the condensation temperatures for Venus, earth, and Mars, no condensed volatiles are possible. The atmospheres of these planets and the earth's oceans are thus an embarrassment to equilibrium condensation models, as is the presence of Fe^{+3} in terrestrial rocks. Despite these problems, the equilibrium condensation model is very useful as one of several end members of idealized scenarios for planet formation.

Heterogeneous Accretion

Another model for planet formation is based on the premise that all solids were vaporized in a hot solar nebula. During cooling, condensation ensued. As each new phase formed, it was immediately accreted. The equilibrium condensation sequence was, in a sense, beheaded, because the removal of each solid phase prevented its subsequent reaction with the vapor. This is somewhat analogous to perfect fractional crystallization, which we discussed in Chapter 11, but in a cosmic setting. At some point, the process was interrupted as the nebula was dissipated. Because temperatures in the nebula decreased radially outward, the progress of condensation was not everywhere the same. Mercury might have passed through only the stages at which Ca, Al-rich material, metallic iron, and some magnesium silicates condensed. Planets further out may have acquired veneers of phyllosilicate- and organic-rich material before condensation was halted.

This model is called *heterogeneous accretion* (Turekian and Clark, 1969; Anderson, 1972), because what are produced are layered planets, with the most refractory material in cores and more volatile phases at the surfaces. In practice, the model cannot be applied exclusively, because it leads to bizarre results. For example, the earth should contain no FeO or FeS, due to the fact that all of the iron was buried in the core and isolated from further reaction through mantling by magnesian silicates. The advocates of this model do not practice undeviating loyalty, however; they ask only that equilibrium condensation be modified to allow some heterogeneous accretion to take place. This flexibility makes it impossible to derive from this model a unique composition for a planet.

Heterogeneous accretion does, however, suggest a source for volatile compounds in planetary atmospheres and the earth's hydrosphere. Volatiles were accreted late in the condensation sequence onto planetary surfaces. The model also explains why rocks of the earth's mantle contain higher siderophile element abundances than expected for rocks in equilibrium with a metallic core, as well as quantities of Fe^{+2} and Fe^{+3}, which are not stable in the presence of metallic iron. If reduced and oxidized iron had ever been mingled, they should have reacted to a uniform, intermediate oxidation state. A major difficulty with this model is that the requirement for total vaporization of dust in the nebula and for very rapid accretion are at odds with current conceptions of nebular conditions.

The Chondrite Model

If we accept the proposition that chondrites are leftover planetary building blocks, it becomes unnecessary to rely on abstract concepts of nebular condensate material to construct models. In this case, we have actual samples that can be scrutinized to understand cosmochemical behavior. Chondrites seem to tell us that planets were probably

Estimating Bulk Compositions of Planets

formed from an assortment of materials radically out of equilibrium. After all, chondrites consist of components processed at both high (Ca, Al-rich inclusions and chondrules) and low (matrix) temperatures, as well as under oxidizing (ferrous silicates) and reducing (metallic iron) conditions. This disequilibrium assemblage is intimately mixed on a millimeter scale.

Our task would be simple if the compositions of the planets matched those of individual chondrite groups. As appealing a model as this is, it doesn't quite work. Planets seem to show the same kinds of volatile element depletions that some chondrite groups do, but planetary fractionations are more extreme. Mixing various classes of chondrites provides better matches for planetary compositions, but even this is not enough.

A better procedure is to use the *components* of chondrites to make planets. Early chondrite models employed only two components, refractory-rich chondrules and volatile-rich matrix, but later models have become increasingly complex. One formulation (Anders, 1977; Morgan and Anders, 1979) identifies seven components: refractory-rich material, metal, silicates, troilite (FeS), volatile-rich material, remelted metal, and remelted silicates. The first five components can be considered condensates, and these incorporate elements according to their volatility. The volatility group assignment for each element is given in Figure 13.14. The last two components formed by melting of some of the others, leading to other fractionations. Although these are

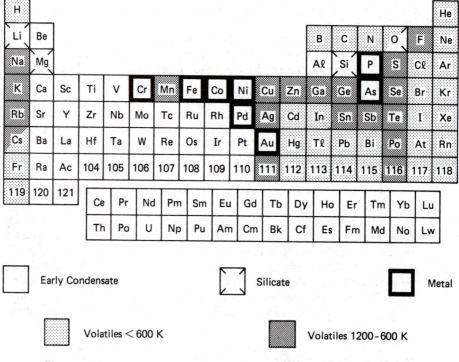

Figure 13.14 Elements condensing from the solar nebula are separated into groups according to their volatility, as illustrated by this Periodic Table. The condensation temperatures for these groups are refractory, > 1400 K; metal, 1400–1200 K; silicate, 1200–600 K; moderately volatile, 1200–600 K; highly volatile, <600 K. (After J. W. Morgan and E. Anders, Chemical composition of Earth, Venus, and Mercury. *Proc. Natl. Acad. Sci. USA* 77: 6973–77. Copyright 1980. Reprinted with permission of the authors.)

defined as chemical components, most of them have physical counterparts in chondrites. For example, Ca, Al-rich inclusions qualify as refractory-rich material, matrix is volatile-rich material, and chondrules are remelted silicates.

The chondrite model is based on the postulate that the materials that made the planets experienced exactly the same fractionations as recorded by chondrites. Because the seven components can be mixed in different proportions to reproduce the chemistry of the various chondrite groups, they clearly were fractionated to some extent in the nebula. Thus, these same components can be blended to make planets, though more extreme proportions of some may be required. Each component carries its own suite of elements in approximately cosmic proportions, but the components have been fractionated from each other to varying degrees. Therefore, this model requires knowledge of the abundance of only a few *index elements* to predict the composition of a planet. The index elements and their condensation ranges are as follows: U, > 1400 K; Fe, $1400-1200$ K; K, $1200-600$ K; and T1, < 600 K.

Worked Problem 13-5

What is the bulk chemical composition of Mars, based on the chondrite model? We will outline the procedure used by Morgan and Anders (1979) in solving this problem. They first derived compositions for the components listed above using limited temperature intervals of the equilibrium condensation sequence. These compositions are not tabulated here, but may be found in Table 2 of Morgan and Anders' paper. The compositions of silicate components were calculated on a FeO-, MnO-, and Cr_2O_3-free basis, eliminating the need to worry about oxidation state and iron content.

We will use the following symbols in working this problem: c = refractory condensate, s = silicate, m = metal, t = troilite, and v = volatile-rich material; subscripts u and r refer to unmelted and remelted materials, respectively. We can write a mass balance reaction such that

$$c + s_u + s_r + m + t + v + FeO + MnO + Cr_2O_3 = 1. \qquad (13\text{-}6)$$

The index element for the refractory component is uranium, assigned a value of 28 ppb for Mars based on Soviet orbiter measurements and thermal evolution calculations. The proportion of refractory condensate c in Mars relative to cosmic abundance (in ppb) is then

$$c = U_{Mars}/U_c = 28/197 = 0.142.$$

Volatile elements are usually scaled to thallium in this model, but no Tl data are available for Mars. The abundance of this element (0.14 ppb) was estimated from measured atmospheric ^{36}Ar, because of an observed correlation between these volatile elements in chondrites. The volatile-rich component v is then

$$v = T1_{Mars}/T1_v = 0.14/143 = 9.80 \times 10^{-4}.$$

The unmelted silicate condensate component s_u is indicated by potassium, and a Martian value of 62 ppm based on K/U ratio from orbital γ-ray data was used. Correction must be made to account for the fact that component v above also contains potassium. Therefore,

$$s_u = [K_{Mars} - vK_v]/K_c = (62 - 1.3)/1460 = 4.17 \times 10^{-2}.$$

The other components—remelted silicate s_r, metal m, and troilite t—require a more elaborate approach that will not be treated in detail here. The FeO content can be estimated from Fe/(Fe + Mg) in mantle silicates. For Mars, a value of 0.77 was determined by progressively varying this ratio until the mantle density matched that estimated from geophysical constraints. This ratio allows the calculation of FeO = 0.13. MnO = 0.0012 and Cr_2O_3 = 0.0053 are calculated from other relationships that gauge redox state.

The metal and troilite fractions are given by

$$Fe_{Mars}[1 + (Ni + Co + P)_{cosmic}/Fe_{cosmic}] = 0.777 \, FeO + m + 0.635t. \qquad (13\text{-}7)$$

TABLE 13-3 MODEL BULK COMPOSITIONS FOR THE TERRESTRIAL PLANETS*

	Mercury EC(1)	Mercury CH(2)	Venus EC(1)	Venus CH(2)	Earth EC(1)	Earth CH(2)	Mars EC(1)	Mars CH(3)	Moon CH(4)
Mantle + Crust									
SiO_2	40.8	47.1	52.9	49.8	48.5	47.9	43.6	41.6	43.3
TiO_2	0.49	0.33	0.20	0.21	0.20	0.20	0.16	0.33	0.40
Al_2O_3	9.6	6.4	3.8	4.1	3.5	3.9	3.1	6.39	7.6
Cr_2O_3	-	3.3	-	0.87	-	0.9	-	0.65	0.3
MgO	40.5	33.7	37.6	35.5	34.5	34.1	31.0	29.8	29.1
FeO	0.0	3.7	0.24	5.4	8.3	8.9	17.2	15.9	13.0
MnO	-	0.06	-	0.09	-	0.14	-	0.15	0.15
CaO	8.6	5.2	3.6	3.3	3.3	3.2	2.9	5.2	6.1
Na_2O	0	0.08	1.6	0.28	1.5	0.25	1.4	0.10	0.15
K (ppm)	0	69	1442	221	1324	200	1190	59	78
U (ppm)	0.053	0.034	0.021	0.022	0.019	0.021	0.017	0.033	0.041
Th (ppm)	0.19	0.122	0.073	0.079	0.067	0.076	0.060	0.113	0.145
Core									
Fe	94.1	93.5	94.4	88.6	68.9	88.8	59.8	88.1	61.5
Ni	5.9	5.4	5.6	5.5	5.0	5.8	5.9	8.0	32.5
S	0	0.35	0	5.1	26.1	4.5	34.3	3.5	4.0
Relative masses									
Mantle + Crust	61.6	32.0	69.8	68.0	69.2	63.9	74.7	81.0	98.0
Core	38.4	68.0	30.2	32.0	30.8	36.1	25.3	19.0	2.0

*For each planet, model *EC* is based on equilibrium condensation and model *CH* is based on chondrites.
References: (1) Basaltic Volcanism Study Project, 1981; (2) Morgan and Anders, 1980; (3) Morgan and Anders, 1979; (4) Morgan et al., 1978.

The proportion of troilite is also limited by the available sulfur, which can be expressed in terms of the cosmic S/K ratio,

$$t = K_{Mars}(S/K)_{cosmic} + 2.742p, \qquad (13-8)$$

where p is an adjustable parameter. We are now left with three equations (13-6, 13-7, 13-8) in three unknowns (s_r, m, and t). Solving these simultaneously provides values that can be substituted into the following equation to obtain Fe, which is the final index element:

$$1.065 \, Fe_{Mars} = \alpha + \beta s_r + m + 0.635t,$$

where α and β are constants. In this case, $Fe_{Mars} = 0.268$.

Combining the compositions of these components in their calculated proportions gives the bulk composition of the planet. From the index elements, it is also possible to calculate the abundances of trace elements. All trace elements that define a component are assumed to be fractionated to the same degree as the index element relative to its cosmic abundance. For example, the abundance of any refractory element (say, zirconium) can be determined by multiplying its cosmic abundance by the refractory component depletion factor $c = 0.142$. The major element composition for Mars calculated in this way is given in Table 13-3.

This procedure seems to work reasonably well for most elements, but breaks down for elements that do not partition cleanly into one component. For example, significant errors in volatile element abundances may result from this approach, because some of the elements may be only partly condensed. A major advantage of the chondrite model is that it can be applied to unsampled or incompletely sampled planets by using orbital geophysical data and by making plausible assumptions. The fact that the abundance of practically the entire periodic table in a planet can be estimated from just a few elements is testimony to the cosmochemical importance of chondritic meteorites.

GEOCHEMICAL CONSTRAINTS ON THE ORIGIN OF THE MOON

Traditionally, the following three principal alternative hypotheses have been favored as possible explanations for the origin of the earth's moon:

1. *Intact Capture*—The moon formed elsewhere in the solar system and was gravitationally captured intact by the earth.
2. *Binary Accretion*—The moon accreted as a companion to the earth from materials captured in geocentric orbit.
3. *Rotational Fission*—The moon separated from an early earth that was rotating so rapidly that it was unstable.

Two hybrid models have been added more recently:

4. *Collisional Ejection*—Off-center collision of a very large projectile with the earth ejected debris into geocentric orbit, from which the moon ultimately accreted.
5. *Disintegrative Capture*—A very large body that barely missed the earth was tidally disrupted. Debris were retained in a geocentric orbit and accreted to form the moon.

Aside from dynamic arguments, how do we choose among these possibilities?

Geochemical data do not provide a definitive answer, but they at least serve as important constraints for these hypotheses. Let us first list some geochemical predictions that each hypothesis makes. Intact Capture is largely untestable, but it seems likely that any body formed elsewhere in the solar system might have a composition, both elemental and isotopic, that was distinct from the composition of the earth. Disintegrative Capture is a variant of the Intact Capture hypothesis, from a geochemical point of view. Collisional Ejection is also similar if the incoming projectile provides most of the ejecta that is accreted to form the moon. In current models, the ejecta contributions of these colliding bodies compared to the contributions from the earth range from roughly equal to predominantly projectile. During the collision, most of the ejecta is vaporized, so that recondensation might result in volatile element depletion in the moon. Binary Accretion suggests that the compositions of earth and moon should be very similar, because they were made from similar accreted materials. Rotational Fission is normally envisioned as occurring after core formation in the earth. Thus, the moon should be depleted in siderophile elements to the same degree as the earth's mantle.

The oxygen isotopic compositions of lunar and terrestrial rocks lie along the same mass fractionation line, previously illustrated in Figure 13.11. This is probably a very significant observation, because most other extraterrestrial materials define mass fractionation lines displaced from the earth-moon line. Isotopic data are most compatible with hypotheses (2) and (3).

The Mg/(Mg + Fe) ratio is an important indicator of the extent of geochemical differentiation. A value of the ratio for the earth's mantle of 0.89 seems fairly well-constrained, but that for the moon is less certain. A ratio of 0.80 is consistent with Fe and Mg abundances in lunar basalts, which are more iron-rich than terrestrial basalts. If the moon really does have a different Mg/(Mg + Fe) ratio, this would argue against its derivation from the earth, such as envisioned by hypothesis (3).

The low uncompressed mean density of the moon (3.40 g cm^{-3} at 10 kbar) is very different from that of the earth (4.07 g cm^{-3}). Variations in oxidation state cannot account for density differences of this magnitude, and it is generally accepted that the moon is impoverished in iron metal relative to the earth. Siderophile trace elements are severely depleted in both the earth's mantle and the moon. The earth's situation is easy enough to understand: elements like nickel and cobalt are now sequestered in the core. But the moon has only a small core. Hypotheses (3) and (4) can readily explain this constraint if fission or collision occurred after core formation. Preferential accretion of the moon from the silicate parts of differentiated planetisimals could allow hypotheses (2) and (5) to be consistent. Hypothesis (1) must explain how two independent bodies with such different iron contents have such similar siderophile trace element abundances.

Volatile trace elements are much more depleted in the moon than in the earth. This compositional distinction could be an inherent property of the materials that accreted to form these bodies, or it could result from some later, high-temperature process. This observation is difficult to explain if hypothesis (2) is correct. Compositional differences are expected for hypothesis (1) and (5). Most of the ejecta formed during collision would be vaporized, and the observed volatile element depletion might result from recondensation in hypothesis (3). Because a rapidly rotating earth would probably be partially molten, hypothesis (4) may also be consistent with this constraint.

These few examples obviously do not answer the question of the moon's origin, but they are sufficient to illustrate how geochemical data can be utilized to test possible solutions for extremely complex problems of vast scale. The geochemical data, plus the observation that the earth-moon system has very high angular momentum, suggest to many workers that a Collisional Ejection model is likely (e.g. Taylor, 1987). To learn about more constraints (geochemical and otherwise) on the origin of the moon, consult Hartmann et al. (1986).

PLANETARY MODELS: CORES AND MANTLES

The techniques we have just discussed provide ways to estimate the bulk chemical compositions of planets. It is also of obvious value to know how these elements are distributed within a planet, that is, which are partitioned into crust, mantle, and core. Because planetary crusts contain such a minor portion of the mass, it is common practice in constructing planetary models to combine mantle plus crust as one differentiated component. The problem then is reduced to that of how to estimate the composition of either the mantle-plus-crust or the core.

Estimating Core Compositions

Calculation of the bulk chemical composition of the earth by the methods already discussed leads to the conclusion that it must contain large quantities of iron. The mean density of the earth is consistent with a large, metallic iron core in the interior, and moment of inertia and free oscillations specify how this mass is concentrated. A major seismic discontinuity at 2900 km depth marks the mantle-core transition. Variations in compressional wave velocities through the core have led to the conclusion that the outer core is liquid and the inner core is solid. Most of what we know about cores is derived from these kinds of indirect observations. Despite the rigorous aspects of this discipline, one geophysicist has described the inversion of geophysical data to yield geological information as "like trying the reconstruct the inside of a piano from the sound it makes crashing down stairs." Can geochemistry shed any light on this complex problem?

Geophysical methods applied to study of the core are capable of detecting elements with very different atomic weights, but they cannot distinguish among elements similar in weight to iron. One such element, nickel, is thought to be an important constituent of the core, based on its siderophile character and high cosmic abundance. To first order, the core can thus be visualized as a mass of Fe-Ni alloy, part liquid and part solid. We can also infer that other siderophile elements, such as cobalt and phosphorus, might be important constituents.

The inferred density of the outer core is about 8% less than for the inner core. This is commonly interpreted to reflect the presence of an additional component, an element of low atomic weight, in the outer core (Brett, 1976). Considerable support has developed for the hypothesis that this element is sulfur. Ahrens (1979) studied the seismic velocities of sulfide and metal at high pressures, and concluded that 9 to 12 weight % S would be necessary to account for the lower density of the outer core. This is equivalent to 2.9–3.9 % of sulfur in the whole planet, implying that the earth captured 28–38 % of the cosmic proportion of this element. The only problem with this interpretation is that sulfur is highly volatile, condensing as FeS below 650 K. We should then expect that the terrestrial abundances of other elements of similar volatility would be equivalent. We can define a *depletion factor* as the abundance of a particular element relative to its cosmic abundance. The general sequence of depletion factors for volatile elements in the earth's mantle corresponds to the volatility sequence presented earlier in Figure 13-7.[5] However, many nonsiderophile elements that are less volatile than sulfur (for example, Mn, Na, K, Rb, Cs, Zn) have depletion factors (generally 0.3 to 0.03) in the bulk earth that are equal to or greater than those of sulfur. How can this be? Was sulfur somehow retained while less volatile elements were lost?

Based partly on this reasoning, other workers have argued for another candidate for the light element in the outer core. Bonding in FeO is expected to become metallic, rather than ionic, at high pressure, and an outer core of about 60 percent FeO and 40 percent metallic Fe-Ni would satisfy the density constraint. But does this really solve the problem? Iron does not begin to oxidize in the solar nebula until 750 K, a temperature not much greater than that required for FeS formation. Thus, it seems difficult to incorporate large amounts of FeO into the accreting earth at temperatures that would be required to account for the observed depletions of volatile elements.

---

[5] This generalization does not hold for volatile siderophile elements, which are, of course, depleted in the mantle because they are alloyed with metallic iron in the core.

Actually, neither of these arguments is as cogent as it seems. The abundances of specific volatile elements are extremely difficult to predict, because most are not fully condensed at the temperatures in question. Observed fractionations in chondrites suggest that FeS should be treated as an independent component in mixing models, so its abundance may not be strictly tied to volatility. We have also argued that the planets probably accreted from a disequilibrium assemblage of components. Studies of similar components in chondrites indicate that these differ in oxidation state. Reactions of metal with oxidized species would produce more FeO in addition to primary FeO carried in silicate components. Thus, from a geochemical perspective, either (or both) of these elements may be suitable candidates for the light element in the outer core.

One way to estimate the composition of a core directly is to assume that siderophile elements are present in cosmic abundances relative to iron. An example of this approach is shown in the following Worked Problem. The mantle plus crust composition can then be derived from the bulk planet composition if the core mass fraction is known. Planetary models obtained in this way are given in Table 13-3, for bulk compositions calculated from condensation and chondrite models.

Worked Problem 13-6

What is the composition of the earth's core? One way to estimate core composition is to assume that siderophile elements accompany iron in cosmic proportions. Of course, we must also include major amounts of a light element to account for the density of the outer core, and we have just seen that this element may not be in cosmic proportion relative to iron.

Let us assume a value of 9.0 weight % sulfur in the core, based on sulfur abundance from density constraints. The first step, then, is to determine how much iron must be combined with sulfur to make FeS:

$$\frac{9.0}{32.05 \text{ g mol}^{-1}} = \frac{x}{55.85 \text{ g mol}^{-1}}$$

$$x = 15.68 \text{ weight } \% \text{ Fe.}$$

The weight % FeS in the core is then $15.68 + 9.0 = 24.68$. By difference, the core must consist of 75.32 weight % metal.

The metal phase also contains other siderophile elements in addition to iron. Siderophile elements with the highest cosmic abundances are Ni, Co, and P, and these are prominent in iron meteorites. A mass balance equation for metal is thus

$$\text{Fe}_{\text{metal}} + \text{Ni} + \text{Co} + \text{P} = 75.32\%,$$

or

$$\text{Fe}_{\text{metal}} = 75.32 - \text{Ni} - \text{Co} - \text{P}, \qquad (13\text{-}9)$$

where the abundance of each element is expressed in weight %. Ni can be determined from cosmic weight ratio of Ni and Fe:

$$\text{Ni} = \left(\frac{\text{Ni}_{\text{cosmic}}}{\text{Fe}_{\text{cosmic}}}\right) \text{Fe}_{\text{core}}$$

$$= \left(\frac{4.93 \times 10^4}{58.77 \text{ g mol}^{-1}}\right)\left(\frac{9 \times 10^5}{55.85 \text{ g mol}^{-1}}\right) \text{Fe}_{\text{core}}$$

$$= 0.0521 \text{ Fe}_{\text{core}},$$

where Fe_{core} is the total abundance of iron in the core, equal to $\text{Fe}_{\text{metal}} + 15.68$. Similarly, $\text{Co} = 0.0024 \text{ Fe}_{\text{core}}$ and $\text{P} = 0.0208 \text{ Fe}_{\text{core}}$. Substitution of these values into equation 13-9 gives

$$\text{Fe}_{\text{metal}} = 75.32 - 0.0521 \text{ Fe}_{\text{core}} - 0.0024 \text{ Fe}_{\text{core}} - 0.0208 \text{ Fe}_{\text{core}}$$

= 68.94 weight %.

Thus, the total amount of iron in the core is 68.94 + 15.68 = 84.62 weight %. Ni, Co, and P weight percentages are obtained by multiplying their cosmic weight ratios by 84.62; the respective values are 4.41, 0.20, and 1.76 weight %.

Estimating Mantle Compositions

Core composition can also be determined indirectly if the compositions of the bulk planet and mantle are known. The core composition is presumably complementary to that of the mantle from which it was extracted, so that depletions of siderophile and chalcophile elements relative to the bulk planet abundance can provide insight into the relative abundances of core-forming elements.

Treiman et al. (1986) provided an interesting example of this approach in calculating core compositions for the earth and the shergottite parent body (SPB), an igneous meteorite parent body inferred to be Mars (arguments reviewed by McSween, 1985). They used the measured elemental abundances in terrestrial basalts and shergottites to constrain mantle compositions. Because basalts are generally fractionated, Treiman and coworkers considered ratios of siderophile and lithophile elements (for example, Ag/La) that behaved identically, that is, were not fractionated from each other, during igneous processes. The abundance ratios were compared to chondrite values, yielding depletion or enrichment factors relative to cosmic abundances. In turn, these depletion or enrichment factors were normalized to bulk planet values. Their results for abundances of siderophile elements and chalcophile elements are illustrated in Figure 13.15.

A number of conclusions can be drawn from these patterns. Noble metals like rhenium and gold are greatly depleted in the mantles of both planets, as expected from their strong siderophile affinities. Phosphorus and tungsten are both moderately siderophile; their affinity for iron is partly controlled by redox state. Both of these elements are more depleted in the earth's mantle than the SPB, suggesting that the SPB is more highly oxidized than the earth. Nickel and cobalt are depleted in both mantles, but the Ni/Co ratios are distinct, implying some signficant (but not presently understood) difference in planetary evolution. Chalcophile elements also show depletions in both mantles. Sulfur and selenium occur chiefly in sulfide phases, and both elements are depleted to the same degree in both mantles. The other chalcophile elements have some lithophile affinities, and are variably depleted. The SPB shows consistently greater chalcophile element depletions than the earth's mantle, suggesting that its core has a higher ratio of sulfide to metal.

Geochemical Models for Planetary Differentiation

Core formation is one of the most important geochemical events in the evolution of a planet. A number of hypotheses have been advocated to explain how this occurs. The most straightforward is a model in which equilibrium was maintained between an Fe-S-O metallic liquid and silicate during accretion and core separation. This model is testable if mantle depletion factors for siderophile and chalcophile elements match those predicted from experimentally determined metal-silicate distribution coefficients. *Inefficient core formation* refers to a condition in which separation of solid and liquid metal phases from silicate mantle is not perfect. Small amounts of trapped metal would thus dominate the siderophile and chalcophile element abundances of the mantle, giv-

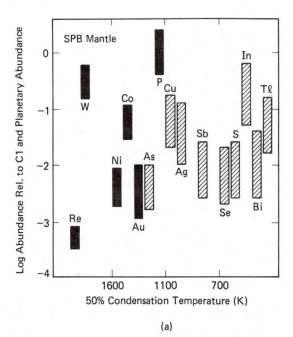

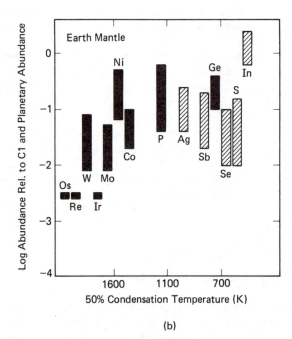

Figure 13.15 Abundances of siderophile elements (black bars) and chalcophile elements (hatchured bars) in the mantles of the earth and SPB (inferred to be Mars), relative to C1 chondrite and bulk planet abundances. The depletions from 0 are due to fractionation of metal and sulfide during core formation. The horizontal axis is the temperature at which 50% of the element would be condensed in the nebula.

ing chondritic relative proportions. Another model is *heterogeneous accretion,* in which the core is chemically isolated from the mantle, which accreted later.

These models are difficult to test. Treiman et al. (1986) suggested that elemental abundances in the SPB mantle and core could be modeled adequately by assuming equilibrium between silicate and metal/sulfide. However, none of these three processes explains the composition of the earth's mantle and core. Additional complications could be the unknown effects of high pressure or a compositionally layered mantle.

SUMMARY

Cosmochemistry is the application of geochemical principles to systems of vast scale. Fundamental to this discipline is the determination of the cosmic abundance of the elements. At the beginning of this chapter, we saw how elemental abundance patterns were controlled by nucleosynthesis in stars. The mixing of the products of fusion, nuclear statistical equilibrium, neutron capture, and proton capture reactions in earlier generations of stars accounts for the composition of the sun.

Chondritic meteorites are materials surviving from the earliest stages of solar system history. Their chemical compositions are essentially cosmic, although small fractionations occur in some chondrite groups. Element fractionations in the solar nebula, inferred from chondrites, were governed by volatility and by geochemical affinity, reflected in siderophile, chalcophile, and lithophile behavior.

The equilibrium condensation sequence provides a conceptual framework with which we can examine the effects of volatility. Although it is unlikely that the solar nebula was ever hot enough to vaporize completely all solids, many of the features in chondrites can be understood by using this tool. Thermal processes in the solar nebula and in stars that preceded it have imprinted the effects of both condensation and evaporation on chondrite components.

We have also seen that the isotopic compositions of the most refractory components of chondrites point to a supernova origin. The infusion of now-extinct radionuclides indicates that a stellar explosion must have occurred in the vicinity of the nebula. It has been proposed that the shock wave from such a supernova triggered the collapse of gas and dust to form the solar system.

Some of the most volatile materials in chondrites are organic compounds, synthesized by catalytic reactions in the nebula, possibly mixed with surviving interstellar compounds. Among these are most of the raw materials for life, suggesting that the earth may not have had to synthesize organic compounds from scratch. Ices were the last materials to condense in the nebula. Highly volatile materials are concentrated in the outer solar system—in the Jovian planets, and in comets.

The compositions of the terrestrial planets can be modelled based on condensation theory, assuming that a radial temperature and pressure distribution controlled the stability of elements and compounds. The heterogeneous accretion model assumes that materials were aggregated and isolated from the nebula as soon as they condensed. The basis for a third planetary model is that fractionations in protoplanetary materials were the same as those observed in chondrites, so that chondritic chemical components can be used to construct planets. The compositions of planetary cores can also be estimated from cosmochemical considerations. The fact that these different techniques lead to similar results suggests that cosmochemical behavior offers a valid mechanism for understanding planetary-scale problems.

SUGGESTED READINGS

Because chondrites provide most of the data for cosmochemistry, much of the literature in this field is embedded in books and papers on meteorites. Many of the references below are technically demanding, but they provide a thorough background in this subject.

ANDERS, E. and EBIHARA, M. 1982. Solar-system abundances of the elements. *Geochim. Cosmochim. Acta* 46: 2363–80. (An important paper in which cosmic abundances are derived and discussed.)

Basaltic Volcanism Study Project (BSVP). 1981. *Basaltic volcanism on the terrestrial planets.* Elmsford, N.Y.: Pergamon. (An exhaustive reference book on aspects of planetary geochemistry. Section 4.3 gives an excellent review of cosmochemical constraints on planetary compositions.)

DODD, R. T. 1980. *Meteorites, A chemical-petrologic synthesis*. New York: Cambridge Univ. Press. (One of the best current books on meteorites. Chapters 2, 3, and 4 describe chondrite petrology and chemistry.)

GROSSMAN, L., and LARIMER, J. W. 1974. Early chemical history of the solar system. *Rev. Geophys. Space Phys.* 12: 71–101. (A comprehensive review of condensation and other nebular processes.)

HENDERSON, P. 1982. *Inorganic geochemistry*. Oxford: Pergamon Press. (Chapter 2 of this interesting book gives a particularly fine account of nucleosynthesis in stars and the cosmic abundance pattern.)

McSWEEN, H. Y., JR. 1987. *Meteorites and their parent planets*. New York: Cambridge Univ. Press. (A nontechnical introduction to meteorites and what can be inferred about their parent bodies.)

RINGWOOD, A. E. 1979. *Origin of the earth and moon*. New York: Springer-Verlag. (A thoughtful book about processes that shaped the planets: Chapter 3 describes core formation and Chapters 6 and 8 consider nebula processes and accretion.)

SUESS, H. E. 1987. *Chemistry of the solar system*. New York: John Wiley and Sons. (A nice summary at a basic level. Chapter 1 gives the isotopic composition of the elements, and chapter 2 discusses nucleosynthesis.)

WASSON, J. T. 1985. *Meteorites, their record of early solar-system history*. New York: Freeman. (A meteorite monograph that is rich in cosmochemical information. Chapter 7 and 8 describe nebula fractionation in detail.)

WOOD, J. A. 1979. *The solar system*. Englewood Cliffs, N.J.: Prentice-Hall. (Possibly the most lucid introduction to planetary geology and stellar processes available.)

WOOD, J. A., and CHANG, S., eds. 1985. *The cosmic history of the biogenic elements and compounds*. Washington, D.C.: NASA. (An up-to-date account of nucleosynthesis of biogenic elements in stars and in the interstellar medium, as well as an explanation of nebula processes, compiled by 12 contributors.)

The following additional papers are referenced in this chapter. The interested student may wish to explore them.

ANDERS, E. 1977. Chemical composition of the moon, earth, and eucrite parent body. *Phil. Trans. Roy. Soc. London* A295: 23–40.

ANDERSON, D. L. 1972. Implications of the inhomogeneous planetary accretion hypothesis. *Comments Earth Sci: Geophys.* 2: 93–98.

AHRENS, T. J. 1979. Equations of state of iron sulfide and constraints on the sulfur content of the earth. *J. Geophys. Res.* 84: 985–1008.

BARSHAY, S. S., and LEWIS, J. S. 1976. Chemistry of primitive solar material. *Ann. Rev. Astron. Astrophys.* 14: 81–94.

BRETT, R. 1976. The current status of speculations on the composition of the core of the earth. *Rev. Geophys. Space Phys.* 14: 375–83.

BURBIDGE, E. M.; BURBIDGE, G. R.; FOWLER, W. A.; and HOYLE, F. 1957. Synthesis of the elements in stars. *Rev. Mod. Phys.* 29: 547–650.

CAMERON, A. G. W. 1978. Physics of the primitive solar accretion disk. *The Moon and Planets* 18: 5–40.

———, and TRURAN, J. W. 1977. The supernova trigger for formation of the solar system. *Icarus* 30: 447–61.

CLAYTON, R. N.; GROSSMAN, L.; and MAYEDA, T. K. 1973. A component of primitive nuclear composition in carbonaceous meteorites. *Science* 182: 485–88.

GOETTEL, K. A., and BARSHAY, S. S. 1978. The chemical equilibrium model for condensation in the solar nebula: Assumptions, implications, and limitations. In *The origin of the solar system*, ed. S. F. Dermott, 611–27. New York: John Wiley and Sons.

GROSSMAN, L. 1972. Condensation in the primitive solar nebula. *Geochim. Cosmochim. Acta* 36: 597–619.

HARTMANN, W. K.; PHILLIPS, R. J.; and TAYLOR, G. J., eds. 1986. *Origin of the moon*. Houston: Lunar and Planetary Institute.

JANAF. 1971. *Thermochemical Tables*. 2d ed. U.S. Nat. Stand. Ref. Data Ser., Nat. Bur. Stand. Publ. 37. Plus supplements in *J. Phys. Chem. Ref. Data*. (See Appendix B for complete references.)

LEE, T.; PAPANASTASSIOU, D.; and WASSERBURG, G. J. 1977. Aluminum-26 in the early solar system: Fossil or fuel? *Astrophys. J.* 211: L107–10.

LEWIS, J. S. 1972. Low temperature condensation in the solar nebula. *Icarus* 16: 241–52.

McSWEEN, H. Y., JR. 1979. Are carbonaceous chondrites primitive or processed? A review. *Rev. Geophys. Space Phys.* 17: 1059–78.

———. 1985. SNC meteorites: Clues to Martian petrologic evolution? *Rev. Geophys.* 23: 391–416.

MORGAN, J. W.; HERTOGEN, J.; and ANDERS, E. 1978. The moon: Composition determined by nebular processes. *Moon and Planets* 18: 465–478.

———, and ANDERS E. 1979. Chemical composition of Mars. *Geochim. Cosmochim. Acta* 43: 1601–10.

———, and ANDERS E. 1980. Chemical composition of Earth, Venus, and Mercury. *Proc. Natl. Acad. Sci. USA* 77: 6973–77.

RINGWOOD, A. E. 1959. On the chemical evolution and densities of planets. *Geochim. Cosmochim. Acta* 15: 257–83.

TREIMAN, A. H.; DRAKE, M. J.; JANSSENS, M. J.; WOLF, R.; and EBIHARA, M. 1986. Core formation in the earth and shergottite parent body (SPB): Chemical evidence from basalts? *Geochim. Cosmochim. Acta* 50: 1071–91.

TUREKIAN, K., and CLARK, S. P. 1969. Inhomogeneous accretion of the earth from the primitive solar nebula. *Earth Planet. Sci. Lett.* 6: 346–48.

UREY, H. C. 1952. *The planets*. New Haven: Yale University Press.

VAN SCHMUS, W. R., and WOOD, J. A. 1967. A chemical-petrologic classification for the chondritic meteorites. *Geochim. Cosmochim. Acta* 31: 747–65.

WEIDENSCHILLING, S. J. 1976. Accretion of the terrestrial planets II. *Icarus* 27: 161–70.

PROBLEMS

1. Discuss why the comparison of terrestrial Nd-Sm isotopic data with a chondritic (ϵ_{Nd}) evolution curve provides useful information.

2. Using Figure 13.10 and the cosmic abundances in Table 13-1, estimate the bulk composition of the refractory nebular condensate that forms in the temperature interval from 1900–1400 K.

3. What are the relative weight proportions of condensable matter (rock plus ices) versus non-condensable gases (He and leftover H) in a gas of solar composition? (Part of the solution has already been calculated in Worked Problem 13-3.)

4. Calculate the composition of the earth's core in terms of Fe, Ni, Co, P, S, and O, assuming that half of the light element is sulfur and half is oxygen. (See Worked Problem 13-7.)

5. List the chemical fractionations observed in chondrites, and suggest a plausible mechanism for each.

6. Using Figure 13.12, calculate the proportion of an exotic oxygen component consisting of pure ^{16}O that would have to be added to "normal" solar-system oxygen (lying along the terrestrial mass fractionation line) to produce a Ca, Al-rich inclusion plotting along the mixing line at $\delta^{17}O = -25$ per mil.

Appendix A
Mathematical Methods

Most of the problems in this book require mathematics that is no more complicated than standard algebra or first-year calculus. We have found, however, that students often need a refresher in some of these basics to boost their confidence. We strongly suggest that you use the problems in this text as an excuse to dust off your old math books and get some practice. If you do not already own a handbook of standard functions or a guide to mathematical methods in the sciences, you should probably add one to your shelf. We have found Burington (1973), Boas (1983), Potter (1978) and Dence (1975) to be particularly useful.

This short appendix is intended to introduce a few concepts not generally included in math courses required for the geology curriculum, but that are very useful in geochemistry. We do not pretend that it is an in-depth presentation; our goal is purely pragmatic. Each of these topics appears in one or more of the problems in this text, and should become familiar as you advance in geochemistry.

PARTIAL DIFFERENTIATION

If a function, f, has values that depend on only one physical parameter, x, then you know from your first calculus course that the derivative, $df(x)/dx$, represents the rate of change of $f(x)$ with respect to x. In graphical terms, $df(x)/dx$ (if it exists in the range of interest) is the instantaneous slope of the curve defined by $y = f(x)$. Strictly, $df(x)/dx$ is given by

$$\frac{df(x)}{dx} = \lim_{h \to 0} \frac{f(x + h) - f(x)}{h}.$$

Equations involving rates of change in the physical world are extremely common, but it is rare to find one in which the value of the function f depends on only one parameter, x. More commonly, we deal with $f(x, y, z, \ldots)$. It is still useful, however, to know how the function f varies if we change the value of only *one* of its controlling parameters at a time. For a function $f(x, y)$, for example, we define *partial derivatives* $(\partial f(x, y)/\partial x)_y$ and $(\partial f(x, y)/\partial y)_x$ by:

$$\left(\frac{\partial f(x, y)}{\partial x}\right)_y = \lim_{h \to 0} \frac{f(x + h, y) - f(x, y)}{h}$$

and

$$\left(\frac{\partial f(x, y)}{\partial y}\right)_x = \lim_{h \to 0} \frac{f(x, y + h) - f(x, y)}{h}.$$

When we write partial derivatives, we use the Greek symbol ∂, rather than d, to remind ourselves that f is a function of more than one variable, and we indicate the variables that are being held fixed by means of subscripts. In Chapter 2, for example, we find that the Gibbs Free Energy (G) of a phase is a function of temperature (T), pressure (P), and the number of moles of each of the components $(n_1, n_2, \ldots, n_i)$ in the phase. To consider how G changes as a function of T alone, we look for the value of $(\partial G/\partial T)_{P, n_1, n_2, \ldots, n_i}$.

Because there can be many interrelationships among variables that describe a system, we can write many different partial derivatives. If pressure, temperature, and volume all influence the state of a system, for example, we may be interested in $(\partial P/\partial T)_V$, $(\partial V/\partial P)_T$, $(\partial T/\partial V)_P$, or any of their reciprocals. It should be clear, though, that these are not independent of each other. The Maxwell relations in Chapter 2 can be used to investigate some of the relationships among partial derivatives of thermodynamic functions. These can also be derived more generally by learning four simple rules for working with partial derivatives. These follow from the standard Chain Rule, which states that if $w = f(x,y,z)$ and if x, y, and z are each functions of u, and v, then

$$\left(\frac{\partial w}{\partial u}\right)_v = \left(\frac{\partial w}{\partial x}\right)_{y,z}\left(\frac{\partial x}{\partial u}\right)_v + \left(\frac{\partial w}{\partial y}\right)_{x,z}\left(\frac{\partial y}{\partial u}\right)_v + \left(\frac{\partial w}{\partial z}\right)_{x,y}\left(\frac{\partial z}{\partial u}\right)_v.$$

This has many nonobvious consequences. For example, if $w = f(x, y)$ and $x = g(y, z)$, then

$$\left(\frac{\partial w}{\partial z}\right)_y = \left(\frac{\partial w}{\partial x}\right)_y\left(\frac{\partial x}{\partial z}\right)_y + \left(\frac{\partial w}{\partial y}\right)_x\left(\frac{\partial y}{\partial z}\right)_y,$$

but, $(\partial y/\partial z)_y = 0$, so

1.
$$\left(\frac{\partial w}{\partial x}\right)_y\left(\frac{\partial x}{\partial z}\right)_y = \left(\frac{\partial w}{\partial z}\right)_y.$$

It can also be shown that

2.
$$\left(\frac{\partial w}{\partial x}\right)_y = \frac{1}{(\partial x/\partial w)_y},$$

3.
$$\left(\frac{\partial x}{\partial y}\right)_z\left(\frac{\partial y}{\partial z}\right)_x\left(\frac{\partial z}{\partial x}\right)_y = -1,$$

and

4.
$$\left(\frac{\partial w}{\partial y}\right)_x = \left(\frac{\partial w}{\partial y}\right)_z + \left(\frac{\partial w}{\partial z}\right)_y\left(\frac{\partial z}{\partial y}\right)_x.$$

We commonly describe changes in some property of a system by writing *differential equations*, in which an infinitesimal change df is related to infinitesimal changes in one or more of the variables that control the property. For a function of a single variable, $f(x)$, we can write

$$df = \left(\frac{df}{dx}\right) dx .$$

If more than one variable can potentially influence the value of f, the equivalent expression is

$$df = \left(\frac{\partial f}{\partial x}\right)_{y,...} dx + \left(\frac{\partial f}{\partial y}\right)_{x,...} dy + \ldots$$

The expression on the right side of this equation is called a *total differential*, because it accounts for the total change df by summing the changes due to each independent variable separately. A familiar property that you may already consider in this way is land elevation, which varies as a function of both latitude and longitude on the earth's surface. If you were to move a short distance, the total change in elevation would be equal to the slope of the land surface in a north-south direction, times the distance you actually moved in that direction, plus the slope in an east-west direction, times the distance you moved in that direction.

In many geochemical applications, we find ourselves working with a special type of total differential, called an *exact* or *perfect* differential. The differentials of each of the energy functions (dE, dH, dF, dG) introduced in Chapter 2, for example, belong in this category. To define what we mean, suppose that we know of three properties of a system, each of which is a function of the same set of three independent variables. Identify these properties as $P(x,y,z)$, $Q(x,y,z)$, and $R(x,y,z)$. Can these three properties be related by some function $f(x,y,z)$ in such a way that $df = P\,dx + Q\,dy + R\,dz$? If so, then the expression on the right side is not only the total differential of f, it is an exact differential. We say that f is a *function of state*, with the following particularly useful characteristics:

1. Not only are P, Q, and R the partial derivatives of f with respect to x, y, and z, but their own partial derivatives (2nd derivatives of f) are also interrelated. The following *cross-partial* reciprocity expressions are true:

$$\left(\frac{\partial P}{\partial y}\right)_{x,z} = \left(\frac{\partial Q}{\partial x}\right)_{y,z} ; \left(\frac{\partial P}{\partial z}\right)_{x,y} = \left(\frac{\partial R}{\partial x}\right)_{y,z} ; \left(\frac{\partial Q}{\partial z}\right)_{x,y} = \left(\frac{\partial R}{\partial y}\right)_{x,z} .$$

2. We can determine the total change in f between two states of the system by integrating df between (x_1, y_1, z_1) and (x_2, y_2, z_2) along some reaction pathway, C. An integral of this type is known as a *line integral*. For an illustration of how this operation is performed, see Worked Problem 2–1. When we do this integration for a function of state, we discover that

$$\oint_{x_1,y_1,z_1}^{x_2,y_2,z_2} df = \oint_{x_1,y_1,z_1}^{x_2,y_2,z_2} (P\,dx + Q\,dy + R\,dz) = f(x_2, y_2, z_2) - f(x_1, y_1, z_1) .$$

Put simply, this means that we always get the same answer regardless of how we get from state 1 to state 2. The integral of df is said to be independent of path.

3. This implies that if we integrate df around a closed loop from state 1 to state 2 and back again, the net change in f will be zero.

MATRIX OPERATIONS

A *matrix* is a rectangular array of numbers. For example, these might correspond to the row and column entries in a data table or the stoichiometric coefficients in a set of linear equations. Mathematical operations which apply to matrices are important to physical scientists because they permit us to manipulate large amounts of information simultaneously and to explore fundamental relationships that characterize data sets. A familiarity with linear algebra, which includes the study of matrix operations, is essential for those who write computer programs for data analysis.

The size of a matrix is defined by specifying the number of rows (m) and columns (n) in it. In this brief overview, we will use a standard convention and refer to a matrix by labeling it with a capital letter, and will identify its *entries* with the corresponding lower-case letter. Individual entries in matrix A, therefore, are designated as elements a_{ij}, where i is the particular row and j is the column where entry a_{ij} lies. An endless variety of types of matrices are distinguished from one another on the basis of size, relative magnitudes of m and n, degree of symmetry of entries, and the type of values used as entries. If either m or n (but not both) is equal to 1, for example, we refer to A as a *vector*. If $m = n$, A is said to be a *square matrix* of order n. Several types of vectors and square matrices are important in geochemistry.

Arithmetic operations on matrices are similar in many respects to operations on scalar quantities. In particular, we see that

1. Matrices of the same size can be added by simply adding their corresponding entries, so that $A + B = C$ implies that for each element c_{ij}, $c_{ij} = a_{ij} + b_{ij}$. This operation is both commutative and associative, so $A + B = B + A$ and $A + (B + C) = (A + B) + C$.
2. A matrix can be multiplied by a scalar quantity k by scaling each of its entries accordingly. Thus $kA = C$ implies that $c_{ij} = ka_{ij}$. This operation is commutative, so $kA = Ak$. Also, scalar multiplication and matrix addition are distributive, so that $k(A + B) = kA + kB$.

   This familiar behavior does not extend to operations which multiply matrices by each other. Here, we find that
3. If A is an m by r matrix and B is an r by n matrix, the product AB, which equals C, is found by computing each entry c_{ij} from $c_{ij} = \Sigma\, a_{iq}b_{qj}$, where q ranges from 1 to r. For example, in the operation

$$\begin{pmatrix} a_{11} & a_{12} \\ a_{21} & a_{22} \\ a_{31} & a_{32} \end{pmatrix} \begin{pmatrix} b_{11} & b_{12} \\ b_{21} & b_{22} \end{pmatrix} = \begin{pmatrix} c_{11} & c_{12} \\ c_{21} & c_{22} \\ c_{31} & c_{32} \end{pmatrix},$$

the entry c_{12} is the sum $(a_{11}b_{12} + a_{12}b_{22})$, and $c_{21} = (a_{21}b_{11} + a_{22}b_{21})$. Clearly, in this case BA is not defined, because the number of rows in A is not the same as the number of columns in B. If both were square matrices, however, we might be tempted to compute the product BA. We would discover, then, that matrix multiplication is not commutative. That is, usually $BA \neq AB$. Matrix multiplication is associative, however, so that $A(BC) = (AB)C$. You should verify these statements by experimentation.

Another very common matrix operation involves "repackaging" A:

4. The *transpose* of matrix A, referred to as A^T, results from swapping rows and columns in A. Thus, if $C = A^T$, $c_{ij} = a_{ji}$.

   There is one other basic operation, applicable to square matrices, with which you should be familiar:

5. If the *inverse* of matrix A, referred to as A^{-1}, exists, it can be shown that $AA^{-1} = A^{-1}A = I$, where I is a square matrix (known as the *identity matrix*) in which all elements for which $i = j$ are equal to 1, and all others are equal to zero. This operation is particularly useful in problems which seek to solve sets of simultaneous linear equations. It can be shown that if A is an invertible n by n matrix, then for each n by 1 vector B, the system of equations $AX = B$ has exactly one solution, namely, $X = A^{-1}B$.

Virtually any computer program (like CYCLE) that solves sets of equations, therefore, is built around a matrix inversion routine. Unfortunately, finding A^{-1} typically involves performing a very large number of calculations and, for some matrices, can involve significant uncertainties due to cumulative round-off errors. Rather than discuss any of the various techniques for matrix inversion here, we suggest that you consult an elementary text in linear algebra.

It is beyond the scope of this appendix to discuss most of the specific applications of matrix algebra that are scattered through this text. We will make a brief exception, however, to discuss one fundamental property of square matrices that is central to the analysis of geochemical cycles, discussed in Chapters 1 and 4.

An *eigenvector*, Ψ, of A is a nonzero column vector that satisfies the equation $A\Psi = E\Psi$, where E is a scalar value, called an *eigenvalue*. At first glance, it might seem that a large number of vectors and scalar values could satisfy this condition, but this is not so. Consider the particular matrix

$$A = \begin{pmatrix} 3 & 0 \\ 8 & -1 \end{pmatrix},$$

and pick a vector

$$B = \begin{pmatrix} 2 \\ 3 \end{pmatrix}$$

at random. Their product, AB, is equal to

$$\begin{pmatrix} 6 \\ 13 \end{pmatrix}.$$

There is no value, E, however, for which $EB = AB$, so B is not an eigenvector of A. We can show that only two independent eigenvectors of A exist by solving the matrix equation

$$\begin{pmatrix} 3 & 0 \\ 8 & -1 \end{pmatrix}\begin{pmatrix} x \\ y \end{pmatrix} = E\begin{pmatrix} x \\ y \end{pmatrix},$$

which is equivalent to solving the simultaneous equations

$$3x = Ex$$

and

$$8x - y = Ey.$$

The first of these can be true if either x or $(3 - E)$ is equal to zero. If $x = 0$, then substitution into the second equation tells us that $E = -1$ and that y can take any value. In the other case, where $(3 - E) = 0$, we find that $E = 3$ and $2x = y$. The two eigenvalues of A, then, are $E_1 = 3$ and $E_2 = -1$, and the eigenvectors are multiples of

$$\Psi_1 = \begin{pmatrix} 0 \\ 1 \end{pmatrix}$$

and

$$\Psi_2 = \begin{pmatrix} 1 \\ 2 \end{pmatrix}.$$

You should verify these results by substituting them in the original matrix equation.

The two important observations here are that (1) any multiples of Ψ_1 or Ψ_2 (like $5\Psi_1$ or $2.8\pi\Psi_2$) are also eigenvectors of A, although they are obviously not independent of Ψ_1 and Ψ_2; and (2) there are only a very limited number of eigenvalues (in this case, two). The values of E can be thought of as the fundamental harmonic modes of the matrix. In fact, where the elements in a matrix A represent the terms in a set of wave equations (for oscillating springs, or interatomic bonds), the set of eigenvalues is referred to as the *spectrum* of the matrix. In Chapters 1 and 4, we have discussed Lasaga's use of eigenvector analysis to determine the response times of reservoirs in an "oscillating" geochemical system, where A is a matrix of time constants in differential equations describing those reservoirs.

Elementary techniques for finding the eigenvectors and eigenvalues of a matrix are discussed in any linear algebra text, although a dismally small number of them discuss their application to problems you may recognize as useful. In the context of this book, you may want to read the discussion of molecular vibration modes in Dence (1975, pages 293-303), which is particularly clear. As with matrix inversion, most problems of this type today are solved numerically by computer. A large number of programs for this purpose are available.

ROOT FINDING

Matrix methods are particularly useful for finding the roots of sets of first-order equations. It is very common, however, to encounter problems in the physical sciences which yield equations in which a key variable x is raised to a higher power or appears in a transcendental function like $\sin x$. A familiar equation, from which you learned to extract x when you were in high school, is the polynomial

$$0 = ax^2 + bx + c .$$

In this example, x can be determined by applying the quadratic formula

$$x = \frac{-b \pm \sqrt{b^2 - 4ac}}{2a} .$$

Unfortunately, there are no simple solutions of this type for most equations you will run across. By adding a cubic term to the polynomial above, for example, we would make the task of root-finding considerably more difficult. There are many ways around this problem, most of which are well-suited to analysis by computer. We will briefly discuss one, the Newton-Raphson method, which is mentioned as a means for solving several problems in Chapters 3, 4, and 12.

Consider the curve for the function $y = f(x)$ in Figure A.1. Suppose that $f(x)$ has a real root at $x = a_0$ that we want to find. One way to do this would be simply to guess random successive values of x until we find the one where $f(x) = 0$. By examining Figure A.1, though, we can see a way to make guesses in an educated way. Suppose that our first guess is that $x = a_1$, reasonably close to a_0 but not correct. The error in this guess is $|a_1 - a_0|$, shown by the distance AC. Assuming that this error is unacceptably large, we now make a better guess by drawing the tangent BP to the curve at point P. For $x = a_2$, the error is only $|a_2 - a_0|$, or the distance AB. We repeat the process, each time estimating the improved value of a_{i+1} by drawing a tangent to $y = f(x)$ at the point corresponding to a_i, until the error $|a_i - a_0|$ is acceptably small.

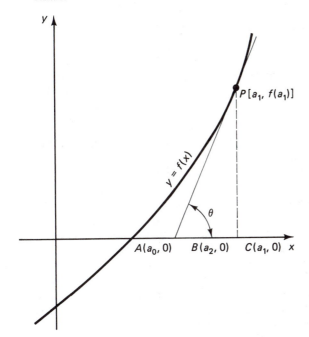

Figure A.1 Geometrical basis for the Newton-Raphson method for finding a root of $f(x)$. BP is the tangent to $f(x)$ at point P.

By comparing the first two guesses geometrically, we see that

$$AB = AC - BC$$

or

$$(a_2 - a_0) = (a_1 - a_0) - PC/\tan\theta$$

from which $a_2 = a_1 - f(a_1)/f'(a_1)$, where $f'(a_1)$ is the instantaneous slope of $f(x)$ at a_1. This simple rule gives us a powerful means for making consecutive guesses, provided that (1) $f(x)$ has a derivative in the vicinity of a_0, (2) a_1 is reasonably close to a_0, so the method doesn't converge on some other root, and (3) $f'(a_{i+1})$ is not very close to zero, so that we avoid introducing a large numerical error in a_{i+1}.

Most texts that discuss the Newton-Raphson method fail to point out how easily it can be applied to *systems* of equations. Because geochemical problems commonly involve functions of more than one variable, you may find it useful to know how this is done. Suppose that several variables $(x_1, x_2, \ldots, x_n)$, related by a set of n independent equations, describe a system. To make life frustrating, each of the equations is a complicated function in which the variables appear in several terms, so that finding the values of x_1 through x_n by substitution or simple matrix methods is impossible. If each of the functions is differentiable with respect to each of the variables, however, there is hope for use of the Newton-Raphson method.

As in the simple case above, begin by guessing initial values for x_1 through x_n. Using these, calculate the numerical values of each of the functions $(f_1, f_2, \ldots, f_n)$ and place them in a column vector F. Also, calculate each of the partial derivatives $(\partial f/\partial x)$ and build a matrix J that looks like this:

$$
J = \begin{pmatrix}
(\partial f_1/\partial x_1) & (\partial f_1/\partial x_2) & \cdots & (\partial f_1/\partial x_n) \\
(\partial f_2/\partial x_1) & (\partial f_2/\partial x_2) & \cdots & (\partial f_2/\partial x_n) \\
\vdots & \vdots & \cdots & \cdots \\
(\partial f_n/\partial x_1) & (\partial f_n/\partial x_2) & \cdots & (\partial f_n/\partial x_n)
\end{pmatrix}
$$

The column vector $X = (x_1, x_2, \ldots, x_n)$ can then be found by successive approximations from the matrix version of the Newton-Raphson formula, which we present without proof:

$$
X_{k+1} = X_k - (J_k)^{-1} F_k .
$$

This is the basis for the algorithm that computes binary mole fractions in the program ASYM.BAS on the program diskette.

FITTING A FUNCTION TO DATA

In a typical research problem, a geochemist gathers observations on a natural or experimental system and then tries to make sense of them by looking for functional relationships between the data and system variables which control them. Sometimes the function that best describes the data has a form that is based on theoretical considerations. At other times, you may choose an empirical function (a polynomial or exponential equation, for example). In any case, the task of fitting that function to your data set involves finding the values of one or more coefficients in the equation, so that it predicts results that are as close to the observed data as possible.

We will demonstrate a common approach to the problem by considering the case in which measured values of some property y appear to depend on some other property x in such a way that a graph of y against x is a straight line. That is, $y = f(x)$ has the form

$$
y = a_0 + a_1 x .
$$

What values of a_0 and a_1 are most appropriate for your data? If all of the observed values of y lay precisely on a line, the answer would be easy to find. Unfortunately, however, there are always random errors in any data set, due to sampling technique or to undiagnosed complexities in the system. (See Figure A.2.) The task, then, is to find values of a_0 and a_1 such that the distance between $f(x)$ and each of the measured values of y is as small as possible. In practice, we generally go one step farther and look for a_0 and a_1 such that the sum of the squares of the deviations from $f(x)$ is minimized. This overall approach, therefore, is known as the *method of least squares*.

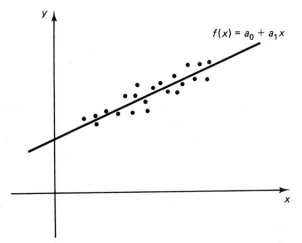

Figure A.2 The measured values of property x, in this example, are apparently related to values of property y by some function $y = f(x)$. Because of random errors in the data set, however, the observed values are scattered around the most probable linear function.

Mathematically, the problem involves minimizing the function

$$Q(a_0, a_1) = \sum_i (y_i - a_0 - a_1 x_i)^2$$

or, if we have some reason to trust some observations more than others,

$$Q(a_0, a_1) = \sum_i w_i(y_i - a_0 - a_1 x_i)^2 \, ,$$

where w_i is a weighting factor for observation i. The summations are each taken over all values of i from 1 to N, the total number of observations. To minimize Q, we need to find the two partial derivatives of Q with respect to a_0 and a_1, and set each equal to zero:

$$(\partial Q / \partial a_0)_{a_1} = 2\left(\sum w_i y_i - a_0 \sum w_i - a_1 \sum w_i x_i \right) = 0$$

and

$$(\partial Q / \partial a_1)_a = 2\left(\sum w_i x_i y_i - a_0 \sum w_i x_i - a_1 \sum w_i x_i^2 \right) = 0.$$

These can be solved simultaneously to yield

$$a_0 = \frac{\left(\sum w_i x_i^2 \right)\left(\sum w_i y_i \right) - \left(\sum w_i x_i \right)\left(\sum w_i x_i y_i \right)}{\left(\sum w_i \right)\left(\sum w_i x_i^2 \right) - \left(\sum w_i x_i \right)^2}$$

$$a_1 = \frac{\left(\sum w_i \right)\left(\sum w_i x_i y_i \right) - \left(\sum w_i x_i \right)\left(\sum w_i y_i \right)}{\left(\sum w_i \right)\left(\sum w_i x_i^2 \right) - \left(\sum w_i x_i \right)^2} \, .$$

The same procedure can be followed for any polynomial expression.

Fitting a Function to Data

THE ERROR FUNCTION

You are probably familiar, but perhaps only in a loose way, with the function

$$f(x) = \frac{1}{\sqrt{2\pi}} e^{-t^2/2},$$

where $t = (x - \bar{x})/\sigma$. This is the standard form of the *normal frequency function*, which describes the frequency distribution of events (x) around a mean, $\bar{x}$. The quantity σ is the *standard deviation* of x, and is a measure of the spread of values around the mean. This is the "curve" that has been used to calculate the grade distribution in classes since you were in grade school, and describes the distributions of a very large number of other events in nature. In this text, we have used it indirectly as a means of finding the distribution of mobile ions diffusing across a boundary between adjacent phases.

It can be shown that the area under this curve between zero and some penetration distance x is equal to the probability that an ion lies within that range. This, in fact, is the context in which the normal distribution appears in Chapter 5. The *error function*, $erf(x)$, can be defined as

$$erf(x) = \frac{1}{\sqrt{2\pi}} \int_0^x e^{-t^2/2} dt$$

in order to calculate this probability. This form of the error function is commonly used by statisticians. Unfortunately, several other forms are also used by physical scientists. The definition most common in discussions of transport equations is the one we have applied in Chapter 5:

$$erf(x) = \frac{2}{\sqrt{\pi}} \int_0^x e^{-t^2} dt.$$

It is a simple matter to relate one form of the error function to another by making an appropriate substitution for t and adjusting the integration limits accordingly. Because several subtly different versions exist, you should always check to see how $erf(x)$ is defined for the particular application you have in mind.

Unfortunately, the integration of e^{-t^2} cannot be performed analytically (that is, in neat, closed form). Because $erf(x)$ is a widely used function, however, there are many popular ways to evaluate it. The easiest is simply to look it up in a table. The error function and its complement, $erfc(x) = 1 - erf(x)$, are included in most standard volumes of mathematical functions (for example, Burington, 1973). There are also a large number of approximate solutions that are reliable for calculations with pencil and paper. One which yields less than 0.7% error is the function $y = \sqrt{[1 - \exp(-4x^2/\pi)]}/2$. In most applications today, however, $erf(x)$ is calculated by numerical integration, as we have done in the program SEMIINF.

REFERENCES

BOAS, M.L. 1983. *Mathematical methods in the physical sciences*. 2nd ed. New York: John Wiley and Sons.

BURINGTON, R.S. 1973. *Handbook of mathematical tables and formulas*. 5th ed. New York: McGraw-Hill.

DENCE, J.B. 1975. *Mathematical techniques in chemistry*. New York: John Wiley and Sons.

POTTER, M.C. 1978. *Mathematical methods in the physical sciences*. Englewood Cliffs, N.J.: Prentice-Hall.

Appendix B
Finding and Evaluating Geochemical Data

SELECTED DATA SOURCES

The data of geochemical interest are scattered through a large number of journals and research reports which focus on the results of narrowly defined studies. In most cases, a search for data on specific systems should begin with the major abstracting journals, *Chemical Abstracts* and *Mineralogical Abstracts,* which regularly review the major journals and provide a topical summary of their contents. Limited literature surveys, such as the excellent summary of thermodynamic data sources in Nordstrom and Munoz (1985), are also useful guides to the primary literature. Rather than provide another survey of this type, we have compiled in this appendix a short list of general references that are widely used by geochemists. In almost all cases, these are tabulations, or "data bases," built of information gathered from the primary research literature. They offer the advantage of being a quick overview of values that might otherwise be hard to find except in obscure corners of the library. More importantly, a large number of these (but not all!) are critical evaluations of the data, checked for internal consistency.

General Reference

DeBievre, P.; Gallet, M.; Holden, N.E.; and Barnes, I.L. 1984. Isotopic abundances and atomic weights of the elements. *J. Phys. Chem. Ref. Data* 13: 809–891.

Greenwood, N.N., and Earnshaw, A. 1984. *Chemistry of the elements.* Oxford: Pergamon Press.

Ronov, A.B., and Yaroshevsky, A.A. 1969. Chemical composition of the earth's crust. In *The earth's crust and upper mantle,* ed. Hart, P.J. 35–57. Monograph 13. Washington, D.C.: Amer. Geophys. Union.

Wedepohl, K.H., ed. 1969. *Handbook of geochemistry.* 2 Vols. New York: Springer-Verlag.

Elements and Inorganic Compounds

CODATA 1978. CODATA recommended key values for thermodynamics 1977. *CODATA Bull.* 28.

———. 1987. 1986 adjustment of the fundamental physical constants. *CODATA Bull.* 63.

HULTGREN, R.; DESAI, P.D.; HAWKINS, D.T.; GLEISER, M.; KELLY, K.K.; and WAGMAN, D.D. 1973. *Selected values of the thermodynamic properties of binary alloys,* Metals Park, Ohio: Amer. Soc. for Metals.

MERRILL, L. 1982. Behavior of the AB_2-type compounds at high pressures and high temperatures. *J. Phys. Chem. Ref. Data* 11: 1005–64.

NAUMOV, G.B.; RYZHENKO, B.N.; and KHODAKOVSKII, I.L. 1974. *Handbook of thermodynamic data.* NTIS Doc. Pb-226, 722/7GA. Washington, D.C.: U.S. Dept. of Commerce.

PARKER, V.B.; WAGMAN, D.D.; and EVANS, W.H. 1971. *Selected values of chemical thermodynamic properties: Tables for the alkaline earth elements (elements 92 through 97 in the standard order of arrangement).* U.S. Nat. Bur. Standards Tech. Note 270–6. Washington, D.C.

SCHUMM, R.H.; WAGMAN, D.D.; BAILEY, S.; EVANS, W.H.; and PARKER, V.B. 1973. *Selected values of chemical thermodynamic properties: Tables for the lanthanide (rare earth) elements (elements 62 through 76 in the standard order of arrangement).* U.S. Nat. Bur. Standards Tech. Note 270–7. Washington, D.C.

STULL, D.R., and PROPHET, H. 1971. *JANAF Thermochemical Tables.* U.S. Nat. Bur. Standards, NSRDS-NBS 37. Washington, D.C.: U.S. Dept. of Commerce.

(Supplement 1974 by CHASE, M.W.; CURNUTT, J.L.; HU, A.T.; PROPHET, H.; and WALKER, L.C.. *J. Chem. Phys. Ref. Data* 3: 311–480.)

(Supplement 1975 by CHASE, M.W.; CURNUTT, J.L.; PROPHET, H.; MCDONALD, R.A.; and SYVERUD, A.N.. *J. Phys. Chem. Ref. Data* 4: 1–175.)

(Supplement 1978 by CHASE, M.W., JR.; CURNUTT, J.L.; MCDONALD, R.A.; and SYVERUD, A.N.. *J. Phys. Chem. Ref. Data* 7: 793–940.)

(Supplement 1982 by CHASE, M.W., JR.; CURNUTT, J.L.; DOWNEY, J.R., JR.; MCDONALD, R.A.; SYVERUD, A.N.; and VALENZUELA, E.A.. *J. Phys. Chem. Ref. Data* 11: 695–940.)

WAGMAN, D.D.; EVANS, W.H.; PARKER, V.B.; HALOW, I.; BAILEY, S.M.; and SCHUMM, R.H. 1968. *Selected values of chemical thermodynamic properties: Tables for the first thirty-four elements in the standard order of arrangement.* U.S. Nat. Bur. Standards Tech. Note 270–3. Washington, D.C.

———. 1969. *Selected values of chemical thermodynamic properties: Tables for elements 35 through 53 in the standard order of arrangement.* U.S. Nat. Bur. Standards Tech. Note 270–4. Washington, D.C.

———; EVANS, W.H.; PARKER, V.B.; HALOW, I. BAILEY, S.M.; SCHUMM, R.H.; and CHURNEY, K.L.; 1971. *Selected values of chemical thermodynamic properties: Tables for elements 54 through 61 in the standard order of arrangement.* U.S. Nat. Bur. Standards Tech. Note 270–5. Washington, D.C.

———; EVANS, W.H.; PARKER, V.B.; and SCHUMM, R.H. 1976. *Chemical thermodynamic properties of sodium, potassium, and rubidium: An interim tabulation of selected material.* U.S. Nat. Bur. Standards Interim Rept. NBSIR 76-1034. Washington, D.C.

———; EVANS, W.H.; PARKER, V.B.; SCHUMM, R.H.; and NUTTALL, R.L. 1981. *Selected values of chemical thermodynamic properties: Compounds of uranium, protactinium, thorium, actinium, and the alkaline metals.* U.S. Nat. Bur. Standards Tech. Note 270–8. Washington, D.C.

———; EVANS, W.H.; PARKER, V.B.; SCHUMM, R.H.; HALOW, I.; BAILEY, S.M.; CHURNEY, K.L.; and NUTTALL, R.L. 1982. *The NBS tables of chemical thermodynamic properties: Selected values for inorganic and C_1 and C_2 organic substances in SI units.* J. Phys. Chem. Ref. Data 11 (Suppl. 2): 1–392.

Minerals

HAAS, J.L., JR.; ROBINSON, G.R.; and HEMINGWAY, B.S. 1981. Thermodynamic tabulations for selected phases in the system $CaO-Al_2O_3-SiO_2-H_2O$ at 101.325 kPa (1 atm) between 273.15 and 1800K. *J. Phys. Chem. Ref. Data* 10: 575–669.

HELGESON, H.C.; DELANY, J.M.; NESBITT, H.W.; and BIRD, D.K. 1978. Summary and critique of the thermodynamic properties of rock-forming minerals. *Amer. J. Sci.* 278–A: 1–229.

ROBIE, R.A.; HEMINGWAY, B.S.; and FISHER, J.R. 1978. Thermodynamic properties of minerals and related substances at 298.15K and one atmosphere pressure and at higher temperatures. *U.S. Geol. Survey Bull.* 1259.

ROBINSON, G.R.; HAAS, J.L.; JR., SCHAFER, C.M.; AND HAZELTON, H.T.; JR. 1982. *Thermodynamic and thermophysical properties of selected phases in the MgO-SiO_2-H_2O-CO_2, CaO-Al_2O_3-SiO_2-H_2O-CO_2, and Fe-FeO-Fe_2O_3-SiO_2 chemical systems, with special emphasis on the properties of basalts and their mineral components.* U.S. Geol. Survey Open-File Rept. 83–79, 429 pp.

WOODS, T.L., and GARRELS, R.M. 1987. *Thermodynamic values at low temperature for natural inorganic materials.* New York: Oxford Univ. Press.

See also Helgeson (1982, 1985) under *Aqueous Species,* the next classification of references in this section.

Aqueous Species

BURNHAM, C.W.; HOLLOWAY, J.R.; and DAVIS, N.F. 1969. *Thermodynamic properties of water to 1000°C and 10,000 bars*. Special Paper 132. Boulder, Colo.: Geol. Soc. Amer.

CRISS, C.M., and COBBLE, J.W. 1964. The thermodynamic properties of high temperature aqueous solutions. IV. Entropies of the ions up to 200° and the correspondence principle. *J. Amer. Chem. Soc.* 86: 5385–90.

———. 1964. The thermodynamic properties of high temperature aqueous solutions. V. The calculation of ionic heat capacities up to 200°. Entropies and heat capacities above 200°. *J. Amer. Chem. Soc.* 86: 5390–93.

HAMER, W.J. 1968. *Theoretical mean activity coefficients of strong electrolytes in aqueous solutions from 0 to 100°C.*. U.S. Nat. Bur. Standards, NSRDS–NBS 24. Washington, D.C.: U.S. Dept. of Commerce.

HELGESON, H.C. 1967. Thermodynamics of complex dissociation in aqueous solution at elevated temperatures. *J. Phys. Chem.* 71: 3121–36.

———. 1969. Thermodynamics of hydrothermal systems at elevated temperatures and pressure. *Amer. J. Sci.* 267: 729–804.

———. 1982. Errata: Thermodynamics of minerals, reactions, and aqueous solutions at high temperatures and pressures. *Amer. J. Sci.* 282: 1144–49.

———. 1985. Errata II: Thermodynamics of minerals, reactions, and aqueous solutions at high pressures and temperatures. *Amer. J. Sci.* 285: 845–55.

——— and KIRKHAM, D.H. 1974. Theoretical prediction of the thermodynamic behavior of aqueous electrolytes at high pressures and temperatures: I. Summary of the thermodynamic/electrostatic properties of the solvent. *Amer. J. Sci.* 274: 1089–1198.

———. 1974. Theoretical prediction of the thermodynamic behavior of aqueous electrolytes at high pressures and temperatures. II. Debye–Hückel parameters for activity coefficients and relative partial molal properties. *Amer. J. Sci.* 274: 1199–1261.

———. 1976. Theoretical prediction of the thermodynamic behavior of aqueous electrolytes at high pressures and temperatures. III. Equation of state for aqueous species at infinite dilution. *Amer. J. Sci.* 276, 97–240.

——— and FLOWERS, G.C. 1982. Theoretical prediction of the thermodynamic behavior of aqueous electrolytes at high pressures and temperatures. IV. Calculation of activity coefficients, osmotic coefficients, and partial molal and standard and relative partial molal properties to 600°C and 5 kb. *Amer. J. Sci.* 281: 1249–1516.

HOGFELDT, E. (1982) *Stability constants of metal ion complexes: Part A. inorganic ligands*. IUPAC Chemical Data Series No. 21. Oxford: Pergamon Press.

HORNE, R.A., ED. 1972. *Water and aqueous solutions: Structure, thermodynamics, and transport properties*. New York: Wiley Interscience.

SILLÉN, L.G., and MARTELL, A.E. 1964. *Stability constants of metal-ion complexes*. Spec. Pub. No. 17. London: The Chemical Society. U.K. (Supplement 1971.)

Reaction Kinetics

COHEN, N., and WESTBERG, K.R. 1983. Chemical kinetic data sheets for high-temperature chemical reactions. *J. Phys. Chem. Ref. Data* 12: 531–590.

Phase Relations

EHLERS, E. G. 1972. *The interpretation of geological phase diagrams*. San Francisco: Freeman.

LEVIN, E.M.; ROBBINS, C.R.; and MCMURDIE, H.F. 1964. *Phase diagrams for ceramists*. Vol. 1. Columbus, Ohio: Amer. Ceram. Soc.

USING THE DATA

Before you use the data in any tabulation, be sure to read the text material that accompanies it, describing the criteria for data selection, the reactions that were investigated experimentally to obtain the data, the choice of standard and reference states, and the methods that were used to calculate interpolated, or "derived," values. Because data are gathered for many purposes, these criteria may not be the same from one source to another. Also, tests for the internal consistency of data compilations may be applied in different ways by various teams of compilers. You should not assume, therefore, that you can safely mix values garnered from different sources in a single problem. For a very good discussion of the potential problems of dealing with multiple data sources, consult Chapter 12 of Nordstrom and Munoz (1985).

Despite these problems, it is usually advisable and often necessary to consult several sources as you compile data for use in your calculations. You will find that teams of researchers have made idiosyncratic choices that emphasize different types of information for the same substances. The JANAF tables (Stull and Prophet, 1971 et seq.), for example, indicate the temperatures at which phase changes take place in the substance being reported, but do not indicate phase changes in the reference state elements used to determine $\Delta \overline{H}^0_f$ and $\Delta \overline{G}^0_f$. Robie et al. (1978) report both. In addition, Robie and co-workers provide pairs of tables for silicate minerals which report free energies and enthalpies of formation from both the elements and from the oxides. The JANAF tables, on the other hand, include a short summary of the data sources consulted during compilation, and in most cases a critical justification for the particular choices that were made in preparing their tables. This is a useful feature rarely found in other compendia. Helgeson et al. (1978) carry this approach to an extreme by producing a document in which evaluation of the data is the major concern. The data tables are only a few pages of the text and are in a compressed form. It is left to the reader, for example, to calculate thermodynamic values above 298 K by laboriously integrating heat capacity power functions.

Most of the data in standard tabulations should be familiar to students who have read Chapters 2, 3, and 8 of this text. The lone exception is the *free energy function,* defined as $(\Delta G^0 - \Delta H^0_{298})/T$. Values of this function do not change very much with temperature, so that it is usually very safe to interpolate linearly to find values between tabulated temperatures. For reactions among various substances, then, it can be shown that

$$\left(\frac{\Delta G^0 - \Delta H^0_{298}}{T} \right)_r = \frac{\Delta G^0_r}{T} - \frac{\Delta H^0_{r,298}}{T}$$

so

$$\Delta G^0_{r,T} = T \left(\frac{\Delta G^0 - \Delta H^0_{298}}{T} \right)_r + \Delta H^0_{r,298}$$

where we have followed our convention that quantities with the subscript r are the stoichiometric sum of values for the product phases minus the reactant phases. When the free energy function is available, it serves as a labor-saving alternative to calculating $\Delta G^0_{r,T}$ from heat capacity data and reference state values for enthalpy and entropy.

For those times when it is preferable to use heat capacity data, it is convenient to have them in the form of a power law. We have written a program called FITS.EXE to calculate coefficients for the expression.

$$C_p = R(a + bT + cT^2 + dT^3 + eT^4),$$

in which R (= 1.98726 cal mole^{-1} K^{-1}) is the gas constant. The values are determined by a technique known as a cubic spline fit to tabulated values which you provide. The program also performs goodness-of-fit statistics by comparing heat capacity, entropy, and enthalpy values calculated from the power law with those which you provide. Because it is a large program, we have not been able to include FITS.EXE on the diskette available for this book. Interested readers can obtain a copy from the first author, however. For further information, run the WELCOME program on the diskette.

Appendix C
Numerical Values
of Geochemical Interest

Dimensions of the earth

Mass of the earth . . . . 5.973×10^{24} kg
Mass of the atmosphere . . . . 5.1×10^{18} kg
Mass of the oceans . . . . 1.4×10^{21} kg
Mass of the crust . . . . 2.6×10^{22} kg
Mass of the mantle . . . . 4.0×10^{24} kg
Mass of the outer core . . . . 1.85×10^{24} kg
Mass of the inner core . . . . 9.7×10^{22} kg
Equatorial radius . . . . 6.378139×10^{6} m
Polar radius . . . . 6.35675×10^{6} m
Surface area of the earth . . . . 5.1×10^{14} m^2
Surface area of the oceans . . . . 3.62×10^{14} m^2
Surface area of the continents . . . . 1.48×10^{14} m^2
Volume of the oceans . . . . 1.37×10^{21} l
Mean depth of the oceans . . . . 3.8×10^{3} m
Continental runoff rate . . . . 3.6×10^{16} l yr^{-1}

Physical constants

Avogadro's number . . . . $N = 6.022094 \times 10^{23}$ mol^{-1}
Gas constant . . . . $R = 1.98717$ cal mol^{-1} K^{-1}
$\qquad = 8.31433$ J mol^{-1} K^{-1}
$\qquad = 82.06$ cm^3 atm mol^{-1} K^{-1}
Faraday's constant . . . . $F = 96,487.0$ coulomb equiv^{-1}
$\qquad = 23,060.9$ cal volt^{-1} equiv^{-1}
Gravitational constant . . . . $G = 6.6732 \times 10^{-11}$ m^3 kg^{-1} sec^{-2}
$\qquad = 6.6732 \times 10^{-11}$ nt m^2 kg^{-2}
Boltzmann's constant . . . . $k = 1.380622 \times 10^{-23}$ J K^{-1}
Planck's constant . . . . $h = 6.626176 \times 10^{-34}$ J sec
Base of natural logarithms . . . . $e = 2.71828$

Conversion factors

Distance . . . . 1 centimeter (cm) $= 10^8$ Ångström (Å)

$= 0.3937$ inch

Time . . . . 1 year $= 3.154 \times 10^7$ sec

Mass . . . . 1 gram (g) $= 2.20462 \times 10^{-3}$ lb

Temperature . . . . Kelvins (K) $= °C + 273.15$

$= 5(°F - 32)/9 + 273.15$

Pressure (mass length^{-1} time^{-2}) 1 atm $= 1.013250 \times 10^6$ dyne cm^{-2}

$= 1.013250$ bar

$= 1.013250 \times 10^5$ pascal (Pa)

$= 1.013250 \times 10^5$ nt m^2

$= 14.696$ lb in^2

Energy (mass length^2 time^{-2})

1 Joule (J) $= 10^7$ erg

$= 2.389 \times 10^{-1}$ cal

$= 9.868 \times 10^{-3}$ liter atm

$= 6.242 \times 10^{18}$ eV

$= 2.778 \times 10^{-7}$ kw-hr

$= 9.482 \times 10^{-4}$ Btu

Entropy (mass length^2 time^{-2} deg^{-1})

1 Gibbs $= 1$ cal K^{-1}

Viscosity (mass length^{-1} time^{-1})

1 poise $= 1$ g sec^{-1} cm^{-1}

$= 1$ dyne sec cm^{-1}

$= 0.1$ Pa sec

Radioactive Decay . . . . 1 Curie (Ci) $= 3.7 \times 10^{10}$ sec^{-1}

Logarithmic values . . . . ln $x = 2.303$ log x

Units of concentration in solutions

Molarity (M) = moles liter^{-1} of solution

Normality (N) = equiv. liter^{-1} of solution

Molality (m) = moles kg^{-1} of H_2O*

Parts per million (ppm) = g/10^6 g (= mg/kg)

*Concentrations are commonly reported in the geochemical literature as millimoles kg^{-1} of *solution*. To convert this measure to molal units requires that the density of the solution be known. Except in highly concentrated brines, however, little error is introduced if this difference is ignored.

Periodic table of the elements

1 H 1.00797																	2 He 4.0026
3 Li 6.939	4 Be 9.0122											5 B 10.811	6 C 12.011	7 N 14.0067	8 O 15.9994	9 F 18.998	10 Ne 20.183
11 Na 22.990	12 Mg 24.312											13 Al 26.982	14 Si 28.086	15 P 30.974	16 S 32.064	17 Cl 35.453	18 Ar 39.948
19 K 39.102	20 Ca 40.08	21 Sc 44.956	22 Ti 47.90	23 V 50.942	24 Cr 51.996	25 Mn 54.938	26 Fe 55.847	27 Co 58.933	28 Ni 58.71	29 Cu 63.54	30 Zn 65.37	31 Ga 69.72	32 Ge 72.59	33 As 74.922	34 Se 78.96	35 Br 79.909	36 Kr 83.80
37 Rb 85.47	38 Sr 87.63	39 Y 88.905	40 Zr 91.22	41 Nb 92.906	42 Mo 95.94	43 Tc (99)	44 Ru 101.07	45 Rh 102.91	46 Pd 106.4	47 Ag 107.870	48 Cd 112.40	49 In 114.82	50 Sn 118.69	51 Sb 121.75	52 Te 127.60	53 I 126.90	54 Xe 131.30
55 Cs 132.91	56 Ba 137.34	*	72 Hf 178.49	73 Ta 180.95	47 W 183.85	75 Re 186.2	76 Os 190.2	77 Ir 192.2	78 Pt 195.09	79 Au 196.97	80 Hg 200.59	81 Tl 204.37	82 Pb 207.19	83 Bi 208.98	84 Po (210)	85 At (210)	86 Rn (222)
87 Fr (223)	88 Ra (226)	**															

*Lanthanide Series (Rare Earths)	57 La 138.91	58 Ce 140.12	59 Pr 140.91	60 Nd 144.24	61 Pm (147)	62 Sm 150.35	63 Eu 151.96	64 Gd 157.25	65 Tb 158.92	66 Dy 162.50	67 Ho 164.93	68 Er 167.26	69 Tm 168.93	70 Yb 173.04	71 Lu 174.97
**Actinide Series	89 Ac (227)	90 Th 232.04	91 Pa (231)	92 U 238.03	93 Np (237)	94 Pu (244)	95 Am (243)	96 Cm (247)	97 Bk (247)	98 Cf (251)	99 Es (254)	100 Fm (253)	101 Md (256)	102 No (254)	103 Lw (257)

Appendix D
Computers and Geochemistry

PROGRAMS ACCOMPANYING THIS TEXT

Our Goal

Most of the topics to which you have been introduced in this book were first explored well before modern computers became available. Furthermore, it will probably always be true that much of the practical work of geochemistry can be performed with a pad of paper and a sharp pencil. On the other hand, the advent of inexpensive desktop computers has relieved geochemists from the drudgery of repetitive calculations and has made it possible to consider in a routine way complex chemical systems and large compendia of data that would have been discouraging to deal with a decade or two ago. Although it would be incorrect to claim that geochemists *must* be familar with how to use computers, it is probably fair to claim that unfamiliarity has become a professional handicap. For this reason, we assume that you have easy access to a personal computer, and we encourage you to consider using it to solve occasional problems in this text.

The diskette available for this text can be read by any IBM®-compatible machine with a single low-density $5\frac{1}{4}''$ disk drive.[1] It contains programs which should allow you to consider, even at the introductory level of this book, some cumbersome problems that have routinely been left out of geochemistry texts in the past.

Our goal in providing this diskette is twofold. First, we want to provide a set of tested routines that you can apply directly to geochemical problems in the classroom and in research. Students who have had little or no previous experience with writing programs should find these simple to use without instructions.

[1] It is difficult to anticipate what type of computer will be available for you. Our guess is that you have an MS-DOS machine of some kind. If not, you may obtain source listings directly from the authors and enter them from your keyboard. We have tried as much as possible to avoid using syntax that is peculiar to IBM®-compatible systems, so the programs should run with few necessary changes. We apologize for the inconvenience this may cause for owners of Apple® or other non-compatible systems.

1. Programs with the extension .EXE will run if you simply enter the program name when you see the system prompt. When you see $A>$, for example, type WEL-COME to run the program WELCOME.EXE.

2. BASIC programs (.BAS) run through your system's BASIC interpreter (BA-SICA, GWBASIC, or equivalent) or compiler (QuickBasic or equivalent). Consult the user's manual for your BASIC language program if you are not sure how to do this.

3. FORTRAN programs (.FOR) will need to be compiled with a FORTRAN compiler if you want to run them. This should not be necessary unless you want to make changes to one of the FORTRAN programs, because we have included .EXE versions of each on the diskette.

Each program begins by displaying a short paragraph on the computer screen to describe its purpose and the type of data it will expect. The program will then ask a series of leading questions to find out which conditions you want to specify for the problem at hand. The results of the computation will be displayed on the screen or, in some cases, as printed output.

Our greater purpose in providing this diskette is to show by example how easy it is to write programs for use in geochemistry. A few programs have been provided in compiled form (as .EXE files) in order to speed them up or because you might not have the appropriate compiler on your machine. Most of these, however, are also included in their uncompiled (source) version. All programs were written in generic versions of BASIC or FORTRAN and are provided as alphanumeric (ASCII) files. This makes it easy for you to examine a source code by using the DOS TYPE or PRINT command, by reading it with your word processor, or (for BASIC programs) by loading it with your BASIC interpreter. You will find that we have used comment lines liberally to clarify program logic, and that we have occasionally included sample data files (with .DAT file extensions) or documentation files (with .DOC extensions) which should help you see how the longer programs are intended to work. In general, the longer a program is, the more effort we have expended to design a sophisticated algorithm. For short problems, we have tended to use "brute force," rather than trying to optimize a program's speed. We expect you to experiment with these programs and to "improve" them for your own purposes. In short, we hope that you use this diskette to learn how to use your computer as a tool for geochemistry.

An Appeal to Common Sense

Important! Before you begin using your diskette, we urge you to heed two warnings:

1. This diskette is not copy-protected, and can easily be damaged by mishandling or accidental erasure. Even if neither of these happens to you, it would be very easy to modify one of these programs beyond recognition while you experiment with it and to not be able to recover the original version. Neither the authors nor Prentice-Hall, Inc. can afford to replace diskettes damaged by the owner. Therefore, PLEASE make a backup copy of the diskette to ensure against disaster.

2. We have tested these programs on several machines and with a reasonable range of input values. We do not claim, however, that they are flawless, or that the results they produce will necessarily be reliable in all applications. The diskette is provided, in other words, in "as is" condition. It is your responsibility to verify

that the results of any computation with these programs are credible before applying them to your own research. Also, neither the authors nor Prentice-Hall, Inc. is liable in any way for damage to computer hardware or software that may arise from use of these programs.

The Programs

ASYM.BAS   This BASIC program uses an asymmetric solution model to calculate temperature, excess mixing parameters, or the compositions of coexisting phases on a binary solvus.

CYCLE.EXE   A compiled FORTRAN program, based on the algorithm of Lasaga (1980), to examine the time-dependent behavior of a multireservoir geochemical system. Users should examine the companion files CYCLE.DOC and CARBON.DAT for detailed instructions and a sample data file. The FORTRAN source file CYCLE.FOR, its subroutines, and its eigenvalue library routines are too large to fit on your program diskette, but may be obtained from the first author on request. Run the program WELCOME.EXE for further information.

DELTAG.BAS   A simple BASIC program to calculate the free energy for a specified reaction among pure, ideal phases as a function of temperature and total pressure. Water or carbon dioxide (as ideal gases) may be included in the reaction.

FETIOX.EXE   A compiled BASIC program based on an algorithm by Lindsley (1977) for calculating temperature and oxygen fugacity from the compositions of coexisting iron-titanium oxides. The source code, FETIOX.BAS, is also included. It was written for the Microsoft® QuickBasic compiler, however, and cannot be run with a BASIC interpreter.

LOOP.BAS   A BASIC program to calculate the liquidus and solidus curves for a binary system from thermodynamic data.

PBPB.BAS   A simple BASIC program for calculating the age of a sample based on its ratio of radiogenic $^{207}Pb/^{206}Pb$.

PLAG.BAS   A BASIC program to demonstrate the construction of liquidus and solidus curves for plagioclase feldspar.

PT.EXE   A compiled FORTRAN program to calculate temperatures and pressures, using selected silicate geothermometers and geobarometers. Users should examine the companion files PT.DOC and PT.DAT for further information and a sample data set. The FORTRAN source code, PT.FOR, is also included. The original version of this program was written by Jonathan Shireman. It is reproduced with his permission.

RAINWATR.BAS   A BASIC program which calculates the *pH* of rainwater in equilibrium with P_{CO_2} at a specified temperature. Results are calculated for values of P_{CO_2} between 3.4×10^{-4} atm and 10 atm.

REGULAR.BAS   A BASIC program which calculates the free energy of mixing for a regular binary solution as a function of temperature and excess mixing parameter.

SEMIINF.BAS   A BASIC program to calculate composition-distance or composition-time profiles in one half of a semiinfinite diffusion couple.

SOLUBLE.BAS   A BASIC program to calculate mineral solubilities in a simple electrolyte solution, calculating ion activity coefficients by iterative approximations with the extended Debye-Hückel equation.

TITRATE.BAS   A BASIC program to illustrate the procedure for calculating solution pH during an alkalinity titration of a dibasic salt solution (Na_2CO_3).

WELCOME.BAS   A BASIC program that contains information about your diskette. An executable version, WELCOME.EXE, is also included.

OTHER GEOCHEMICAL COMPUTER PROGRAMS

Most of the programs we have included on the diskette are "convenience" programs. That is, they address problems which are time-consuming but which, given a lot of patience and a sharp pencil, you could solve without a computer. A large number of programs, however, have been written to investigate geochemical problems that would be unthinkable to do if you had to perform them by hand. The greatest number of these model chemical equilibria among phases in multicomponent systems. The problem to be solved may take many different forms, but generally involves the simultaneous consideration of a large number of potentially stable phases to determine which should actually be present under a specified set of thermodynamic conditions.

Some, but not most, of these large-scale programs are available in a format that will run successfully on a desktop computer. Most, but not all, can be obtained at little or no cost by writing to the author of the program. We will not attempt to list programs in this appendix, but will refer you instead to the excellent compendia by Nordstrom et al. (1979) and Nordstrom and Ball (1984). Also, most of the journals familiar to geochemists regularly publish articles describing algorithms for the numerical evaluation of equilibria or of reaction paths (see, for example, Harvie et al., 1987; Blander and Pelton, 1987; Perkins et al., 1987).

As with the programs on our own diskette, we advise you to be aware of the limitations of programs that you obtain from outside sources. First, read any documentation, including published articles that have reported use of the program for other research problems. Be sure you understand the general approach taken by the program, and pay particular attention to those numerical conditions under which it may give ambiguous or incorrect answers. A free energy minimization routine, for example, may converge on a metastable equilibrium under some circumstances.

Second, be sure that the program you have chosen is appropriate to the problem at hand. Consider, in particular, any boundary conditions that may make your system different from the one being modelled by the program. Is your system open or closed? Isothermal? Adiabatic? Are the phases in your system really in equilibrium?

Third, evaluate the data set used by the program carefully. Most of the major programs in use today work with data that are internally consistent, but may not be compatible with other values that you may add later. You are ultimately responsible for verifying that the output from a program you use for your research is reliable.

Finally, if you are inspired to write your own programs for geochemistry, take advantage of the large number of subroutine libraries that are now available for performing standard mathematical operations like matrix inversion and numerical integration. Some of these, like LINPACK, can be obtained through many university computation centers. Routines for desktop computers have also been published in several good books. If you write in FORTRAN or Pascal, for example, we recommend Press et al. (1986). If your language is BASIC, try Ruckdeschel (1980, 1981).

REFERENCES

BLANDER, M., and PELTON, A.D. 1987. Thermodynamic analysis of binary liquid silicates and prediction of ternary solution properties by modified quasichemical equations. *Geochim. Cosmchim. Acta* 51:85–96.

HARVIE, C.E.; GREENBURG, J.P.; and WEARE, J.H. 1987. A chemical equilibrium algorithm for highly non-ideal multiphase systems: Free energy minimization. *Geochim. Cosmochim. Acta* 51:1045–57.

LASAGA, A.C. 1980. The kinetic treatment of geochemical cycles. *Geochim. Cosmochim. Acta* 44:815–28.

LINDSLEY, D. 1977. Magnetite-ilmenite geothermometer-oxybarometer: An evaluation of old and new data. *EOS* 58:519.

NORDSTROM, D.K., and BALL, J.W. 1984. Chemical models, computer programs, and metal complexation in natural waters. In *International symposium on trace metal complexation in natural waters.*, ed. Kramer, C.J.M., and Duinker, J.C., 149–69. Dordrecht, Netherlands: Martinus Mijhoff/Dr J.W. Junk Publishing Co..

———; PLUMMER, L.N.; WIGLEY, T.M.L.; WOLERY, T.J.; BALL, J.W.; JENNE, E.A.; BASSETT, R.L.; CRERAR, D.A.; FLORENCE, T.M.; FRITZ, B.; HOFFMAN, M.; HOLDREN, G.R.; JR., LAFON, G.M.; MATTIGOD, S.V.; MC-

DUFF, R.E.; MOREL, F.; REDDY, M.M.; SPOSITO, G.; and THRAILKILL, J. 1979. A comparison of computerized chemical models for equilibrium calculations in aqueous systems. In *Chemical modelling in aqueous systems*, ed. Jenne, E.A., 857–92. Symposium Ser. 93, Amer. Chem. Soc..

PERKINS, D.; ESSENE, E.J.; and WALL, V.J. 1987. THERMO: A computer program for calculation of mixed-volatile equilibria. *Amer. Mineral.* 72:446–47.

PRESS, W.H.; FLANNERY, B.P.; TEUKOLSKY, S.A.; and VETTERLING, W.T. 1986. *Numerical recipes*. Cambridge: Cambridge Univ. Press.

RUCKDESCHEL, F.R. 1980. *BASIC scientific subroutines.* Vol. 1. Peterborough, N.H.: BYTE Books.

RUCKDESCHEL, F.R. 1981. *BASIC scientific subroutines.* Vol. 2. Peterborough, N.H.: BYTE Books.

Programs Accompanying this Text

Answers to Selected Problems

CHAPTER 1

4. (a) $3/t$
   (b) $(2\pi a/\alpha t^2) \exp(-2\pi a/\alpha t)$
   (c) $2\alpha t \cos(\alpha t^2)$
   (d) $-1/(2.3\,t)$
   (e) $\ln t - \ln(1-t)$
   (f) $\Sigma_i(E_i/(kt^2)) \exp(-E_i/(kt))$
   (g) $(1-\alpha)\sin(t-\alpha t) - \sin t$
5. (a) $2x\,dx + 2y\,dy + 2z\,dz$
   (b) $yx \cos(xyz)\,dx + xz \cos(xyz)\,dy + xy \cos(xyz)\,dz$
   (c) $(3x^2y - y^2)\,dx + (x^3 - 2xy)\,dy$

8. (a) $\begin{pmatrix} 7 \\ -5 \\ 13 \end{pmatrix}$
   (b) $\begin{pmatrix} -19 & 17 & 16 \\ 5 & 22 & 7 \\ -10 & 20 & 24 \end{pmatrix}$
   (c) $\begin{pmatrix} 2 + 2x^2 - x^4 \\ 1 - 4x^2 - 2x^3 \\ 3 + 8x^2 - 2x^4 \end{pmatrix}$

   (d) -171
   (e) $\begin{pmatrix} -\frac{1}{3} & \frac{1}{3} & \frac{2}{3} \\ \frac{1}{3} & \frac{1}{6} & -\frac{1}{6} \\ \frac{2}{3} & -\frac{2}{3} & -\frac{1}{3} \end{pmatrix}$

CHAPTER 2

3. 8878.7 J mol^{-1}; 8362.7 J mol^{-1}
4. (a) 2382.5 cal   **(b)** 9968.7 J
7. $(\partial V/\partial T)_P = -(\partial S/\partial P)_T$ and $(\partial P/\partial T)_V = (\partial S/\partial V)_T$
   so $\dfrac{-(\partial V/\partial T)_P}{(\partial P/\partial T)_V} = \dfrac{(\partial S/\partial P)_T}{(\partial S/\partial V)_T} = (\partial V/\partial P)_T$

CHAPTER 3

2. $K_{sp} = 4.48 \times 10^{-5}$; $a_{Ca^{2+}} = 6.7 \times 10^{-3}$

4. 1.4×10^{-3} g

5. 2.33×10^{-3} g

6. 1.24×10^{-3}

9. $x_{Fs} = 0.00291$; $x_{En} = 0.831$; $x_{Di} = 0.00746$; $x_{CrCATS} = 7.35 \times 10^{-7}$; $x_{Py} = 0.0308$.

CHAPTER 4

2. 10.84

4. 0.67×10^{-8} molal

5. $NaHCO_3$ increases alkalinity; NaCl causes no change; HC1 decreases alkalinity; $MgCO_3$ increases alkalinity.

CHAPTER 5

1. 7.94 m

3.

Depth (m)	ϕ	ω (cm yr^{-1})	U (cm yr^{-1})
5	0.644	0.051	-0.049
10	0.477	0.034	-0.031
20	0.277	0.025	-0.019
30	0.178	0.022	-0.013
40	0.141	0.021	-0.01
60	0.092	0.020	-0.003
80	0.083	0.020	-0.0007

CHAPTER 6

1. 1.1×10^{-12}

3. At pH 9, 3.2 ppm; at pH 10, 8.1 ppm; at pH 11, 216 ppm

4. $m_{H_4SiO_4} = 2 \times 10^{-3}$ molal; $m_{H_3SiO_4^-} = 1.26 \times 10^{-8}$ molal; $m_{H_2SiO_4^{2-}}$ $= 3.16 \times 10^{-14}$ molal.

6. 0.866 volt

8. 0.181 ppm

10. 579; 4

CHAPTER 7

1. **(a)** Mass 44 = 98.424%, mass 45 = 1.1688%, mass 46 = 0.4032%.

(b) Mass 28 = 98.66%, mass 29 = 1.134%, mass 30 = 0.202%.

4. $1000 \ln \alpha_{SiO_2\text{-}UO_2} = -0.5 \times 10^{-6} T^{-1} + 13.29 \times 10^3 T - 7.36$.

5. $3.23°/_{oo}$

7. $\delta^{18}O = 0.695 \, T(°C) - 13.603$; $\delta D = 5.55 \, T(°C) - 100.95$.

8. $\delta^{18}O = 1.46°/_{oo}$; $\delta D = -207.5°/_{oo}$.

CHAPTER 8

6. (a) $\Delta \overline{H}^{\circ}_{800} = 508.63$ cal mol^{-1}; $\Delta \overline{S}^{\circ}_{800} = -8.11$ cal mol^{-1} deg^{-1};
$\Delta \overline{G}^{\circ}_{800} = 6996.63$ cal mol^{-1}.

(b) $\Delta \overline{G}^{\circ}_{800} = 6735.63$ cal mol^{-1}. ΔC_P is a function of temperature.

8. (a) 836 K; **(b)** 503.3 K

12. 597°C, 6 Kbar

CHAPTER 9

4. The first drop of liquid is very rich in Ab. As melting proceeds, the liquid becomes more An-rich. The final liquid has $X_{An} = 0.2$.

11. At 900°C, Tr:Kspar = 11/76 ; at 1100°C, Liq:Or = 11/18; at 1300°C, Liq:Lc = 67:12.

CHAPTER 10

1. Approx. 250 kJ or 60 kcal

4. (b) 28.405×10^{-7} cal cm^{-2}

6. (a) $h_c = r_c - (2(\sigma_{LS} - \sigma_{CS})/\Delta G_v)$; $V_c = (\pi h_c/6)(3x^3 + h_c^2)$,
where $x = r_c \left[1 - \left(\frac{r_c - h_c}{r_c} \right)^2 \right]^{1/2}$.

(b) $V_{hetero} = 1.6 \times 10^{-18}$ cm^3; $V_{hetero}/V_{homo} = 1.9 \times 10^{-5}$.

CHAPTER 11

1. At about 1850°C, Fo starts to crystallize and is removed from the system. Liquid composition changes along the liquidus. At the peritectic point, En begins to crystallize. Liquid composition moves continuously to the eutectic point, where En+Cr crystallize.

2. At about 1850°C, the homogeneous liquid separates into two liquids. The SiO$_2$-poor liquid is volumetrically more important, but the proportion of SiO$_2$-rich liquid increases slightly with decreasing temperature. At 1695°C, the SiO$_2$-rich liquid reacts to form Cr. Because there are three coexisting phases, temperature cannot change further until all of the SiO$_2$-rich liquid is consumed. Below 1695°C, Cr continues to crystallize until the eutectic temperature (1543°C), where En+Cr crystallize until all liquid is exhausted.

3. *Equilibrium melting*: At about 1410°C, olivine begins to melt, generating a liquid with Fo$_{20}$. As T increases, both solid and liquid become more magnesian. Melting is complete at about 1700°C, where the liquid is Fo$_{50}$. The last solid to melt has Fo$_{80}$. *Fractional melting*: Melting begins at about 1410°C, but melt is removed as it is formed, so late melt fractions are not mixed with early ones. Melting continues to almost 1900°C, where the final liquid is nearly pure Fo$_{100}$.

CHAPTER 12

1. $^{87}Sr_0/^{86}Sr = 0.7045$; $t = 397.7$ m.y.

2. (a) $t = \frac{1}{\lambda} \ln \left[\frac{^{40}Ar}{0.124\ ^{40}K} + 1 \right]$; **(b)** 1.66×10^5 years

3. 1.9×10^{21} atoms of ^{207}Pb

6. 7760 years

CHAPTER 13

2. Relative to 10^6 atoms of Si: 3.22×10^6 Mg, 1.22×10^6 Ca, 4.8×10^4 Ti, 1.7×10^4 Al, 2.14×10^2 Zr

3. 0.023

4. 84.714% Fe, 4.5% O, 4.5% S, 4.41% Ni, 1.762% P, 0.203% Co

Glossary

Accretion. The accumulation of materials in space to form larger objects. This process may occur homogeneously (producing bodies with uniform composition) or heterogeneously (producing layered bodies).

Activation energy. The amount of energy that must be provided in order to overcome some kinetic barrier.

Activity. The concentration of a component, adjusted for any effects of nonideality; also, a measurement of the number of radioactive decay events per unit of time.

Activity coefficient. The ratio of the activity of a species to its concentration; a measure of the degree of nonideal behavior of a chemical species in solution.

Adiabatic process. A process that occurs without exchange of heat with the surroundings.

Advection. The transport of ions or molecules within a moving medium.

Alkalinity. The charge deficit between the sum of dissolved conservative cations and anions in an electrolyte solution.

Assimilation. Incorporation of solid material into a magma.

Atomic number. The number of protons in an atomic nucleus.

Biopolymer. A complex organic molecule synthesized by plants or animals. Examples of these molecules include carbohydrates, proteins, lignin, and lipids.

Buffer. An assemblage of chemical species whose coexistence allows a system to resist change in some intensive property, such as *pH* or oxygen fugacity.

Calorimetry. An experimental measurement of heat evolved or absorbed during a specific reaction.

Carbonate compensation depth (CCD). The depth in the oceans at which the downward flux of carbonate minerals is balanced by their rate of dissolution.

Chalcophile. An element with an affinity for sulfide phases.

Chelate. A large molecule or complex that can enclose weakly bound atoms or ions.

Chemical potential (μ_i). Partial molar free energy, which describes the way in which total free energy for a phase responds to a change in the amount of component i in the phase $[= (\partial G/\partial n_i)_{P,T,n_{j\neq i}}]$.

Chondrite. A common type of meteorite, with nearly cosmic elemental abundances; thought to be a sample of the earliest solar system material.

Closed system. A system that can exchange energy, but not matter, with its surroundings.

Colloid. A stable electrostatic suspension of small particles in a liquid.

Components. Abstract chemical entities, independently variable within a system, which collectively describe all of the potential compositional variations within it.

Compressibility. The measure of the relationship between phase volume and lithostatic pressure, $[\beta_T = -1/V\,(\partial \overline{V}/\partial P)_T;\ \beta_S = -1/V\,(\partial \overline{V}/\partial P)_S]$.

Condensation. The formation of solids or liquids from a gas phase during cooling.

Congruent reaction. A reaction in which one phase melts or dissolves to form another phase of the same composition.

Conservative species. Any dissolved species the concentrations of which are not affected in solution in any way, other than dilution, by variations in the abundance of other species in solution.

Cosmic abundance of the elements. The relative elemental abundance pattern in the sun.

Cosmogenic nuclide. Any isotope formed by interaction of matter with cosmic rays.

Cotectic. A boundary curve on an n-component phase diagram, along which two or more phases crystallize simultaneously.

Delta notation. The notation used for stable isotope data, equal to $[(R_{sample} - R_{standard})/R_{standard}] \cdot 1000$, where R is the ratio of heavy to light isotope of the element of interest.

Diagenesis. The set of processes other than weathering or metamorphism that change the texture or mineral composition of sediments; these include compaction, cementation, recrystallization, and authigenesis.

Differentiation. The separation within a homogeneous planet of crust, mantle, and core; also, any process by which magma can give rise to rocks of contrasting composition.

Diffusion. The dispersion of ions or molecules through a medium that is not moving, due to a gradient in some intensive property across the system.

Diffusion coefficent. The proportionality constant relating flux to gradient in a diffusion equation.

Distribution coefficient (K_D). The ratio of concentrations of a component in two coexisting phases.

e-folding time. The time it takes for a compositional variable to return to within a factor of $1/e$ of its steady state value following some perturbing event, numerically equal to the inverse of the kinetic rate constant.

Eh. Redox potential of a cell, expressed in terms of voltage.

Electrolyte. Any substance which dissociates into ions when dissolved in an appropriate medium.

Elementary reaction. Any chemical reaction which occurs at the molecular level among species as they appear in the written representation.

Endothermic reaction. A reaction that absorbs heat, resulting in a decrease in enthalpy.

Enthalpy. A thermodynamic function of state, H; a measure of the energy that is irretrievably converted to heat during any natural process.

Entropy. A thermodynamic function, S, the change of which is defined as the change in heat gained by a body at a certain temperature in a reversible process, $[dS = (d\,Q/T)_{rev}]$.

Equation of state. A function that defines interrelationships among intensive properties of a system.

Equilibrium. The condition in which the properties of a system do not change with time.

Equilibrium constant (K_{eq}). The constant relating the composition of a system at equilibrium to its Gibbs free energy.

Eutectic. The point on a phase diagram at which the liquidus touches the solidus, where two or more solids crystallize simultaneously from a melt.

Exothermic reaction. A reaction that evolves heat, resulting in an increase in enthalpy.

Exsolution. The unmixing of two phases in the solid state.

Extended defects. Dislocations and planar defects in crystals that provide effective pathways for diffusion.

Extensive property. A variable whose value is a measure of the size of the system.

Extinct radionuclide. Any unstable nuclide that decays so rapidly that virtually none remains at the present time.

Fluid inclusion. Fluid trapped within crystals. These may be primary (trapped as the crystal grew) or secondary (introduced at some later time).

Flux. The amount of material moving from one location to another per unit time.

Fractional crystallization. Physical separation of crystals and liquid, preventing equilibrium between phases.

Fractionation. Separation of any two entities, such as isotopes (by mass), elements (by geochemical or cosmochemical properties), or crystals and melt (by fractional crystallization or fractional melting).

Fractionation factor. The ratio of a heavy to a light isotope of an element in one phase divided by the same ratio in another coexisting phase.

Fugacity. Gas partial pressure that has been adjusted for nonideality.

Geobarometer. A calibrated mineral exchange reaction that is a strong function of pressure but not temperature.

Geochronology. The determination of the ages of rocks using radiometric dating techniques.

Geopolymer. Any of several complex organic compounds of high molecular weight, formed from biopolymers during diagenesis; these include kerogen and bitumen.

Geothermal gradient. The rate of change of temperature with depth in the earth; also called *geotherm.*

Geothermometer. A calibrated mineral exchange reaction that is a strong function of temperature but not pressure.

Gibbs free energy. A thermodynamic function of state G, which is a measure of the energy available for changing chemical bonds in a system.

Half-life. The time required for one-half of a given number of atoms of a radionuclide to decay.

Half-reaction. A hypothetical reaction that illustrates either gain or loss of electrons, useful for understanding redox equilibria.

Heat. The transfer of energy that results in an increase in temperature.

Heat capacity. The functional relationship between enthalpy and temperature, $[C_P = (\partial H/\partial T)_P; C_V = (\partial H/\partial T)_V]$.

Helmholtz free energy. A thermodynamic function of state F, which, for an isothermal process, is a measure of the energy transferred as work; also called the *work function.*

Henry's law. An expression of solution behavior in which the activities of dissolved species become directly proportional to their concentrations at high dilution.

Ideal solution. Any solution in which the end-member constituents behave as if they were independent.

Immiscibility. The condition that results in spontaneous phase separation.

Incongruent reaction. A reaction in which one phase melts or dissolves to form phases of different composition.

Intensive property. A variable whose value is independent of the size of the system.

Internal energy. A thermodynamic function of state E, the change of which is a measure of the energy transferred as heat or work between the system and its surroundings.

Isochron. A line on an isotope plot whose slope determines the age of a rock system. Isochrons may be defined by the isotopic compositions of constituent minerals or whole rocks.

Isolated system. A system that cannot exchange either matter or energy with the universe beyond the system's limits.

Isotopes. Atoms of the same element but with different numbers of neutrons and hence different atomic weights. Isotopes may be stable or unstable; also called *nuclides*.

Kinetics. The description of a system's behavior in terms of its rates of change.

Liquidus. The boundary curve on a phase diagram representing the onset of crystallization of a liquid with lowering temperature.

Lithophile. An element with an affinity for silicate phases.

Local equilibrium. Equilibrium that is attained on a small scale, but not on a larger scale; also called *mosaic equilibrium*.

Lysocline. The depth in the oceans at which the effects of carbonate dissolution are first discernable.

Mass number. The number of protons + neutrons in a nucleus.

Mean residence time. The average residence time for a chemical species in a system before it is removed by some loss process.

Metamorphism. A collection of processes that change the texture or phase composition of a rock through the action of temperature, pressure, and fluids. Metamorphism can be prograde (with increasing temperature and pressure) or retrograde (with falling temperature and pressure).

Metasomatism. Metamorphism with accompanying change in chemical composition, usually as a result of fluid migration.

Metastable system. A system that appears to be stable because it is observed over a short time compared to the rates of reactions that alter it.

Mobile component. A component free to migrate into or out of a system.

Nuclear statistical equilibrium. The steady state in which disintegrative and constructive nuclear reactions balance in an evolved star.

Nucleation. The initiation of a small volume of a new phase. This process may occur homogeneously (without a substrate) or heterogeneously (on a substrate).

Nucleosynthesis. The production of nuclides, primarily in stars.

Nuclides. *See* Isotopes.

Open system. A system that can exchange matter and energy with its surroundings.

Overall reaction. A chemical reaction that describes a net change involving several intermediate steps and competing pathways.

Partial melting. The fusion of something less than an entire rock, a process that may occur under equilibrium or fractional (continuous separation of melt and crystals) conditions.

Peritectic. The point on a phase diagram at which an incongruent reaction occurs.

pH. A measure of acidity in solution [$pH = -log_{a_{H^+}}$].

Phase. A substance with continous physical properties.

Phase diagram. A graphical summary of how a system reacts to changing conditions or composition.

Point defects. Crystal imperfections caused by the absence of atoms at lattice sites; such defects may be intrinsic or extrinsic.

Pyrolysis. The distillation of gases to produce refractory organic compounds.

Radioactive decay. The spontaneous transformation of one unstable nuclide into another isotope; this occurs by alpha or beta decay, positron decay, electron capture, or spontaneous fission.

Radiogenic nuclide. An isotope formed by decay of some parent radionuclide.

Raoult's law. An expression of solution behavior in which the activities of dissolved species are equal to their concentrations.

Refractory. Having a high melting temperature or condensing from a gas at a high temperature.

Salinity. The total dissolved salt content.

Saturated hydrocarbons. Organic molecules consisting only of carbon and hydrogen atoms linked by single bonds. These may form chain structures called aliphatic compounds, or cyclical structures called alicyclic compounds.

Secular equilibrium. The condition in which the decay rate for daughter radionuclides in a decay series equals that of the parent.

Siderophile. An element with an affinity for metallic phases.

Solidus. The boundary curve on a phase diagram representing complete solidification of a system with lowering temperature.

Solubility product constant (K_{sp}). The equilibrium constant for a dissolution reaction.

Solute. The dissolved species in a solution.

Solvent. The host species in a solution.

Solvus. The region of unmixing on a phase diagram.

Spinodal. The region of unmixing inside the solvus on a phase diagram; such unmixing involves no nucleation barrier.

Standard state. An arbitrarily selected state of a system

used as a reference against which changes in thermodynamic properties can be compared.

Steady state.   The condition in which rates of change within a system balance one another, so that there is no net change in its appearance for an indefinite time period, even though various parts of the system are not in thermodynamic equilibrium.

Stereoisomers.   Symmetrically distinct forms of the same chemical compound.

Stoichiometric.   Summing to the whole, as in stoichiometric equations or coefficients.

Supernova.   A massive stellar explosion, important for nucleosynthesis.

Surface tension.   Tensional force applied perpendicular to any line on a droplet surface.

System.   That portion of the universe that is of interest for a particular problem; a system can be open, closed, or isolated.

Thermal expansion.   The measure of the relationship between phase volume and temperature, $[\propto_P = - V (\partial \overline{V}/\partial T)_P]$.

Thermocline.   The transitional range between the warm surface zone and the cold deeper zone in the oceans.

Thermodynamics.   A set of laws that predict the equilibrium configuration of a system and how it will change if its environmental parameters are changed.

Titration.   An experiment in which a reaction is allowed to proceed incrementally, so that proportions of reactants can be determined.

Trace elements.   Elements that occur in rocks with concentrations of a few tenths of a percent or less by weight; these may exhibit compatible or incompatible behavior.

Transition elements.   Elements with inner orbitals incompletely filled by electrons.

Troposphere.   The lower portion of the atmosphere that contains the bulk of its mass.

Uncompressed mean density.   The mean density of a planet, corrected for the effects of gravitational self-compression.

Undercooling.   The difference between the equilibrium temperature for the appearance of a phase and the temperature at which it actually appears.

Unsaturated hydrocarbons.   Organic molecules with double or triple bonds between carbon atoms; these may form chain structures, called aliphatic compounds, or cyclic structures, called aromatic compounds.

Variance.   The number of independent parameters that must be defined in order to specify the state of a system at equilibrium; also called *degrees of freedom.*

Volatile.   A term indicating that a given element or compound commonly occurs in the gaseous state at the temperature of interest, or indicating that a given element condenses from a gas at low temperature.

Weathering.   Those processes occurring at the earth's surface that cause decomposition of rocks.

Work.   A transfer of energy that causes a mechanical change in a system or its surroundings; the integral of force $\times$ displacement.

Xenolith.   A foreign rock fragment trapped in igneous rock.

Index